An Introduction to Geophysical Exploration

GEOSCIENCE TEXTS

SERIES EDITOR

A. HALLAM

Lapworth Professor of Geology
University of Birmingham

1 Principles of Mineral Behaviour
A. PUTNIS & J.D.C. McCONNELL

2 An Introduction to Ore Geology
ANTHONY M. EVANS

3 Sedimentary Petrology: an Introduction
MAURICE E. TUCKER

GEOSCIENCE TEXTS VOLUME 4

An Introduction to Geophysical Exploration

P. KEAREY
BSc, PhD
Lecturer in Geophysics
Department of Geology
University of Bristol

M. BROOKS
BSc, PhD
Professor, Department of Geology
University College, Cardiff

BLACKWELL SCIENTIFIC PUBLICATIONS

OXFORD LONDON EDINBURGH

BOSTON PALO ALTO MELBOURNE

Editorial offices:
Osney Mead, Oxford, OX2 0EL
8 John Street, London, WC1N 2ES
9 Forrest Road, Edinburgh, EH1 2QH
52 Beacon Street, Boston,
Massachusetts 02108, USA
706 Cowper Street, Palo Alto
California 94301, USA
99 Barry Street, Carlton
Victoria 3053, Australia

First published 1984

Photoset by Enset (Photosetting)
Midsomer Norton, Bath, Avon
and printed and bound in Great Britain by
Adlard & Son Ltd, Dorking.

DISTRIBUTORS

USA and Canada
Blackwell Scientific Publications Inc
PO Box 50009, Palo Alto
California 94303

Australia
Blackwell Scientific Book Distributors
31 Advantage Road, Highett
Victoria 3190

British Library
Cataloguing in Publication Data

Kearey, P.
An introduction to geophysical exploration.——
(Geoscience texts; V.4)
1. Prospecting——Geophysical methods
I. Title II. Brooks, M. III. Series
622′.15 TN269

ISBN 0-632-01049-5

Contents

Preface

This book provides a general introduction to the most important methods of geophysical exploration. These methods represent a primary tool for investigation of the subsurface and are applicable to a very wide range of problems. Although their main application is in prospecting for natural resources the methods are also used, for example, as an aid to geological surveying, as a means of deriving information on the Earth's internal physical properties, and in engineering or archaeological site investigations. Consequently, geophysical exploration is of importance not only to geophysicists but also to geologists, physicists, engineers and archaeologists. The book covers the physical principles, methodology, interpretational procedures and fields of application of the various survey methods. The main emphasis has been placed on seismic methods because these represent the most extensively used techniques, being routinely and widely employed by the oil industry in prospecting for hydrocarbons. Since this is an introductory text we have not attempted to be completely comprehensive in our coverage of the subject. Readers seeking further information on any of the survey methods described should refer to the more advanced texts listed at the end of each chapter. Limitations of space have forced us to omit certain of the more specialized methods, in particular, geophysical borehole logging, radiometric surveying and remote sensing.

We hope that the book will serve as an introductory course text for students in the above-mentioned disciplines and also as a useful guide for specialists who wish to be aware of the value of geophysical surveying to their own disciplines. In preparing a book for such a wide possible readership it is inevitable that problems arise concerning the level of mathematical treatment to be adopted. Geophysics is a highly mathematical subject and although we have attempted to show that no great mathematical expertise is necessary for a broad understanding of geophysical surveying, a full appreciation of the more advanced data processing and interpretational techniques does require a reasonable mathematical ability. Our approach to this problem has been to keep the mathematics as simple as possible and to restrict full mathematical analysis to relatively simple cases. We consider it important, however, that any user of geophysical surveying should be aware of the more advanced techniques of analysing and interpreting geophysical data since these can greatly increase the amount of useful information obtained from the data. In discussing such techniques we have adopted a semi-quantitative or qualitative approach which allows the reader to assess their scope and importance, without going into the details of their implementation.

Acknowledgements

We thank our friends and colleagues Dr P.F. Ellis, Dr J. Shaw and Dr R.G. Pearce for their helpful comments on early versions of the manuscript. The text-figures were drafted by Mrs J. Bees, Mrs A. Gregory and Mrs M. Millen. Mr D.G. Hilton is thanked for extensive photographic assistance. We also thank Miss J. Cole, Mrs J. Rowlands, Mrs M. Rundle and Miss P. Westall for their patience in typing so many drafts of the manuscript.

1

The Principles and Limitations of Geophysical Exploration Methods

1.1 Introduction

This chapter is provided for readers with no prior knowledge of geophysical exploration methods and is pitched at an elementary level. It may be passed over by readers already familiar with the basic principles and limitations of geophysical surveying.

The science of geophysics applies the principles of physics to the study of the Earth. Geophysical investigations of the interior of the Earth involve taking measurements at or near the Earth's surface that are influenced by the internal distribution of physical properties. Analysis of these measurements can reveal how the physical properties of the Earth's interior vary vertically and laterally.

By working at different scales, geophysical methods may be applied to a wide range of investigations from studies of the entire Earth (global geophysics) to exploration of a localized region of the upper crust. In the geophysical exploration methods (also referred to as geophysical surveying) discussed in this book, measurements within geographically restricted areas are used to determine the distributions of physical properties at depth that reflect the local subsurface geology.

An alternative method of investigating subsurface geology is, of course, by drilling boreholes, but these are expensive and provide information only at discrete locations. Geophysical surveying, although sometimes prone to major ambiguities or uncertainties of interpretation, provides a relatively rapid and cost-effective means of deriving areally distributed information on subsurface geology. In the exploration for subsurface resources the methods are capable of detecting and delineating local features of potential interest that could not be discovered by any realistic drilling programme. Geophysical surveying does not dispense with the need for drilling but, properly applied, it can optimize exploration programmes by maximizing the rate of ground coverage and minimizing the drilling requirement. The importance of geophysical exploration as a means of deriving subsurface geological information is so great that the basic principles and scope of the methods and their main fields of application should be appreciated by any practicing earth scientist. This book provides a general introduction to the main geophysical methods in widespread use.

1.2 The survey methods

There is a broad division of geophysical surveying methods into those that make use of natural fields of the Earth and those that require the input into the ground of artificially generated energy. The natural field methods utilize the gravitational, magnetic, electrical and electromagnetic fields of the Earth, searching for local perturbations in these naturally occurring fields that may be caused by concealed geological features of economic or other interest. Artificial source methods involve the generation of local electrical or electromagnetic fields that may be used analogously to natural fields, or, in the most important single group of geophysical surveying methods, the generation of seismic waves whose propagation velocities and transmission paths through the subsurface are mapped to provide information on the distribution of geological boundaries at depth. Generally, natural field methods can provide information on Earth properties to significantly greater depths and are logistically more simple to carry out than artificial source methods. The latter, however, are capable of producing a more detailed and better resolved picture of the subsurface geology.

Several geophysical surveying methods can be used at sea or in the air. The higher capital and operating costs associated with marine or airborne work and the problems of accurate position fixing are offset by the increased speed of operation and the benefit of being able to survey areas where ground access is difficult or impossible.

A wide range of geophysical surveying methods exists, for each of which there is an 'operative' physical property to which the method is sensitive. The methods are listed in Table 1.1.

Table 1.1. Geophysical surveying methods.

Method	Measured parameter	'Operative' physical property
Seismic	Travel times of reflected/ refracted seismic waves	Density and elastic moduli, which determine the propagation velocity of seismic waves
Gravity	Spatial variations in the strength of the gravitational field of the Earth	Density
Magnetic	Spatial variations in the strength of the geomagnetic field	Magnetic susceptibility and remanence
Electrical		
Resistivity	Earth resistance	Electrical conductivity
Induced Polarization	Polarization voltages or frequency-dependent ground resistance	Electrical capacitance
Self Potential	Electrical potentials	Electrical conductivity
Electromagnetic	Response to electromagnetic radiation	Electrical conductivity and inductance

The type of physical property to which a method responds clearly determines its range of applications. Thus, for example, the magnetic method is very suitable for locating buried magnetite ore bodies because of their high magnetic susceptibility. Similarly, seismic or electrical methods are suitable for the location of a buried water table because saturated rock may be distinguished from dry rock by its higher seismic velocity and higher electrical conductivity.

Other considerations also determine the type of methods employed in a geophysical exploration programme. For example, reconnaissance surveys are often carried out from the air because of the high speed of operation. In such cases the electrical or seismic methods are not applicable, since these require physical contact with the ground for the direct input of energy.

Geophysical methods are often used in combination. Thus, the initial search for metalliferous mineral deposits often utilizes airborne magnetic and electromagnetic surveying. Similarly, routine reconnaissance of continental shelf areas often includes simultaneous gravity, magnetic, and seismic surveying. At the interpretation stage, ambiguity arising from the results of one survey method may often be removed by consideration of results from a second survey method.

Geophysical exploration commonly takes place in a number of stages. For example, in the offshore search for oil and gas, an initial gravity reconnaissance survey may reveal the presence of a large sedimentary basin that is subsequently explored using seismic methods. A first round of seismic exploration may highlight areas of particular interest where further detailed seismic work needs to be carried out.

The main fields of application of geophysical surveying, together with an indication of the most appropriate surveying methods for each application, are listed in Table 1.2.

Table 1.2. Geophysical surveying applications.

Application	Appropriate survey methods*
Exploration for fossil fuels (oil, gas, coal)	S, G, M, (EM)
Exploration for metalliferous mineral deposits	M, EM, E, SP, IP
Exploration for bulk mineral deposits (sand & gravel)	S, (E), (G)
Exploration for underground water supplies	E, S, (G)
Engineering/construction site investigation	E, S, (G), (M)

*G: gravity; M: magnetic; S: seismic; E: electrical resistivity; SP: self potential; IP: induced polarization; EM: electromagnetic. Subsidiary methods in brackets.

Exploration for hydrocarbons and for metalliferous minerals represent the most important geophysical surveying activities. In terms of the amount of money expended annually, seismic methods are the most important technique because of their routine and widespread use in the exploration for hydrocarbons. Seismic methods are particularly well suited to the investigation of the layered sequences in sedimentary basins that are the primary targets for oil or gas. On the other hand, seismic methods are quite unsuited to the exploration of igneous and metamorphic terrains for the near-surface, irregular ore bodies that represent the main source of metalliferous minerals. Exploration for ore bodies is mainly carried out using electromagnetic and magnetic surveying methods.

In several geophysical survey methods it is the local variation in a measured parameter, relative to some normal background value, that is of primary interest.

Such variation is attributable to a localized subsurface zone of distinctive physical property and possible geological importance. A local variation of this type is known as a *geophysical anomaly*. For example, the Earth's gravitational field, after the application of certain corrections, would everywhere be constant if the subsurface were of uniform density. Any lateral density variation associated with a change of subsurface geology results in a local deviation in the gravitational field. This local deviation from the otherwise constant gravitational field is referred to as a gravity anomaly.

Although many of the geophysical methods require complex methodology and relatively advanced mathematical treatment in interpretation, much information may be derived from a simple assessment of the survey data. This is illustrated in the following section where a number of geophysical surveying methods are applied to the problem of detecting and delineating a specific geological feature, namely a salt dome. No terms or units are defined here, but the examples serve to illustrate the way in which geophysical surveys can be applied to the solution of a particular geological problem.

Salt domes are emplaced when a buried salt layer, because of its buoyancy, rises through overlying denser strata in a series of approximately cylindrical bodies. The rising columns of salt pierce the overlying strata or arch them into a domed form. A salt dome has physical properties that are different from the surrounding sediments and which enable its detection by geophysical methods. These properties are (1) a relatively low density, (2) a negative magnetic sus-

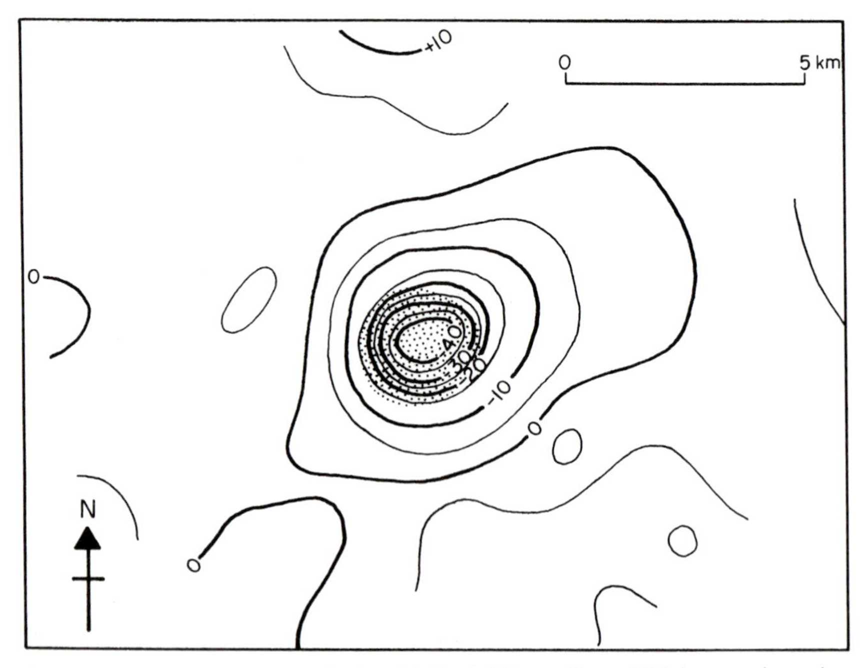

Fig. 1.1. The gravity anomaly over the Grand Saline Salt Dome, Texas, USA (contours in gravity units – see Chapter 6). The stippled area represents the subcrop of the dome. (Redrawn from Peters & Dugan 1945.)

ceptibility, (3) a relatively high propagation velocity for seismic waves, and (4) a high electrical resistivity (specific resistance).

(1) The relatively low density of salt with respect to its surroundings renders the salt dome a zone of anomalously low mass. The Earth's gravitational field is perturbed by subsurface mass distributions and the salt dome therefore gives rise to a gravity anomaly that is negative with respect to surrounding areas. Fig. 1.1 presents a contour map of gravity anomalies measured over the Grand Saline Salt Dome in east Texas, USA. The gravitational readings have been corrected for effects which result from the Earth's rotation, irregular surface relief and regional geology so that the contours reflect only variations in the shallow density structure of the area resulting from the local geology. The location of the salt dome is known from both drilling and mining operations and its subcrop is indicated. It is readily apparent that there is a well-defined negative gravity anomaly centred over the salt dome and the circular gravity contours reflect the circular outline of the dome. Clearly, gravity surveys provide a powerful method for the location of features of this type.

(2) A less familiar characteristic of salt is its negative magnetic susceptibility, full details of which must be deferred to Chapter 7. This property of salt causes a local decrease in the strength of the Earth's magnetic field in the vicinity of a salt dome. Fig. 1.2 presents a contour map of the strength of the magnetic field over the Grand Saline Salt Dome covering the same area as Fig. 1.1. Readings have been corrected for the large scale variations of the magnetic field with latitude, longitude and time so that, again, the contours reflect only those variations resulting

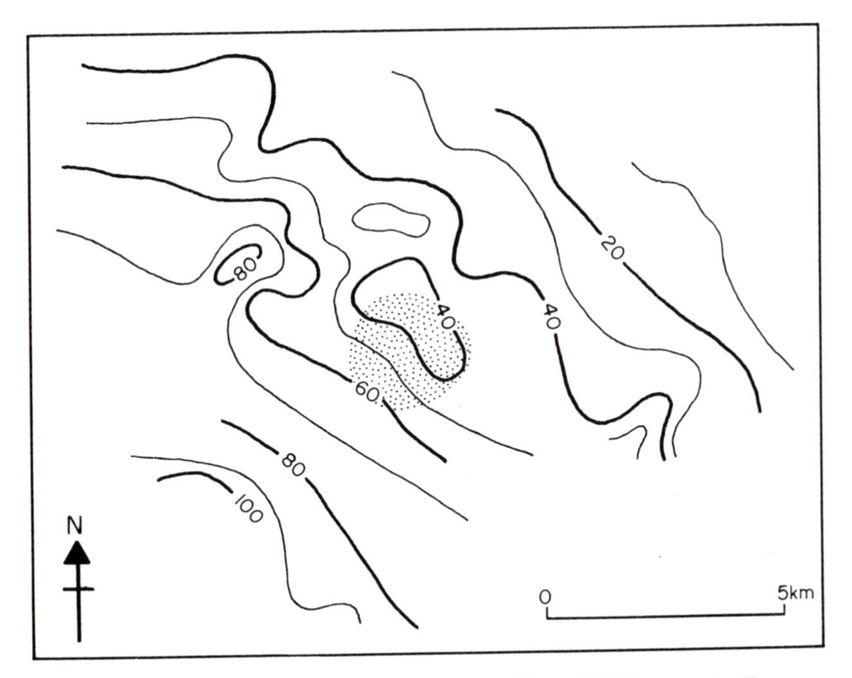

Fig. 1.2. Magnetic anomalies over the Grand Saline Dome, Texas, USA (contours in nT – see Chapter 7). The stippled area represents the subcrop of the dome. (Redrawn from Peters & Duggan 1945.)

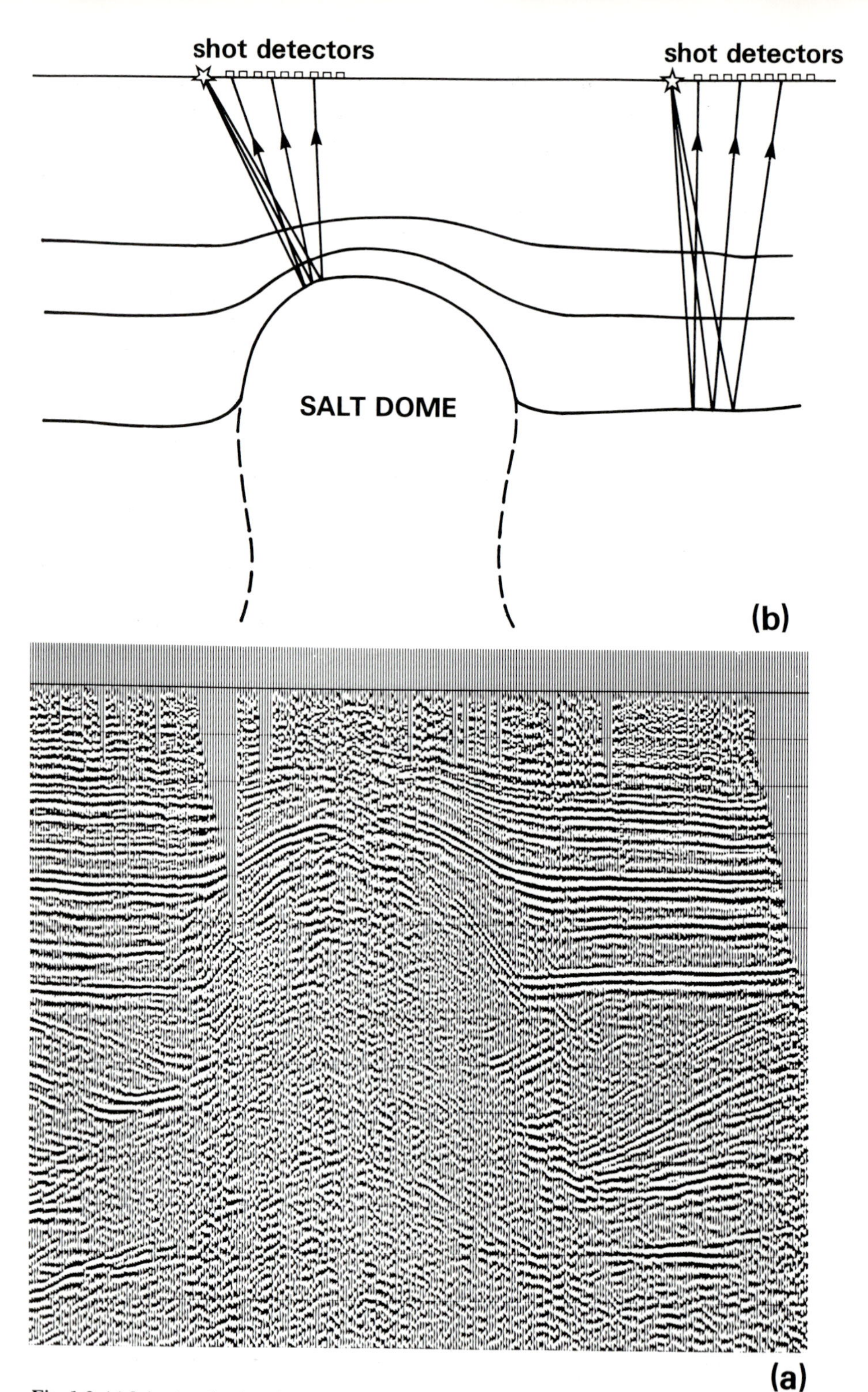

Fig. 1.3. (a) Seismic reflection section across a buried salt dome (courtesy Prakla-Seismos GMBH). (b) Simple structural interpretation of the seismic section, illustrating some possible raypaths for reflected rays.

from variations in the magnetic properties of the subsurface. As expected, the salt dome is associated with a negative magnetic anomaly although the magnetic low is displaced slightly from the centre of the dome. This example illustrates that salt domes may be located by magnetic surveying but the technique is not widely used as the associated anomalies are usually very small and therefore difficult to detect.
(3) Seismic rays normally propagate through salt at a higher velocity than through the surrounding sediments. A consequence of this velocity difference is that any seismic energy incident on the boundary of a salt body is partitioned into a refracted phase that is transmitted through the salt and a reflected phase that travels back through the surrounding sediments (Chapter 3). These two seismic phases provide alternative means of locating a concealed salt body.

For a series of seismic rays travelling from a single shot point into a fan of seismic detectors (Fig. 5.21), rays transmitted through any intervening salt dome will travel at a higher average velocity than in the surrounding medium and, hence, will arrive relatively early at the recording site. By means of this 'fan-shooting' it is possible to delineate sections of ground which are associated with anomalously short travel times and which may therefore be underlain by a salt body.

An alternative, and more effective, approach to the seismic location of salt domes utilizes energy reflected off the salt, as shown schematically in Fig. 1.3. A

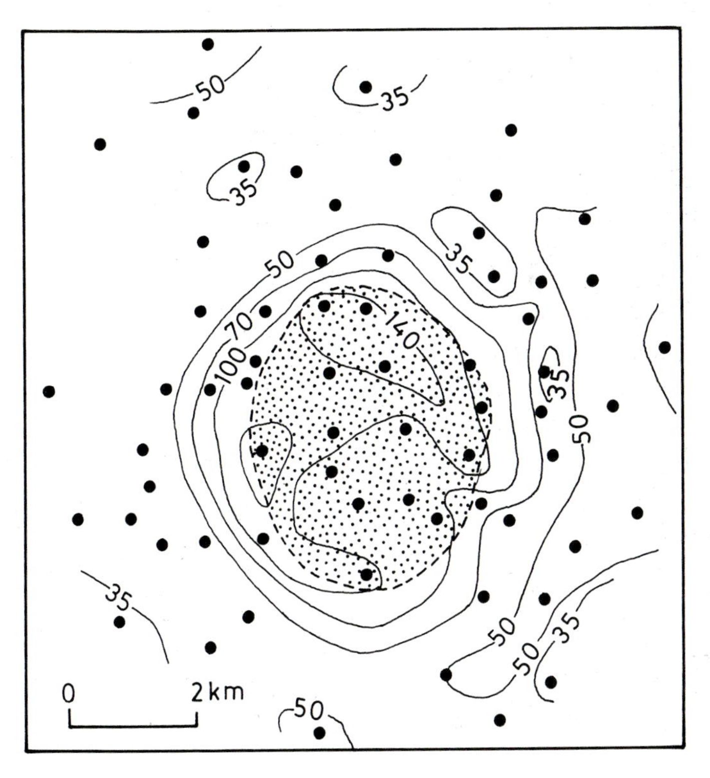

Fig. 1.4. Perturbation of telluric currents over the Haynesville Salt Dome, Texas, USA (for explanation of units see Chapter 9). The stippled area represents the subcrop of the dome. (Redrawn from Boissonas & Leonardon 1948.)

survey configuration of closely-spaced shots and detectors is moved systematically along a profile line and the travel times are measured of rays reflected back from any subsurface geological interfaces. If a salt dome is encountered, rays reflected off its top surface will delineate the shape of the concealed body.
(4) Earth materials with anomalous electrical resistivity may be located using either electrical or electromagnetic geophysical techniques. Shallow features are normally investigated using artificial field methods in which an electrical current is introduced into the ground and potential differences between points on the surface are measured to reveal anomalous material in the subsurface (Chapter 8). However, this method is restricted in its depth of penetration by the limited power that can be introduced into the ground. Much greater penetration can be achieved by making use of the natural Earth currents (telluric currents) generated by the motions of charged particles in the ionosphere. These currents extend to great depths within the Earth and, in the absence of any electrically anomalous material, flow parallel to the surface. A salt dome, however, possesses an anomalously high electrical resistivity and electric currents preferentially flow around and over the top of such a structure rather than through it. This pattern of flow causes distortion of the constant potential gradient at the surface that would be associated with a homogeneous subsurface and indicates the presence of the anomalous salt. Fig. 1.4 presents the results of a telluric current survey of the Haynesville Salt Dome, Texas, USA. The contour values represent quantities describing the extent to which the telluric currents are distorted by subsurface phenomena and their configuration reflects the shape of the subsurface salt dome with some accuracy.

1.3 The problem of ambiguity in geophysical interpretation

If the internal structure and physical properties of the Earth were precisely known, the magnitude of any particular geophysical measurement taken at the Earth's surface could be predicted uniquely. Thus, for example, it would be possible to predict the travel time of a seismic wave reflected off any buried layer or to determine the value of the gravity or magnetic field at any surface location. In geophysical surveying the problem is the converse of the above, namely, to deduce some aspect of the Earth's internal structure on the basis of geophysical measurements taken at (or near to) the Earth's surface. The former type of problem is known as a *direct* problem, the latter as an *inverse* problem. Whereas direct problems are theoretically capable of unambiguous solution, inverse problems suffer from an inherent ambiguity, or non-uniqueness, in the conclusions that can be drawn.

To exemplify this point a simple analogy to geophysical surveying may be considered. In *echo-sounding*, high frequency acoustic pulses are transmitted by a transducer mounted on the hull of a ship and echoes returned from the sea bed are detected by the same transducer. The travel time of the echo is measured and converted into a water depth, multiplying the travel time by the velocity with which sound waves travel through water, i.e. 1500 m/s. Thus an echo time of 0.10 s indicates a path length of $0.10 \times 1500 = 150$ m, or a water depth of

$150/2 = 75$ m, since the pulse travels down to the sea bed and back up to the ship.

Using the same principle, a simple seismic survey may be used to determine the depth of a buried geological interface (e.g. the top of a limestone layer). This would involve generating a seismic pulse at the Earth's surface and measuring the travel time of a pulse reflected back to the surface from the top of the limestone. However, the conversion of this travel time into a depth requires knowledge of the velocity with which the pulse travelled along the reflection path and, unlike the velocity of sound in water, this information is generally not known. If a velocity is assumed, a depth estimate can be derived but it represents only one of many possible solutions. And since rocks differ significantly in the velocity with which they propagate seismic waves, it is by no means a straightforward matter to translate the travel time of a seismic pulse into an accurate depth to the geological interface from which it was reflected.

The solution to this particular problem, as discussed in Chapter 4, is to measure the travel times of reflected pulses at several offset distances from a seismic source because the variation of travel time as a function of range provides information on the velocity distribution with depth. However, although the degree of uncertainty in geophysical interpretation can often be reduced to an acceptable level by the general expedient of taking additional field measurements, the problem of inherent ambiguity cannot be circumvented.

The general problem in geophysical surveying is that significant differences from an actual subsurface geological situation may give rise to insignificant, or immeasurably small, differences in the quantities actually measured during the survey. Thus the problem of ambiguity in the geophysical interpretation of subsurface geology arises because many different geological configurations could reproduce the observed measurements. This basic limitation arises directly from the unavoidable fact that geophysical surveying attempts to solve a difficult inverse problem. In spite of this limitation, however, geophysical surveying is an invaluable tool for the investigation of subsurface geology and occupies a key role in exploration programmes for geological resources.

1.4 The structure of the book

The above introductory sections illustrate in a simple way the very wide range of approaches to the geophysical investigation of the subsurface and warn of inherent limitations in geophysical interpretations.

In Chapters 3 to 9 the individual survey methods are treated systematically in terms of their basic principles, survey procedures, interpretation techniques and types of application. These accounts of the individual methods are preceded, in Chapter 2, by a short account of the more important data processing techniques of general applicability in geophysics. All the above chapters contain suggestions for further reading which provide a much more extensive treatment of the material presented in these chapters. To complete the book, in Chapter 10 the various survey methods are placed in the context of their major fields of application, illustrated by a series of case histories.

2

Geophysical Data Processing

2.1 Introduction

Much of geophysical surveying is concerned with the measurement and analysis of waveforms that express the variation of some measurable quantity as a function of distance or time. The quantity may, for example, be the strength of the Earth's gravitational or magnetic field along a profile line across a geological structure; or it may be the displacement of the ground surface as a function of time associated with the passage of seismic waves from a nearby explosion. The analysis of waveforms such as these represents an essential aspect of geophysical data processing and interpretation. The fundamental principles on which the various methods of data analysis are based are brought together in this chapter, along with a discussion of the techniques of digital data processing by computer that are routinely used by geophysicists.

Throughout this chapter waveforms are referred to as functions of time, but all the principles discussed, relating to spectral analysis and digital filtering, are equally applicable to functions of distance. In the latter case, frequency (number of waveform cycles per unit time) is replaced by spatial frequency or *wavenumber* (number of waveform cycles per unit distance).

2.2 Digitization of geophysical data

Waveforms of geophysical interest generally represent continuous (analogue) functions of time or distance. The quantity of information and, in some cases, the complexity of data processing to which these waveforms are subjected are such that the processing can only be accomplished effectively and economically by digital computers. Consequently, the data often need to be expressed in digital form for input to a computer, whatever the form in which they were originally recorded.

A continuous, smooth function of time or distance can be expressed digitally by sampling the function at a fixed interval and recording the instantaneous value of the function at each sampling point. Thus the analogue function of time $f(t)$ shown in Fig. 2.1(a) can be represented as the digital function $g(t)$ shown in Fig. 2.1(b) where the continuous function has been replaced by a series of discrete values at fixed intervals of time τ.

The two basic parameters of a digitizing system are the sampling precision (dynamic range) and the sampling frequency.

Dynamic range is an expression of the ratio of the largest measurable amplitude

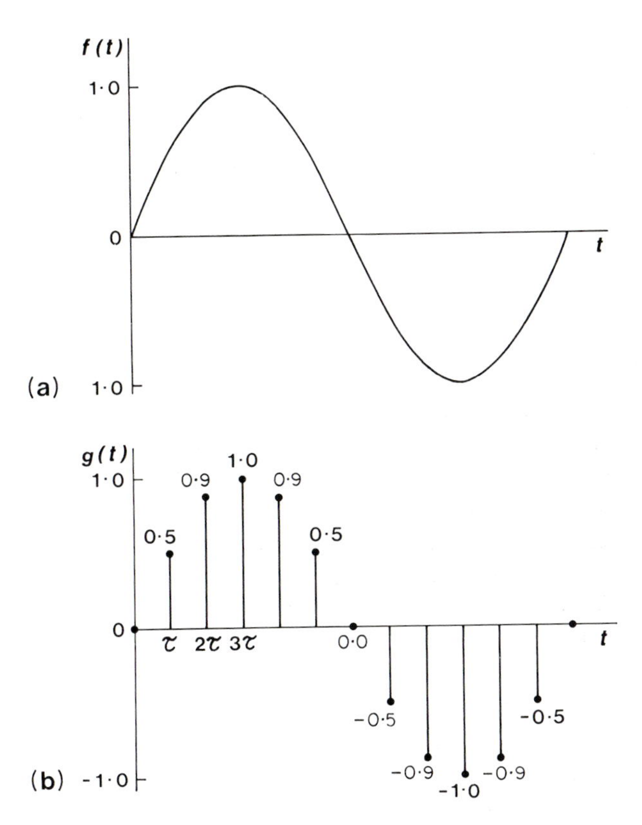

Fig. 2.1. (a) Analogue representation of a sinusoidal function. (b) Digital representation of the same function.

A_{max} to the smallest measurable amplitude A_{min} in a sampled function. The higher the dynamic range, the more faithfully will amplitude variations in the analogue waveform be represented in the digitized version of the waveform. Dynamic range is normally expressed in the *decibel* (dB) scale used to define electrical power ratios: the ratio of two power values P_1 and P_2 is given by $10 \log_{10} (P_1/P_2)$ decibels. Since electrical *power* is proportional to the square of *signal amplitude* A, $10 \log_{10} (P_1/P_2) = 10 \log_{10} (A_1/A_2)^2 = 20 \log_{10} (A_1/A_2)$. Thus, if a digital sampling scheme measures amplitudes over the range from 1 to 1024 units of amplitude, the dynamic range is given by $20 \log_{10} (A_{max}/A_{min}) = 20 \log_{10} 1024 \doteq 60$ dB.

For convenience of handling in digital computers, digital samples are expressed in binary form (i.e. they are composed of a sequence of digits that have the value of either 0 or 1). Each binary digit is known as a *bit* and the sequence of bits representing the sample value is known as a *word*. The dynamic range of a digitized waveform is determined by the number of bits in each word. For example, a dynamic range of 60 dB requires 10-bit words since the appropriate amplitude ratio of 1024 ($= 2^{10}$) is rendered as 1000 000 000 in binary form. A dynamic range of 84 dB represents an amplitude ratio of 2^{14} and, hence, requires sampling with 14-bit words. Thus, increasing the number of bits in each word in digital sampling increases the dynamic range of the digital function.

Intuitively, it may appear that the digital sampling of a continuous function inevitably leads to a loss of fidelity in the resultant digital function, since the latter is only specified by discrete values at a series of spaced points. In fact, as discussed below, there is no significant loss of information content as long as the frequency of sampling is at least twice as high as the highest frequency component in the sampled function.

Sampling frequency is the number of sampling points in unit time or unit distance. Thus if a waveform is sampled every two milliseconds (sampling interval), the sampling frequency is 500 samples per second (or 500 Hz). Sampling at this rate will preserve all frequencies up to 250 Hz in the sampled function. This frequency of half the sampling frequency is known as the *Nyquist frequency* (f_N) and the *Nyquist interval* is the frequency range from zero up to f_N.

$$f_N = 1/2\Delta t \qquad (2.1)$$

where Δt = sampling interval.

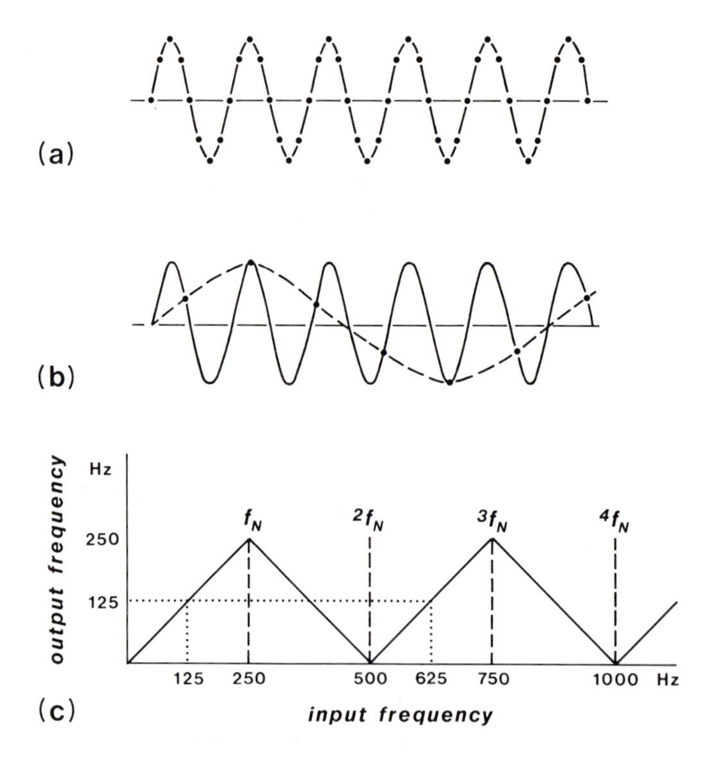

Fig. 2.2. (a) Sine wave frequency less than Nyquist frequency. (b) Sine wave frequency greater than Nyquist frequency showing the fictitious frequency that is generated by aliasing. (c) Relationship between input and output frequencies for a sampling frequency of 500 Hz (Nyquist frequency f_N = 250 Hz).

If frequencies above the Nyquist frequency are present in the sampled function, a serious form of distortion results known as *aliasing*, in which the higher frequency components are 'folded back' into the Nyquist interval. Consider the example illustrated in Fig. 2.2 in which a sine wave is sampled at different sampling frequencies. At the higher sampling rate (Fig. 2.2(a)) the waveform is accurately reproduced but at the lower rate (Fig. 2.2(b)) it is rendered as a fictitious frequency within the Nyquist interval. The relationship between input and output frequencies in the case of a sampling frequency of 500 Hz is shown in Fig. 2.2(c). It is apparent that an input frequency of 125 Hz, for example, is retained in the output but that an input frequency of 625 Hz is folded back to be output at 125 Hz also.

To overcome the problem of aliasing, either the sampling frequency must be at least twice as high as the highest frequency component present in the sampled function, or the function must be passed through an *antialias filter* prior to digitization. The antialias filter is a high cut filter with a sharp cut-off that removes frequency components above the Nyquist frequency, or attenuates them to an insignificant amplitude level.

2.3 Spectral analysis

A distinction may be made between *periodic waveforms* (Fig. 2.3(a)), that repeat themselves at a fixed time period T, and *transient waveforms* (Fig. 2.3(b)), that are non-repetitive.

By means of *Fourier analysis* any periodic waveform, however complex, may be decomposed into a series of sine (or cosine) waves whose frequencies are

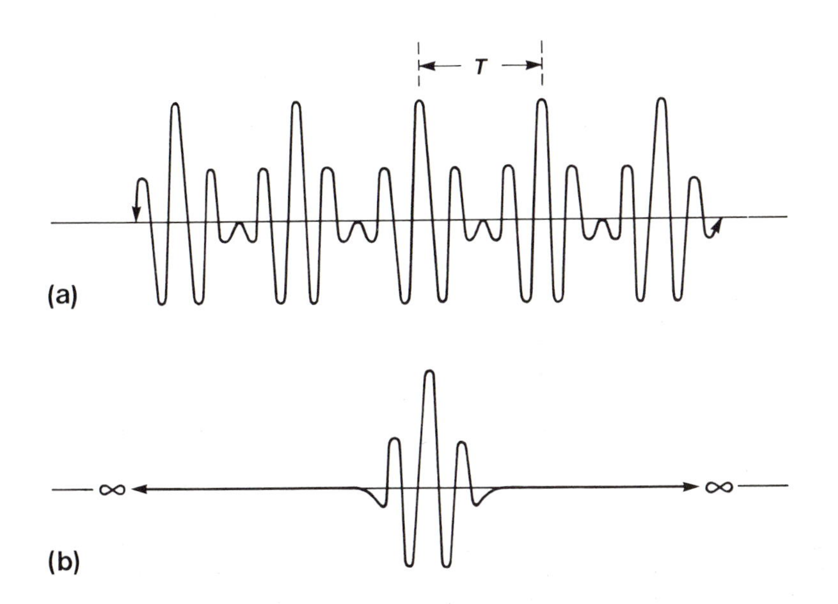

Fig. 2.3. (a) Periodic and (b) transient waveforms.

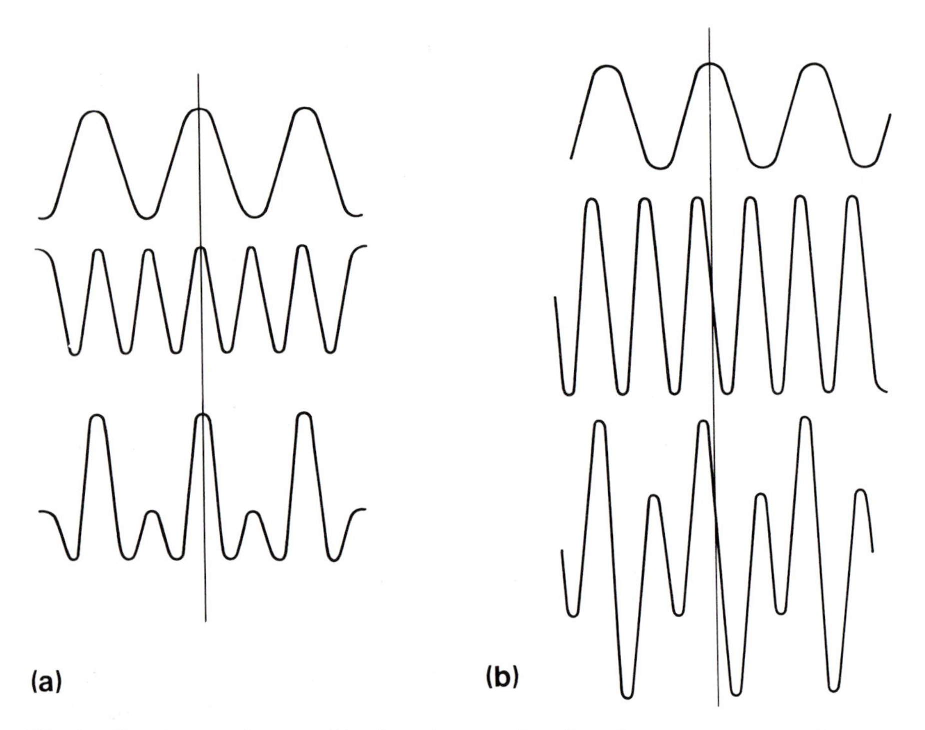

Fig. 2.4. Complex waveforms resulting from the summation of two sine wave components of frequency f and $2f$. (a) The two sine wave components are of equal amplitude and in phase. (b) The higher frequency component has twice the amplitude of the lower frequency component and is $\pi/2$ out of phase. (After Anstey 1965.)

integral multiples of the basic repetition frequency $1/T$, known as the fundamental frequency. The higher frequency components, at frequencies of n/T ($n = 1, 2, 3 \ldots$), are known as harmonics. Thus the complex waveform of Fig. 2.4(a) is built up from the addition of the two individual sine wave components shown. To express any waveform in terms of its constituent sine wave components, it is necessary to define not only the frequency of each component but also its amplitude and phase. If in the above example the relative amplitude and phase relations of the individual sine waves are altered, summation can produce the quite different waveform illustrated in Fig. 2.4(b).

From the above it follows that a periodic waveform can be expressed in two different ways: in the *time domain*, expressing wave amplitude as a function of time, and in the *frequency domain*, expressing the amplitude and phase of its constituent sine waves as a function of frequency. The waveforms shown in Fig. 2.4(a) and (b) are represented in Fig. 2.5(a) and (b) in terms of their *amplitude and phase spectra*. These spectra, being composed of a series of discrete amplitude and phase components, are known as line spectra.

Transient waveforms do not repeat themselves, that is, they have an infinitely long period. They may thus be regarded, by analogy with a periodic waveform, as having an infinitesimally small fundamental frequency ($1/T \rightarrow 0$) and, consequently, harmonics that occur at infinitesimally small frequency spacings to give continuous amplitude and phase spectra rather than the line spectra of

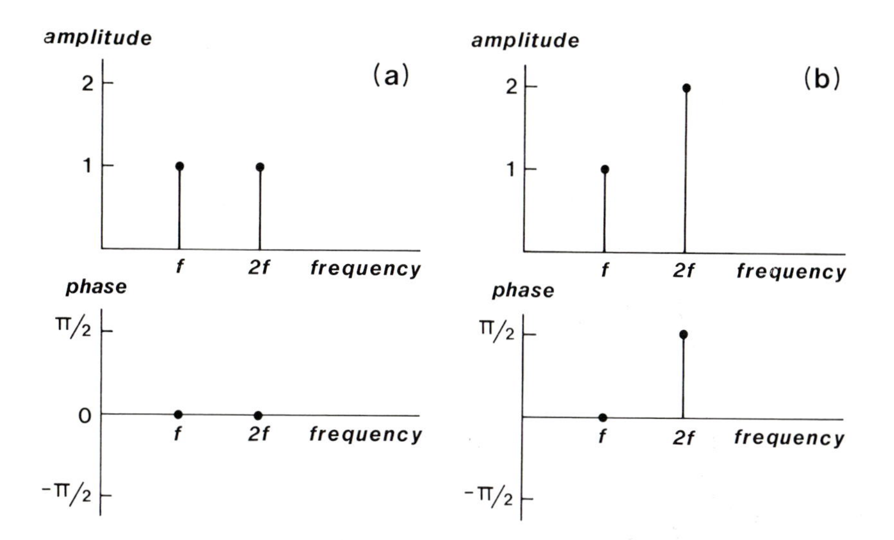

Fig. 2.5. Representation in the frequency domain of the waveforms illustrated in Fig. 2.4, showing their amplitude and phase spectra.

periodic waveforms. Digitization provides a means of dealing with the continuous spectra of transient waveforms. Clearly it is impossible to cope analytically with a spectrum containing an infinite number of sine wave components and the continuous amplitude and phase spectra are therefore subdivided into a number of thin frequency slices, giving each slice a frequency equal to the mean frequency of the slice and an amplitude and phase proportional to the area of the slice of the appropriate spectrum (Fig. 2.6). This digital expression of a continuous spectrum

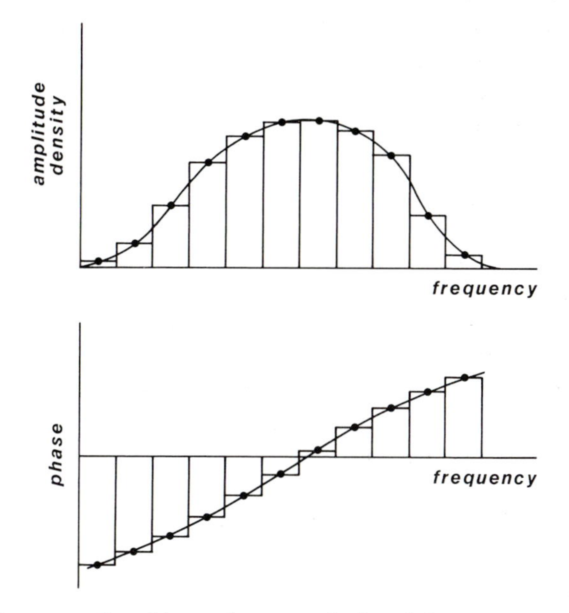

Fig. 2.6. Digital representation of the continuous amplitude and phase spectra associated with a transient waveform.

in terms of a finite number of discrete frequency components provides an approximate representation in the frequency domain of a transient waveform in the time domain. Increasing the number of frequency slices improves the accuracy of the approximation.

Fourier transformation may be used to convert a time function $g(t)$ into its equivalent amplitude and phase spectra $A(f)$ and $\phi(f)$, or into a complex function of frequency $G(f)$ known as the frequency spectrum, where

$$G(f) = A(f)e^{i\phi(f)} \tag{2.2}$$

The time and frequency domain representations of a waveform, $g(t)$ and $G(f)$, are known as a *Fourier pair*, represented by the notation

$$g(t) \longleftrightarrow G(f) \tag{2.3}$$

Components of a Fourier pair are interchangeable, such that, if $G(f)$ is the Fourier transform of $g(t)$, then $g(t)$ is the Fourier transform of $G(f)$.

Fig. 2.7 illustrates Fourier pairs for various waveforms of geophysical significance. All the examples illustrated have zero phase spectra, that is, the individual sine wave components of the waveforms are in phase at zero time. In this case $\phi(f) = 0$ for all values of f.

Fig. 2.7(a) shows a spike function (also known as a Dirac function), which is the shortest possible transient waveform. Fourier transformation shows that the spike function has a continuous frequency spectrum of constant amplitude from zero to infinity; thus, a spike function contains all frequencies from zero to infinity at equal amplitude. The 'DC bias' waveform of Fig. 2.7(b) has, as would be expected, a line spectrum comprising a single component at zero frequency. Note that Fig. 2.7(a) and (b) demonstrate the principle of interchangeability of Fourier pairs stated above (equation (2.3)).

Fig. 2.7(c) and (d) illustrate transient waveforms approximating the shape of seismic pulses, together with their amplitude spectra. Both have a band-limited amplitude spectrum, the spectrum of narrower bandwidth being associated with the longer transient waveform. In general, the shorter a time pulse the wider is its frequency bandwidth and in the limiting case a spike pulse has an infinite bandwidth.

Waveforms with zero phase spectra such as those illustrated in Fig. 2.7 are symmetrical about the time axis and, for any given amplitude spectrum, produce the maximum peak amplitude in the resultant waveform. If phase varies linearly with frequency, the waveform remains unchanged in shape but is displaced in time; if the phase variation with frequency is non-linear the shape of the waveform is altered. A particularly important case in seismic data processing is the phase spectrum associated with *minimum delay* in which there is a maximum concentration of energy at the front end of the waveform. Analysis of seismic pulses sometimes assumes that they exhibit minimum delay (see Chapter 4).

Fourier transformation of digitized waveforms is readily enacted by computers, using a 'fast Fourier transform' (FFT) algorithm as in the Cooley-Tukey method (Brigham 1974). FFT subroutines can thus be routinely built into data processing programmes in order to carry out spectral analysis of geophysical waveforms.

Fourier transformation can be extended into two dimensions (Rayner 1971), and can thus be applied to areal distributions of data such as gravity and magnetic contour maps. In this case the time variable is replaced by horizontal distance and the frequency variable by wavenumber (number of waveform cycles per unit distance). The application of two-dimensional Fourier techniques to the interpretation of potential field data is discussed in Chapters 6 and 7.

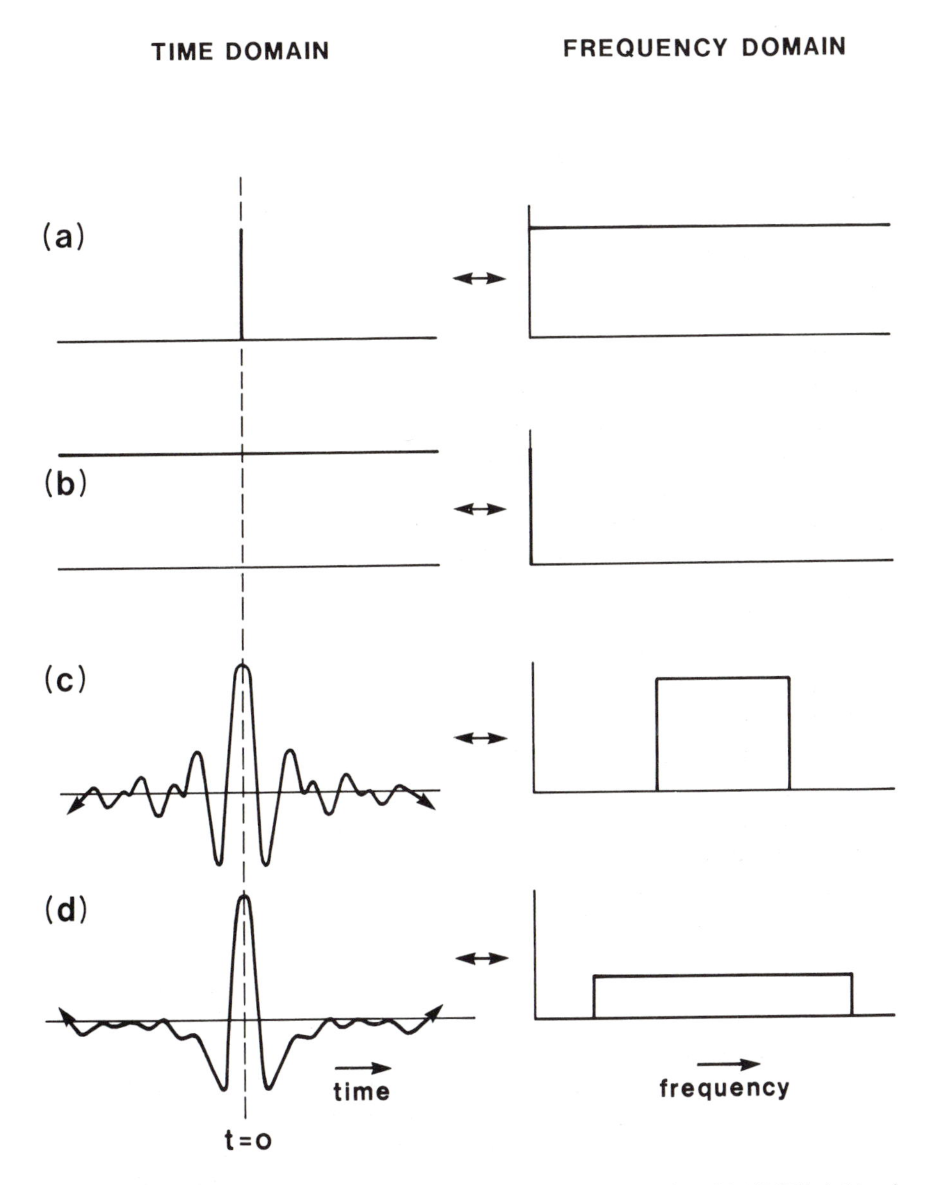

Fig. 2.7. Fourier transform pairs for various waveforms: (a) a spike function; (b) a 'DC bias'; (c) and (d) transient waveforms approximating seismic pulses.

2.4 Waveform processing

The principles of convolution, deconvolution and correlation form the common basis for many methods of geophysical data processing, especially in the field of seismic reflection surveying. They are introduced here in general terms and are referred to extensively in later chapters.

2.4.1 Convolution

Convolution (Kanasewich 1981) is a mathematical operation defining the change of shape of a waveform resulting from its passage through a filter. Filtering modifies a waveform by discriminating against its constituent sine wave components to alter their relative amplitudes or phase relations or both. Filtering is an inherent characteristic of any transmission system. Thus, for example, a seismic pulse generated by an explosion is altered in shape by filtering effects, both in the ground and in the recording system, so that the recorded pulse (the filtered output) differs significantly from the initial pulse (the input).

As a simple example of filtering, consider a weight suspended from the end of a vertical spring. If the top of the spring is perturbed by a sharp up-and-down movement (the input), the response of the weight (the filtered output) is a series of damped oscillations out of phase with the initial perturbation (Fig. 2.8).

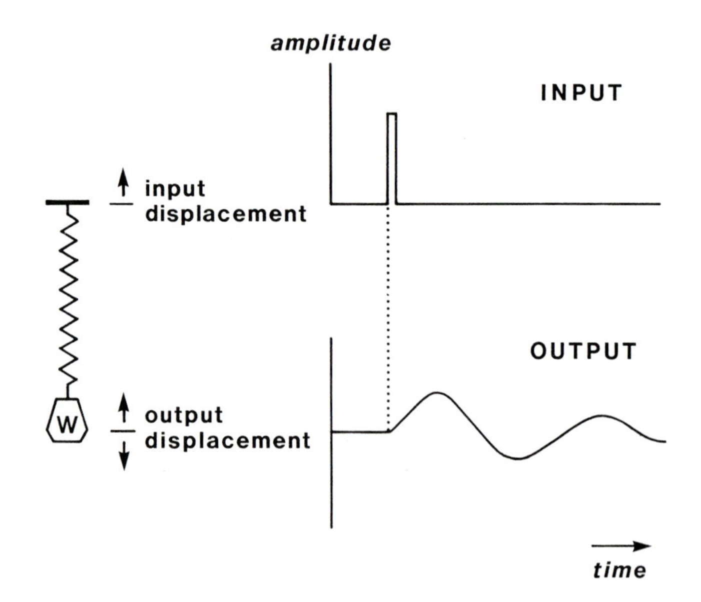

Fig. 2.8. The principle of filtering illustrated by the perturbation of a suspended weight system.

The effect of a filter may be categorized by its *impulse response* which is defined as the output of the filter when the input is a spike function (Fig. 2.9). The Fourier transform of the impulse response is known as the *transfer function* and this specifies the amplitude and phase response of the filter, thus defining its operation completely.

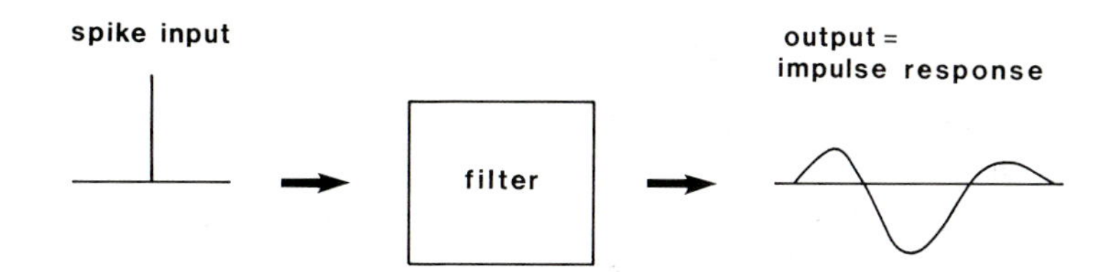

Fig. 2.9. The impulse response of a filter.

The effect of a filter is described mathematically by a convolution operation such that, if the input signal $g(t)$ to the filter is *convolved with* the impulse response $f(t)$ of the filter, known as the *convolution operator*, the filtered output $y(t)$ is obtained

$$y(t) = g(t) \star f(t) \tag{2.4}$$

where the asterisk denotes the convolution operation.

Fig. 2.10(a) shows a spike function input to a filter whose impulse response is given in Fig. 2.10(b). Clearly the latter is also the filtered output since, by definition, the impulse response represents the output for a spike input. Fig. 2.10(c) shows an input comprising two separate spike functions and the filtered

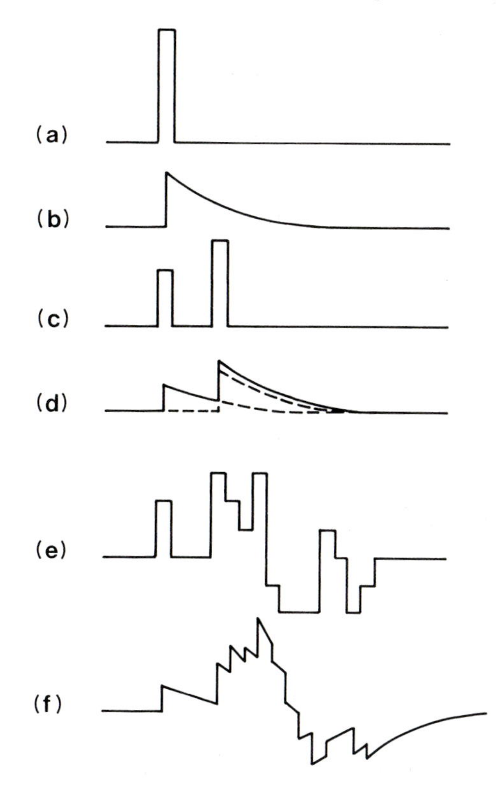

Fig. 2.10. Examples of filtering: (a) a spike input; (b) filtered output equivalent to impulse response of filter; (c) an input comprising two spikes; (d) filtered output given by summation of two impulse response functions offset in time; (e) a complex input represented by a series of contiguous spike functions; (f) filtered output given by the summation of a set of impulse responses. (After Anstey 1965.)

output (Fig. 2.10(d)) is now the superimposition of the two impulse response functions offset in time by the separation of the spikes and scaled according to the individual spike amplitudes. Since any transient wave can be represented as a series of spike functions (Fig. 2.10(e)), the general form of a filtered output (Fig. 2.10(f)) can be regarded as the summation of a set of impulse responses related to a succession of spikes simulating the overall shape of the input wave.

In Fig. 2.11 the individual steps in the convolution process are shown for two digital functions, a double spike function given by $g_i = g_1, g_2, g_3 = 2, 0, 1$ and an impulse response function given by $f_i = f_1, f_2, f_3, f_4 = 4, 3, 2, 1$, where the numbers refer to discrete amplitude values at the sampling points of the two functions.

From Fig. 2.11 it can be seen that the convolved output $y_i = y_1, y_2, y_3, y_4, y_5, y_6 = 8, 6, 8, 5, 2, 1$. Note that the convolved output is longer than the input waveforms: if the functions to be convolved have lengths of m and n, the convolved output has a length of $(m+n-1)$.

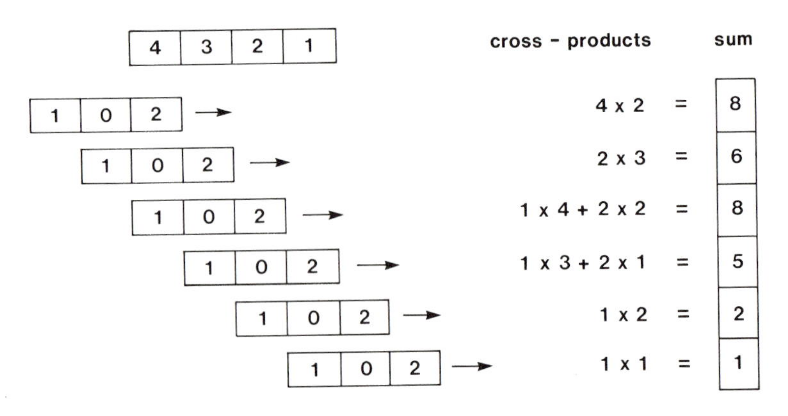

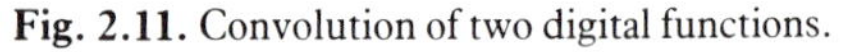
Fig. 2.11. Convolution of two digital functions.

Convolution involves time inversion (or folding) of one of the functions and its progressive sliding past the other function, the individual terms in the convolved output being derived by summation of the cross-multiplication products over the overlapping parts of the two functions.

In general, if $g_i (i = 1, 2, \ldots, m)$ is an input function and $f_j (j = 1, 2, \ldots, n)$ is a convolution operator, then the convolution output function y_k is given by

$$y_k = \sum_{i=1}^{m} g_i f_{k-i} \quad (k = 1, 2, \ldots, m+n-1)$$

It can be shown that the convolution of two functions in the time domain is mathematically equivalent to multiplication of their amplitude spectra and addition of their phase spectra in the frequency domain. The operation of convolution can thus be performed by transforming the time functions into the frequency domain, multiplying their amplitude spectra, summing their phase spectra and taking the inverse transform of the resultant frequency spectrum. Thus, digital filtering can be enacted in either the time domain or the frequency domain. With large data sets, filtering by computer is more efficiently carried out in the frequency domain as less mathematical operations are involved.

Convolution, or its equivalent in the frequency domain, finds very wide application in geophysical data processing, notably in the digital filtering of seismic and potential field data and the construction of synthetic seismograms for comparison with field seismograms (see Chapters 4 and 6).

2.4.2 Deconvolution

Deconvolution, or *inverse filtering* (Kanasewich 1981) is a process that counteracts a previous convolution (or filtering) action. Consider the convolution operation given in equation (2.4)

$$y(t) = g(t) \star f(t)$$

$y(t)$ is the filtered output derived by passing the input waveform $g(t)$ through a filter of impulse response $f(t)$. Knowing $y(t)$ and $f(t)$, the recovery of $g(t)$ represents a deconvolution operation. Suppose that $f'(t)$ is the function that must be convolved with $y(t)$ to recover $g(t)$

$$g(t) = y(t) \star f'(t) \tag{2.6}$$

Substituting for $y(t)$ as given by equation (2.4)

$$g(t) = g(t) \star f(t) \star f'(t) \tag{2.7}$$

Now

$$g(t) = g(t) \star \delta(t) \tag{2.8}$$

where $\delta(t)$ is a spike function (a unit amplitude spike at zero time); that is, a time function $g(t)$ convolved with a spike function produces an unchanged convolution output function $g(t)$. From equations (2.7) and (2.8) it follows that

$$f(t) \star f'(t) = \delta(t)$$

Thus, $f'(t)$ can be derived for application in equation (2.6) to recover the input signal $g(t)$. The function $f'(t)$ represents the *deconvolution operator*.

Deconvolution is an essential aspect of seismic data processing, being used to improve seismic records by removing the adverse filtering effects encountered by seismic waves during their passage through the ground. In the seismic case, referring to equation (2.4), $y(t)$ is the seismic record resulting from the passage of a seismic wave $g(t)$ through a portion of the Earth, which acts as a filter with an impulse response $f(t)$. The particular problem with deconvolving a seismic record is that the input waveform $g(t)$ and the impulse response $f(t)$ of the Earth filter are in general unknown. Thus the 'deterministic' approach to deconvolution outlined above cannot be employed and the deconvolution operator has to be designed using statistical methods. This special approach to the deconvolution of seismic records, known as predictive deconvolution, is discussed further in Chapter 4.

2.4.3 *Correlation*

Cross-correlation of two digital waveforms involves cross-multiplication of the individual waveform elements and summation of the cross-multiplication products over the common time interval of the waveforms. The cross-correlation function involves progressively sliding one waveform past the other and, for each time shift, or lag, summing the cross-multiplication products to derive the cross-correlation as a function of lag value. The cross-correlation operation is similar to convolution but does not involve folding of one of the waveforms.

Given two digital waveforms of finite length, x_i and y_i ($i = 1, 2, \ldots, n$), the cross-correlation function is given by

$$\phi_{xy}(\tau) = \sum_{i=1}^{n-\tau} x_{i+\tau} y_i \, (-m < \tau < +m)$$

where τ is the lag and m is known as the maximum lag value of the function.

It can be shown that cross-correlation in the time domain is mathematically equivalent to multiplication of amplitude spectra and subtraction of phase spectra in the frequency domain.

Clearly if two identical non-periodic waveforms are cross-correlated (Fig. 2.12) all the cross-multiplication products will sum at zero lag to give a maximum positive value. When the waveforms are displaced in time, however, the cross-multiplication products will tend to cancel out to give small values. The cross-correlation function therefore peaks at zero lag and reduces to small values at large time shifts. Two closely similar waveforms will likewise produce a cross-correlation function that is strongly peaked at zero lag. On the other hand, if two

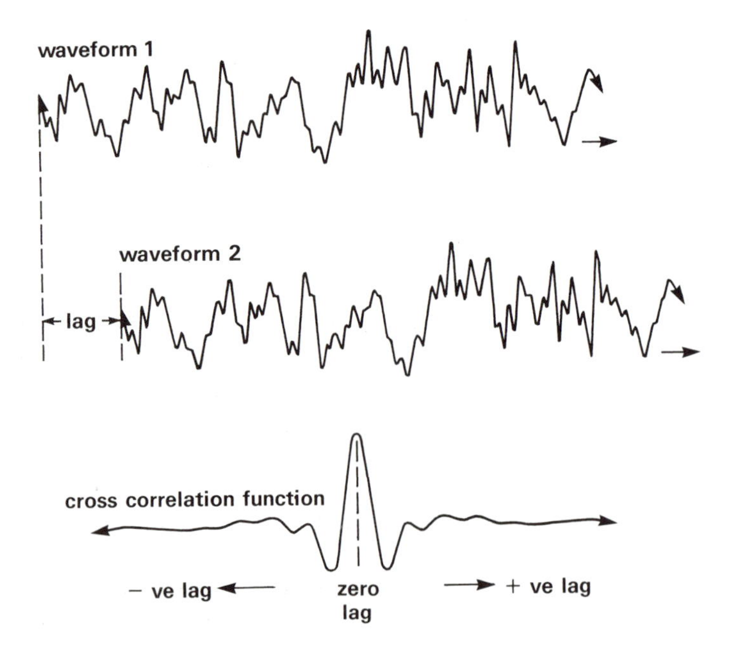

Fig. 2.12 Cross-correlation of two identical waveforms.

dissimilar waveforms are cross-correlated the sum of cross-multiplication products will always be near to zero due to the tendency for positive and negative products to cancel out at all values of lag. In fact, for two waveforms containing only random noise the cross-correlation function $\phi_{xy}(\tau)$ is zero for all values of τ. Thus, the cross-correlation function measures the degree of similarity of waveforms.

An important application of cross-correlation is in the detection of weak signals embedded in noise. If a waveform contains a known signal concealed in noise at unknown time, cross-correlation of the waveform with the signal function will produce a cross-correlation function centred on the time value at which the signal function and its concealed equivalent in the waveform are in phase (Fig. 2.13).

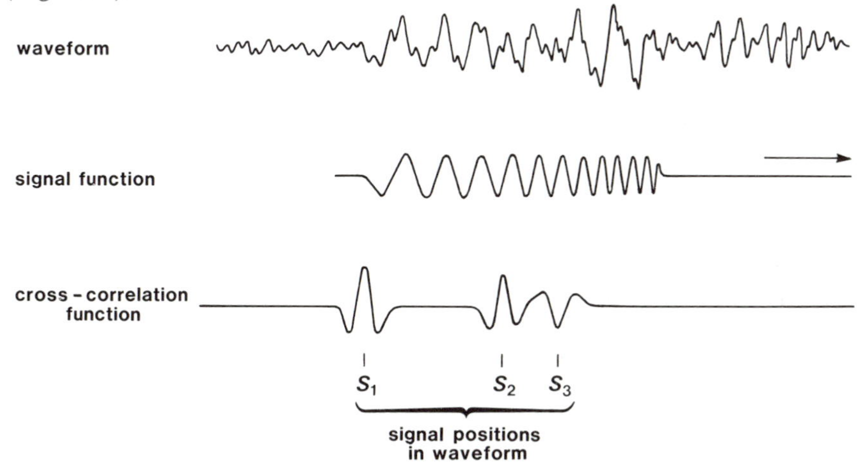

Fig. 2.13. Cross-correlation to detect occurrences of a known signal concealed in noise. (After Sheriff 1973.)

A special case of correlation is that in which a waveform is cross-correlated with itself, to give the *autocorrelation function* $\phi_{xx}(\tau)$. This function is symmetrical about a zero lag position, so that

$$\phi_{xx}(\tau) = \phi_{xx}(-\tau).$$

The autocorrelation function of a periodic waveform is also periodic, with a frequency equal to the repetition frequency of the waveform. Thus, for example, the autocorrelation function of a cosine wave is also a cosine wave. For a transient waveform, the autocorrelation function decays to small values at large values of lag. These differing properties of the autocorrelation function of periodic and transient waveforms determine one of its main uses in geophysical data processing, namely, the detection of hidden periodicities in any given waveform. Side lobes in the autocorrelation function (Fig. 2.14) are an indication of the existence of periodicities in the original waveform, and the spacing of the side lobes defines the repetition period. This property is particularly useful in the detection and suppression of multiple reflections in seismic records (see Chapter 4).

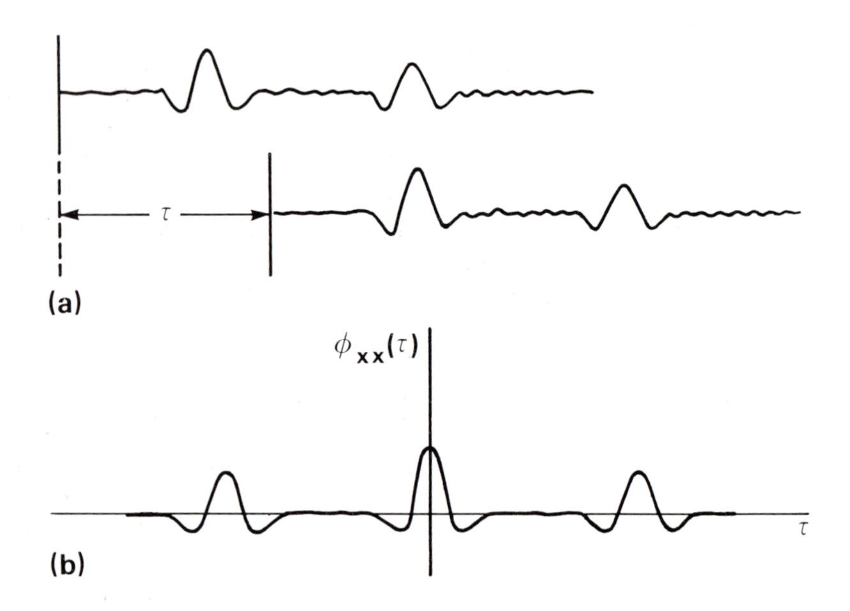

Fig. 2.14. Autocorrelation of the waveform exhibiting periodicity shown in (a) produces the autocorrelation function with side lobes shown in (b). The spacing of the side lobes defines the repetition period of the original waveform.

The autocorrelation function contains all the frequency information of the original waveform but none of the phase information, the original phase relationships being replaced by a zero phase spectrum. In fact, the autocorrelation function and the square of the amplitude spectrum $A(f)$ can be shown to form a Fourier pair

$$\phi_{xx}(\tau) \longleftrightarrow A(f)^2$$

Since the square of the amplitude represents the power term (energy contained in the frequency component) the autocorrelation function can be used to compute the *power spectrum* of a waveform.

2.5 Digital filtering

In waveforms of geophysical interest, the signal is almost invariably superimposed on unwanted noise. In favourable circumstances the signal/noise ratio (SNR) is high, so that the signal is readily identified and extracted for subsequent analysis. Often the SNR is low and special processing is necessary to enhance the information content of the waveforms. Digital filtering is widely employed in geophysical data processing to improve SNR or otherwise improve the signal characteristics.

A very wide range of digital filters is in routine use in geophysical, and especially seismic, data processing. The two main types of digital filter are frequency filters and inverse (deconvolution) filters.

2.5.1 *Frequency filters*

Frequency filters discriminate against selected frequency components of an input waveform and may be low-pass (LP), high-pass (HP), band-pass (BP) or band-reject (BR) in terms of their frequency response. Frequency filters are employed when the signal and noise components of a waveform have different frequency characteristics and can therefore be separated on this basis.

Analogue frequency filtering is still in widespread use and analogue antialias (LP) filters are an essential component of analogue-to-digital conversion systems (Section 2.2). Nevertheless, digital frequency filtering by computer offers much greater flexibility of filter design and facilitates filtering of much higher performance than can be obtained with analogue filters.

To illustrate the design of a digital frequency filter, consider the case of a LP filter whose cut-off frequency is f_c. The desired output characteristics of the ideal LP filter are represented by the amplitude spectrum shown in Fig. 2.15(a). The spectrum has a constant unit amplitude between 0 and f_c and zero amplitude outside this range: the filter would therefore pass all frequencies between 0 and f_c without attenuation and would totally suppress frequencies above f_c. This amplitude spectrum represents the transfer function of the ideal LP filter.

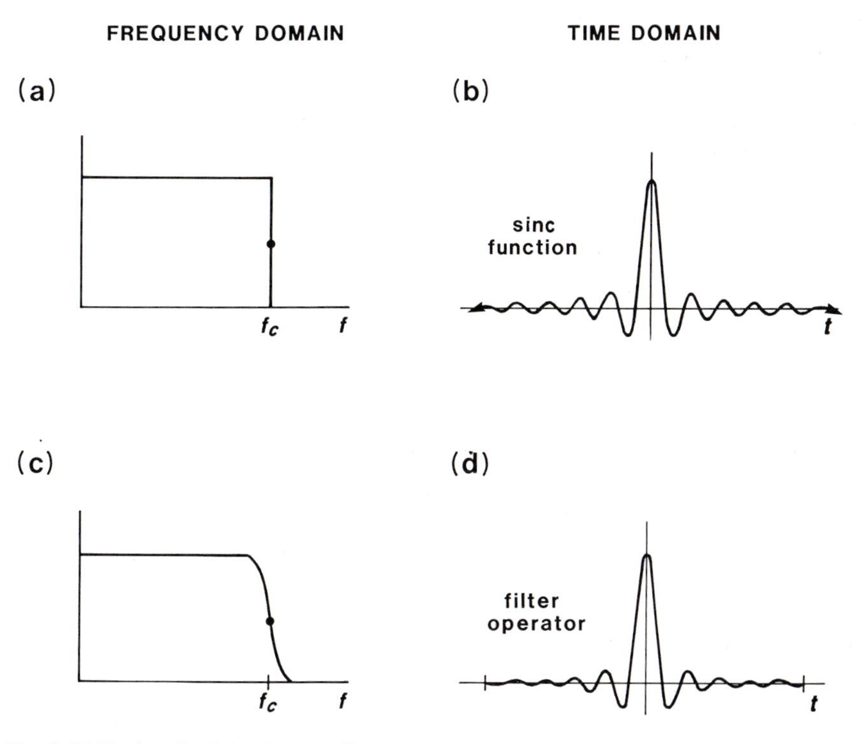

Fig. 2.15. Design of a digital low-pass filter.

Inverse Fourier transformation of the transfer function into the time domain yields the impulse response of the ideal LP filter (see Fig. 2.15(b)). However, this impulse response (a sinc function) is infinitely long and must therefore be truncated for practical use as a convolution operator in a digital filter. Fig. 2.15(c) represents the frequency response of a practically realizable LP filter operator of finite length (Fig. 2.15(d)). Convolution of the input waveform with the latter will result in LP filtering with a ramped cut-off (Fig. 2.15(c)) rather than the instantaneous cut-off of the ideal LP filter.

HP, BP and BR time-domain filters can be designed in a similar way by specifying a particular transfer function in the frequency domain and using this to design a finite-length impulse response function in the time domain. As with analogue filtering, digital frequency filtering generally alters the phase spectrum of the waveform and this effect may be undesirable. However, *zero-phase filters* can be designed that facilitate digital filtering without altering the phase spectrum of the filtered signal.

2.5.2 *Inverse (deconvolution) filters*

The main applications of inverse filtering to remove the adverse effects of a previous filtering operation lie in the field of seismic data processing. A discussion of inverse filtering in the context of deconvolving seismic records is given in Chapter 4.

Further reading

Brigham, E.O. (1974) *The Fast Fourier Transform*. Prentice-Hall, New Jersey.
Claerbout, J.F. (1976) *Fundamentals of Geophysical Data Processing*. McGraw-Hill, New York.
Kanasewich, E.R. (1981) *Time Sequence Analysis in Geophysics*. University of Alberta Press (3rd edn).
Kulhánek, O. (1976) *Introduction to Digital Filtering in Geophysics*. Elsevier, Amsterdam.
Rayner, J.N. (1971) *An Introduction to Spectral Analysis*. Pion, England.
Robinson, E.A. & Trietel, S. (1980) *Geophysical Signal Analysis*. Prentice-Hall, New Jersey.
Sheriff, R.E. & Geldart, L.P. (1983) *Exploration Seismology Vol 2: Data-processing and Interpretation*. Cambridge University Press, Cambridge.

3

Elements of Seismic Surveying

3.1 Introduction

In seismic surveying, seismic waves are propagated through the Earth's interior and the travel times are measured of waves that return to the surface after refraction or reflection at geological boundaries within the ground. These travel times may be converted into depth values and, hence, the distribution of subsurface interfaces of geological interest may be systematically mapped.

Seismic surveying was first carried out in the early 1920s. It represented a natural development of the already long-established methods of earthquake seismology in which the travel times of earthquake waves recorded at seismological observatories are used to derive information on the internal structure of the Earth. Earthquake seismology provides information on the gross internal layering of the Earth, and measurement of the velocity of earthquake waves through the various Earth layers provides major clues as to their composition and constitution. In the same way, but on a smaller scale, seismic surveying provides a clear and, indeed, uniquely detailed picture of subsurface geology. It undoubtedly represents the single most important geophysical surveying method, in terms of the amount of survey activity and the very wide range of its applications.

Many of the principles of earthquake seismology are applicable to seismic surveying. However, the latter is concerned solely with the structure of the ground down to several kilometres at most and utilizes artificial seismic sources such as explosions, whose location, timing and source characteristics are, unlike earthquakes, under the direct control of the geophysicist. Seismic surveying also utilizes specialized recording systems and associated data processing and interpretation techniques.

Seismic methods are widely applied to exploration problems involving the detection and mapping of subsurface boundaries of, normally, simple geometry. The methods are particularly well suited to the mapping of layered sedimentary sequences and are therefore widely used in the search for oil or gas. The methods are also well suited, on a smaller scale, to the mapping of near-surface sediment layers, the location of the water table and, in an engineering context, site investigation of foundation conditions including the determination of depth to bedrock. Seismic surveying can be carried out on land or at sea, and it is used extensively in offshore geological surveys and the exploration for offshore resources.

In this chapter the physical principles on which seismic methods are based are reviewed at an elementary level, starting with a discussion of the nature of seismic waves and going on to consider their mode of propagation through the ground, with particular reference to reflection and refraction at subsurface interfaces between different rock types. To understand the different types of seismic wave that propagate through the ground away from a seismic source some elementary concepts of stress and strain need to be considered.

3.2 Stress and strain

When external forces are applied to a body, balanced internal forces are set up within it. *Stress* is a measure of the intensity of these balanced internal forces. The stress acting on an area of any surface within the body may be resolved into a component of normal stress perpendicular to the surface and a component of shearing stress in the plane of the surface.

At any point in a stressed body three orthogonal planes can be defined on which the components of stress are wholly normal stresses, that is, no shearing stresses act along them. These planes define three orthogonal axes known as the principal axes of stress, and the normal stresses acting in these directions are known as the *principal stresses*. Each principal stress represents a balance of equal-magnitude but oppositely-directed force components. The stress is said to be compressive if the forces are directed towards each other and tensile if they are directed away from each other.

If the principal stresses are all of equal magnitude within a body the condition of stress is said to be *hydrostatic*, since this is the state of stress throughout a fluid body at rest. No shearing stresses exist in a hydrostatic stress field since these cannot be sustained by a fluid body. If the principal stresses are unequal, shearing stresses exist along all surfaces within the stressed body except for the three orthogonal planes intersecting in the principal axes.

A body subjected to stress undergoes a change of shape and/or size known as *strain*. Up to a certain limiting value of stress, known as the yield strength of a material, the strain is linearly related to the applied stress (Hooke's Law). This elastic strain is reversible so that removal of stress leads to a removal of strain. If the yield strength is exceeded the strain becomes non-linear and partly irreversible (i.e. permanent strain results), and is known as plastic or ductile strain. If the stress is increased still further the body fails by fracture. A typical stress-strain curve is illustrated in Fig. 3.1.

The linear relationship between stress and strain in the elastic field is specified for any material by its various *elastic moduli*, each of which expresses the ratio of a particular type of stress to the resultant strain. Consider a rod of original length l and cross-sectional area A which is extended by an increment Δl through the application of a stretching force F to its end faces (Fig. 3.2(a)). The relevant elastic modulus is Young's modulus E, defined by

$$\text{Young's modulus } E = \frac{\text{longitudinal stress } F/A}{\text{longitudinal strain } \Delta l/l}$$

Note that extension of such a rod will be accompanied by a reduction in its

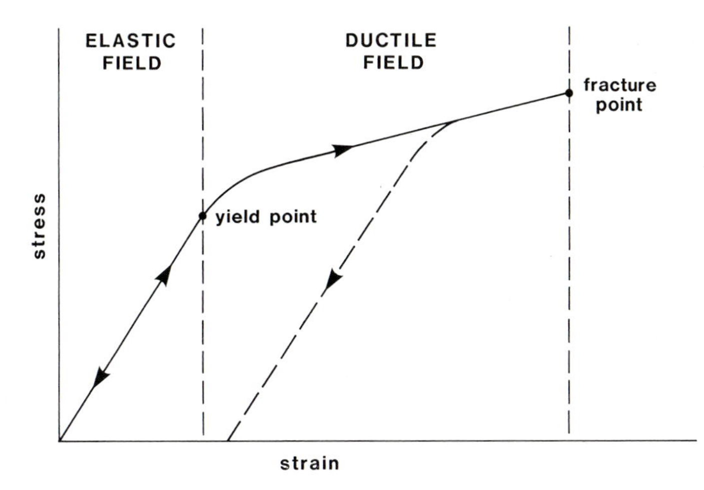

Fig. 3.1. A typical stress-strain curve for a solid body.

diameter, i.e. the rod will suffer lateral as well as longitudinal strain. The ratio of the lateral to the longitudinal strain is known as *Poisson's ratio* (σ).

The *bulk modulus K* expresses the stress-strain ratio in the case of a simple hydrostatic pressure P applied to a cubic element (Fig. 3.2(b)), the resultant

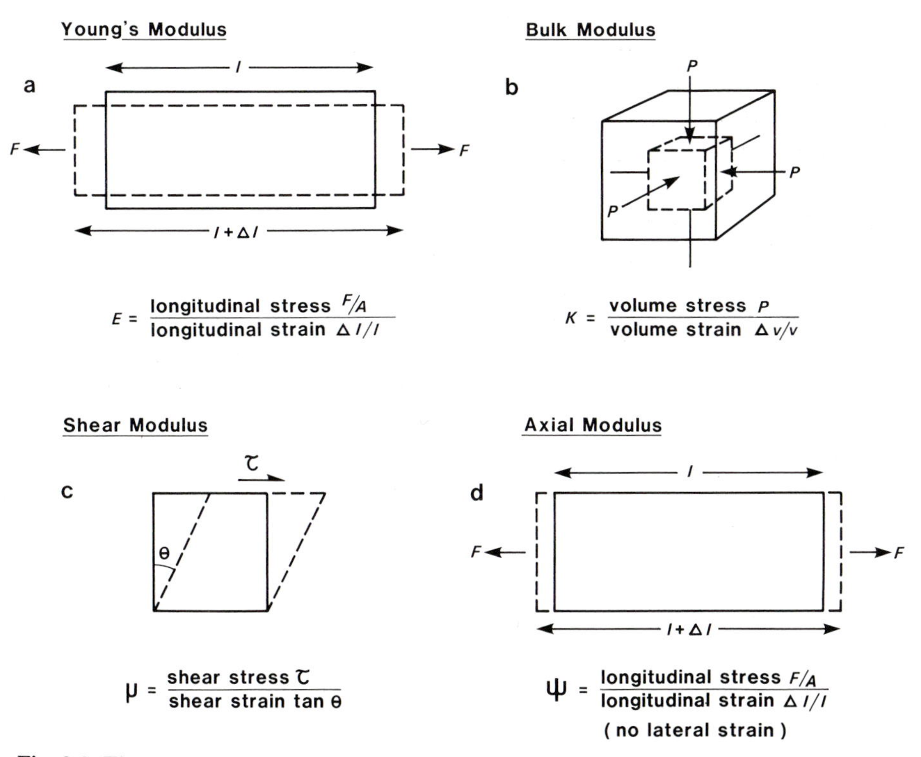

Fig. 3.2. The elastic moduli: (a) Young's modulus E; (b) bulk modulus K; (c) shear modulus μ; (d) axial modulus ψ.

volume strain being the change of volume Δv divided by the original volume v

$$\text{Bulk modulus } K = \frac{\text{volume stress } P}{\text{volume strain } \Delta v/v}$$

In a similar manner the *shear modulus* (μ) is defined as the ratio of shearing stress (τ) to the resultant shear strain tan θ (Fig. 3.2(c))

$$\text{Shear modulus} = \frac{\text{shearing stress } \tau}{\text{shear strain } \tan\theta}$$

Finally, the *axial modulus* ψ defines the ratio of longitudinal stress to longitudinal strain in the case when there is no lateral strain, i.e. when the material is constrained to deform uniaxially (Fig. 3.2(d))

$$\text{Axial modulus } \psi = \frac{\text{longitudinal stress } F/A}{\text{longitudinal strain (uniaxial) } \Delta l/l}$$

3.3 Seismic waves

Seismic waves are parcels of elastic strain energy that propagate outwards from a seismic source such as an earthquake or an explosion. Sources suitable for seismic surveying generate shortlived wave trains known as pulses that typically contain a wide range of frequencies. Except in the immediate vicinity of the source, the strains associated with the passage of a seismic pulse are minute and may be assumed to be elastic. On this assumption the propagation velocities of seismic pulses are determined by the elastic moduli and densities of the materials through which they pass. There are two groups of seismic waves, *body waves* and *surface waves*.

3.3.1 Body waves

Body waves of two types can propagate through the body of an elastic solid. *Compressional waves* (the longitudinal, primary or P-waves of earthquake seismology) propagate by compressional and dilatational uniaxial strains in the direction of wave travel. Particle motion associated with the passage of a compressional wave involves oscillation, about a fixed point, in the direction of wave propagation (Fig. 3.3(a)). *Shear waves* (the transverse, secondary or S-waves of earthquake seismology) propagate by a pure shear strain in a direction perpendicular to the direction of wave travel. Individual particle motions involve oscillation, about a fixed point, in a plane at right angles to the direction of wave propagation (Fig. 3.3(b)). If all the particle oscillations are confined to a plane, the shear wave is said to be plane-polarized.

The velocity of propagation of a body wave in any material is given by

$$v = \left[\frac{\text{appropriate elastic modulus of material}}{\text{density } \rho \text{ of material}}\right]^{1/2}$$

Hence the velocity v_p of a compressional body wave, which involves a uniaxial

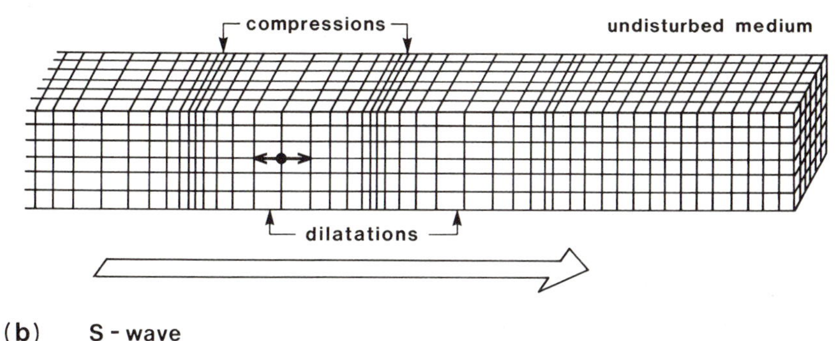

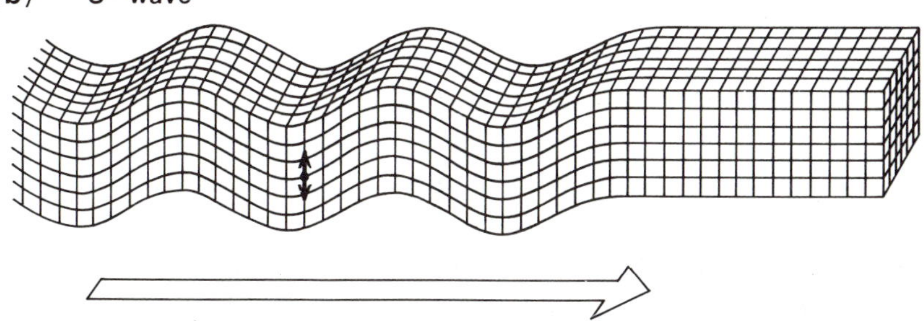

Fig. 3.3. Elastic deformations and ground particle motions associated with the passage of body waves: (a) a P-wave; (b) an S-wave. (From Bolt 1982.)

compressional strain, is given by

$$v_p = \left[\frac{\psi}{\rho}\right]^{1/2}$$

or, since $\psi = K + 4/3\mu$, by

$$v_p = \left[\frac{K + 4/3\mu}{\rho}\right]^{1/2}$$

and the velocity v_s of a shear body wave, which involves a pure shear strain, is given by

$$v_s = \left[\frac{\mu}{\rho}\right]^{1/2}$$

It will be seen from these equations that compressional waves always travel faster than shear waves. The ratio v_p/v_s in any material is determined solely by the value of Poisson's ratio (σ) for that material

$$v_p/v_s = \left[\frac{2(1-\sigma)}{(1-2\sigma)}\right]^{1/2}$$

and since Poisson's ratio for consolidated rocks is typically about 0.25, $v_p \doteq 1.7\, v_s$.

Body waves are non-dispersive, i.e. all frequency components in a wavetrain or pulse travel through any material at the same velocity, determined only by the elastic moduli and density of the material.

3.3.2 Surface waves

In a bounded elastic solid, seismic waves known as surface waves can propagate along the boundary of the solid.

Rayleigh waves propagate along a free surface, or along the boundary between two dissimilar solid media, the associated particle motions being elliptical in a plane perpendicular to the surface and containing the direction of propagation (Fig. 3.4(a)). The orbital particle motion is in the opposite sense to the circular particle motion associated with a oscillatory water wave, and is therefore sometimes described as retrograde. A further major difference between Rayleigh waves and oscillatory water waves is that the former involve a shear strain and are thus restricted to solid media. The amplitude of Rayleigh waves decreases exponentially with distance below the surface. They have a propagation velocity lower than that of shear body waves and in a homogeneous half space they would be non-dispersive. In practice, Rayleigh waves travelling round the surface of the Earth are observed to be dispersive, their waveform undergoing progressive change during propagation as a result of the different frequency components travelling at different velocities. This dispersion is directly attributable to velocity variation with depth in the Earth's interior and, indeed, analysis of the observed pattern of dispersion is a powerful method of studying the velocity structure of the lithosphere and asthenosphere (Knopoff 1983).

In a layered solid a second set of surface waves, known as *Love waves*, appears in the surface layer if its shear body wave velocity v_s is lower than that of the underlying layer. Love waves are polarized shear waves with an associated

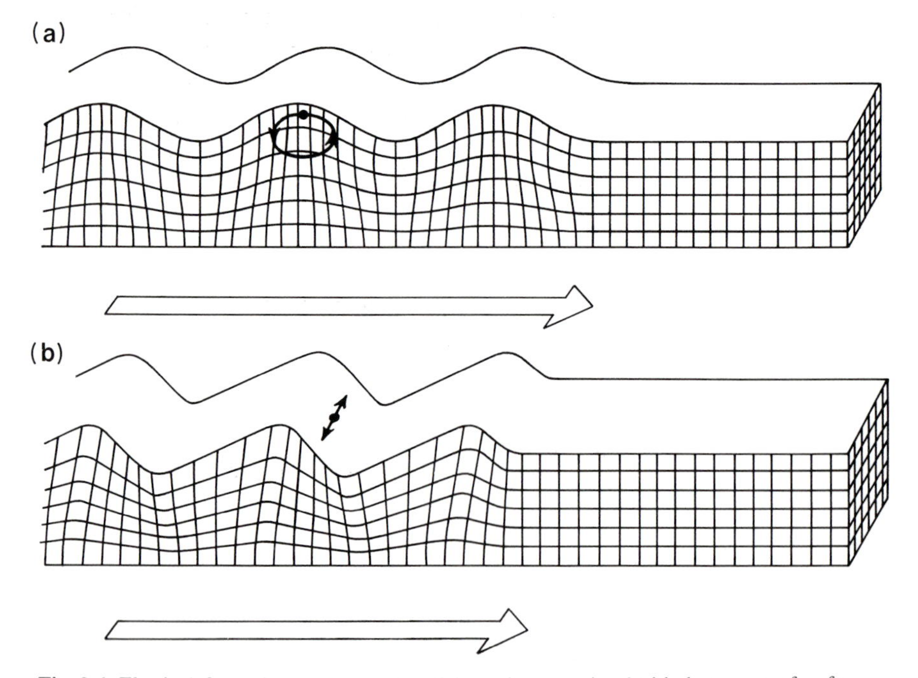

Fig. 3.4. Elastic deformations and ground particle motions associated with the passage of surface waves: (a) a Rayleigh wave; (b) a Love wave. (From Bolt 1982.)

oscillatory particle motion parallel to the free surface and perpendicular to the direction of wave motion (Fig. 3.4(b)). The velocity of Love waves is intermediate between the shear wave velocity of the surface layer and that of deeper layers, and Love waves are inherently dispersive. The observed pattern of Love wave dispersion can be used in a similar way to Rayleigh wave dispersion to study the structure of the lithosphere and asthenosphere (Knopoff 1983).

Although recent experimental surveys of shallow structure have been carried out using shear waves and surface waves (for example, local surface wave dispersion patterns can be used to study the thickness and structure of sedimentary basins), the vast bulk of seismic surveying utilizes only compressional waves and in the following account attention will be concentrated on these waves.

3.3.3 *Waves and rays*

A seismic pulse propagates outwards from a seismic source at a velocity determined by the physical properties of the surrounding rocks. If the pulse travels through a homogeneous rock it will travel at the same velocity in all directions away from the source so that at any subsequent time the wavefront, defined as the locus of all points to which the pulse has reached, will be a sphere. *Seismic rays* are defined as thin pencils of seismic energy travelling along raypaths that, in isotropic media, are everywhere perpendicular to wavefronts (Fig. 3.5). Rays have no physical significance but represent a useful concept in discussing travel paths of seismic energy through the ground.

It should be noted that the propagation velocity of a seismic wave is the velocity with which the seismic energy travels through a medium. This is *not* the same as the velocity of a particle of the medium perturbed by the passage of the wave. In the case of compressional body waves, for example, their propagation velocity through rocks is typically a few thousand metres per second. The associated oscillatory ground motions involve *particle velocities* that depend on the amplitude of the wave. For the weak seismic events routinely recorded in seismic surveys particle velocities may be as small as 10^{-8} m/s and involve ground

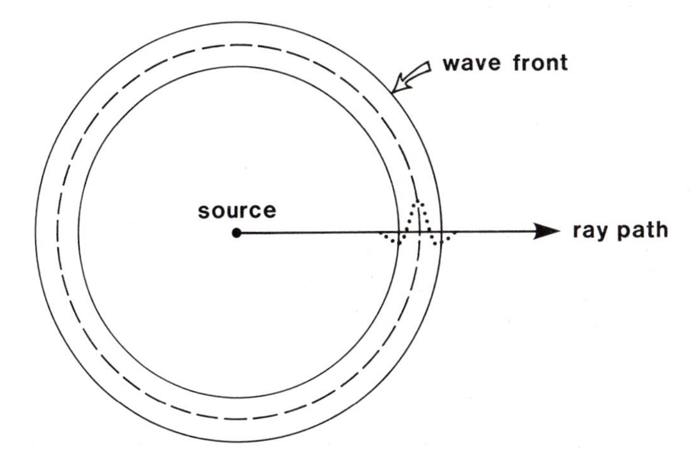

Fig. 3.5. The relationship of a ray path to the associated wavefront.

displacements of only about 10^{-10} m. The detection of seismic waves involves measuring these very small particle velocities.

3.4 Compressional wave velocities of rocks

By virtue of their various compositions, textures (e.g. grain shape and degree of sorting), porosities and contained pore fluids, rocks differ in their elastic moduli and densities and, hence, in their seismic velocities. Information on the compressional wave velocity v_p of rock layers encountered by seismic surveys is important for two main reasons: firstly it is necessary for the conversion of seismic wave travel times into depths; secondly, it provides an indication of the lithology of a rock or, in some cases, the nature of the pore fluids contained within it.

Rock velocities may be measured *in situ* by field measurements, or in the laboratory using suitably prepared rock samples. In the field, seismic surveys yield estimates of velocity for rock layers delineated by reflecting or refracting interfaces, as discussed in detail in Chapters 4 and 5. If boreholes exist in the vicinity of a seismic survey, it may be possible to correlate velocity values so derived with individual rock units encountered within borehole sequences. Velocity may also be measured directly in boreholes using a sonic probe, which emits high frequency pulses and measures the travel time of the pulses through a small vertical interval of wall rock. Drawing the probe up through the borehole yields a *sonic log*, or continuous velocity log (CVL), which is a record of velocity variation through the borehole section (Fig. 3.6).

In the laboratory, velocities are determined by measuring the travel time of high frequency (about 1 MHz) acoustic pulses transmitted through cylindrical rock specimens. By this means, the effect on velocity of varying temperature, confining pressure, pore fluid pressure or composition may be quantitatively assessed. It is important to note that laboratory measurements at low confining pressures are of doubtful validity. The intrinsic velocity of a rock is not normally attained in the laboratory below a confining pressure of about 100 MPa (megapascals), or 1 kbar, at which pressure the original solid contact between grains characteristic of the pristine rock is re-established.

The following general findings of velocity studies are noteworthy:

(1) Compressional wave velocity increases with confining pressure (very rapidly over the first 100 MPa).

(2) Sandstone and shale velocities show a systematic increase with depth of burial and with age, due to the combined effects of progressive compaction and cementation.

(3) For a wide range of sedimentary rocks the compressional wave velocity is related to density, and well-established velocity-density curves have been published (Sheriff & Geldart 1983, chapter 7) (see Section 6.10; Fig. 6.16). Hence, the density of inaccessible subsurface layers may be predicted if their velocity is known from seismic surveys.

(4) The presence of gas in sedimentary rocks reduces the elastic moduli, Poisson's ratio and the v_p/v_s ratio. v_p/v_s ratios greater than 2.0 are characteristic of unconsolidated sand, whilst values less then 2.0 may indicate either a con-

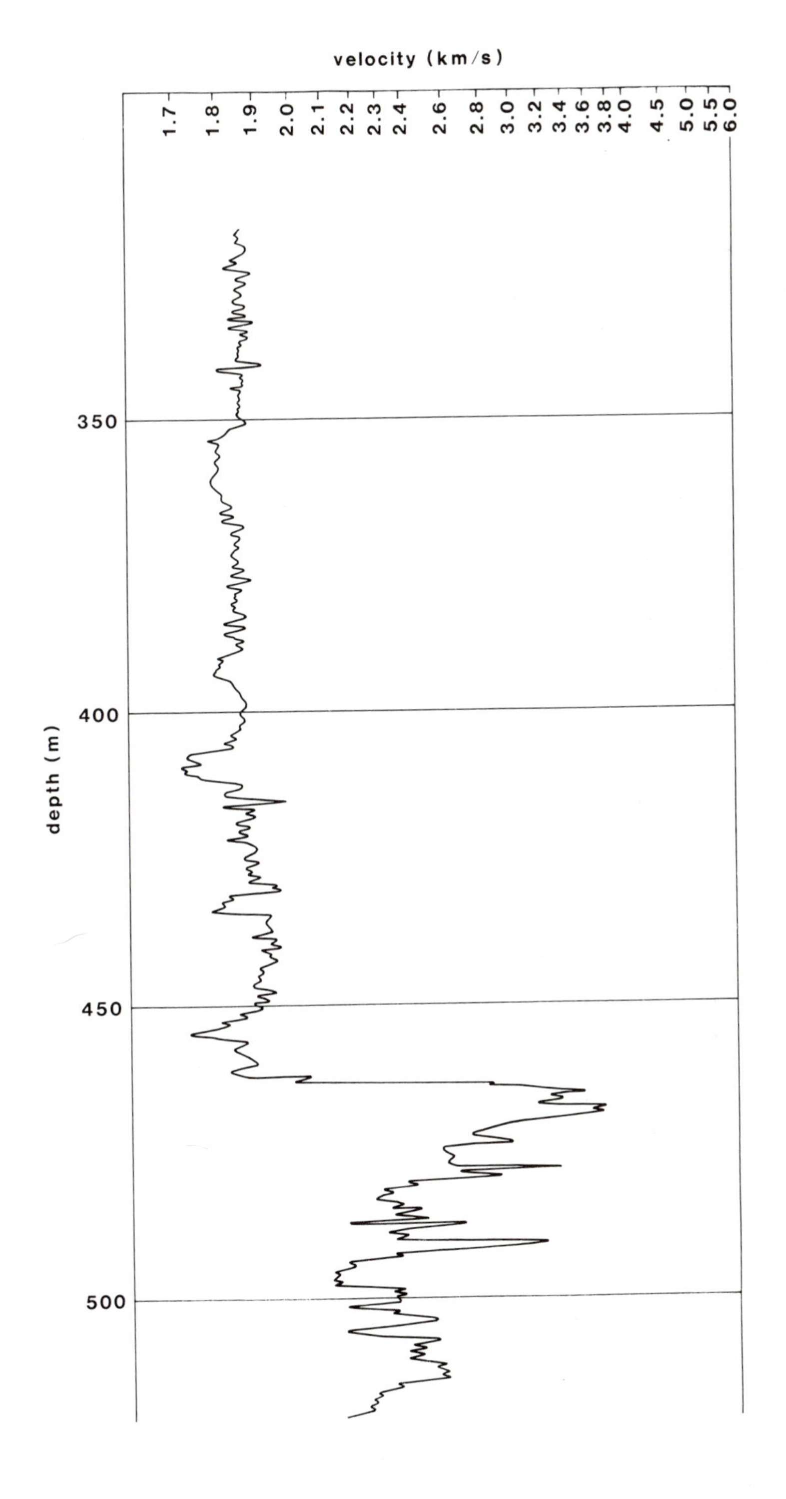

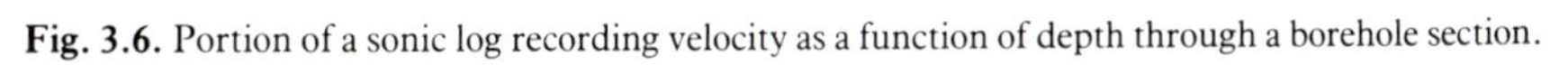
Fig. 3.6. Portion of a sonic log recording velocity as a function of depth through a borehole section.

solidated sandstone or a gas-filled unconsolidated sand. The potential value of v_s in detecting gas-filled sediments accounts for the current interest in shear wave seismic surveying.

Typical compressional wave velocity values and ranges for a wide variety of Earth materials are given in the following table

Table 3.1. Compressional wave velocities in Earth materials.

	v_p (km/s)
Unconsolidated materials	
Sand (dry)	0.2–1.0
Sand (water saturated)	1.5–2.0
Clay	1.0–2.5
Glacial till (water saturated)	1.5–2.5
Permafrost	3.5–4.0
Sedimentary rocks	
Sandstones	2.0–6.0
Tertiary sandstone	2.0–2.5
Pennant sandstone (Carboniferous)	4.0–4.5
Cambrian quartzite	5.5–6.0
Limestones	2.0–6.0
Cretaceous chalk	2.0–2.5
Jurassic oolites and bioclastic limestones	3.0–4.0
Carboniferous limestone	5.0–5.5
Dolomites	2.5–6.5
Salt	4.5–5.0
Anhydrite	4.5–6.5
Gypsum	2.0–3.5
Igneous/Metamorphic rocks	
Granite	5.5–6.0
Gabbro	6.5–7.0
Ultramafic rocks	7.5–8.5
Serpentinite	5.5–6.5
Pore fluids	
Air	0.3
Water	1.4–1.5
Ice	3.4
Petroleum	1.3–1.4
Other materials	
Steel	6.1
Iron	5.8
Aluminium	6.6
Concrete	3.6

3.5 Attenuation of seismic energy

As a seismic pulse propagates, the original energy E transmitted outwards from the source becomes distributed over a spherical shell of expanding radius. If the radius of the shell is r, the amount of energy contained within a unit area of the shell is $E/4\pi r^2$. Along a raypath, therefore, the energy contained in the ray falls off as r^{-2} due to the effect of the *geometrical spreading* of the energy. Wave amplitude which, within a homogeneous material, is proportional to the square root of the wave energy, therefore falls off as r^{-1}.

A further cause of energy loss along a raypath arises because the ground is imperfectly elastic in its response to the passage of seismic waves. Elastic energy is gradually absorbed into the medium by internal frictional losses, leading eventually to the total disappearance of the seismic disturbance. *The absorption coefficient* (α) expresses the proportion of energy lost during transmission through a distance equivalent to a complete wavelength λ. Values of α for common Earth materials range from 0.25 to 0.75 decibels per λ.

Over the range of frequencies utilized in seismic surveying the absorption coefficient is normally assumed to be independent of frequency. If the amount of absorption per wavelength is constant, it follows that higher frequency waves attenuate more rapidly than lower frequency waves as a function of time or distance. To illustrate this point, consider two waves with frequencies of 10 Hz and 100 Hz to propagate through a rock in which $v_p = 2.0$ km/s and $\alpha = 0.5$ dB/λ. The 100 Hz wave ($\lambda = 20$ m) will be attenuated due to absorption by 5 dB over a distance of 200 m whereas the 10 Hz wave ($\lambda = 200$ m) will be attenuated by only 0.5 dB over the same distance. The shape of a seismic pulse with a broad

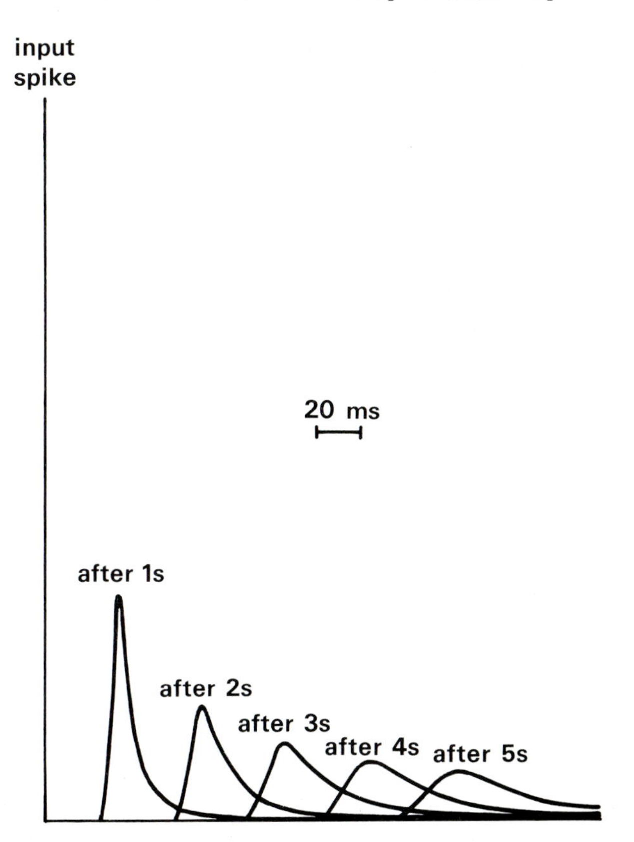

Fig. 3.7 The progressive change of shape of an original spike pulse during its propagation through the ground due to the effects of absorption. (After Anstey 1977.)

frequency content therefore changes continuously during propagation due to the progressive loss of the higher frequencies. In general, the effect of absorption is to produce a progressive lengthening of the seismic pulse (see Fig. 3.7).

3.6 Raypaths in layered media

At an interface between two rock layers there is generally a change of propagation velocity resulting from the difference in physical properties of the two layers. At such an interface the energy within an incident seismic pulse is partitioned into transmitted and reflected pulses. The relative amplitudes of the transmitted and reflected pulses, in terms of the velocities and densities of the two layers, are given by Zoeppritz's equations (Telford *et al.* 1976).

3.6.1 Reflection and transmission of normally incident seismic rays

Consider a compressional ray of amplitude A_0 normally incident on an interface between two media of differing velocity and density (Fig. 3.8). A transmitted ray of amplitude A_2 travels on through the interface in the same direction as the incident ray and a reflected ray of amplitude A_1 returns back along the path of the incident ray.

The total energy of the transmitted and reflected rays must, of course, equal the energy of the incident ray. The relative proportions of energy transmitted and reflected are determined by the contrast in *acoustic impedance* Z across the interface. The acoustic impedance of a rock is the product of its density and its compressional wave velocity, i.e. $Z = \rho v$. It is difficult to relate acoustic impedance to a tangible rock property but, in general, the harder a rock the higher is its acoustic impedance.

Acoustic impedance is closely analogous to electrical impedance and, just as the maximum transmission of electrical energy requires a matching of electrical impedances, so the maximum transmission of seismic energy requires a matching of acoustic impedances. Hence, the smaller the contrast in acoustic impedance across a rock interface the greater is the proportion of energy transmitted through the interface.

The *reflection coefficient* R is the ratio of the amplitude A_1 of the reflected ray to

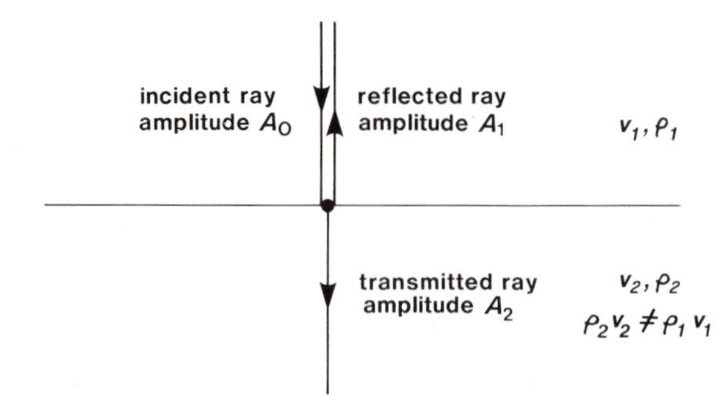

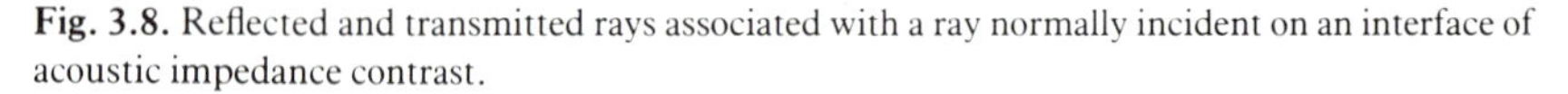
Fig. 3.8. Reflected and transmitted rays associated with a ray normally incident on an interface of acoustic impedance contrast.

the amplitude A_0 of the incident ray

$$R = A_1/A_0$$

For a normally incident ray this is given, from solution of Zoeppritz's equations, by

$$R = \frac{\rho_2 v_2 - \rho_1 v_1}{\rho_2 v_2 + \rho_1 v_1} = \frac{Z_2 - Z_1}{Z_2 + Z_1}$$

where ρ_1, v_1, Z_1 and ρ_2, v_2, Z_2 are the density, *P*-wave velocity and acoustic impedance values in the first and second layers, respectively. From this equation it follows that $-1 \leq R \leq +1$. A negative *R*-value signifies a phase change of π (180°) in the reflected ray.

The *transmission coefficient* T is the ratio of the amplitude A_2 of the transmitted ray to the amplitude A_0 of the incident ray

$$T = A_2/A_0$$

For a normally incident ray this is given, from solution of Zoeppritz's equations, by

$$T = \frac{2Z_1}{Z_2 + Z_1}$$

Reflection and transmission coefficients are sometimes expressed in terms of energy rather than wave amplitude. If energy intensity I is defined as the amount of energy flowing through a unit area normal to the direction of wave propagation in unit time, so that I_0, I_1 and I_2 are the intensities of the incident, reflected and transmitted rays respectively, then

$$R' = I_1/I_0 = \left[\frac{Z_2 - Z_1}{Z_2 + Z_1}\right]^2$$

and

$$T' = I_2/I_0 = \frac{4Z_1 Z_2}{(Z_2 + Z_1)^2}$$

where R' and T' are the reflection and transmission coefficients expressed in terms of energy.

If R or $R' = 0$, all the incident energy is transmitted. This is the case when there is no contrast of acoustic impedance across an interface (i.e. $Z_1 = Z_2$), even if the density and velocity values are different in the two layers. If R or $R' = +1$ or -1, all the incident energy is reflected. A good approximation to this situation occurs at the free surface of a water layer: rays travelling upwards from an explosion in a water layer are almost totally reflected back from the water surface with a phase change of π ($R = -0.9995$).

Values of reflection coefficient R for interfaces between different rock types rarely exceed ± 0.5 and are typically less than ± 0.2. Thus, normally, the bulk of seismic energy incident on a rock interface is transmitted and only a small proportion is reflected.

3.6.2 *Reflection and refraction of obliquely-incident rays*

When a P-ray is obliquely incident on an interface of acoustic impedance contrast, reflected and transmitted P-rays are generated as in the case of normal incidence. Additionally, some of the incident compressional energy is converted into reflected and transmitted S-rays (Fig. 3.9) that are polarized in a vertical plane. Zoeppritz's equations show that the amplitudes of the four phases are a function of the angle of incidence θ. The converted rays may attain a significant magnitude at large angles of incidence; they are, however, of only minor interest in seismic surveying and are not considered further here.

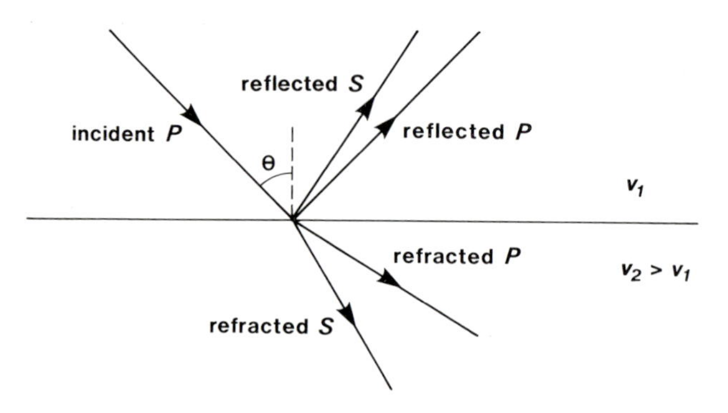

Fig. 3.9. Reflected and refracted P- and S-rays generated by a P-ray obliquely incident on an interface of acoustic impedance contrast.

In the case of oblique incidence, the transmitted P-ray travels through the lower layer with a changed direction of propagation (Fig. 3.10) and is referred to as a *refracted ray*. The situation is directly analogous to the behaviour of a light ray obliquely incident on the boundary between, say, air and water and Snell's Law of Optics applies equally to the seismic case. The generalized form of Snell's Law states that for any ray the quantity sin i/v remains a constant, known as the *ray parameter* p, where i is the angle of inclination of the ray in a layer in which it is travelling with a velocity v.

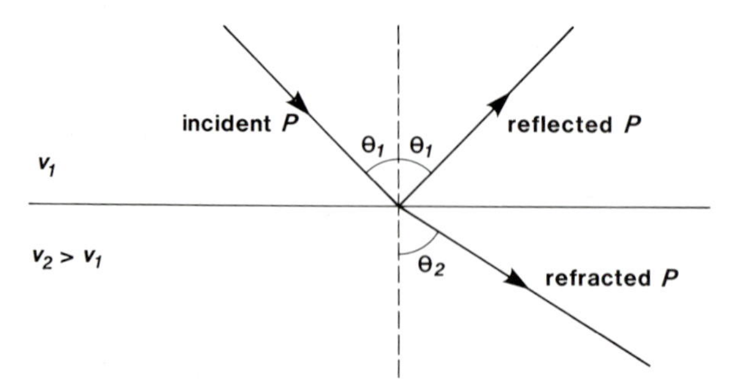

Fig. 3.10 Reflected and refracted P-rays associated with a P-ray obliquely incident on an interface of acoustic impedance contrast.

For the refracted P-ray shown in Fig. 3.10, therefore

$$\frac{\sin\theta_1}{v_1} = \frac{\sin\theta_2}{v_2}$$

or

$$\frac{\sin\theta_1}{\sin\theta_2} = \frac{v_1}{v_2}$$

Note that if $v_2 > v_1$ the ray is refracted away from the normal to the interface, i.e. $\theta_2 > \theta_1$.

Snell's Law applies to the reflected ray also, from which it follows that the angle of reflection equals the angle of incidence (Fig. 3.10).

3.6.3 *Critical refraction*

When the velocity is higher in the underlying layer there is a particular angle of incidence, known as the *critical angle* θ_c, for which the angle of refraction is 90°. This gives rise to a critically refracted ray that travels along the interface at the higher velocity v_2. At any greater angle of incidence there is total internal reflection of the incident energy (apart from converted S-rays over a further range of angles). The critical angle is given by

$$\frac{\sin\theta_c}{v_1} = \frac{\sin 90°}{v_2} = \frac{1}{v_2}$$

so that

$$\theta_c = \sin^{-1}(v_1/v_2)$$

The passage of the critically refracted ray along the top of the lower layer causes a perturbation in the upper layer that travels forward at the velocity v_2, which is greater than the seismic velocity v_1 of the layer. The situation is analogous to that of a projectile travelling through air at a velocity greater than the velocity of sound in air and the result is the same, namely the generation of a shock wave. This wave is known as a *head wave* in the seismic case, and it passes up obliquely through the upper layer towards the surface (see Fig. 3.11). Any ray associated with the head wave is inclined at the critical angle θ_c. By means of the head wave, seismic energy is returned to the surface after critical refraction in an underlying layer of higher velocity.

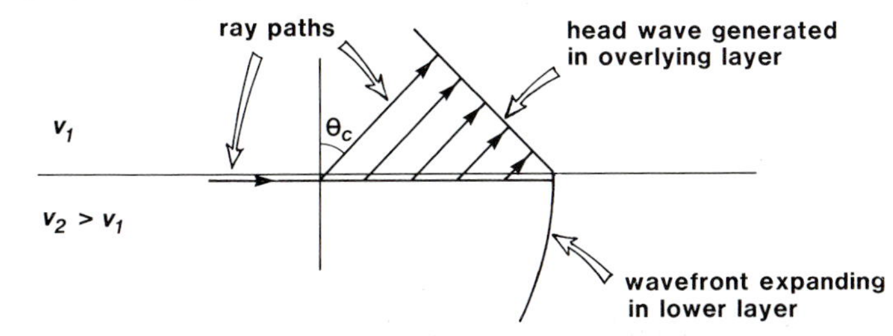

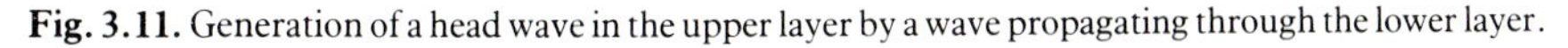
Fig. 3.11. Generation of a head wave in the upper layer by a wave propagating through the lower layer.

3.6.4 *Diffraction*

In the above discussion of the reflection and transmission of seismic energy at interfaces of acoustic impedance contrast it was implicitly assumed that the interfaces were continuous and of low curvature. At abrupt discontinuities in interfaces, or structures whose radius of curvature is shorter than the wavelength of incident waves, the laws of reflection and refraction no longer apply. Such phenomena give rise to a radial scattering of incident seismic energy known as diffraction. Common sources of diffraction in the ground include the edges of faulted layers (Fig. 3.12) and small isolated objects, such as boulders, in an otherwise homogeneous layer.

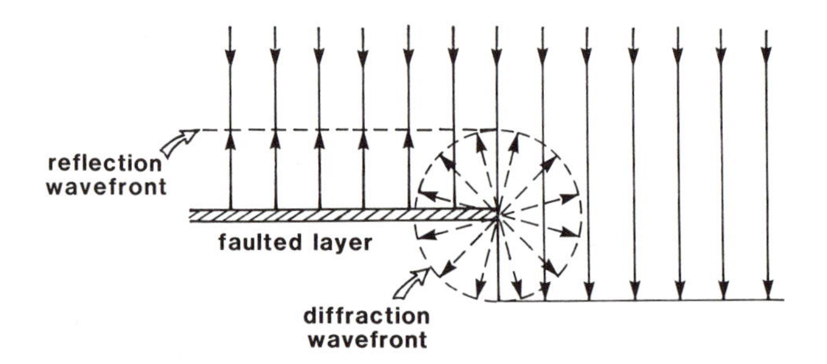

Fig. 3.12. Diffraction caused by the truncated end of a faulted layer.

Diffracted phases are commonly observed in seismic recordings and are sometimes difficult to discriminate from reflected and refracted phases, as discussed in Chapter 4.

3.7 Reflection and refraction surveying

Consider the simple geological section shown in Fig. 3.13 involving two homogeneous layers of seismic velocity v_1 and v_2 separated by a horizontal interface at a depth z, the compressional wave velocity being higher in the underlying layer (i.e. $v_2 > v_1$).

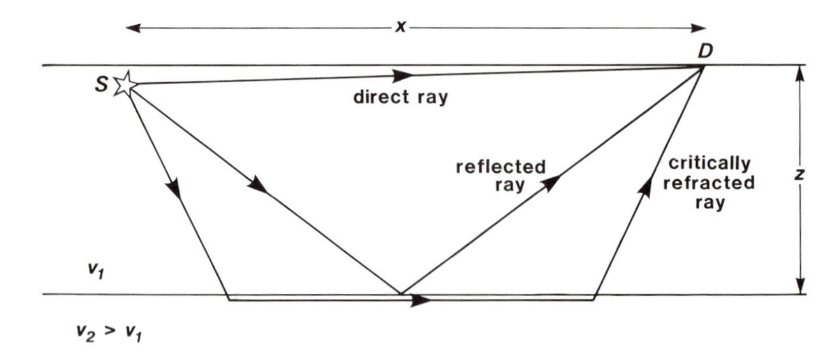

Fig. 3.13. Direct, reflected and refracted ray paths from a near surface source to a surface detector in the case of a simple two-layer model.

From a near-surface seismic source S there are three types of ray path by which energy reaches the surface at a distance from the source, where it may be recorded by a suitable detector as at D, a horizontal distance x from S. The *direct ray* travels along a straight line through the top layer from source to detector at velocity v_1. The *reflected ray* is obliquely incident on the interface and is reflected back through the top layer to the detector, travelling along its entire path at the top layer velocity v_1. The *refracted ray* travels obliquely down to the interface at velocity v_1, along a segment of the interface at the higher velocity v_2, and back up through the upper layer at v_1.

The travel time of a direct ray is given simply by

$$t_{DIR} = x/v_1$$

which defines a straight line of slope $1/v_1$ passing through the time-distance origin.

The travel time of a reflected ray is given by

$$t_{RFL} = (x^2+4z^2)^{1/2}/v_1$$

which, as discussed in Chapter 4, is the equation of an hyperbola.

The travel time of a refracted ray (for derivation see Chapter 5) is given by

$$t_{RFR} = x/v_2 + \frac{2z\,(v_2^2-v_1^2)^{1/2}}{v_1 v_2}$$

which is the equation of a straight line having a slope of $1/v_2$ and an intercept on the time axis of $2z\,(v_2^2-v_1^2)^{1/2}/v_1v_2$.

Travel-time curves, or time-distance curves, for direct, refracted and reflected rays are illustrated in Fig. 3.14. By suitable analysis of the travel-time curve for reflected *or* refracted rays it is possible to compute the depth to the underlying layer. This provides two independent seismic surveying methods for locating and mapping subsurface interfaces, namely, *reflection surveying* and

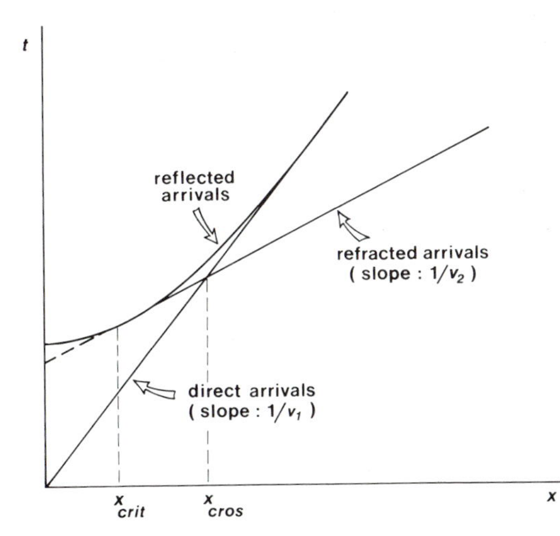

Fig. 3.14. Travel-time curves for direct, reflected and refracted rays in the case of a simple two-layer model.

refraction surveying. These have their own distinctive methodologies and fields of application and they are discussed separately in detail in Chapters 4 and 5. However, some general remarks about the two methods may be made here with reference to the travel-time curves of Fig. 3.14. The curves are more complicated in the case of a multilayered model, but the following remarks still apply.

The first arrival of seismic energy at a surface detector offset from a surface source is always a direct ray or a refracted ray. The direct ray is overtaken by a refracted ray at the *crossover distance* x_{cros}. Beyond this offset distance the first arrival is always a refracted ray. Since critically refracted rays travel down to the interface at the critical angle there is a certain distance, known as the *critical distance* x_{crit}, within which refracted energy will not be returned to surface. At the critical distance, the travel times of reflected rays and refracted rays coincide because they follow effectively the same path. Reflected rays are never first arrivals: they are always preceded by direct rays and, beyond the critical distance, by refracted rays also.

The above characteristics of the travel-time curves determine the methodology of refraction and reflection surveying. In refraction surveying, recording ranges are chosen to be sufficiently large to ensure that the crossover distance is well exceeded in order that refracted rays may be detected as first arrivals of seismic energy. Indeed, some types of refraction survey consider only these first arrivals, which can be detected with unsophisticated field recording systems. In general this approach means that the deeper a refractor, the greater is the range over which recordings of refracted arrivals need to be taken.

In reflection surveying, by contrast, reflected phases are sought that are never first arrivals and are normally of very low amplitude because geological reflectors tend to have small reflection coefficients. Accordingly, reflections are normally concealed in seismic records by higher amplitude events such as direct or refracted body waves and surface waves. Reflection surveying methods therefore have to be capable of discriminating between reflected energy and many types of synchronous noise. Recordings are normally restricted to small offset distances, well within the critical distance for the reflecting interfaces of main interest. Since reflections may always be detected at small offset distances there is, in general, no necessity to increase the overall recording range to investigate deeper reflectors.

3.8 Seismic sources and the seismic/acoustic spectrum

A seismic source is a localized region within which the sudden release of energy leads to a rapid stressing of the surrounding medium. Most seismic sources preferentially generate the compressional wave energy that is mainly utilized in seismic surveying.

There is a very wide variety of seismic sources, characterized by differing energy levels and frequency characteristics. In general a seismic source contains a wide range of frequency components within the range from 1 Hz to a few hundred hertz, though the energy is often concentrated in a narrower frequency band. In addition to the seismic sources there are also several acoustic sources that generate acoustic waves (i.e. sound waves in water or air) which are useful in marine

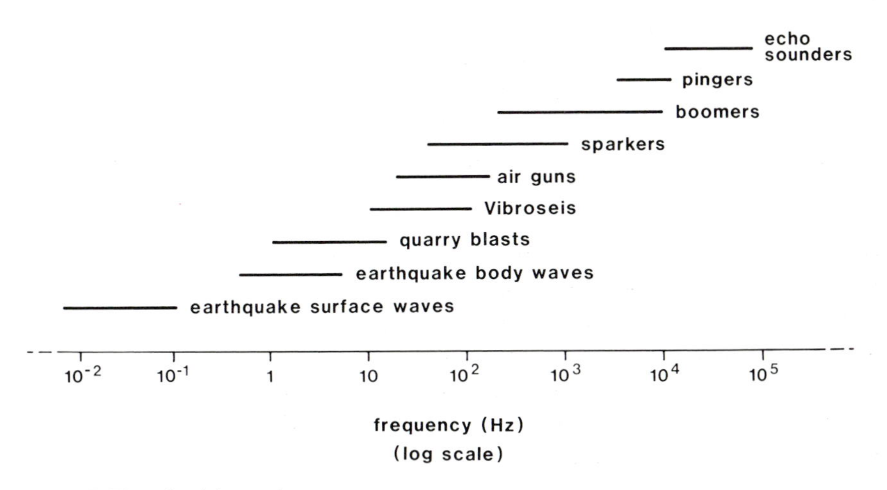

Fig. 3.15. The seismic/acoustic spectrum.

seismic surveying. The complete seismic/acoustic spectrum is shown in Fig. 3.15.

Many considerations govern the selection of a suitable seismic source for a particular survey application. The general problem in seismic surveying is to recognize a seismic signal that has been markedly attenuated by propagation through the ground and which is embedded in the general background level of seismic noise that characterizes any recording site. There are many sources of noise in the seismic spectrum including microseisms (weak natural sources such as wind or water waves), industrial activity and traffic vibration. There is, therefore, an inherent problem of signal : noise ratio (SNR) in seismic surveying. This problem becomes extreme when the SNR reduces below unity.

Source characteristics can be modified by the use of several sources in an array designed, for example, to improve the frequency spectrum of the transmitted pulse. This matter is taken up in Chapter 4 when discussing the design parameters of seismic reflection surveys. In this section the various types of seismic/acoustic source in common use are introduced.

3.8.1 *Explosive sources*

On land, explosives are normally detonated in shallow shot holes to improve the coupling of the energy source with the ground and to minimize surface damage. An inherent problem of explosions at sea is the generation of *bubble pulses* caused by oscillation of the high-pressure gas bubble resulting from the initial explosion. Bubble pulses have the effect of unduly lengthening the seismic pulse (Fig. 3.16). Steps can, however, be taken to suppress the effect of the bubble pulse by detonating near to the water surface so that the gas bubble escapes into the air.

Explosives offer a reasonably cheap and highly efficient seismic source with a wide frequency spectrum but their use normally requires special permitting and presents logistical difficulties of storage and transportation. They are slow to use on land because of the need to drill shot holes. Their main shortcoming, however, is that they do not provide the type of precisely repeatable source signature required by some modern processing techniques, nor can the detonation of explosives be repeated at fixed and precise time intervals as required for efficient reflection profiling at sea carried out by survey vessels underway.

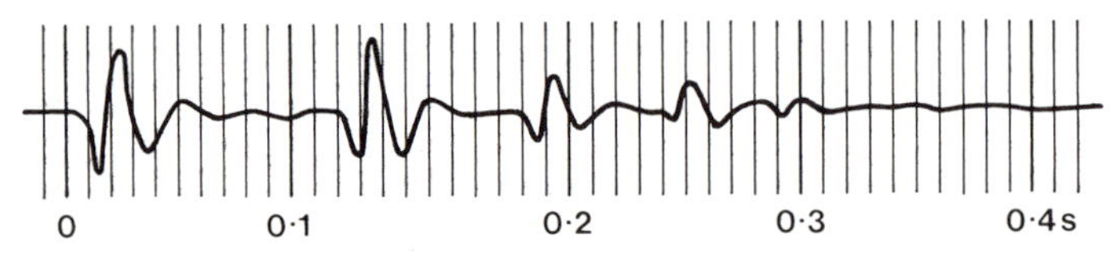

Fig. 3.16. The train of bubble pulses associated with the detonation of an explosive charge in water. (After Sheriff 1973.)

3.8.2 *Non-explosive sources*

(a) Land sources

The most common method of reflection surveying on land utilizing a non-explosive source is the Vibroseis® method. This uses tractor-mounted vibrators to pass into the ground an extended vibration of low amplitude and continuously varying frequency, known as a *sweep signal*. A typical sweep signal lasts for seven or more seconds and varies in frequency between limits of about 10 and 80 Hz. The field recordings consist of overlapping reflected wavetrains of very low amplitude concealed in seismic noise, but cross-correlation (see Section 2.4.3) of the recordings with the known sweep signal enhances the seismic signals and yields a seismic recording of similar appearance to that which would be obtained with an explosive source (Fig. 3.17).

The Vibroseis® source is quick and convenient to use and produces a precisely known and repeatable signal. The vibrator unit needs a firm base on which to operate, such as a tarmac road, and it will not work well on soft ground. The peak

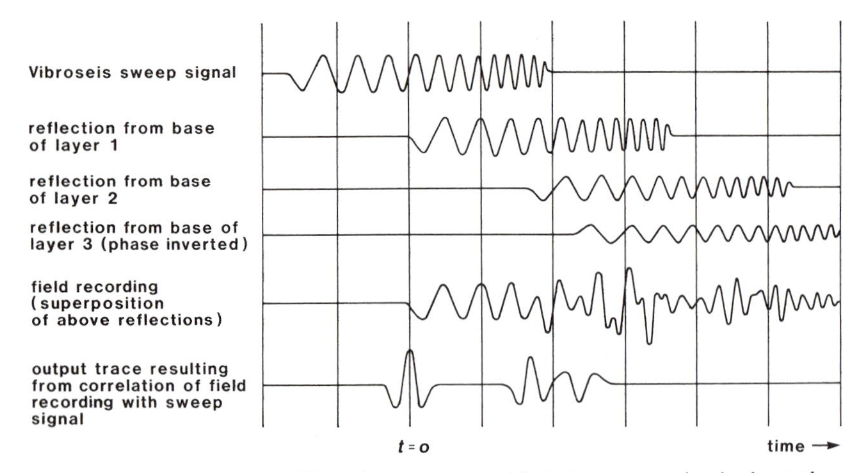

Fig. 3.17. Cross-correlation of a Vibroseis® seismogram with the input sweep signal to locate the positions of occurrence of reflected arrivals.

force of a vibrator is only about 10^5N and to increase the transmitted energy for deep penetration surveys, vibrators are typically employed in groups with a phase-locked response. A particular advantage of vibrators is that they can be used in towns since they cause no damage or significant disturbance to the environment. The cross-correlation method of extracting the signal is also capable of coping with the inherently high noise levels of urban areas.

The principle of using a precisely known source signature of long duration is extended with the *Mini-Sosie* source. A pneumatic hammer delivers a random sequence of impacts to a base plate, thus transmitting a pulse-encoded signal of low amplitude into the ground. The source signal is recorded by a detector on the base plate and used to cross-correlate with the field recordings of reflected arrivals of the pulse-encoded signal from buried interfaces. Peaks in the cross-correlation function reveal the positions of reflected signals in the recordings.

The horizontal impact of a weight on to one side of a vertical plate partially embedded in the ground can be used as a source for shear wave seismology, in which special shear wave detectors are used. One application of shear wave seismology is in engineering site investigation where the separate measurement of v_p and v_s for near-surface layers allows direct calculation of Poisson's ratio and estimation of the elastic moduli, which provide valuable information on the *in situ* geotechnical properties of the ground.

(b) Marine sources

Several sources, having different energy levels and frequency characteristics, are available for marine reflection surveying. Before describing these sources it should be noted that seismic reflection surveys are normally required to achieve a specific depth of penetration and a specific resolution (i.e. ability to resolve individual, closely-spaced reflectors) and that a source must be chosen appropriate to the specified task.

The resolution is basically determined by the pulse length: for a pulse of any particular length there is a minimum separation of reflectors below which the reflected pulses will overlap in time in the seismic recording. Although the pulse length can be shortened at the processing stage by deconvolution (see Section 4.9.2) many seismic sources are used in conjunction with simple seismic profiling systems in which the analogue signal from the receiver is amplified, band-pass filtered and fed directly to a chart recorder (see Chapter 4). In such systems the resolution of the resulting seismic record is inherently limited by the recorded pulse length. Since the higher energy sources necessary for deeper penetration are characterized by lower dominant frequencies and longer pulse lengths there is generally a trade-off between penetration and resolution: the deeper the penetration the lower will be the resolving power. The common types of marine sources are described briefly below.

Air guns (Fig. 3.18(a)) are pneumatic sources in which a chamber is charged with compressed air fed through a hose from a shipboard compressor and the air is released, by electrical triggering, through side vents into the water in the form of a high-pressure bubble. The operating pressure is typically 10–15 MPa. A wide

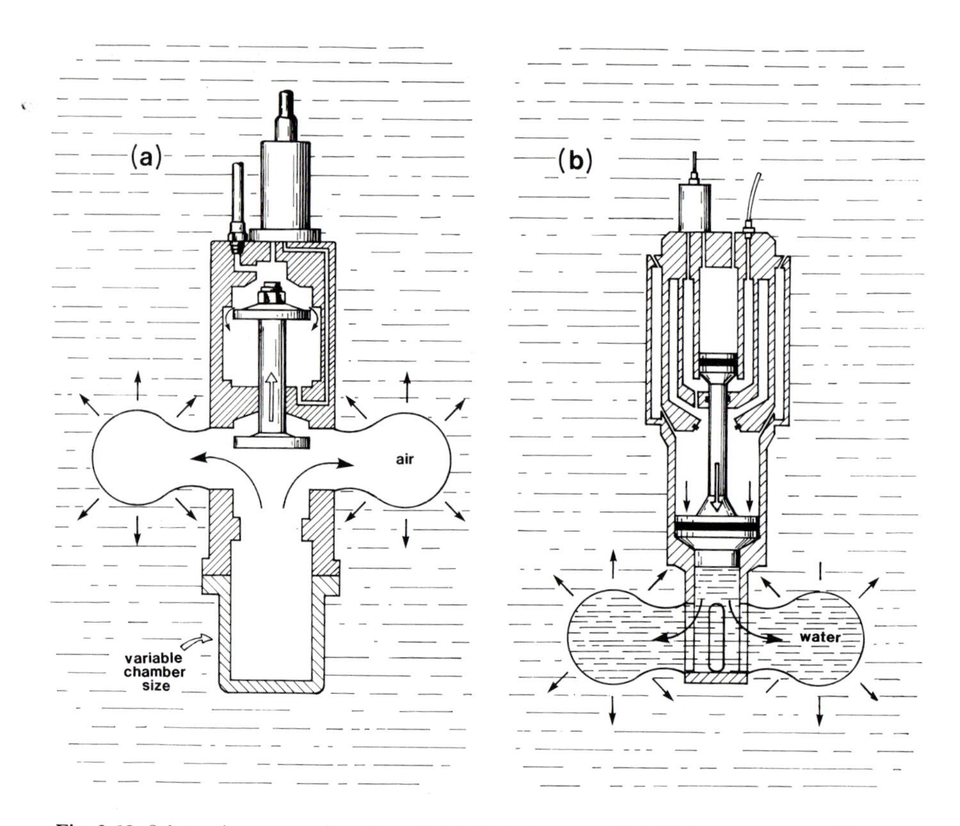

Fig. 3.18. Schematic cross-sections through (a) a Bolt air gun and (b) a Sodera water gun to illustrate the principles of operation. (Redrawn with permission of Bolt Associates and Sodera Ltd.)

range of chamber sizes is available, leading to different energy outputs and frequency characteristics. For deep penetration surveys the total energy transmitted may be increased by the use of arrays of air guns mounted on a frame that is towed behind the survey vessel. The primary pulse generated by an air gun is followed by a train of bubble pulses that increase the overall length of the pulse. With some loss of peak energy output, the growth of bubble pulses can be effectively suppressed by reducing the rate at which the compressed air is released into the water layer. Arrays of guns of differing dimensions and, therefore, different bubble pulse periods can be used to produce a high-energy source in which primary pulses interfere constructively whilst bubble pulses interfere destructively (Fig. 3.19).

Water guns (Fig. 3.18(b)) are pneumatic sources in which the compressed air, rather than being released into the water layer, is used to drive a piston that ejects a water jet into the surrounding water. A vacuum cavity is created behind the advancing water jet and this implodes under the influence of the ambient hydrostatic pressure generating a strong acoustic pulse free of bubble oscillations. Since the implosion represents collapse into a vacuum no gaseous material is compressed to 'bounce back' as a bubble pulse. The resulting short pulse length offers a potentially higher resolution than is achieved with air guns.

Several marine sources utilize explosive mixtures of gases. In *sleeve exploders* propane and oxygen are piped into a submerged flexible rubber sleeve where the gaseous mixture is fired by means of a spark plug. The products of the resultant explosion cause the sleeve to expand rapidly generating a shock wave in the surrounding water. The exhaust gases are vented to surface through a valve that opens after the explosion, thus attenuating the growth of bubble pulses.

Sparkers, boomers and *pingers* are devices for converting electrical energy into acoustic energy. The sparker pulse is generated by the discharge of a large capacitor bank directly into the sea water through an array of electrodes towed in a frame behind the survey vessel. Operating voltages are typically 3.5 to 4.0 kV and peak currents may exceed 200 A. This electrical discharge leads to the formation and rapid growth of a plasma bubble and the consequent generation of an acoustic

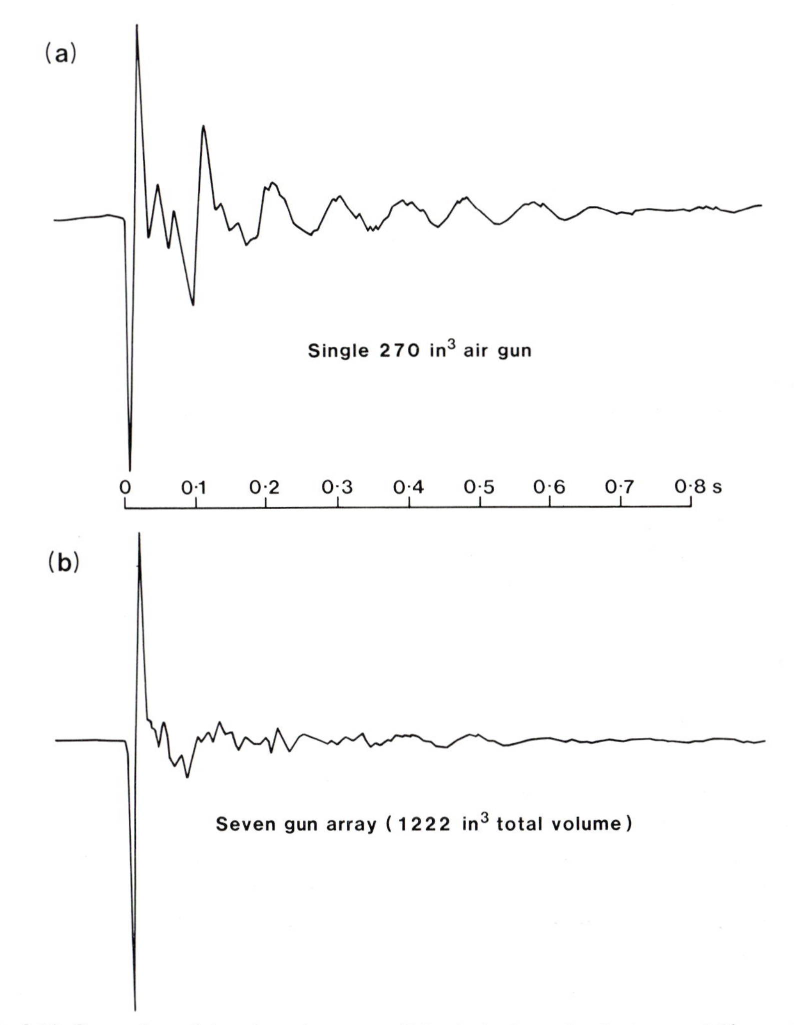

Fig. 3.19. Comparison of the source signatures of (a) a single air gun (peak pressure: 4.6 bar metres) and (b) a seven-gun array (peak pressure: 19.9 bar metres). Note the effective suppression of bubble pulses in the latter case. (Redrawn with permission of Bolt Associates.)

pulse. The boomer source comprises a rigid aluminium plate attached by a spring-loaded mounting to a resin block in which is embedded a heavy-duty spiral coil. A capacitor bank is discharged through the coil and the electromagnetic field thus generated sets up eddy currents in the aluminium plate. These currents generate a secondary field that opposes the primary field and the plate is rapidly repulsed, setting up a compressional wave in the water. The device is typically towed behind the survey vessel in a catamaran mounting. Sparkers and boomers generate broad band acoustic pulses and can be operated over a wide range of energy levels so that the source characteristics can to some extent be tailored to the needs of a particular survey. In general, boomers offer better resolution (down to 0.5 m) but more restricted depth penetration (a few hundred metres maximum).

Pingers consist of small ceramic piezoelectric transducers, mounted in a towing fish, which when activated by an electrical impulse emit a very short, high frequency acoustic pulse of low energy. They offer a very high resolving power (down to 0.1 m) but limited penetration (a few tens of metres in mud, much less in sand or rock). They are useful in offshore engineering applications such as surveys of proposed routes for submarine pipelines.

Further discussion of the use of air guns, sparkers, boomers and pingers in single-channel seismic reflection profiling systems is given in Section 4.12.

3.9 Seismic data acquisition systems

The basic field activity in seismic surveying is the collection of *seismograms* which may be defined as analogue or digital time series that register the amplitude of ground motions as a function of time during the passage of a seismic wavetrain. The acquisition of seismograms involves conversion of the seismic ground motions into electrical signals, amplification and filtering of the signals and their registration on a chart recorder and/or tape recorder. The conventional seismic survey procedure is to monitor ground motions at a large number of surface locations; thus multichannel recording systems are usually employed with, exceptionally, up to several hundred separate recording channels. Except in the simplest recording systems the data are tape recorded to facilitate subsequent processing. Modern recording systems utilize digital tape recording so that the data are available in a suitable form for input to computers. A block diagram of a seismic recording system is shown in Fig. 3.20.

3.9.1 Seismic detectors

The detectors used in seismic surveying are electromechanical transducers that convert a mechanical input (the seismic pulse) into an electrical ouput. Devices used on land to detect seismic ground motions are known as *seismometers* or *geophones*. In water, the passage of a compressional seismic wave is marked by transient pressure changes and these are detected by *hydrophones* towed or suspended in the water column or, in very shallow water, laid on the sea bed. Hydrophones may also be used in the water-saturated ground conditions encountered in swamps or marshland. Detectors may comprise individual geo-

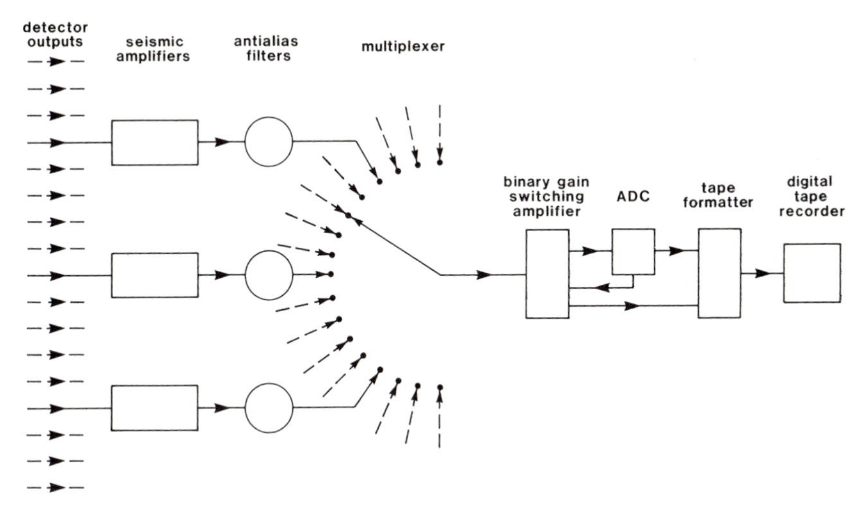

Fig. 3.20. Schematic block diagram of a seismic recording system. (After Telford *et al.* 1976.)

phones or hydrophones, or arrays of these devices connected together in series/parallel to provide a summed output.

Although there are several types of geophone the most common is the *moving-coil* geophone (Fig. 3.21). In this instrument a cylindrical coil is suspended from a spring support in the field of a permanent magnet which is attached to the instrument casing. The magnet has a cylindrical pole piece inside the coil and an annular pole piece surrounding the coil. The suspended coil represents an

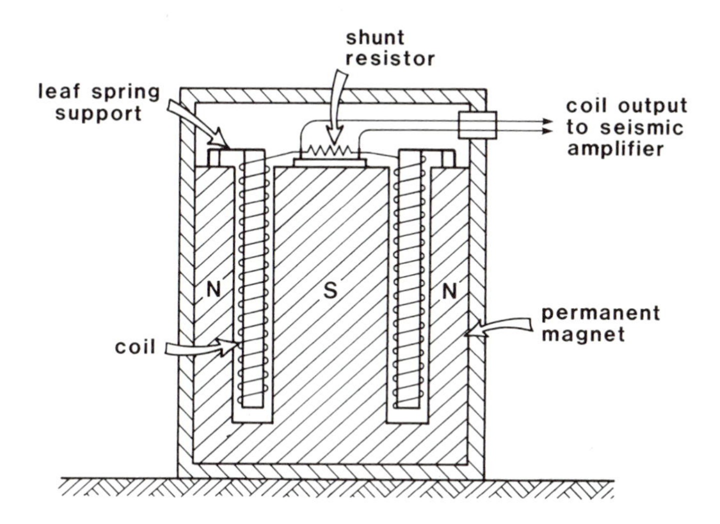

Fig. 3.21. Schematic cross-section through a moving-coil geophone.

oscillatory system with a resonant frequency determined by the mass of the coil and the stiffness of its spring suspension.

The geophone is fixed by a spike base into soft ground or mounted firmly on hard ground. It moves in sympathy with the ground surface during the passage of a seismic wave, causing relative motion between the suspended coil and the fixed magnet. Movement of the coil in the magnetic field generates a voltage across the terminals of the coil. The oscillatory motion of the coil is inherently damped because the current flowing in the coil induces a magnetic field that interacts with the field of the magnet to oppose the motion of the coil. The amount of this damping can be altered by connecting a shunt resistance across the coil terminals to control the amount of current flowing in the coil. Additional damping is arranged by winding the coil on a metal former. The magnetic field induced by eddy currents flowing in the metal former also opposes the coil motion.

Ideally, the output waveform of a geophone closely mirrors the ground

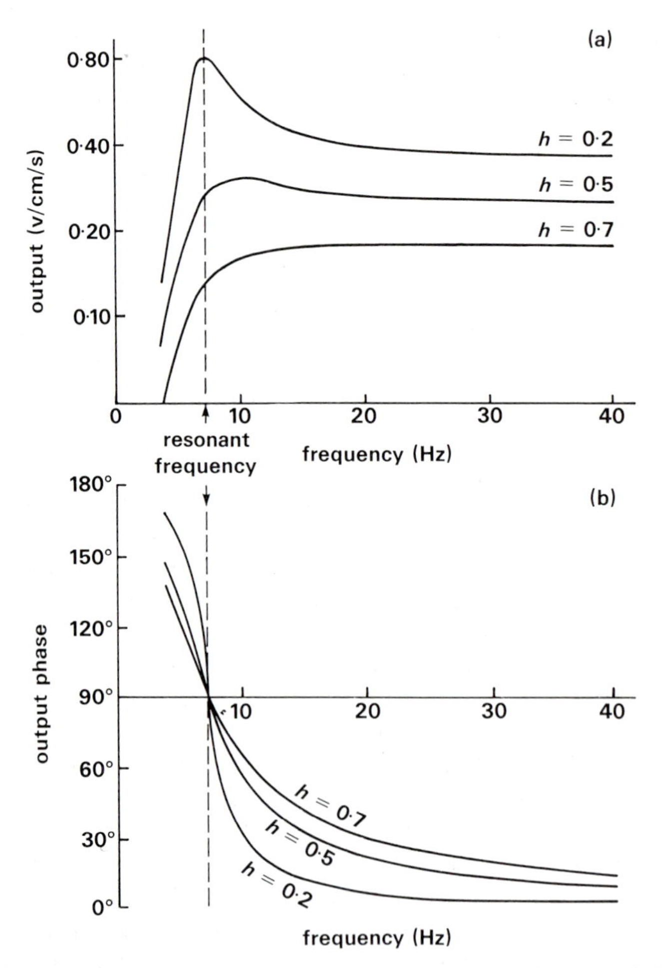

Fig. 3.22. Amplitude and phase responses of a geophone with a resonant frequency of 7 Hz, for different damping factors h. Output phase is expressed relative to input phase. (After Telford *et al.* 1976.)

motion and this is arranged by careful selection of the amount of damping. Too little damping results in an oscillatory output at the resonant frequency, whilst overdamping leads to a reduction of sensitivity. Damping is typically arranged to be about 0.7 of the critical value at which oscillation would just fail to occur for an impulsive mechanical input such as sharp tap. With this amount of damping the frequency response of the geophone is effectively flat above the resonant frequency. The effect of differing amounts of damping on the frequency and phase response of a geophone is shown in Fig. 3.22.

To preserve the shape of the seismic waveform, geophones should have a flat frequency response and minimal phase distortion within the frequency range of interest. Consequently, geophones should be arranged to have a resonant frequency well below the main frequency band of the seismic signal to be recorded. Most commercial seismic reflection surveys employ geophones with a resonant frequency between 4 and 15 Hz.

Above the resonant frequency, the output of a moving-coil geophone is proportional to the velocity of the coil. Note that the coil velocity is related to the very low particle velocity associated with a seismic ground motion and not to the much higher propagation velocity of the seismic energy (see p. 33). The sensitivity of a geophone, measured in output volts per unit of velocity, is determined by the number of windings in the coil and the strength of the magnetic field, hence, instruments of larger and heavier construction are required for higher sensitivity. The miniature geophones used in commercial reflection surveying typically have a sensitivity of about 10 V/m/s.

Moving coil geophones are sensitive only to the component of ground motion along the axis of the coil. Vertically travelling compressional waves from subsurface reflectors cause vertical ground motions and are therefore best detected by geophones with an upright coil as illustrated in Fig. 3.21. The optimal recording of seismic phases that involve mainly horizontal ground motions, such as horizontally polarized shear waves, requires geophones in which the coil is mounted and constrained to move horizontally. As discussed in Chapter 4, geophones are typically deployed in linear or areal arrays containing several geophones whose individual outputs are summed. Such arrays provide detectors with a directional response that facilitates the enhancement of signal and the suppression of certain types of noise (see p. 68).

In seismic surveying the outputs of several detectors are fed to a multichannel recording system mounted in a recording vehicle. The individual detector outputs may be fed along a multicore cable or multiplexed at the detector location and transmitted along a lighter cable containing far fewer conductors. Some modern systems utilize lightweight fibre-optic cables or telemetry links to transmit the detector outputs to the recording vehicle.

Hydrophones are composed of ceramic piezoelectric elements which produce an output voltage proportional to the pressure variations associated with the passage of a compressional seismic wave through water. The sensitivity is typically 0.1 mV/Pa. For multichannel seismic surveying at sea, large numbers of individual hydrophones are made up into hydrophone *streamers* by distributing them along an oil-filled plastic tube. The tube is arranged to have neutral

3

buoyancy and is manufactured from materials with an acoustic impedance close to that of water to ensure good transmission of seismic energy to the hydrophone elements. Since piezoelectric elements are sensitive to accelerations, hydrophones are often composed of two elements mounted back to back and connected in series so that the effects of accelerations of the streamer as it is towed through the water are cancelled out in the hydrophone outputs. As with geophone deployment in land seismic surveying, groups of hydrophones may be connected together into linear arrays to produce detectors with a directional response.

3.9.2 *Seismic amplifiers and tape recorders*

Seismic amplifiers are required to amplify signals in the frequency range from a few hertz to a few hundred hertz (or, in some marine systems, up to a few thousand hertz), and have to cope with a very wide range of signal amplitudes. The amplitude of ground motions near to a seismic source may reduce by a factor of a million or more between the early arrivals of strong direct waves and surface waves and the later arrivals of very weak waves reflected back to the surface from deep interfaces. An amplitude ratio of one million is equivalent to a dynamic range of 120 dB. A maximum dynamic range for geophones of about 140 dB and an inherent minimum noise level in seismic amplifiers of about 1 microvolt effectively limits the maximum dynamic range of a seismic recording to 120 dB.

Most seismic amplifier systems contain frequency filters for high-pass, low-pass, band-pass or band-reject filtering. Filtering is commonly employed to produce a suitable visual record in the field for monitoring purposes, either at the time of the original recording or subsequently, by playback of a tape recording. The tape is normally recorded broad band (except for antialias filtering in the case of digital recording; see Chapter 2) in order to retain the maximum amount of information in the seismic recording. Optimal frequency filtering can then be carried out digitally as an aspect of the subsequent computer processing of the data.

The approach to seismic amplifier design depends upon whether analogue or digital tape recording is to be employed. The maximum dynamic range of analogue tape recording is about 50 dB so that in analogue recording systems the dynamic range of the seismic signal needs to be reduced prior to the recording stage. This can be accomplished by various means. *Automatic gain control* (AGC)

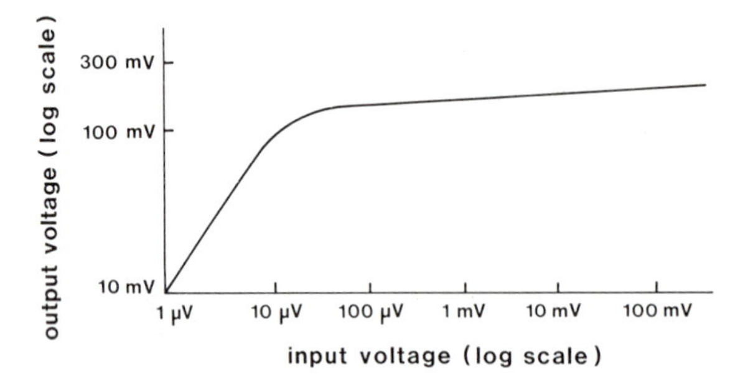

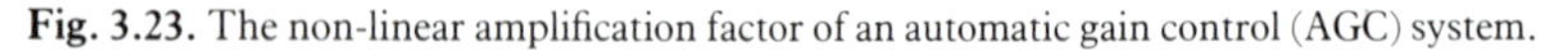
Fig. 3.23. The non-linear amplification factor of an automatic gain control (AGC) system.

alters the amplification factor of the amplifier in accordance with the amplitude of the input signal (Fig. 3.23). Up to a certain input level the gain is approximately constant but it reduces progressively for higher input levels. Thus the stronger signals are relatively attenuated and the overall dynamic range is markedly reduced. Time variable gain can be used to suppress the gain when strong signals are being received (known as *initial suppression* or *presuppression*), and to increase the gain in the later part of the recording, when the seismic signal has reduced to a very low level.

In digital recording systems, the analogue output of the seismic amplifiers has to be passed through an *analogue-to-digital converter* (ADC) (Fig. 3.20). Conventionally, a single ADC is used to digitize all the seismic channels by means of multiplexing. This involves electronic switching of the ADC sequentially through all the channels and, for each channel, sampling the instantaneous output value and registering it in digital form as a binary word (see Chapter 2). Thus one full scan of all the channels produces a sequence of binary words each representing a sample value for an individual channel. The required switching rate of the ADC is determined by the required digital sampling interval and the number of channels to be multiplexed. For example, if each channel of a 50-channel amplifier system is to be sampled every 2.5 ms the ADC must scan all 50 channels in less than 2.5 ms, which requires a switching rate of faster than 0.05 ms. In fact, ADCs operate at much faster switching rates than this.

Tape recording of the seismic data in a multiplexed form means that the initial stage of processing on playback is demultiplexing to recover the form of the outputs of the individual seismic amplifier channels. Demultiplexing by computer is a simple matter of reordering the sequence of binary words recorded on tape into separate one-dimensional arrays representing the outputs of each recording channel.

The multiplexed digital data are recorded in a standard tape format. A common, internationally accepted format is 9-track recording on half-inch tape with a data packing density of 1600 bits per inch, but a higher packing density of 6250 bits per inch is becoming increasingly common.

Since the dynamic range of a digitized waveform is determined solely by the length (i.e. number of bits) of each binary word (see Chapter 2), the only limitation on the dynamic range of a digital recording is the number of bits recorded. However, there are practical limits to the recorded word length because the greater the number of bits the faster the required tape speed and the greater the data storage problem. In addition to the digital word defining the amplitude of the ground motion it is necessary to record an extra bit, known as the sign-bit, to register the associated direction of ground motion (up or down). In conventional seismic recording, 16 to 20 bits are acquired per sample point. For any given number of bits, the effective range of a digital recording can be increased by the use of *floating point* amplifiers. These measure the magnitude of any channel output sampled by the multiplexer and represent the digital output by two numbers, one giving the value of the output to the required number of significant places, the other giving the power of two to which this number has to be raised.

Floating point amplifiers are gradually replacing *binary gain* amplifiers which automatically adjust the gain level of the recording in binary steps (6 dB) through the recording period on the basis of the amplitude of the signal output from the multiplexer (Fig. 3.20). Tape recording the binary gain level as a function of time enables true signal amplitudes to be recovered during subsequent computer processing.

For the visual display of seismic records for monitoring purposes, a multi-channel oscillographic recorder is used either to display the filtered output of the seismic amplifiers or for the field playback of tape recorded data. In some seismic recording systems without tape recording facilities the chart recording produced by the oscillographic recorder represents the only permanent record of the seismic data.

Some oscillographic recorders contain a facility to store records digitally in an internal memory and to sum the results obtained from successive shots prior to display of the seismograms. Summing of the results from a number of shots results in an improved SNR in the resultant seismograms. In such recorders, the content of the memory can typically be inspected on a display screen before being played out as a chart recording to produce a permanent record.

Further reading

Al-Sadi, H.N. (1980) *Seismic Exploration*. Birkhäuser Verlag, Basel.

Anstey, N.A. (1977) *Seismic Interpretation: The Physical Aspects*. IHRDC, Boston.

Anstey, N.A. (1981) *Seismic Prospecting Instruments. Vol 1: Signal Characteristics and Instrument Specifications*. Gebrüder Borntraeger, Berlin.

Gregory, A.R. (1977) Aspects of rock physics from laboratory and log data that are important to seismic interpretation. *In*: Payton, C.E. (ed.), *Seismic Stratigraphy–Applications to Hydrocarbon Exploration*. Memoir 26, American Association of Petroleum Geologists, Tulsa.

Sheriff, R.E. & Geldart, L.P. (1982) *Exploration Seismology Vol 1: History, Theory and Data Acquisition*. Cambridge University Press, Cambridge.

Sheriff, R.E. & Geldart, L.P. (1983) *Exploration Seismology Vol 2: Data-processing and Interpretation*. Cambridge University Press, Cambridge.

Waters, K.H. (1978) *Reflection Seismology–A Tool For Energy Resource Exploration*. Wiley, New York.

4

Seismic Reflection Surveying

4.1 Introduction

In seismic reflection surveys the travel times are measured of arrivals reflected from subsurface interfaces between media of different acoustic impedance. Reflection surveys are most commonly carried out in areas of shallowly dipping sedimentary sequences. In such situations, velocity varies much more as a function of depth, due to the differing physical properties of the individual layers, than horizontally, due to lateral facies changes within the individual layers. For the purposes of initial consideration, the horizontal variations of velocity may be ignored.

Fig. 4.1 shows a horizontally-layered ground with vertical reflected ray paths from the various layer boundaries. This model assumes the subsurface to be composed of a series of depth intervals each characterized by an *interval velocity* v_i, which may be the uniform velocity within a homogeneous geological unit or the average velocity over a depth interval containing more than one unit. If z_i is

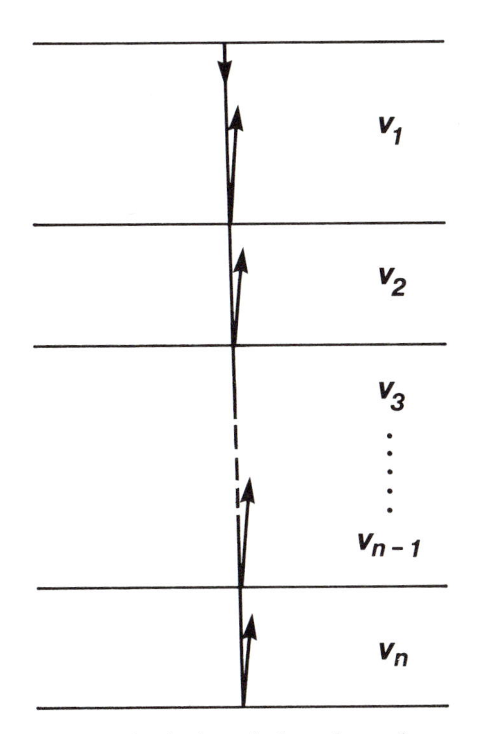

Fig. 4.1. Vertical reflected raypaths in a horizontally-layered ground.

the thickness of such an interval and τ_i is the one-way travel time of a ray through it, the interval velocity is given by

$$v_i = \frac{z_i}{\tau_i}$$

The interval velocity may be averaged over several depth intervals to yield a *time-average velocity* or, simply, *average velocity* $\bar{V}$. Thus the average velocity of the top n layers in Fig. 4.1 is given by

$$\bar{V} = \frac{\sum_{i=1}^{n} z_i}{\sum_{i=1}^{n} \tau_i} = \frac{\sum_{i=1}^{n} v_i \tau_i}{\sum_{i=1}^{n} \tau_i}$$

or, if Z_n is the total thickness of the top n layers and T_n is the total one-way travel time through the n layers

$$\bar{V} = \frac{Z_n}{T_n}$$

4.2 Geometry of reflected ray paths

4.2.1 Single horizontal reflector

The basic geometry of the reflected ray path is shown in Fig. 4.2(a) for the simple case of a single horizontal reflector lying at a depth z beneath a homogeneous top layer of velocity V. The equation for the travel time t of the reflected ray from a shot point to a detector at a horizontal offset, or shot-detector separation, x is given by the ratio of the travel path length to the velocity

$$t = (x^2 + 4z^2)^{1/2}/V \qquad (4.1)$$

In a reflection survey, reflection times t are measured at offset distances x and it is required to determine z and V. If reflection times are measured at different offsets x, equation (4.1) can be solved for these unknowns.

Equation (4.1) can be arranged into the normal hyperbolic form to give

$$\frac{V^2 t^2}{4z^2} - \frac{x^2}{4z^2} = 1 \qquad (4.2)$$

Thus the graph of travel time of reflected rays plotted against offset distance (the *time-distance curve*) is an hyperbola whose axis of symmetry is the time axis (Fig. 4.2(b)). Substituting $x = 0$ in equation (4.1), the travel time t_0 of a vertically reflected ray is obtained. $t_0 = 2z/V$ represents the intercept on the time axis of the time-distance curve (see Fig. 4.2(b)). Equation (4.1) can be written

$$t^2 = \frac{4z^2}{V^2} + \frac{x^2}{V^2} \qquad (4.3)$$

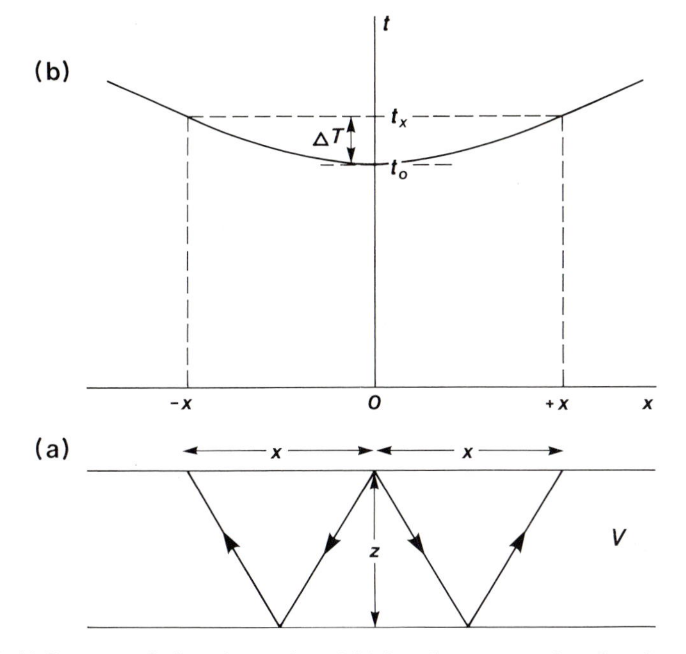

Fig. 4.2. (a) Geometry of reflected raypaths and (b) time-distance curve for reflected rays from a horizontal reflector. ΔT = normal moveout.

Thus

$$t^2 = t_0^2 + \frac{x^2}{V^2} \tag{4.4}$$

From equation (4.3)

$$t = \frac{2z}{V}\left\{1+\left(\frac{x}{2z}\right)^2\right\}^{1/2}$$

$$= t_0\left\{1+\left(\frac{x}{Vt_0}\right)^2\right\}^{1/2} \tag{4.5}$$

Binomial expansion of equation (4.5) gives

$$t = t_0\left\{1+\tfrac{1}{2}\left(\frac{x}{Vt_0}\right)^2 - \tfrac{1}{8}\left(\frac{x}{Vt_0}\right)^4 + \ldots\right\}$$

For small offset-to-depth ratios (i.e. $x/Vt_0 \ll 1$), which is the normal case in reflection surveying, this equation may be truncated after the first term to obtain

$$t \doteq t_0\left\{1+\tfrac{1}{2}\left(\frac{x}{Vt_0}\right)^2\right\} \tag{4.6}$$

This is the most convenient form of the time-distance equation for reflected rays and it is used in various ways in the processing and interpretation of reflection data.

Moveout is defined as the difference between the travel times t_1 and t_2 of

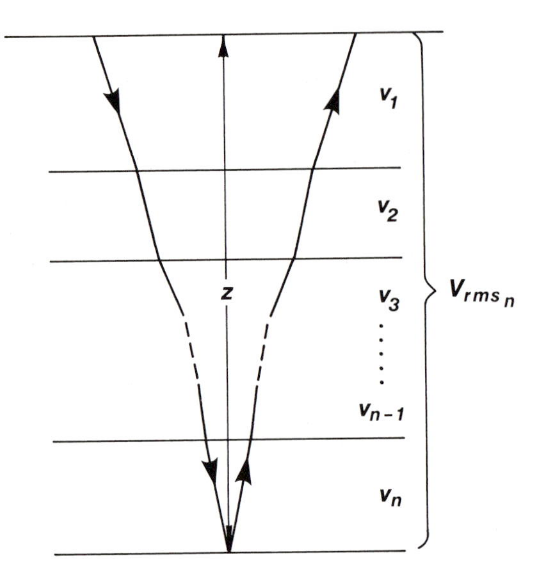

Fig. 4.3. Complex travel path of a reflected ray through a multilayered ground.

reflected ray arrivals recorded at two offset distances x_1 and x_2. From equation (4.6)

$$t_2 - t_1 \doteq \frac{x_2^2 - x_1^2}{2V^2 t_0}$$

Normal moveout (NMO) at an offset distance x is the difference in travel time ΔT between reflected arrivals at x and at zero offset (see Fig. 4.2)

$$\Delta T = t_x - t_0 \doteq \frac{x^2}{2V^2 t_0} \qquad (4.7)$$

Note that NMO is a function of offset, velocity and reflector depth z (since $z = Vt_0/2$). The concept of moveout is fundamental to the recognition, correlation and enhancement of reflection events, and to the calculation of velocities using reflection data. It is implicitly or explicitly used at several stages in the processing and interpretation of reflection data.

To exemplify its use, consider the *T–ΔT method* of velocity analysis. Rearranging the terms of equation (4.7) yields

$$V = \frac{x}{(2t_0 \Delta T)^{1/2}} \qquad (4.8)$$

Using this relationship, the velocity V above the reflector can be computed from knowledge of the zero-offset reflection time t_0 $(= T)$ and the NMO ΔT. In practice, such velocity values are obtained by computer analysis which produces a statistical estimate based upon many such calculations using large numbers of reflected ray paths (see Section 4.8). Once the velocity has been derived, it can be used in conjunction with t_0 to compute the depth z to the reflector using $z = Vt_0/2$.

4.2.2 *Multiple horizontal reflectors*

In a multilayered ground, inclined rays reflected from the nth interface undergo refraction at all higher interfaces to produce a complex travel path (Fig. 4.3). The effect of travel through several layers is to replace the velocity in equations (4.1) and (4.7) by the *average velocity* $\bar{V}$ or, to a closer approximation (Dix 1955), the *root-mean-square velocity* V_{rms} of the layers overlying the reflector.

The root-mean-square velocity of the section of ground down to the nth interface is given by

$$V_{rms_n} = \left[\sum_{i=1}^{n} v_i^2 \tau_i \Big/ \sum_{i=1}^{n} \tau_i \right]^{1/2}$$

where v_i is the interval velocity of the ith layer and τ_i is the one-way travel time of the reflected ray through the ith layer.

Thus the total travel time t_n of the ray reflected from the nth interface, at a depth z, is given by

$$t_n = (x^2+4z^2)^{1/2}/V_{rms_n} \qquad \text{(cf. equation (4.1))}$$

and the NMO for the nth reflector is given by

$$\Delta T_n \doteq \frac{x^2}{2V_{rms_n}^2 t_0} \qquad \text{(cf. equation (4.7))}$$

The individual NMO value associated with each reflection event may therefore be used to derive a root-mean-square velocity value for the layers above the reflector. Values of V_{rms} down to different reflectors can then be used to compute interval velocities using the *Dix formula*. To compute the interval velocity v_n for the nth interval

$$v_n = \left[\frac{V_{rms_n}^2 t_n - V_{rms_{n-1}}^2 t_{n-1}}{t_n - t_{n-1}} \right]^{1/2}$$

where $V_{rms_{n-1}}$, t_{n-1} and V_{rms_n}, t_n are, respectively, the root-mean-square velocity and reflected ray travel times to the $n-1$ th and nth reflectors (Dix 1955).

4.2.3 *Dipping reflector*

In the case of a dipping reflector (Fig. 4.4(a)) the value of dip θ enters the time-distance equation as an additional unknown

$$t = (x^2+4z^2+4xz \sin\theta)^{1/2}/V \qquad \text{(cf. equation (4.1))}$$

or, in the hyperbolic form

$$\frac{V^2t^2}{4z^2\cos^2\theta} - \frac{(x+2z\sin\theta)^2}{4z^2\cos^2\theta} = 1 \qquad \text{(cf. equation (4.2))}$$

The axis of symmetry of the hyperbola is now no longer the time axis (Fig. 4.4(b)).

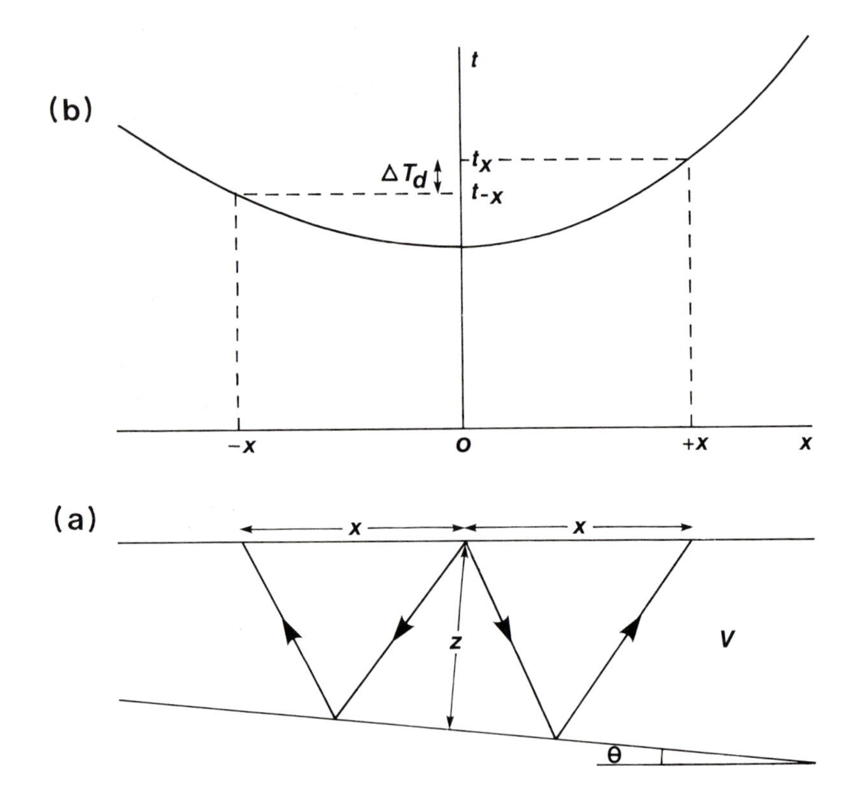

Fig. 4.4. (a) Geometry of reflected raypaths and (b) time-distance curve for reflected rays from a dipping reflector. ΔT_d = dip moveout.

Proceeding as in the case of a horizontal reflector the following truncated binominal expansion is obtained

$$t \doteq t_0 \left\{1+\frac{(x^2+4xz\sin\theta)}{2V^2t_0^2}\right\} \tag{4.9}$$

Consider two receivers at equal offsets x updip and downdip from a central shot point (Fig. 4.4). Because of the dip of the reflector, the reflected ray paths are of different length and the two rays will therefore have different travel times. *Dip moveout* ΔT_d is defined as the difference in travel times t_x and t_{-x} of rays reflected from the dipping interface to receivers at equal and opposite offsets x and $-x$

$$\Delta T_d = t_x - t_{-x}$$

Using the individual travel times defined by equation (4.9)

$$\Delta T_d = \frac{2x \sin \theta}{V}$$

Rearranging terms, and for small angles of dip (when $\sin \theta \doteq \theta$)

$$\theta \doteq V \Delta T_d / 2x$$

Hence the dip moveout ΔT_d may be used to compute the reflector dip θ if V is known. V can be derived via equation (4.8) using the NMO ΔT which, for small dips, may be obtained with sufficient accuracy by averaging the updip and downdip moveouts

$$\Delta T \doteq (t_x + t_{-x} - 2t_0)/2$$

4.2.4 *Ray paths of multiple reflections*

In addition to rays that return to the surface after reflection at a single interface, known as *primary reflections*, there are many paths in a layered subsurface by which rays may return to the surface after reflection at more than one interface. Such rays are called *reverberations, multiple reflections* or simply *multiples*. A variety of possible ray paths involving multiple reflection is shown in Fig. 4.5.

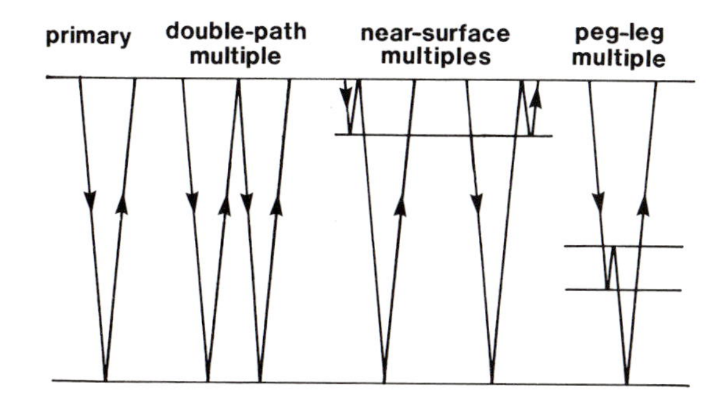

Fig. 4.5. Various types of multiple reflection in a layered ground.

Multiple reflections tend to have lower amplitudes than primary reflections because of the loss of energy at each reflection. However, there are two types of multiple that are reflected at interfaces of high reflection coefficient and therefore tend to have amplitudes comparable with primary reflections: (1) *ghost reflections*, where rays from a buried explosion on land are reflected back from the surface or the base of the weathered layer (see p. 75) to produce a reflection event, known as a *ghost reflection*, that arrives a short time after the primary; and (2) *water layer reverberations*, where rays from a marine source are repeatedly reflected at the sea bed and sea surface.

Multiple reflections that involve only a short additional path length arrive so soon after the primary event that they merely extend the overall length of the recorded pulse. Such multiples are known as *short-path multiples* (or short-period reverberations) and these may be contrasted with *long-path multiples* (long-period reverberations) whose additional path length is sufficiently long that the multiple reflection is a distinct and separate event in the seismic record.

The correct recognition of multiples is essential. Mis-identification of a long-path multiple as a primary event, for example, would lead to serious interpretation error. The arrival times of multiple reflections are predictable, however, from primary reflection times and multiples can therefore be suppressed by suitable data processing techniques to be described later (Section 4.9).

4.3 Multichannel reflection profiling

The basic requirement of a reflection survey is to obtain recordings of reflected pulses at several offset distances from a shot point. As discussed in Chapter 3 this requirement is complicated in practice by the fact that the reflected pulses are never the first arrivals of seismic energy, and they are generally of very low amplitude. Moreover, reflected pulses are typically concealed in noise which includes other, unwanted, seismic phases such as direct and refracted body waves and surface waves. Special procedures to enhance reflected arrivals have to be adopted during field recording and subsequent data reduction and processing.

The requirement to record reflected arrivals at more than one offset distance is met by multichannel recording of seismic arrivals at a spread of detectors located in the vicinity of a shot point. Reflection surveying is normally carried out along profile lines, the shot point and its associated spread of detectors being moved progressively along the line to build up lateral coverage of the underlying geological section. This progression is carried out in a stepwise fashion on land but continuously, by a ship underway, at sea. The gathering of reflection survey data along profile lines permits an imaging, in reflection times, of the geological structure underlying the survey line. The third dimension of geological structure may be studied by implementing an intersecting network of reflection lines.

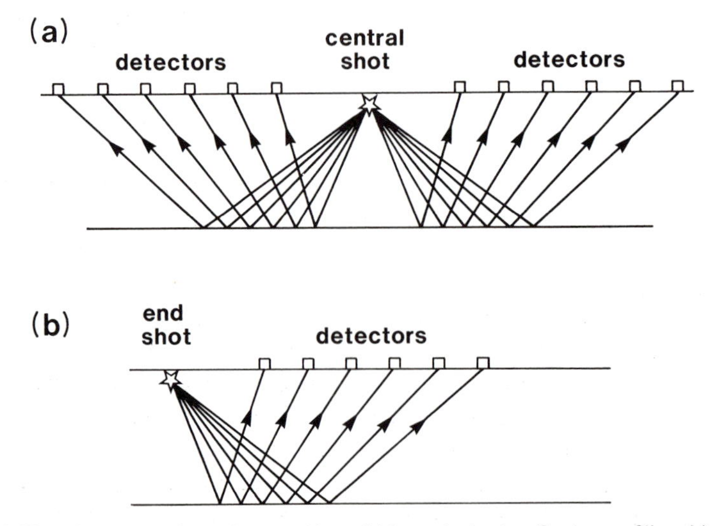

Fig. 4.6. Shot-detector configurations used in multichannel seismic reflection profiling: (a) split spread, or straddle spread; (b) single-ended spread.

The two most common shot-detector configurations in multichannel reflection profiling surveys are the *split spread* (or *straddle spread*) and the *single-ended spread* (Fig. 4.6), with the number of detectors in a spread being normally 24 or more. In split spreads, the detectors are distributed on either side of a central shot point; in single-ended spreads, the shot point is located at one end of the detector spread. Surveys on land are commonly carried out with a split spread geometry, but in marine reflection surveys single-ended spreads are the normal configuration, with a marine source towed ahead of a hydrophone streamer.

A recent major development is *three-dimensional reflection surveying*. 3D data acquisition at sea normally utilizes the conventional single-ended reflection profiling technique described above, but data are collected along very closely spaced parallel profile lines (the *parallel-profiling method*). 3D data acquisition onshore normally utilizes the *crossed-array method* in which detectors are distributed along a profile line or several parallel lines and seismic sources are arranged along another line, thus building up a dense areal array of seismic data. Analysis of reflection times from the areal arrays or closely spaced profile lines allows three-dimensional modelling of the subsurface structure.

3D seismic surveying is an expensive survey method, involving much more laborious data acquisition techniques than normal reflection profiling and requiring special methods of data processing, including 3D migration (see Section 4.10). It is therefore not routinely applied but provides the best possible interpretation of subsurface structure where highly detailed geological information is required.

4.4 The reflection seismogram (seismic trace)

The oscillographic recording of the amplified output of each detector in a reflection spread is a visual representation of the local pattern of vertical ground motion (on land) or pressure variation (at sea) over a short interval of time following the triggering of a nearby seismic source. This *reflection seismogram* or *seismic trace* represents the combined response of the layered ground and the recording system to a seismic pulse.

At each layer boundary a proportion of the incident energy in the pulse is reflected back towards the detector. The detector therefore receives a series of reflected pulses, scaled in amplitude according to the distance travelled and the reflection coefficients of the various layer boundaries. The pulses arrive at times determined by the depths to the boundaries and the velocities of propagation between them.

Assuming that the pulse shape remains unchanged as it propagates through such a layered ground, the resultant seismic trace may be regarded as the convolution of the input pulse with a time series known as a *reflectivity function* composed of a series of spikes. Each spike has an amplitude related to the reflection coefficient of a boundary and a travel time equivalent to the two-way reflection time for that boundary. This time series represents the *impulse response* of the layered ground (i.e. the output for a spike input). The convolution model is illustrated schematically in Fig. 4.7. Since the pulse has a finite length, individual

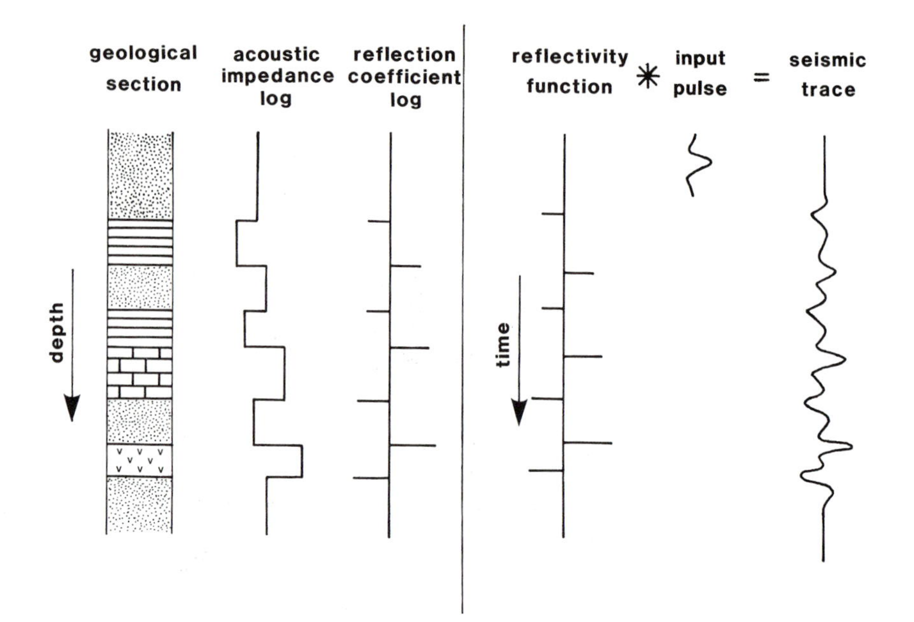

Fig. 4.7. The reflection seismogram viewed as the convolved output of a reflectivity function with an input pulse.

reflections from closely-spaced boundaries are seen to overlap in time on the resultant seismogram.

In practice, as the pulse propagates it lengthens due to the progressive loss of its higher frequency components by absorption. The basic reflection seismogram may then be regarded as the convolution of the reflectivity function with a *time-varying* seismic pulse. The seismogram will be further complicated by the superimposition of various types of noise such as multiple reflections, direct and refracted body waves, surface waves (ground roll), air waves and coherent and incoherent noise unconnected with the seismic source. In consequence of these several effects, reflection seismograms generally have a complex appearance and reflection events are often not recognizable without the application of suitable processing techniques.

4.5 Presentation of reflection survey data: the seismic section

In *seismic sections* the individual seismograms, or seismic traces, are plotted side by side in their correct relative positions and displayed with their time axes arranged vertically in a 'draped' fashion. In these draped sections, recognition of reflection events and their correlation from trace to trace is much assisted if one half of the normal 'wiggly-trace' waveform is blocked out. Fig. 4.8 shows a draped section with this mode of display, derived from a split spread multichannel survey. A short time after the shot instant the first arrival of seismic energy reaches the innermost phones (the central traces) and this energy passes out symmetrically through the two arms of the split spread. The first arrivals are followed by a series of reflection events revealed by their hyperbolic moveout.

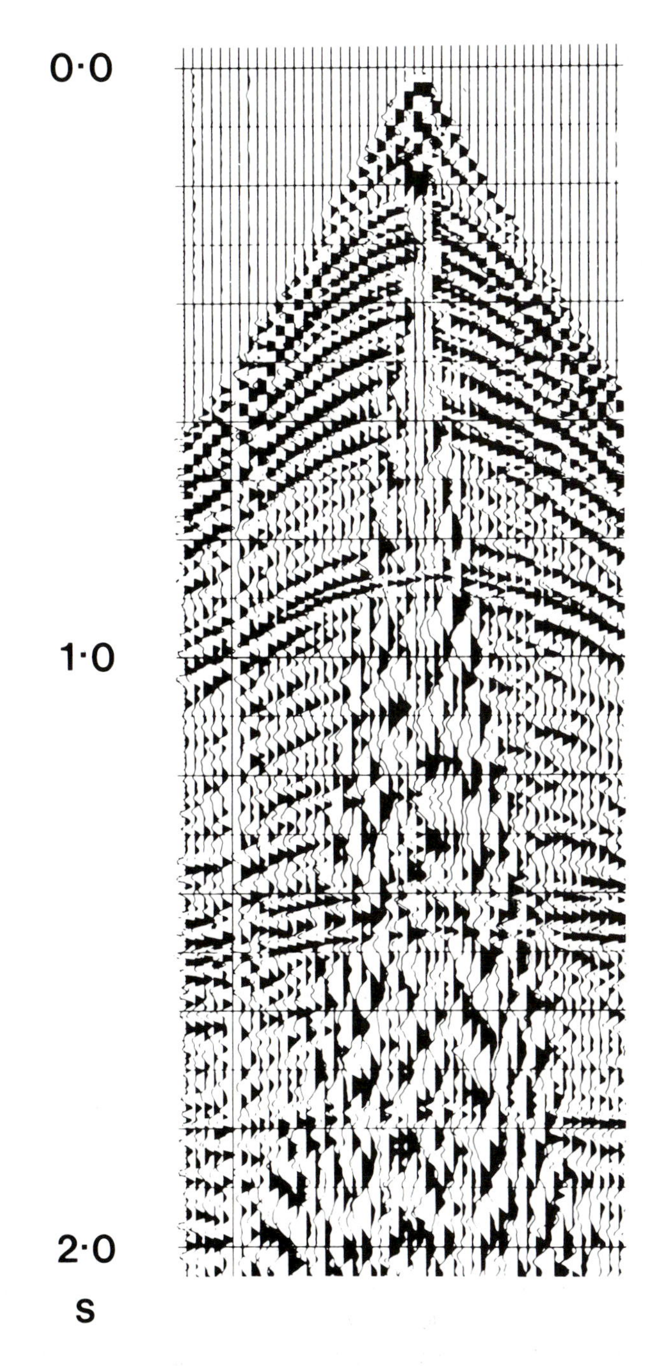

Fig. 4.8. A draped seismic section from a split spread (courtesy Prakla-Seismos GMBH). Sets of reflected arrivals from individual interfaces are recognizable by their characteristic hyperbolic form. The late-arriving, high-amplitude, low-frequency events, defining a triangular-shaped central zone within which reflected arrivals are masked, represent surface waves (ground roll).

4.6 Survey design parameters

Reflection surveys are normally designed to provide a specified depth of penetration and a particular degree of resolution of the subsurface geology. The vertical resolution (i.e. the ability to recognize individual, closely-spaced reflectors) is determined by the pulse length on the seismic section. It may be noted in passing that the vertical resolution of a survey may be improved at the data processing stage by a shortening of the recorded pulse length using inverse (deconvolution) filtering (see Section 4.9). The horizontal resolution is determined by the spacing of the individual depth estimates from which the reflector geometry is reconstructed. From Fig. 4.9 it can be seen that (for a flat-lying reflector) the horizontal resolution is equal to half the detector spacing and, also, that the length of reflector sampled by any detector spread is half the spread length. The spacing of detectors needs to be kept small to ensure that reflections from the same interface can be reliably correlated from trace to trace in areas of complex geology.

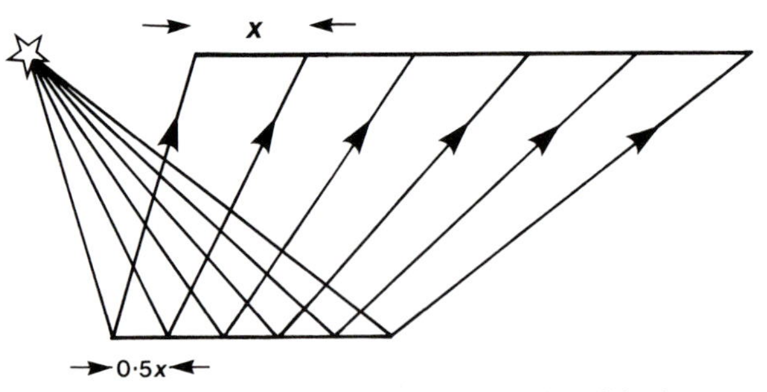

Fig. 4.9. The horizontal resolution of a seismic reflection survey is half the detector spacing.

Each detector in a conventional reflection spread consists of an *array* (or *group*) of several geophones or hydrophones arranged in a specific pattern and connected together in series/parallel to produce a single channel of output. The effective offset of an array is taken to be the distance from the shot to the centre of the array. Arrays of phones provide a directional response and are used to enhance the near-vertically travelling reflected pulses and to suppress several types of horizontally travelling *coherent* noise, i.e. noise that can be correlated from trace to trace as opposed to random noise. To exemplify this, consider a Rayleigh surface wave (a vertically polarized wave travelling along the surface) and a vertically travelling compressional wave reflected from a deep interface to pass simultaneously through two geophones connected in series/parallel and spaced at half the wavelength of the Rayleigh wave. At any given instant, ground motions associated with the Rayleigh wave will be in opposite directions at the two phones and the individual outputs of the phones will therefore be out of phase and cancelled by summing. Ground motions associated with the reflected compressional wave will, however, be in phase at the two phones and the summed outputs of the phones will therefore be twice their individual outputs.

The directional response of any linear array is governed by the relationship between the apparent wavelength λ_a of a wave in the direction of the array, the number of elements n in the array and their spacing Δx. The response is given by a response function R

$$R = \frac{\sin n\beta}{\sin \beta}$$

where

$$\beta = \pi \Delta x / \lambda_a$$

R is a periodic function that is fully defined in the interval $0 \leq \Delta x/\lambda_a \leq 1$ and is symmetrical about $\Delta x/\lambda_a = 0.5$. Typical array response curves are shown in Fig. 4.10.

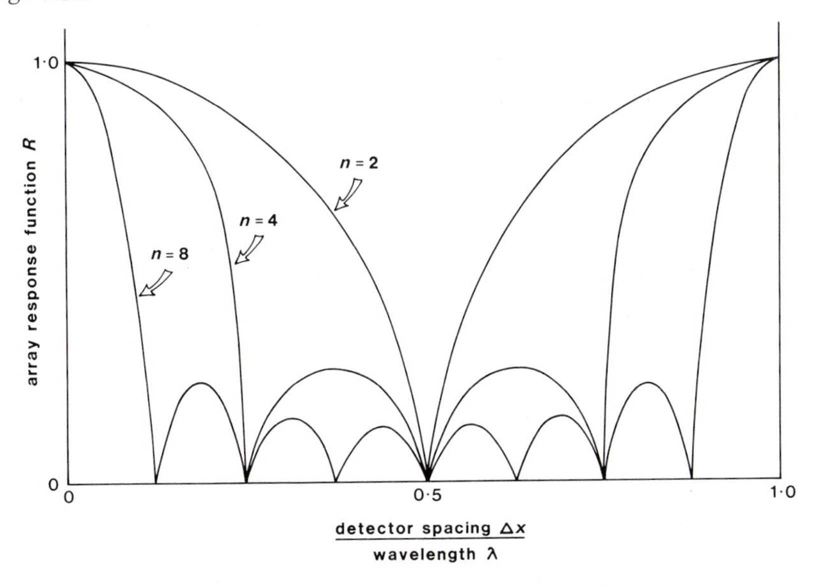

Fig. 4.10. Response functions for different detector arrays. (After Al-Sadi 1980.)

Arrays comprising areal rather than linear patterns of phones may be used to suppress horizontal noise travelling along different azimuths.

The initial stage of a reflection survey involves field trials in the survey area to determine the most suitable combination of source, offset recording range, array geometry and detector spacing (the horizontal distance between the centres of adjacent phone arrays, often referred to as the *group interval*) to produce good seismic data in the prevailing conditions.

Source trials involve tests of the effect of varying, for example, the shot depth and charge size of an explosive source or the number, chamber sizes and trigger delay times of individual guns in an air gun array. The detector array geometry needs to be designed to suppress the prevalent coherent noise events (mostly source-generated). On land, the local noise is investigated by means of a *noise test* in which shots are fired into a spread of closely-spaced detectors (*noise spread*) consisting of individual phones, or arrays of phones clustered together to eliminate their directional response. A series of shots is fired with the noise spread

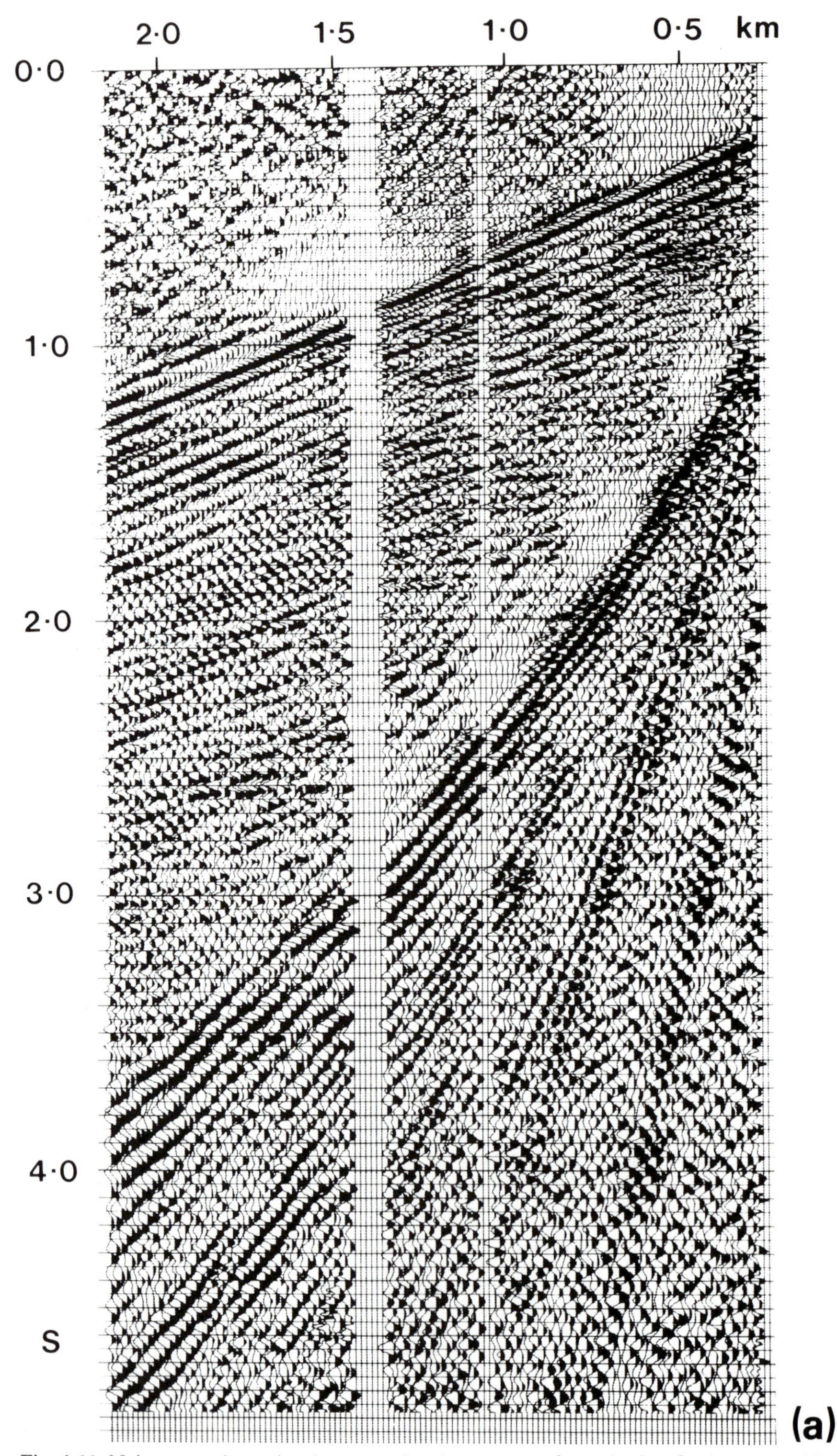

Fig. 4.11. Noise test to determine the appropriate detector array for a seismic reflection survey: (a) draped seismic section obtained with a noise spread composed of clustered (or 'bunched') geophones;

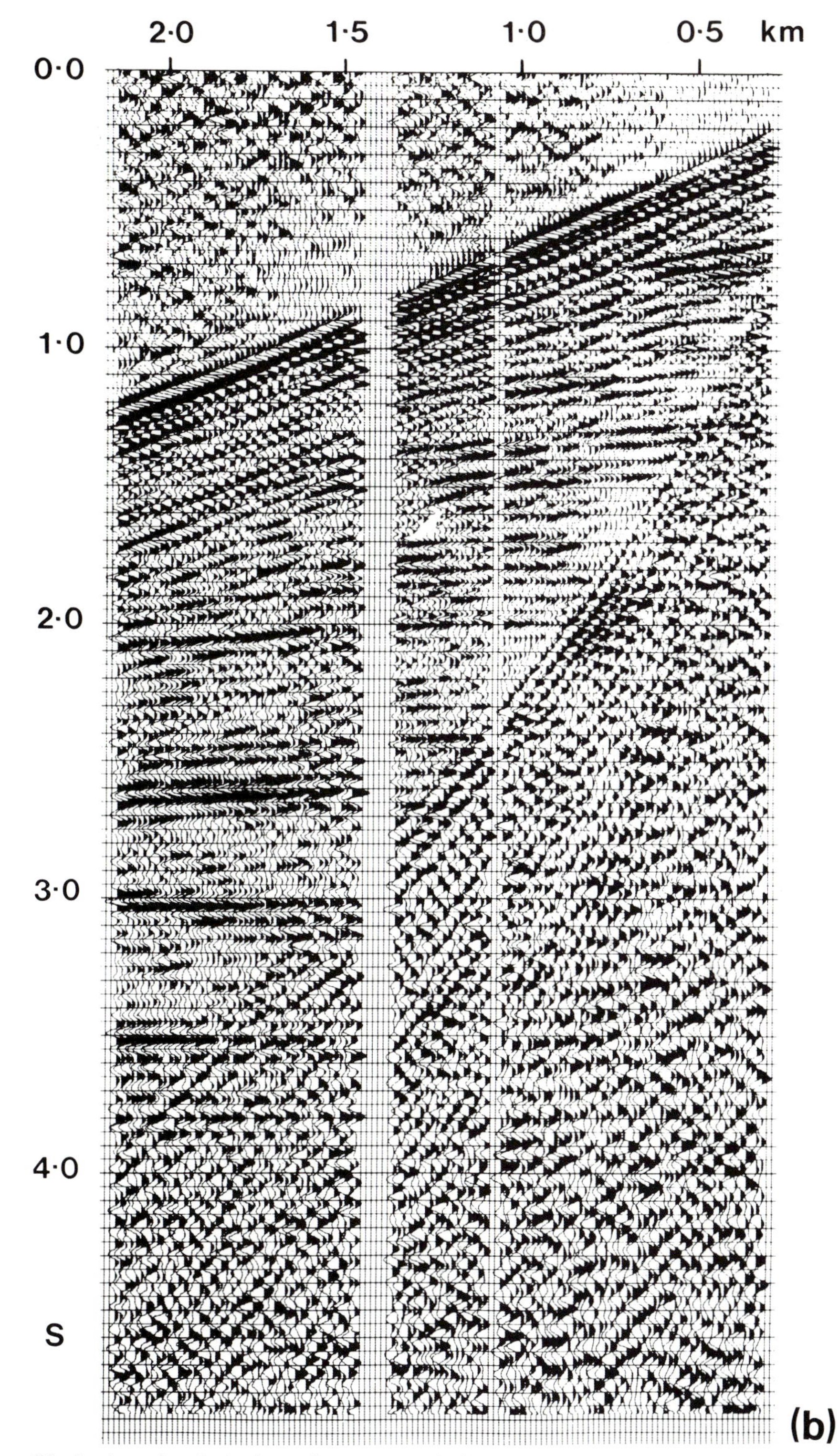

(b) seismic section obtained over the same ground with a spread composed of 140 m-long geophone arrays. (From Waters 1978.)

being moved progressively out to large offset distances. The purpose of the noise test is to determine the characteristics of the coherent noise, in particular, the velocity across the spread and dominant frequency of the air waves, surface waves (ground roll), direct and shallow refracted arrivals, that together tend to conceal the low-amplitude reflections. A typical *noise section* derived from such a test is shown in Fig. 4.11(a) and clearly reveals a number of coherent noise events that need to be suppressed to enhance the SNR of reflected arrivals. Such noise sections provide the necessary information for the optimal design of detector phone arrays. Fig. 4.11(b) shows a time section obtained with a suitable array geometry designed to suppress the local noise events and reveals the presence of reflection events that were totally concealed in the noise section.

It is apparent from the above account that the use of suitably designed arrays can markedly improve the SNR of reflection events on field seismic recordings. Further improvements in SNR and survey resolution are achievable by various types of data processing discussed later in the chapter. However, consideration is firstly given to a field technique, *common depth point* or *common reflection point* profiling, that brings about major improvements in seismic sections and the interpreted geological sections derived from them.

4.7 Common depth point (CDP) profiling and stacking

If the shot-detector spread configuration is moved forward in a multichannel reflection profiling survey in such a way that no two reflected ray paths sample the same point on a reflector, the survey coverage is said to be *single-fold*. Each seismic trace then represents an unique sampling of some point on the reflector. In common depth point profiling (Mayne 1962, 1967), which has become the standard method of multichannel reflection surveying, it is arranged that a set of traces recorded at different offsets contains reflections from a common depth point (CDP) on the reflector (Fig. 4.12(a)). The shot points and detector locations for such a set of traces, known as a CDP *gather*, have a *common midpoint* (CMP) below which the common depth point is assumed to lie.

The advantages of this survey procedure are that (1) the CDP gather represents the best possible data set for computing velocity from the normal moveout (NMO) effect; and (2) with accurate velocity information the moveout can be removed from each trace of a CDP gather to produce a set of traces that may be summed algebraically (i.e. *stacked*) to produce a CDP *stack* in which reflected arrivals are enhanced relative to the seismic noise.

The common depth point principle breaks down in the presence of dip because the common depth point then no longer directly underlies the shot-detector midpoint and the reflection point differs for rays travelling to different offsets (see Fig. 4.12(b)). Nevertheless, the method is sufficiently robust that marked improvements in SNR almost invariably result from CDP stacks as compared with single traces.

The fold of the stacking refers to the number of traces in the CDP gather and may conventionally be 6, 12, 24, 48 or, exceptionally, over 1000. The fold is alternatively expressed as a percentage: single-fold = 100% coverage, six-fold = 600% coverage and so on. The theoretical improvement in SNR brought about by stacking n traces containing a mixture of coherent in-phase signals and random (incoherent) noise is $\sqrt{n}$. Stacking attenuates or even totally suppresses the long-path multiples that have a significantly different moveout from the primary reflections: thus when the latter are stacked in phase the former are not in phase and do not sum.

The fold of a CDP survey is determined by the quantity $N/2n$, where N is the number of phone arrays along a spread and n is the number of phone array spacings by which the spread is moved forward between shots (the *move-up rate*). Thus with a 24-channel spread ($N = 24$) and a move-up rate of two array spacings per shot interval ($n = 2$), the coverage would be $24/4 = 6$-fold. A field procedure for the routine collection of 6-fold CDP coverage using a single-ended 12-channel spread configuration progressively moved forward along a profile line is shown in Fig. 4.13.

In 3D seismic surveying using the crossed-array method the common depth point principle still applies but the CDP gather involves an areal rather than a linear distribution of shot points and detector locations. Thus, for example, a 24-fold areal coverage is obtained if reflected ray paths from four shots along different shot lines to six detectors along different profile lines all have a common reflection point.

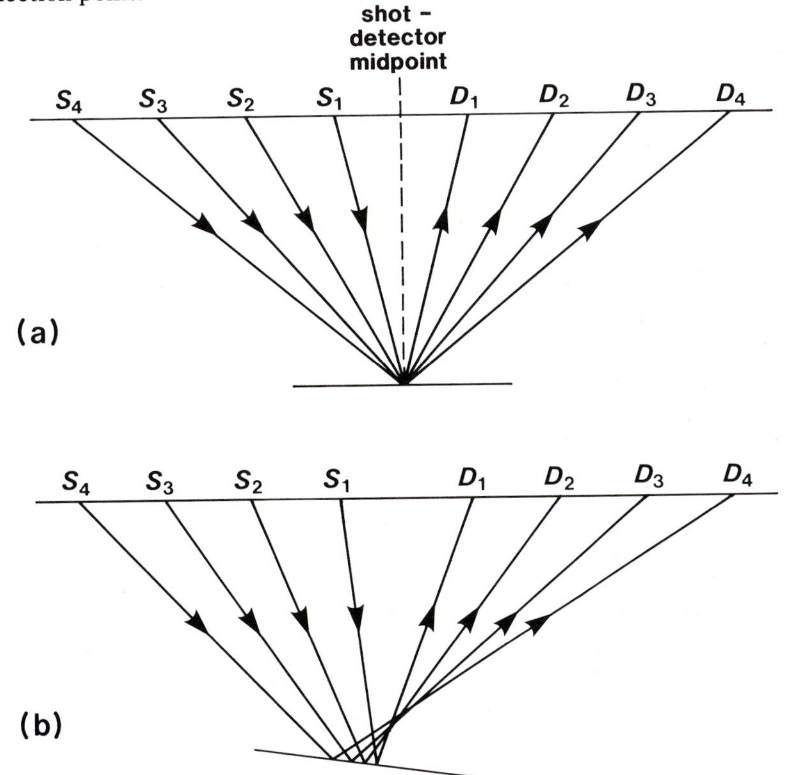

Fig. 4.12. Common depth point (CDP) reflection profiling: (a) a set of rays from different shots to detectors all reflected off a common point on a horizontal reflector; (b) the common depth point is not achieved in the case of a dipping reflector.

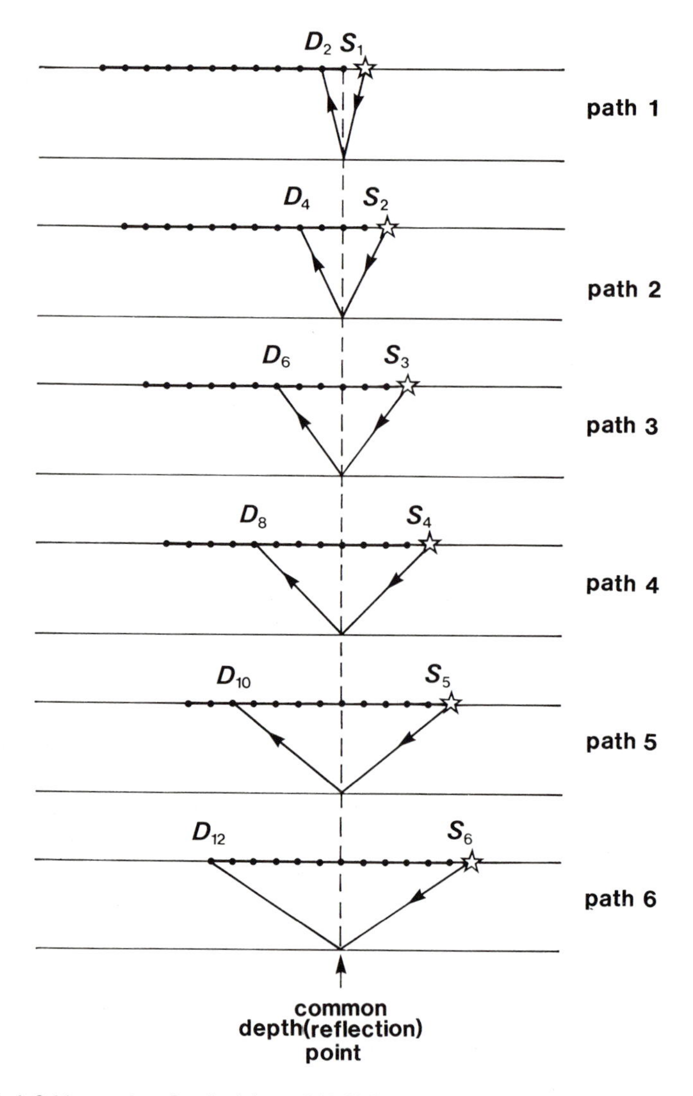

Fig. 4.13. A field procedure for obtaining 6-fold CDP coverage with a single-ended 12-channel detector spread moved progressively along the survey line.

4.8 Time corrections applied to seismic traces

Two main types of correction need to be applied to reflection times on individual seismic traces in order that the resultant seismic sections give a true representation of geological structure. These are the *static* and *dynamic* corrections, so-called because the former is a fixed time correction applied to an entire trace whereas the latter varies as a function of reflection time.

Reflection times on seismic traces recorded on land have to be corrected for time differences introduced by near-surface irregularities. These irregularities have the effect of shifting reflection events on adjacent traces out of their true time relationships. The two major sources of irregularity (Fig. 4.14) are: (1) elevation

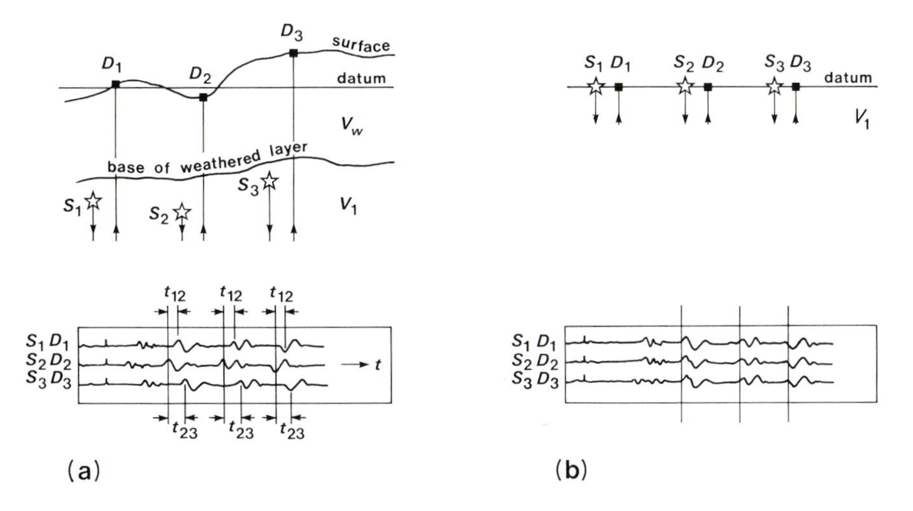

Fig. 4.14. Static corrections. (a) Seismograms showing time differences between reflection events on adjacent seismograms due to the different elevations of shots and detectors and the presence of a weathered layer. (b) The same seismograms after the application of elevation and weathering corrections, showing good alignment of the reflection events. (After O'Brien 1974.)

differences between individual shots and detectors; and (2) the presence of a *weathered layer*, which is a heterogeneous surface layer, a few metres to several tens of metres thick, of abnormally low seismic velocity. The weathered layer is mainly caused by the presence within the surface zone of open joints and micro-fractures and by the unsaturated state of the zone. Although it may be only a few metres thick, its abnormally low velocity causes large time delays to rays passing through it. Thus variations of thickness in the weathered layer may, if not corrected for, lead to false indications of significant structural relief features on underlying reflectors being portrayed on seismic sections. In marine surveys there is no elevation difference between individual shots and detectors but the water layer represents a surface layer of anomalously low velocity in some ways analogous to the weathered layer on land.

The *static correction* is a combined weathering and elevation correction that removes the effects of the low velocity surface layer and reduces all reflection times to a common height datum (Fig. 4.14). The correction is calculated on the assumption that the reflected ray path is effectively vertical immediately beneath any shot or detector. The travel time of the ray is then corrected for the time taken to travel the vertical distance between the shot or detector elevation and the survey datum, there being a component of correction for each end of the ray path. Survey datum may lie above the local base of the weathered layer, or even above the local land surface. In adjusting travel times to datum in such cases the height interval between the base of the weathered layer and datum is effectively replaced by material with the velocity of the main top layer.

Calculation of the static correction requires knowledge of the velocity and thickness of the weathered layer and the velocity of the underlying layer under all shot and detector locations. The first arrivals of energy at detectors in a reflection spread are normally rays that have been refracted along the top of the layer underlying the weathered layer and these arrivals can be used in a refraction interpretation of the top layer geometry using methods discussed in Chapter 5. If the normal reflection spread does not contain recordings at sufficiently small offsets to detect these shallow refracted rays and the direct rays defining the weathered layer velocity v_w, special short refraction spreads may be established for this purpose.

Direct measurements of the weathered layer velocity may also be obtained by *uphole surveys* in which small shots are fired at various depths down boreholes penetrating through the weathered layer and the velocities are measured of rays travelling from the shots to a surface detector. Conversely, a surface shot may be recorded by down-hole detectors. In reflection surveys utilizing buried shots, a

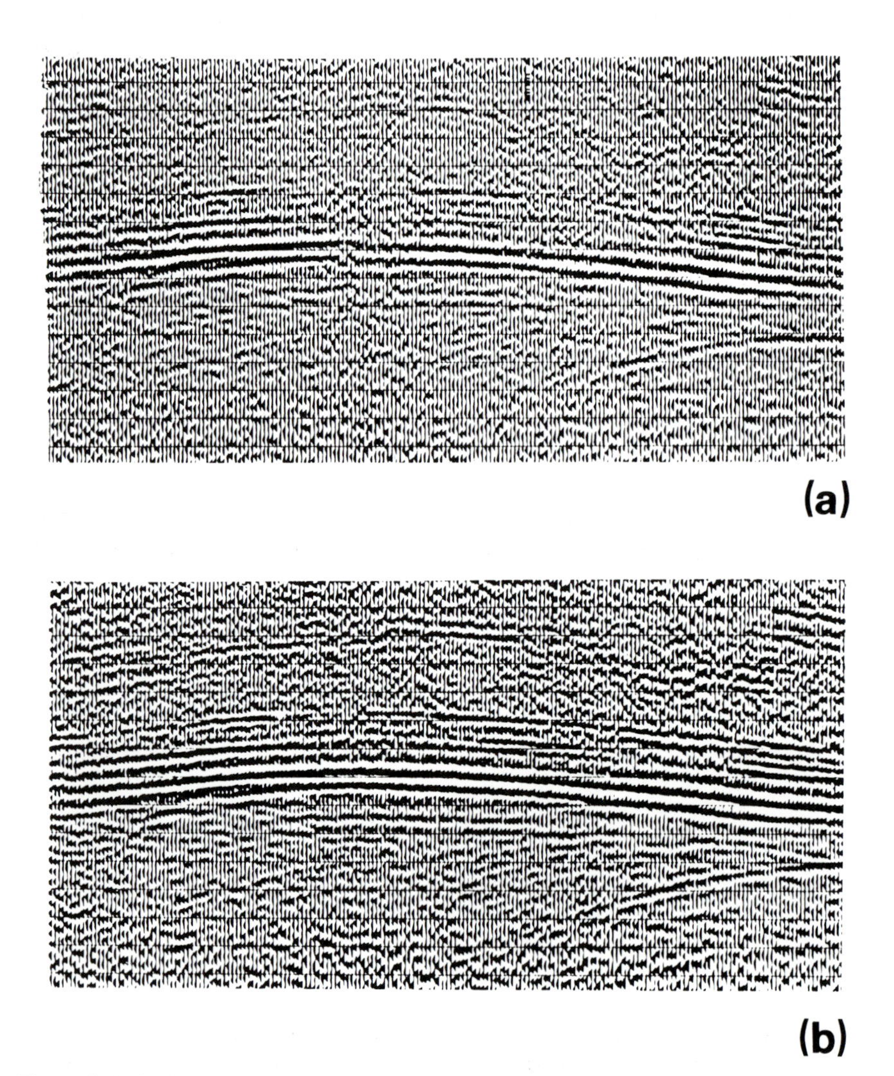

Fig. 4.15. Major improvement to a seismic section resulting from residual static analysis. (a) Manual statics only; (b) After residual static correction. (Courtesy Prakla Seismos GMBH.)

surface detector is routinely located at the surface close to the shot hole to measure the *vertical time* (VT) or *uphole time*, from which the velocity of the surface layer above the shot may be calculated.

A purely empirical but often very effective approach to the computation of the static correction is to assume that the weathered layer and surface relief are the only cause of irregularities in the travel times of rays reflected from a shallow interface and to apply appropriate time adjustments to the individual traces to produce a smooth reflection profile in the time section.

Due to the fact that the velocity and thickness of the weathered layer can never be precisely defined, the static correction always contains errors, or residuals, which have the effect of diminishing the SNR of CDP stacks and reducing the coherence of reflection events on time sections. These residuals can be calculated by computer in a *residual static analysis* by searching for systematic residual effects associated with individual shot and detector locations and applying these as corrections to the time sections. Fig. 4.15 shows the marked improvement in SNR and reflection coherence achievable by the application of these automatically computed residual static corrections.

In marine reflection surveys the static correction is commonly restricted to a conversion of travel times to mean sea level datum, without removing the overall effect of the water layer. Travel times are increased by $(d_s + d_h)v_w$ where d_s and d_h are the depths below mean sea level of the source and hydrophone array and v_w is the velocity of sea water.

The *dynamic correction* is applied to reflection times to remove the effect of normal moveout. The correction is therefore numerically equal to the NMO and, as such, is a function of offset, velocity and reflector depth. Consequently, the correction has to be calculated separately for each time increment of a seismic trace.

Adequate correction for normal moveout is dependent on the use of accurate velocities. In common depth point surveys the appropriate velocity is derived by computer analysis of moveout in the groups of traces (CDP gathers) that contain

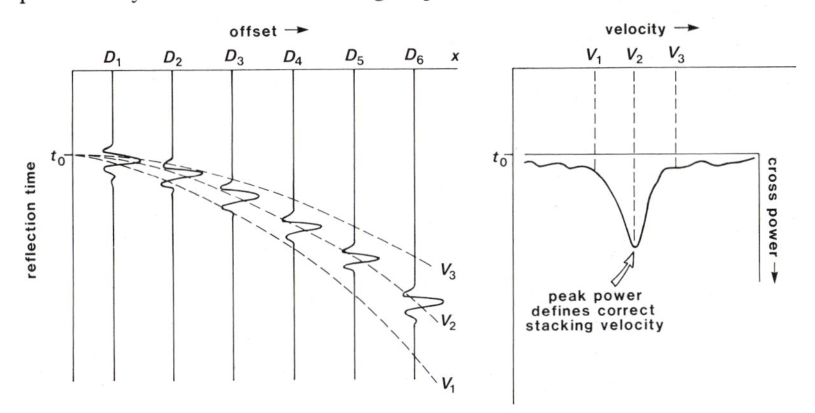

Fig. 4.16. A set of reflection events in a CDP gather is corrected for NMO using a range of velocity values. The stacking velocity is that which produces peak cross-power from the stacked events, i.e. the velocity that most successfully removes the NMO. In the case illustrated, V_2 represents the stacking velocity. (After Taner & Koehler 1969.)

reflections from a common depth point. Prior to this *velocity analysis,* static corrections must be applied to the individual traces to remove the effect of the low velocity surface layer and to reduce travel times to a common height datum. The method is exemplified with reference to Fig. 4.16 which illustrates a set of statically corrected traces containing a reflection event with a zero offset travel time of t_0. Dynamic corrections are calculated for a range of velocity values and the dynamically corrected traces are stacked. The *stacking velocity* V_{st} is defined as that velocity value which produces the maximum amplitude of the reflection event in the stack of traces. This clearly represents the condition of successful removal of NMO. Since the stacking velocity is that which removes NMO, it is given by the equation

$$t^2 = t_0^2 + \frac{x^2}{V_{st}^2} \qquad \text{(cf. equation (4.4))}$$

The stacking velocity closely approximates the root-mean-square velocity V_{rms}, though it is obviously affected by reflector dip, and values of V_{st} for different reflectors can be used in a similar way to derive interval velocities using the Dix

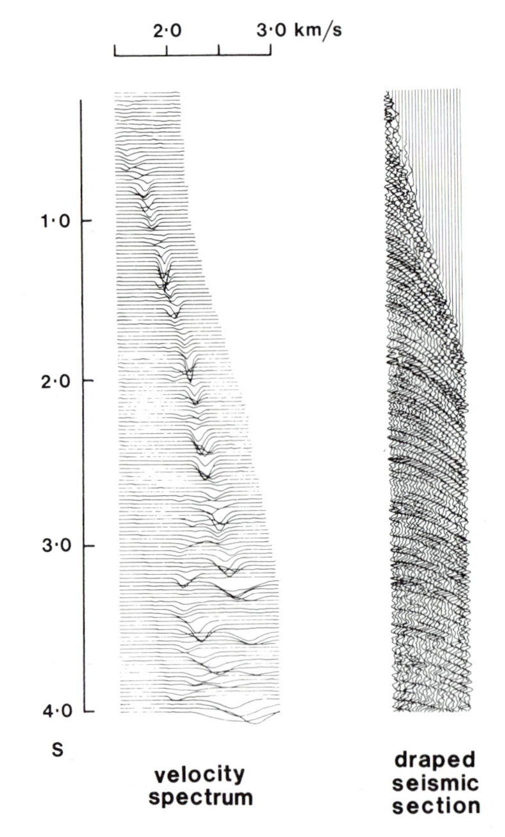

Fig. 4.17. The velocity spectrum defines the stacking velocity as a function of reflection time. The cross-power function used to define the stacking velocity is calculated for a large number of narrow time windows along a seismic section. The velocity spectrum is typically displayed alongside the draped section. (From Taner & Koehler 1969.)

formula (see p. 61). In practice, stacking velocities are computed for narrow time intervals along the entire trace to produce a *velocity spectrum* (Fig. 4.17). Velocity spectra are produced for several locations along a CDP profile to provide stacking velocity values for use in the dynamic correction of local traces.

4.9 Reflection data processing: filtering and inverse filtering of seismic data

In addition to the data *reduction* procedures of static and dynamic time correction and CDP stacking, several digital data *processing* techniques are available for the enhancement of seismic sections. In general, the aim of reflection data processing is to further increase the SNR and improve the vertical resolution of the individual seismic traces by waveform manipulation, in contrast to the simple adjustments of reflection times that characterize data reduction. The two main types of waveform manipulation are frequency filtering and inverse filtering (deconvolution).

4.9.1 Frequency filtering

Any coherent or incoherent noise event whose dominant frequency is different from that of reflected arrivals may be suppressed by frequency filtering (see Chapter 2). Thus, for example, ground roll in land surveys and several types of ship-generated noise in marine seismic surveying can often be markedly attenuated by low-cut filtering, and wind noise by high-cut filtering.

Since the dominant frequency of reflected arrivals reduces with increasing length of travel path, due to the more rapid absorption of the higher frequencies, the characteristics of frequency filters are normally varied as a function of reflection time. For example, the first second of a 3-second seismic trace might typically be band-pass filtered between limits of 15 and 75 Hz whereas the frequency limits for the third second might be 10 and 45 Hz. As the frequency characteristics of reflected arrivals are also influenced by the prevailing geology, the appropriate time-variant frequency filtering may also vary as a function of distance along a seismic profile. The filtering may be carried out by computer in the time domain or the frequency domain (see Chapter 2).

4.9.2 Inverse filtering (deconvolution)

Many components of seismic noise lie within the frequency spectrum of a reflected pulse and therefore cannot be removed by frequency filtering. The suppression of these noise components can, however, be effected by various types of inverse filtering. Inverse filters discriminate against noise and improve signal character using criteria other than simply frequency. They are thus able to suppress types of noise that have the same frequency characteristics as the reflected signal. A wide range of inverse filters is available for reflection data processing, each designed to remove some specific adverse effect of filtering in the ground along the transmission path, such as absorption or multiple reflection.

Deconvolution is the analytical process of removing the effect of some previous filtering operation (convolution). Inverse filters are designed to deconvolve seismic traces by removing the adverse filtering effects associated with the propagation of seismic pulses through a layered ground. Such effects include lengthening of the seismic pulse by the presence of multiple wavetrains and by progressive absorption of the higher frequencies. Mutual interference of extended reflection wavetrains from individual interfaces seriously degrades seismic records since onsets of reflections from deeper interfaces are totally or partially concealed by the wavetrains of reflections from shallower interfaces.

Consider a composite waveform w_k resulting from an initial spike source extended by the presence of short-path multiples near source such as, especially, water layer reverberations. The resultant seismic trace x_k will be given by the convolution of the reflectivity function r_k with the composite input waveform w_k (neglecting the effects of attenuation and absorption)

$$x_k = r_k \star w_k \text{ (plus noise)}$$

Reflected waveforms from closely-spaced reflectors will overlap in time on the seismic trace and, hence, will interfere. Deeper reflections may thus be concealed by the reverberation wavetrain associated with reflections from shallower interfaces, so that only by the elimination of the multiples will all the primary reflections be revealed. Note that short-path multiples have effectively the same normal moveout as the related primary reflection and are therefore not suppressed by CDP stacking, and they have a similar frequency content to the primary reflection so that they cannot be removed by frequency filtering.

Deconvolution has the general aim, not fully realizable, of compressing every occurrence of a composite waveform w_k on a seismic trace into a spike output, in order to reproduce the reflectivity function r_k that would fully define the subsurface layering. This is equivalent to the elimination of the multiple wavetrain. The required deconvolution operator is an inverse filter i_k which, when convolved with the composite waveform w_k, yields a spike function δ_k

$$i_k \star w_k = \delta_k$$

Convolution of the same operator with the entire seismic trace yields the reflectivity function

$$i_k \star x_k = r_k$$

In communications systems where w_k is known, deconvolution can be achieved by the use of *matched filters* which effectively cross-correlate the output with the known input signal (as in the initial processing of Vibroseis® seismic records to compress the long source wavetrain; see Section 3.8.2). *Wiener filters* may also be used when the input signal is known. A Wiener filter (Fig. 4.18) converts the known input signal into an output signal that comes closest, in a least-squares sense, to a desired output signal. The filter optimizes the output signal by arranging that the sum of squares of differences between the actual output and the desired output is a minimum.

However, in reflection surveying both w_k and r_k are generally unknown (the

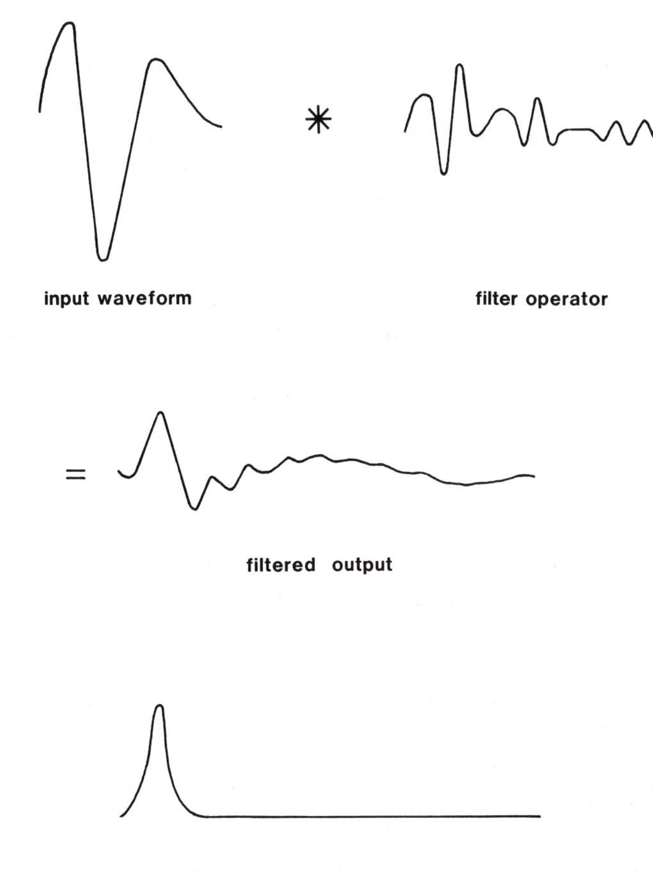

Fig. 4.18. The principle of Wiener filtering.

reflectivity function r_k is, of course, the main target of reflection interpretation) and only the seismic time series x_k is known. A special approach is therefore required to design suitable inverse filters. This approach utilizes statistical analysis of the seismic time series and is known as *predictive deconvolution*. Two important assumptions underlying predictive deconvolution (see e.g. Robinson & Treitel 1980) are: (1) that the reflectivity function r_k represents a random series (i.e. that there is no systematic pattern to the distribution of reflecting interfaces in the ground); and (2) that the composite waveform w_k is minimum-delay (i.e. that its contained energy is concentrated at the front end of the pulse; see Chapter 2). From assumption (1) it follows that the autocorrelation function of the seismic trace represents the autocorrelation function of the composite waveform w_k. From assumption (2) it follows that the autocorrelation function can be used to define the shape of the waveform, the necessary phase information coming from the minimum-delay assumption.

Such an approach allows prediction of the shape of the composite waveform for use in Wiener filtering. A particular case of Wiener filtering in seismic deconvolution is that for which the desired output is a spike function. This is the basis of *spiking deconvolution*, also known as *whitening deconvolution* because a spike has the amplitude spectrum of *white noise* (i.e. all frequency components have the same amplitude).

A wide variety of predictive deconvolution operators can be designed for inverse filtering of real seismic data, facilitating the suppression of multiples (*dereverberation*) and the compression of reflected pulses. The presence of short-period reverberation in a seismogram is revealed by an autocorrelation function with a series of decaying waveforms (Fig. 4.19(a)). Long-period reverberations appear in the autocorrelation function as a series of separate side lobes (Fig. 4.19(b)), the lobes occurring at lag values for which the primary reflection aligns with a multiple reflection. Thus the spacing of the side lobes represents the periodicity of the reverberation pattern. The first multiple is phase-reversed with respect to the primary reflection, due to reflection at the ground surface or the base of the weathered layer. Thus the first side lobe has a negative peak resulting from cross-correlation of the out-of-phase signals. The second multiple undergoes a further phase reversal so that it is in phase with the primary reflection and therefore gives rise to a second side lobe with a positive peak (see Fig. 4.19(b)). Autocorrelation functions such as those shown in Fig. 4.19 form the basis of predictive deconvolution operators for removing reverberation events from seismograms.

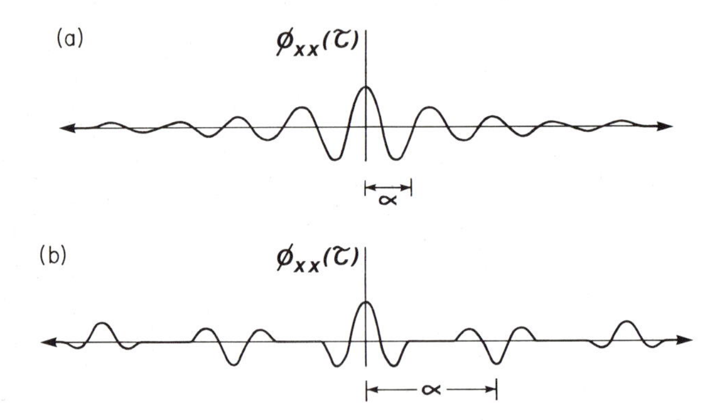

Fig. 4.19. Autocorrelation functions of seismic traces containing reverberations: (a) a gradually decaying function indicative of short-period reverberation; (b) a function with separate side lobes indicative of long-period reverberation.

Practically achievable inverse filters are always approximations to the ideal filter that would produce a reflectivity function from a seismic trace: firstly, the ideal filter operator would have to be infinitely long; secondly, predictive deconvolution makes assumptions about the statistical nature of the seismic time series that are only approximately true. Nevertheless, dramatic improvements to seismic sections, in the way of multiple suppression and associated enhancement of vertical resolution, are routinely achieved by predictive deconvolution. An example of the effectiveness of predictive deconvolution in improving the quality of a seismic section is shown in Fig. 4.20.

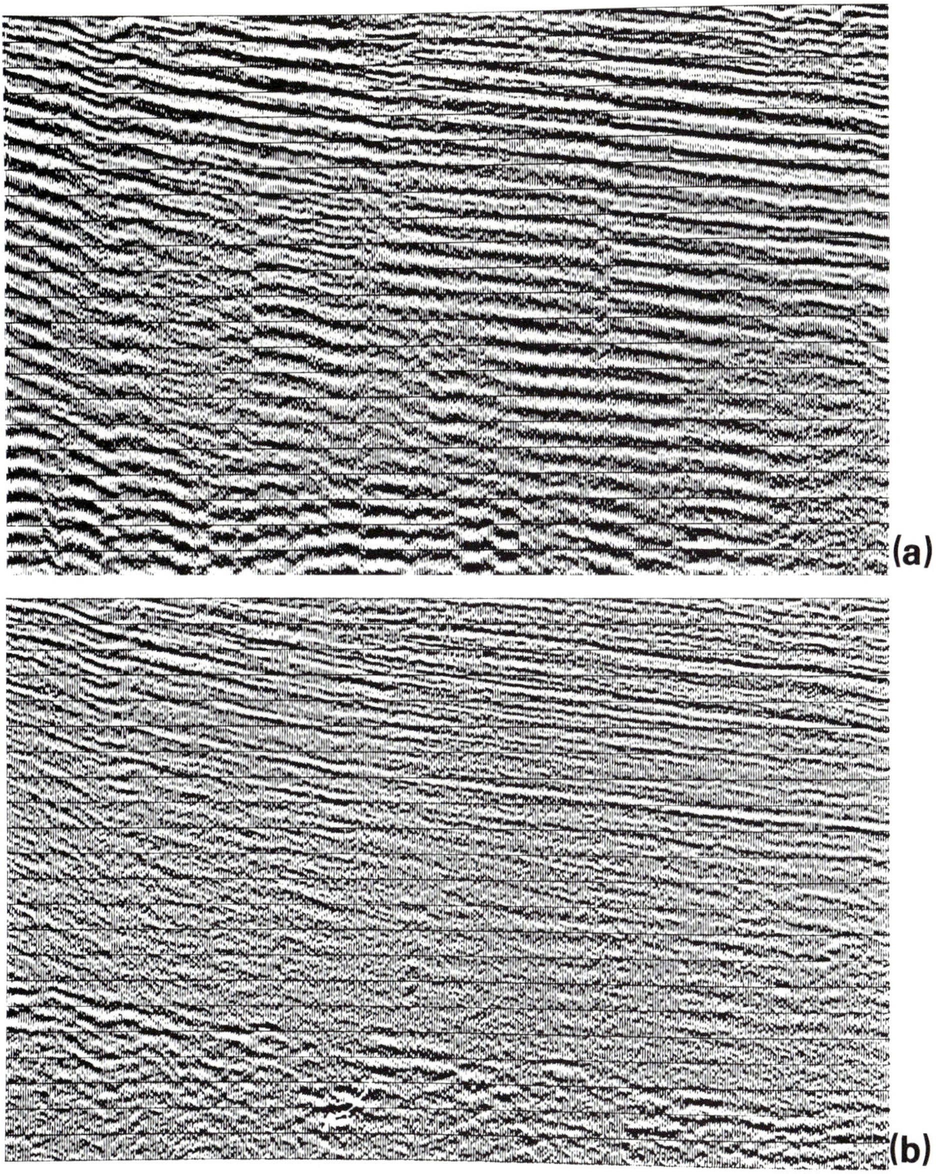

(a)

(b)

Fig. 4.20. Removal of reverberations by predictive deconvolution: (a) seismic record dominated by strong reverberations: (b) same section after spiking deconvolution. (Courtesy Prakla Seismos GMBH.)

4.9.3 Velocity filtering

In velocity filtering (also known as fan filtering or pie slice filtering), seismic signals are discriminated on the basis of the apparent velocity with which they pass across a spread of detectors. Events are passed or rejected on the basis of their value of moveout, coherent events whose moveout falls outside a prescribed range being attenuated relative to those that fall within the range. By this means, seismic phases whose apparent velocity across the spread falls outside the predicted range of reflection events, e.g. shallow refracted arrivals and surface wave arrivals, can be effectively suppressed.

It may be noted that detector arrays operate selectively on seismic arrivals according to their apparent velocity across the array (see Section 4.6), and therefore function as simple velocity filters at the data acquisition stage. Further velocity filtering may be carried out digitally at the data processing stage.

4.10 Migration of reflection data

On seismic sections such as that illustrated in Fig. 4.20 each reflection event is mapped directly beneath the midpoint of the appropriate CDP gather. However, the reflection point is located beneath the midpoint only if the reflector is horizontal. In the presence of a component of dip along the survey line the actual reflection point is displayed in the updip direction; in the presence of a

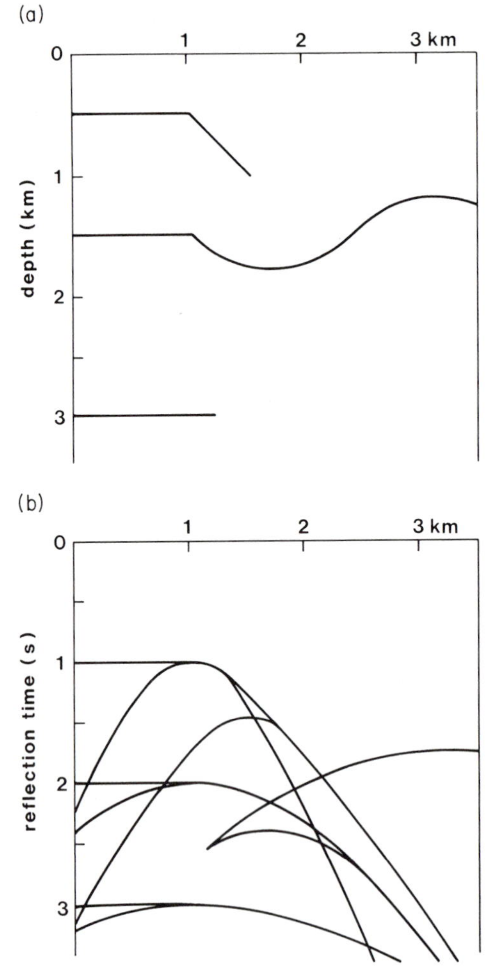

Fig. 4.21. (a) A structural model of the subsurface and (b) the resultant reflection events that would be observed in a non-migrated seismic section, containing numerous diffraction events. (After Sheriff 1978.)

component of dip across the survey line (cross dip) the reflection point is displaced out of the plane of the section. *Migration* is the process of reconstructing a seismic section so that reflection events are repositioned under their correct surface location and at a corrected vertical reflection time. In two-dimensional migration, cross dip is neglected and the migrated reflection point is constrained to lie within the plane of the section; three-dimensional migration, which is analytically more complex, allows the reflection point to migrate out of the plane of the section. Migrated seismic sections can be directly translated into geological depth sections using appropriate velocity information.

The conversion of reflection times to reflector depths using one-way reflection times multiplied by the appropriate velocity yields a reflector geometry known as the *record surface*. This only coincides with the actual *reflector surface* when the latter is horizontal. In the case of dipping reflectors the record surface departs from the reflector surface, i.e. it gives a distorted picture of the reflector geometry. Migration removes the distorting effects of dipping reflectors from seismic sections and their associated record surfaces. Migration also removes the diffracted arrivals resulting from point sources since every diffracted arrival is migrated back to the position of the point source. A variety of geological structures and sources of diffraction are illustrated in Fig. 4.21(a) and the resultant non-migrated seismic section is shown in Fig. 4.21(b). Structural distortion in the

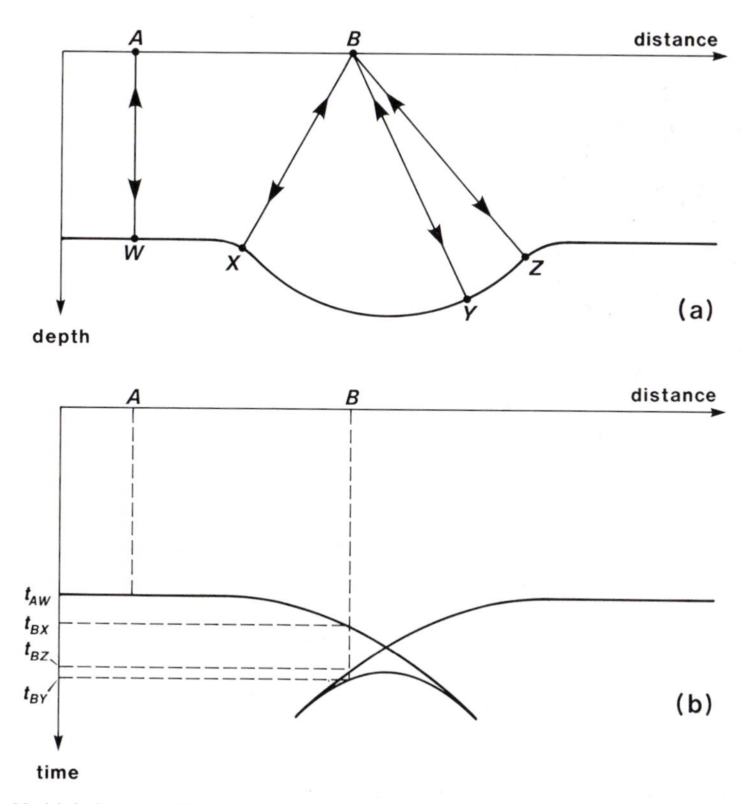

Fig. 4.22. (a) A sharp synclinical feature in a reflecting interface, and (b) the resultant 'bow-tie' shape of the reflection event on the non-migrated seismic section.

4

non-migrated section (and record surfaces derived from it) includes a broadening of anticlines and a narrowing of synclines. The edges of fault blocks act as point sources and typically give rise to strong diffracted phases, represented by hyperbolic patterns of events in the seismic section. Synclines within which the reflector curvature exceeds the curvature of the incident wavefront are represented on non-migrated seismic sections by a 'bow-tie' event resulting from the existence of three discrete reflection points for any surface location (see Fig. 4.22).

Various aspects of migration are discussed below using the simplifying assumption that the source and detector have a common surface position (i.e. the detector has a zero offset, which is approximately the situation involved in CDP gathers). In such a case, the incident and reflected rays follow the same path and the rays are normally incident on the reflector surface. Consider a source-detector on the surface of a medium of constant seismic velocity (Fig. 4.23). Any reflection event is conventionally mapped to lie directly beneath the source-detector but in fact it may lie anywhere on the locus of equal two-way reflection times, which is a semi-circle centred on the source-detector position.

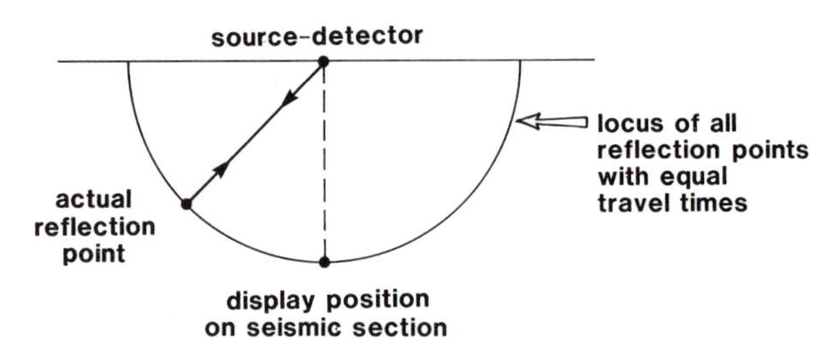

Fig. 4.23. For a given reflection time, the reflection point may lie anywhere on the arc of a circle centred on the source-detection position. On a non-migrated seismic section the point is mapped to lie immediately below the source-detector.

Now consider a series of source-detector positions overlying a planar dipping reflector beneath a medium of uniform velocity (Fig. 4.24). The reflection events are mapped to lie below each source-detector location but the actual reflection points are offset in the updip direction. The construction of arcs of circles

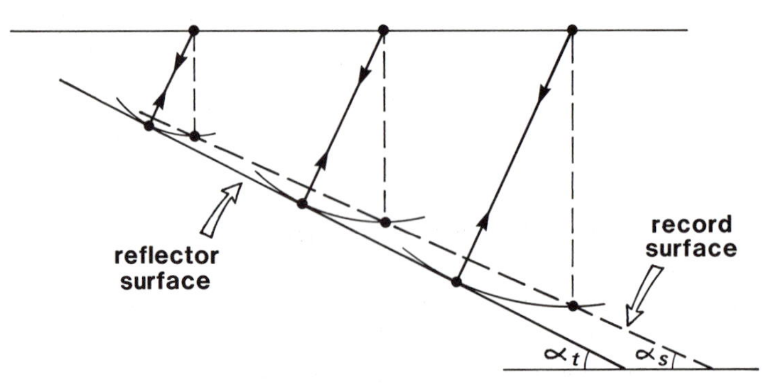

Fig. 4.24. A planar-dipping reflector surface and its associated record surface derived from a non-migrated seismic section.

(wavefront segments) through all the mapped reflection points enables the actual reflector geometry to be mapped. This represents a simple example of migration. The migrated section indicates a steeper reflector dip than the record surface derived from the non-migrated section. In general, if α_s is the dip of the record surface and α_t is the true dip of the reflector, $\sin \alpha_t = \tan \alpha_s$. Hence the maximum dip of a record surface is 45° and represents the case of horizontal reflection paths from a vertical reflector. This *wavefront common-envelope* method of migration can be extended to deal with reflectors of irregular geometry. If there is a variable velocity above the reflecting surface to be migrated, the reflected ray paths are not straight and the associated wavefronts are not circular. In such a case, a *wavefront chart* is constructed for the prevailing velocity-depth relationship and this is used to construct the wavefront segments passing through each reflection event to be migrated.

An alternative approach to migration is to assume that any continuous reflector is composed of a series of closely-spaced point reflectors, each of which is a source of diffractions, and that the continuity of any reflection event results from the constructive and destructive interference of these individual diffraction events. A set of diffracted arrivals from a single point reflector embedded in a uniform velocity medium is shown in Fig. 4.25. The two-way reflection times to different surface locations define an hyperbola. If arcs of circles (wavefront segments) are drawn through each reflection event they intersect at the actual point of diffraction (Fig. 4.25). In the case of a variable velocity above the point reflector the diffraction event will not be an hyperbola but a curve of similar convex shape. No reflection event on a seismic section can have a greater convexity than a diffraction event, hence the latter is referred to as a *curve of maximum convexity*. In *diffraction migration* all dipping reflection events are assumed to be tangential to some curve of maximum convexity. By the use of a wavefront chart appropriate to the prevailing velocity-depth relationship, wavefront segments can be drawn through dipping reflection events on seismic sections and the events migrated back to their diffraction points (Fig. 4.25). Events so migrated will, overall, map the prevailing reflector geometry.

All modern approaches to migration utilize the wave equation, which is a partial differential equation describing the motion of waves within a medium that have been generated by a wave source: in the case under consideration, the motion of seismic waves in the ground generated by a seismic source. The migration problem can be considered in terms of wave propagation through the ground in the following way: for any reflection event, the form of the seismic wavefield at the surface can be reconstructed from the travel times of reflected arrivals to different source-detector locations, and for the purpose of migration it is required to reconstruct the form of the wavefield within the ground, in the vicinity of a reflecting interface. This reconstruction can be achieved by solution of the wave equation to effectively propagate the wave backwards in time. Propagation of the wavefield of a reflection event halfway back to its origin time should place the wave on the reflecting interface, hence, the form of the wavefield at that time should define the reflector geometry.

Migration using the wave equation is known as *wave equation migration*

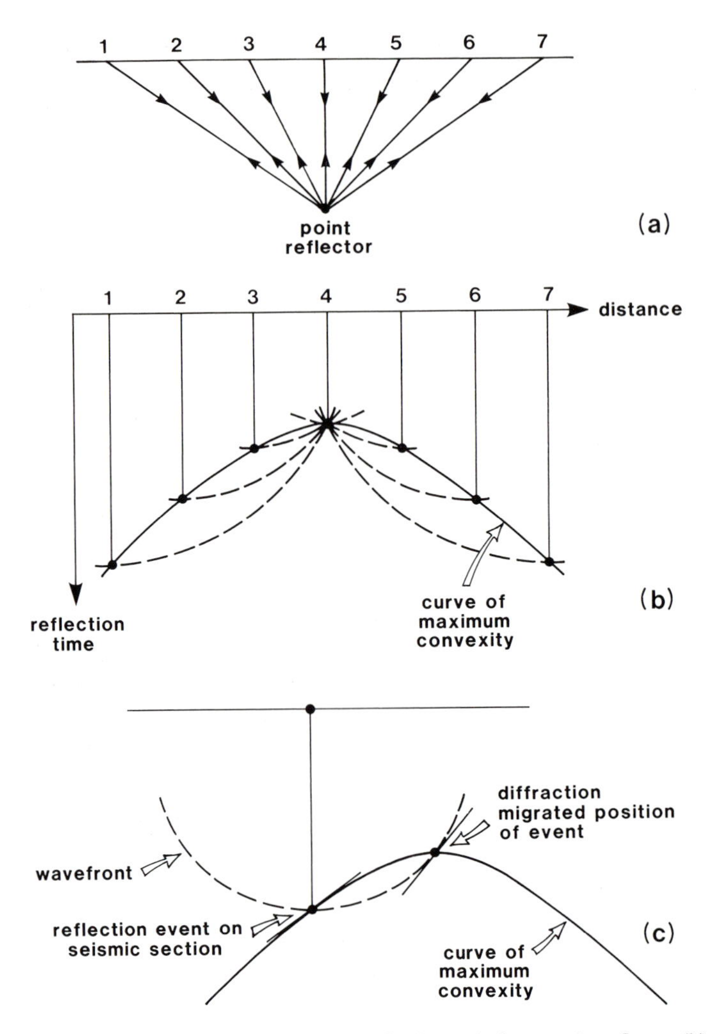

Fig. 4.25. Principles of diffraction migration: (a) reflection paths from a point reflector; (b) migration of individual reflection events back to position of point reflector; (c) use of wavefront chart and curve of maximum convexity to migrate a specific reflection event: the event is tangential to the appropriate curve of maximum convexity, and the migrated position of the event is at the intersection of the wavefront with the apex of the curve.

(Robinson & Treitel 1980). There are several approaches to the problem of solving the wave equation and these give rise to specific types of wave equation migration such as *finite difference migration,* in which the wave equation is approximated by a finite difference equation suitable for solution by computer, and *frequency-domain migration,* in which the wave equation is solved by means of Fourier transformations, the necessary spatial transformations to achieve migration being enacted in the frequency domain and recovered by an inverse Fourier transformation.

Migration by computer can also be carried out by direct modelling of ray paths through hypothetical models of the ground, the geometry of the reflecting interfaces being adjusted iteratively to remove discrepancies between observed and calculated reflection times. Particularly in the case of seismic surveys over highly complex subsurface structures, e.g. those encountered in the vicinity of salt domes and salt walls, this *ray trace migration* method may be the only method capable of successfully migrating the seismic sections.

In order to migrate a seismic section accurately it would be necessary to define fully the velocity field of the ground, i.e. to specify the value of velocity at all points. In practice, for the purposes of migration an estimate of the velocity field is made from prior analysis of the non-migrated seismic section, together with information from borehole logs where available. In spite of this approximation, migration almost invariably leads to major improvement in the seismic imaging of reflector geometry.

Any system of migration represents an approximate solution to the problem of mapping reflecting surfaces into their correct spatial positions and the various methods have different performances with real data. For example, the diffraction method performs well in the presence of steep reflector dips but is poor in the presence of a low SNR. The best all round performance is given by frequency-domain migration. Examples of the migration of seismic sections are illustrated in Figs. 4.26 and 4.27. Note in particular the clarification of structural detail, including the removal of bow-tie effects, and the repositioning of structural features in the migrated sections. Clearly, when planning to test hydrocarbon prospects in areas of structural complexity (as on the flank of a salt dome) it is important that drilling locations are based on interpretation of migrated rather than non-migrated seismic sections. Migration is not routinely carried out, however, because of the relatively high processing cost involved.

4.11 Interpretation

There are two main approaches to the interpretation of seismic sections: *structural analysis* which is the study of reflector geometry on the basis of reflection times, and *stratigraphical analysis* (or *seismic stratigraphy*) which is the analysis of reflection sequences as the seismic expression of lithologically-distinct depositional sequences. Both structural and stratigraphical analyses are greatly assisted by *seismic modelling* in which theoretical (synthetic) seismograms are constructed for layered models in order to derive insight into the physical significance of reflection events contained in seismic sections.

4.11.1 Structural analysis

The main application of structural analysis of seismic sections is in the search for structural traps containing hydrocarbons. Interpretation usually takes place against a background of continuing exploration activity and an associated increase in the amount of information related to the subsurface geology. Reflection events of interest are usually colour coded initially and labelled as, e.g. 'red reflector', 'blue reflector', until their geological significance is established. Whereas an initial interpretation of reflections displayed on seismic sections may lack geological control, at some point the geological nature of the reflectors is likely to

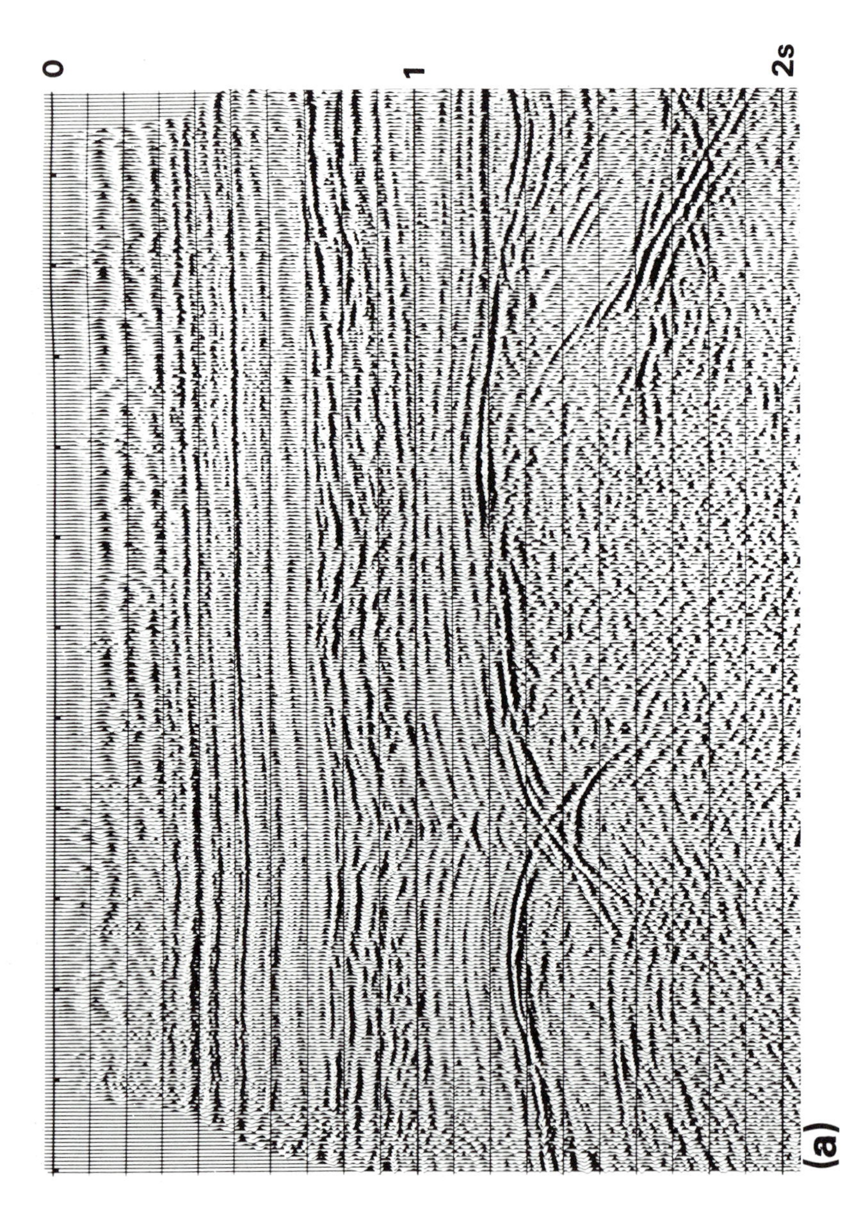
0
1
2s
(a)

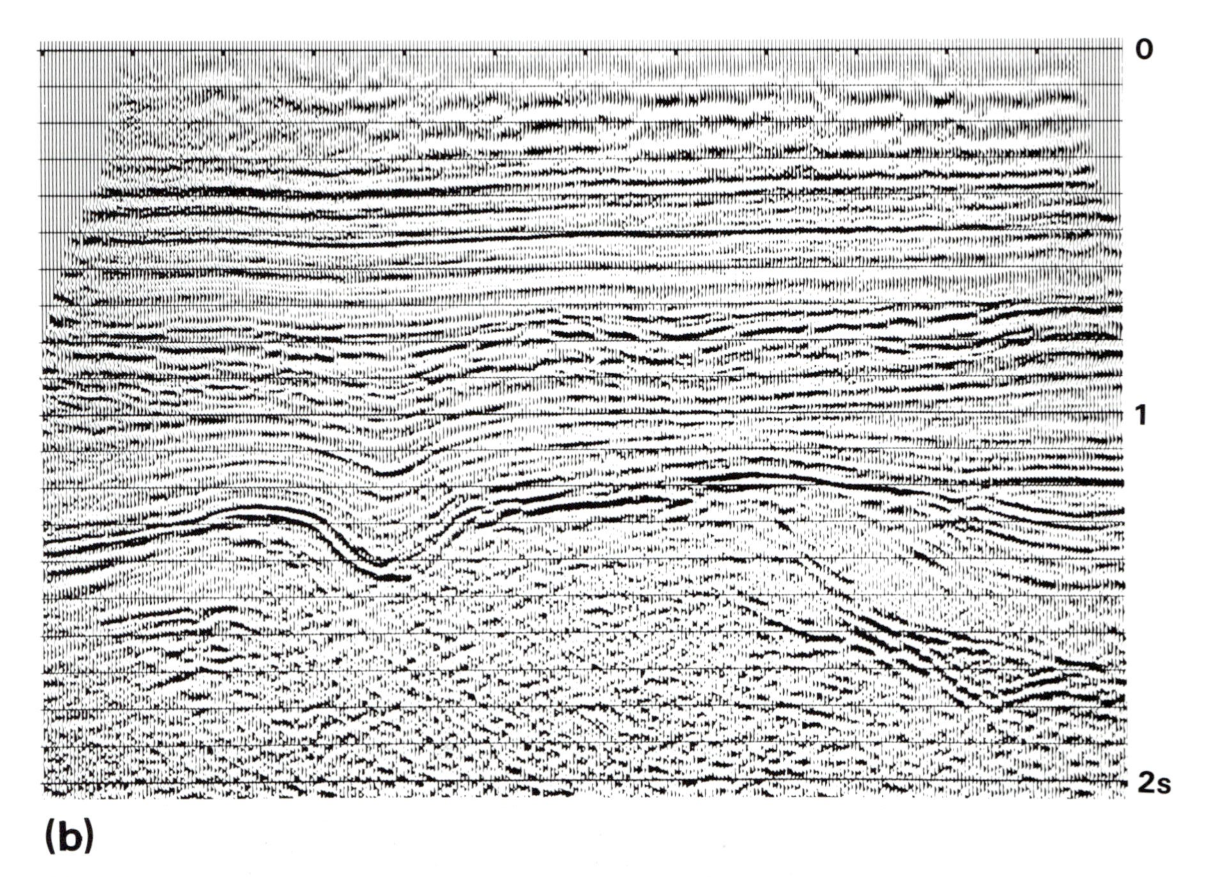

Fig. 4.26. (a) A non-migrated seismic section. (b) The same seismic section after wave equation migration. (Courtesy Prakla-Seismos GMBH.)

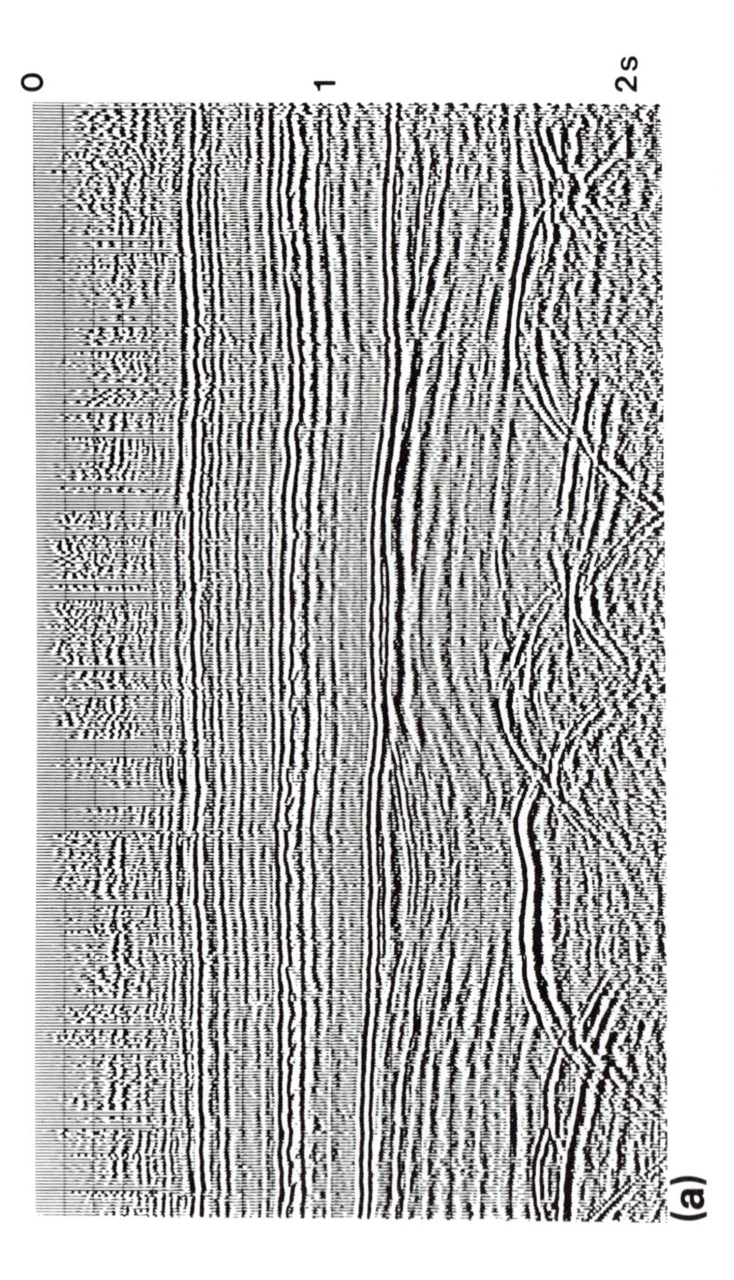
0
1
2s
(a)

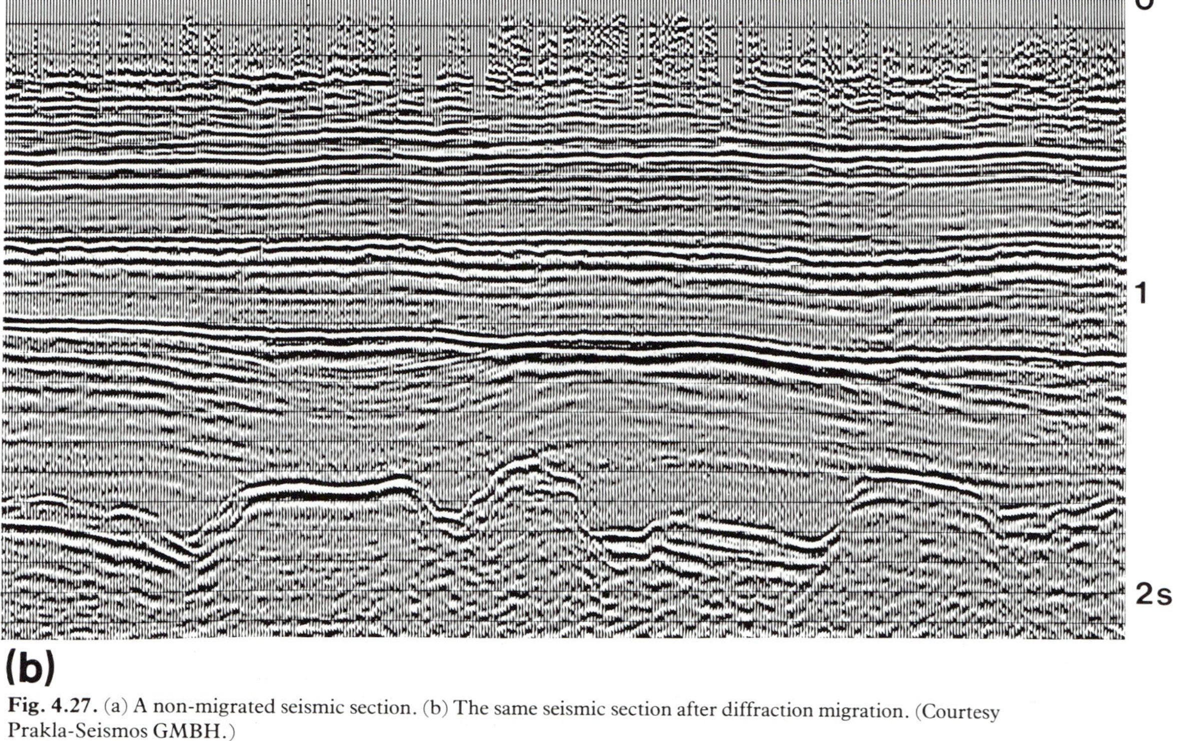

Fig. 4.27. (a) A non-migrated seismic section. (b) The same seismic section after diffraction migration. (Courtesy Prakla-Seismos GMBH.)

become established by tracing reflection events back either to outcrop or to an existing borehole for stratigraphic control. Subsurface reflectors may then be referred to by an appropriate stratigrapical indicator such as 'base Tertiary', 'top Lias'.

Most structural interpretation is carried out in units of two-way reflection time rather than depth, and *time-structure maps* are constructed to display the geometry of selected reflection events by means of contours of equal reflection time (Fig. 4.28). *Stuctural contour maps* can be produced from time-structure maps by conversion of reflection times into depths using appropriate velocity information (e.g. local stacking velocities derived from the reflection survey or sonic log data from boreholes). Time-structure maps obviously bear a close similarity to structural contour maps but are subject to distortion associated with

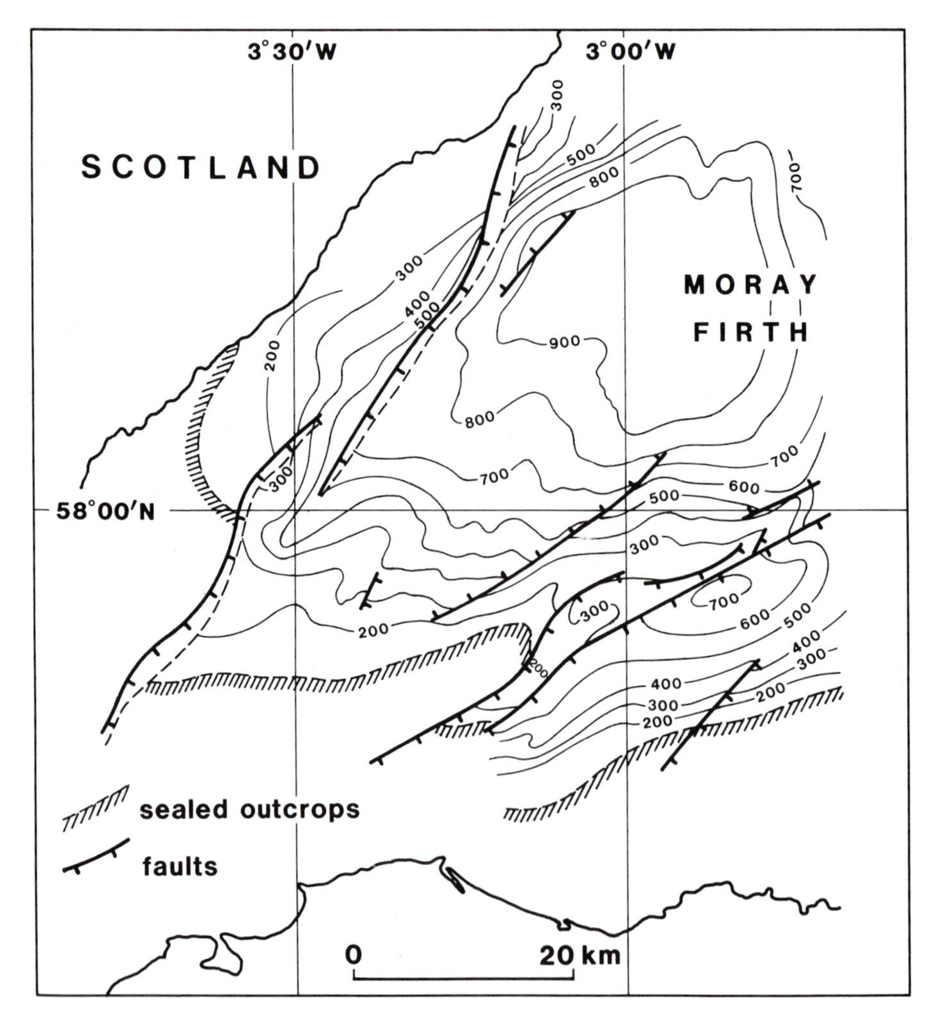

Fig. 4.28. Time-structure map of reflector at the base of the Lower Cretaceous in the Moray Firth off northeast Scotland, UK. Contour values represent two-way travel times of reflection event in milliseconds. (Courtesy British Geological Survey, Edinburgh, UK.)

lateral or vertical changes of velocity in the subsurface interval overlying the reflector. Other aspects of structure may be revealed by contouring variations in the reflection time interval between two reflectors, sometimes referred to as *isochron maps*, and these can be converted into *isopach maps* by the conversion of reflection time intervals into thicknesses using the appropriate interval velocity.

Problems often occur in the production of time-structure or isochron maps. The difficulty of correlating reflection events across areas of poor signal to noise ratio, structural complexity or rapid stratigraphic transition often leaves the disposition of a reflector poorly resolved. Intersecting survey lines facilitate the checking of an interpretation by comparison of reflection times at intersection points. Mapping reflection times around a closed loop of survey lines reveals any errors in the identification or correlation of a reflection event across the area of a seismic survey.

Reprocessing of data, or migration, may be employed to help resolve uncertainties of interpretation, but additional seismic lines are often needed to resolve problems associated with an initial phase of interpretation. It is common for several rounds of seismic exploration to be necessary before a prospective structure is sufficiently well defined to locate the optimal position of an exploration borehole.

4.11.2 *Stratigraphical analysis (seismic stratigraphy)*

Seismic stratigraphy involves the subdivision of seismic sections into sequences of reflections that are interpreted as the seismic expression of genetically-related sedimentary sequences. The principles behind this *seismic sequence analysis* are twofold. Firstly, reflections are taken to define chronostratigraphical units since the types of rock interface that produce reflections are stratal surfaces and unconformities; by contrast, the boundaries of diachronous lithological units tend to be transitional and not to produce reflections. Secondly, genetically-related sedimentary sequences normally comprise a set of concordant strata that exhibit discordance with underlying and overlying sequences, i.e. they are typically bounded by angular unconformities variously representing onlap, downlap, toplap or erosion (Fig. 4.29). A typical seismic sequence is, therefore, a

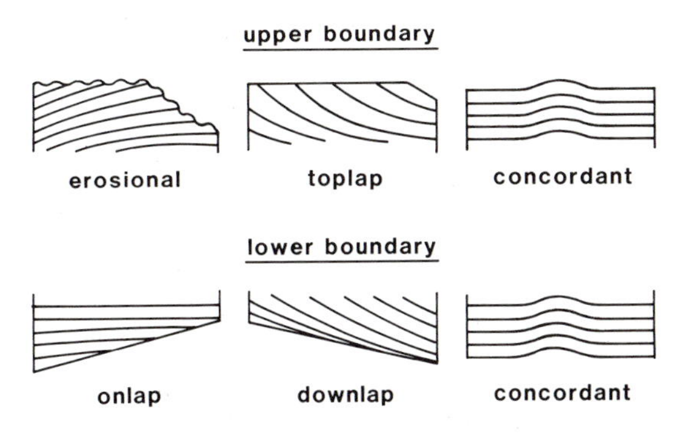

Fig. 4.29. Different types of geological boundary defining seismic sequences. (After Sheriff 1980.)

group of concordant or near-concordant reflection events that terminate against the discordant reflections of adjacent seismic sequences.

Seismic sequence analysis is followed by *seismic facies analysis* in which the identified sequences are interpreted to provide evidence of the environment of deposition and, indirectly, the lithologies of the individual strata that make up the related sedimentary sequences. Different types of reflection configuration (see Fig. 4.30) are diagnostic of different sedimentary environments: e.g. parallel reflections characterize some shallow water shelf environments whilst the deeper water shelf edge and slope environments are often marked by the development of major sigmoidal or oblique cross-bedded units. A simple example of the stratigraphical analysis of a seismic section is shown in Fig. 4.31.

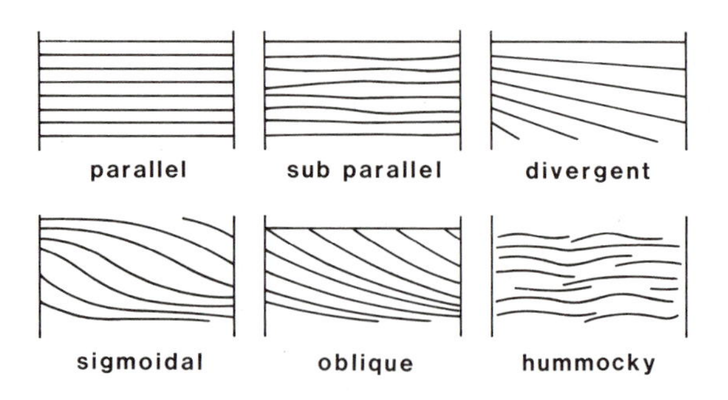

Fig. 4.30. Various internal bedforms that give rise to different seismic facies within sedimentary sequences identified on seismic sections. (After Sheriff 1980.)

Major seismic sequences can often be correlated across broad regions of continental margins and clearly give evidence of being associated with major sea-level changes. The widespread application of seismic stratigraphy in areas of good chronostratigraphical control has led to the development of a model of global cycles of major sea-level change and associated transgressive and regressive depositional sequences throughout the Mesozoic and Cenozoic (Payton 1977). Application of the methods of seismic stratigraphy in offshore sedimentary basins with little or no geological control often enables correlation of locally recognized depositional sequences with the worldwide pattern of sea-level changes (Payton 1977). It also facilitates identification of the major progradational sedimentary sequences which offer the main potential for hydrocarbon generation and accumulation. Stratigraphic analysis therefore greatly enhances the chances of successfully locating hydrocarbon traps in sedimentary basin environments.

Hydrocarbon accumulations are sometimes revealed directly on true-amplitude seismic sections (see below) by localized zones of anomalously strong reflections known as *bright spots*. These high-amplitude reflection events (Fig. 4.32) are attributable to the large reflection coefficients at the top and bottom of gas zones (typically, gas-filled sands) within a hydrocarbon reservoir. In the absence of bright spots, fluid interfaces may nevertheless be directly recognizable by *flat spots* which are horizontal or near-horizontal reflection events discordant to the local geological dip.

4.11.3 Seismic modelling

Conventionally, reflection amplitudes are normalized prior to their presentation on seismic sections so that original distinctions between weak and strong reflections are suppressed. This practice tends to increase the continuity of reflection events across a section and therefore aids their identification and structural mapping. However, much valuable geological information is contained in the true amplitude of a reflection event, which can be recovered from suitably calibrated field recordings. Any lateral variation of reflection amplitude is due to lateral change in the lithology of a rock layer or in its pore fluid content. Thus, whilst the production of normalized-amplitude sections may assist structural mapping of reflectors, it suppresses information that is vital to a full stratigraphic interpretation of the data. With increasing interest centring on stratigraphic interpretation, true-amplitude seismic sections are becoming increasingly important.

In addition to amplitude, the shape and polarity of a reflection event also contain important geological information. Analysis of the significance of lateral changes of shape, polarity and amplitude observed in true-amplitude seismic sections is carried out by *seismic modelling*, often referred to in this context as *stratigraphic modelling*. Seismic modelling involves the production of synthetic seismograms for layered sequences to investigate the effects of varying the model parameters on the form of the resulting seismograms. Synthetic seismograms and synthetic seismic sections can be compared with real data, and models can be manipulated in order to simulate the real data. By this means, valuable insights can be obtained into the subsurface geology responsible for a particular seismic section. The standard type of synthetic seismogram represents the seismic response to vertical propagation of an assumed source wavelet through a model of the subsurface composed of a series of horizontal layers of differing acoustic impedance. Each layer boundary reflects some energy back to the surface, the amplitude and polarity of the reflection being determined by the acoustic impedance contrast. The synthetic seismogram comprises the sum of the individual reflections in their correct travel time relationships (see Fig. 4.33).

In its simplest form, a synthetic seismogram $x(t)$ may be considered as the convolution of the assumed source function $s(t)$ with a reflectivity function $r(t)$ representing the acoustic impedance contrasts in the layered model

$$x(t) = s(t) \star r(t)$$

However, filtering effects along the downgoing and upgoing ray paths and the overall response of the recording system need to be taken into account. Multiples may or may not be incorporated into the synthetic seismogram.

The acoustic impedance values necessary to compute the reflectivity function may be derived directly from sonic log data. This is normally achieved assuming density to be constant throughout the model, but it may be important to derive estimates of layer densities in order to compute more accurate impedance values.

Synthetic seismograms can be derived for more complex models using ray tracing techniques (see Fig. 5.14).

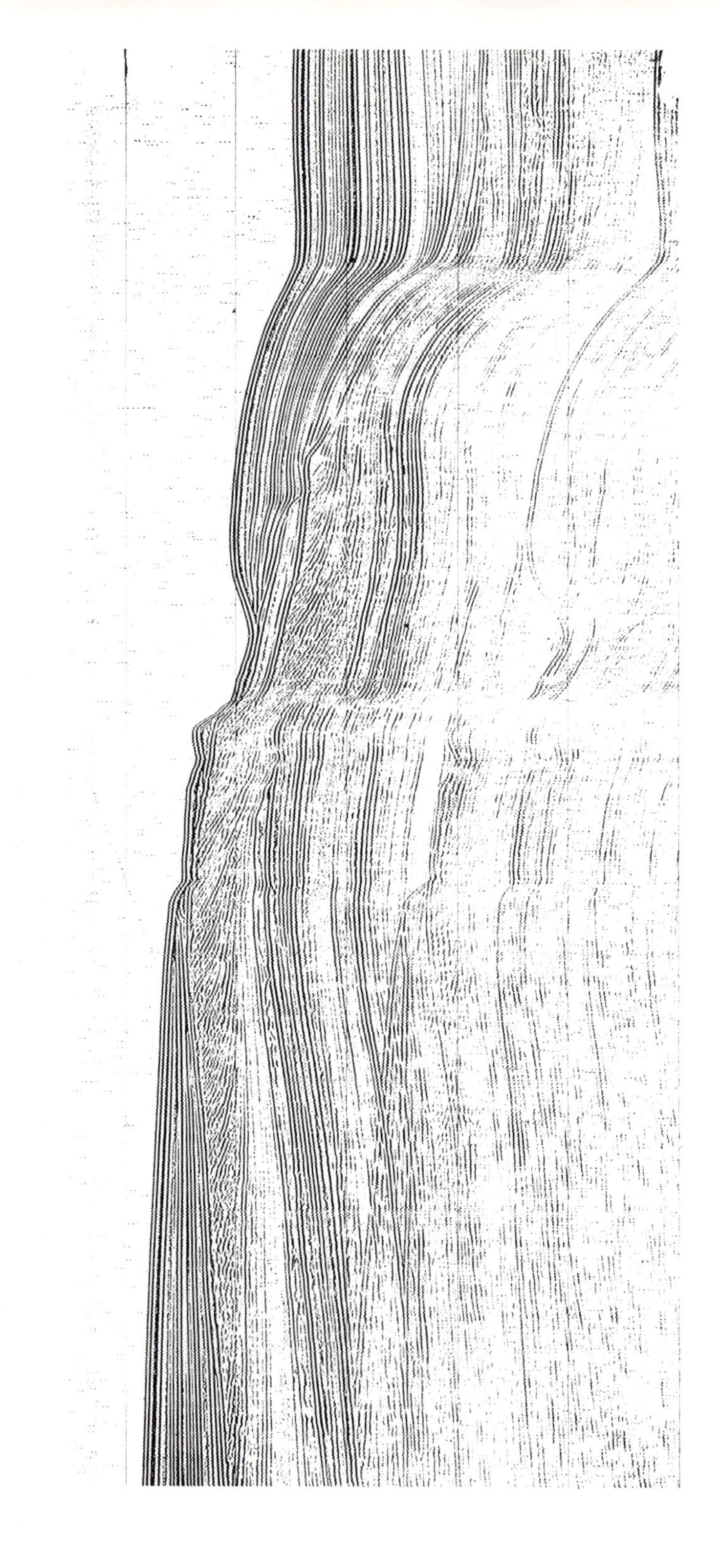

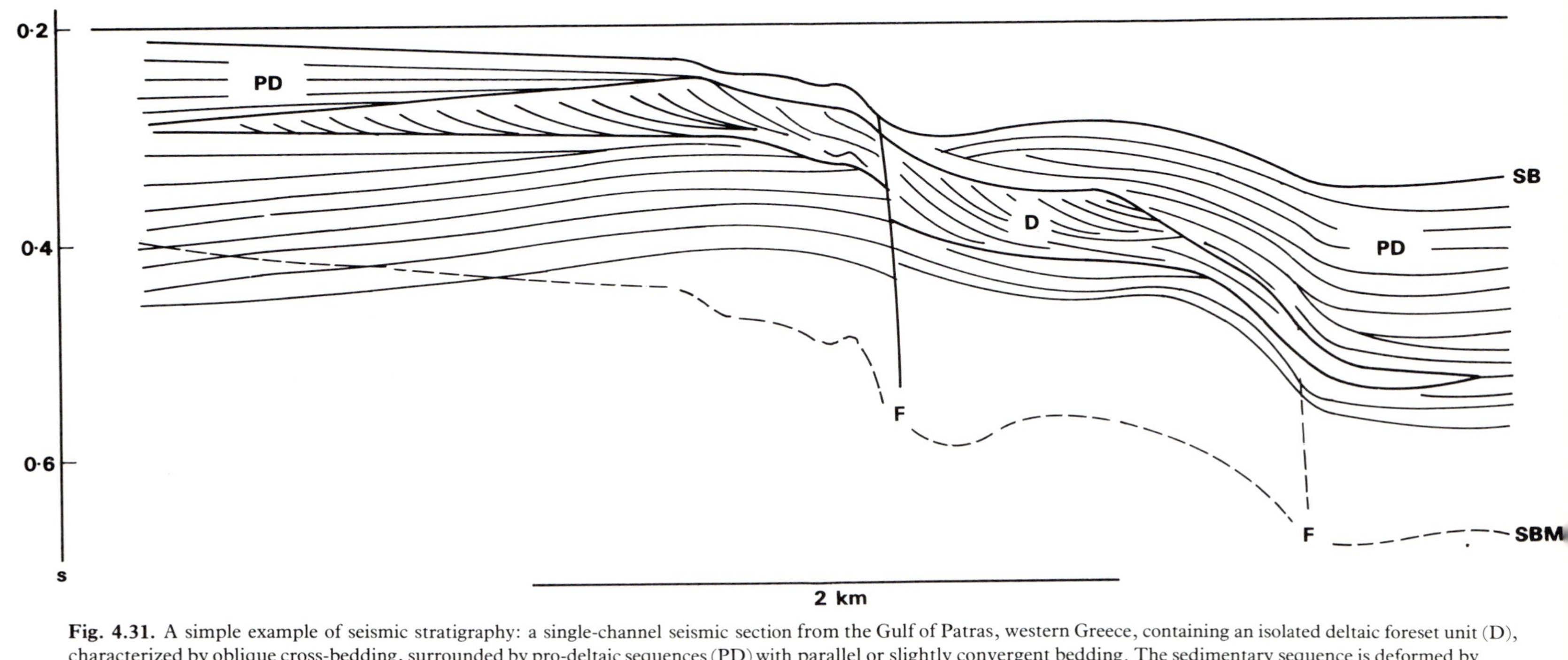

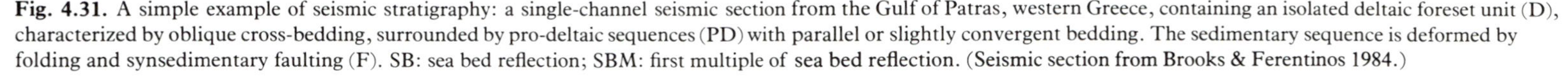

Fig. 4.31. A simple example of seismic stratigraphy: a single-channel seismic section from the Gulf of Patras, western Greece, containing an isolated deltaic foreset unit (D), characterized by oblique cross-bedding, surrounded by pro-deltaic sequences (PD) with parallel or slightly convergent bedding. The sedimentary sequence is deformed by folding and synsedimentary faulting (F). SB: sea bed reflection; SBM: first multiple of sea bed reflection. (Seismic section from Brooks & Ferentinos 1984.)

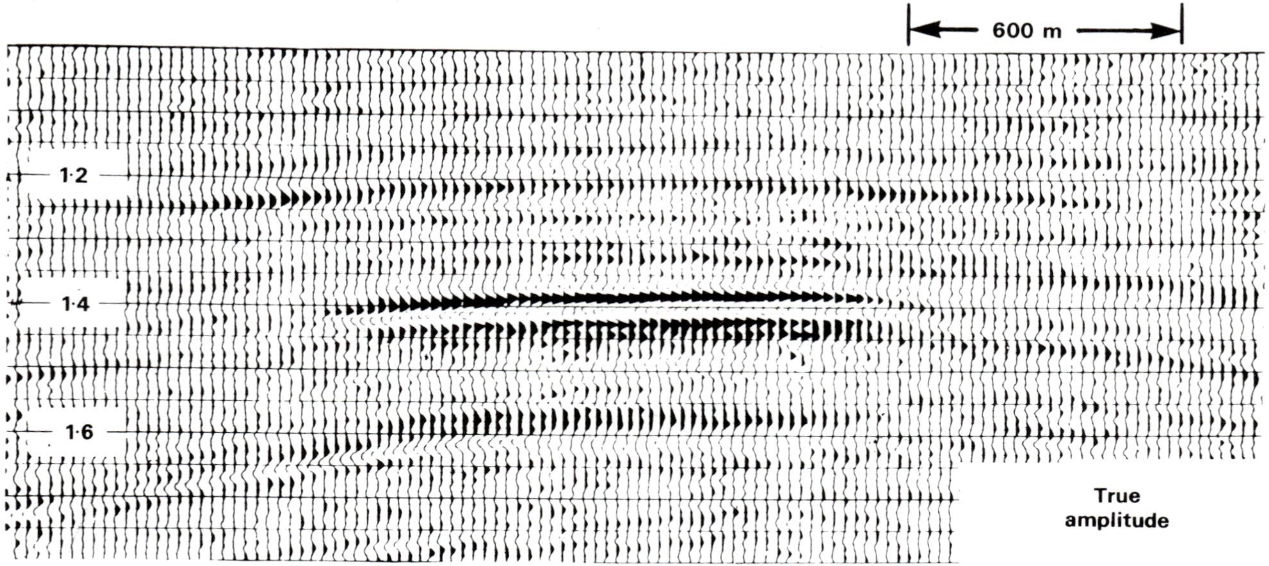

Fig. 4.32. Part of a true-amplitude seismic section containing a seismic bright spot associated with a local hydrocarbon accumulation. (From Sheriff 1980, after Schramm *et al.* 1977.)

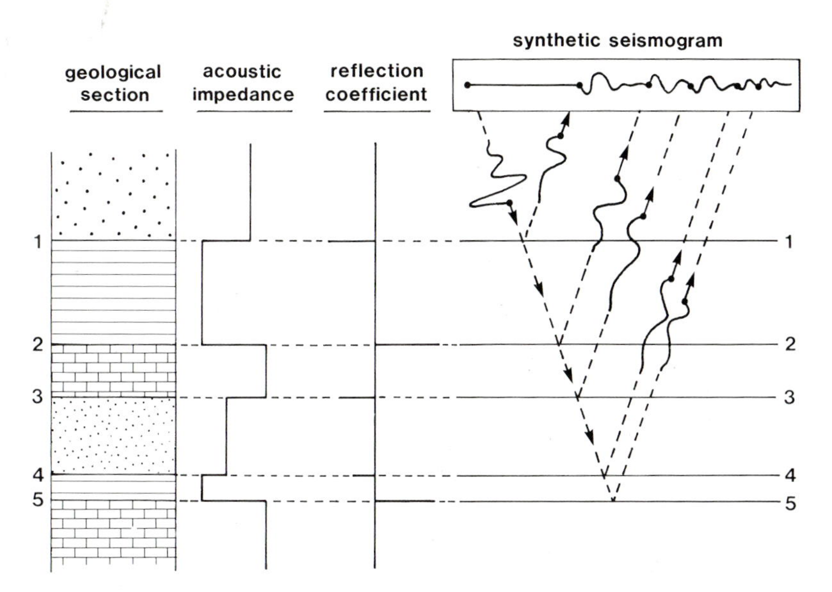

Fig. 4.33. The synthetic seismogram.

Particular stratigraphic features that have been investigated by seismic modelling, to determine the nature of their representation on seismic sections, include thin layers, discontinuous layers, wedge-shaped layers, transitional layer boundaries, variable porosity and type of pore fluid. Fig. 4.34 illustrates synthetic seismograms computed across a section of stratigraphic change. These show how the varying pattern of interference between reflection events expresses itself in lateral changes of pulse shape and peak amplitude.

4.12 Single-channel marine reflection profiling

Single-channel reflection profiling is a simple but highly effective method of seismic surveying at sea that finds wide use in a variety of offshore applications. It represents reflection surveying reduced to its bare essentials: a marine seismic/acoustic source is towed behind a survey vessel and triggered at a fixed firing rate, and signals reflected from the sea bed and from sub-bottom reflectors are detected by a hydrophone streamer towed in the vicinity of the source (Fig. 4.35). The outputs of the individual hydrophone elements are summed and fed to a single-channel amplifier/processor unit and thence to a chart recorder. This survey procedure is not possible on land because only at sea can the source and detectors be moved forward continuously, and a sufficiently high firing rate achieved, to enable surveys to be carried out continuously from a moving vehicle.

The source and hydrophone array are normally towed at shallow depth but some deep water applications utilize deep-tow systems in which the source and receiver are towed close to the sea bed. Deep-tow systems overcome the transmission losses associated with a long water path, thus giving improved penetration of seismic/acoustic energy into the sea bed. Moreover, in areas of rugged bathymetry they produce records that are much simpler to interpret: there is commonly a multiplicity of reflection paths from a rugged sea bed to a surface

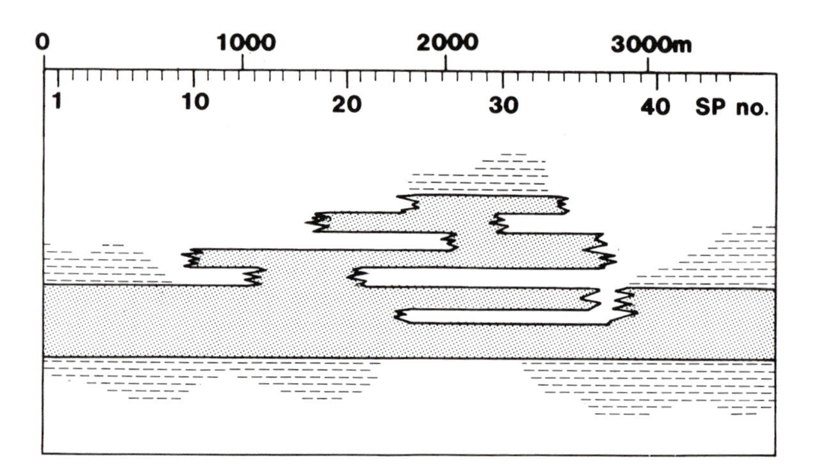

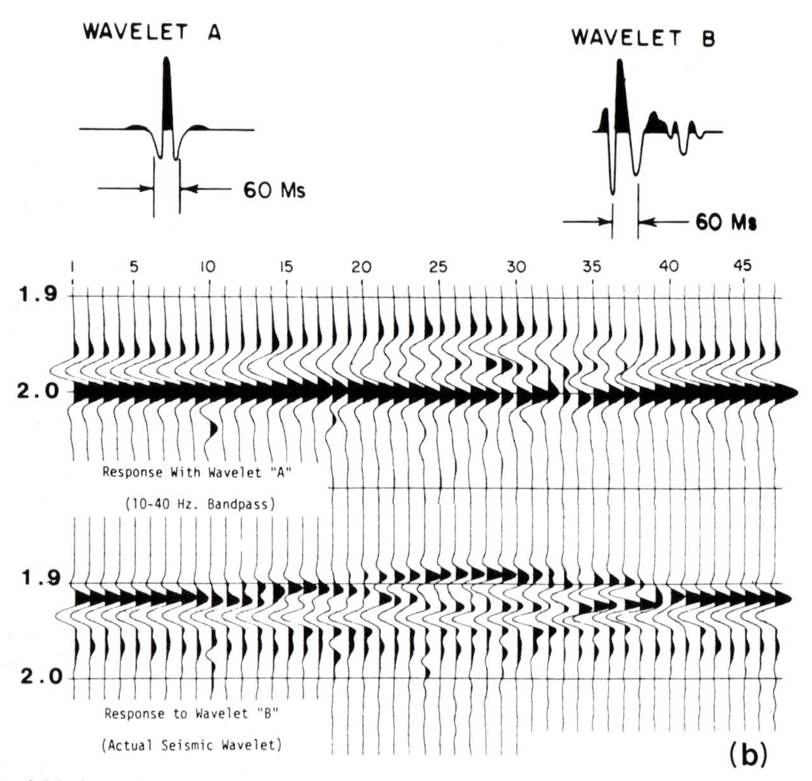

Fig. 4.34. A set of synthetic seismograms simulating a seismic section across a zone of irregular sandstone geometry. (From Neidell & Poggiagliolmi 1977.)

source-detector location, so that records obtained in deep water using shallow-tow systems commonly exhibit hyperbolic diffraction patterns, bow-tie effects and other undesirable features of non-migrated seismic sections.

In place of the oscillographic recorder used in multichannel seismic surveying, single-channel profiling typically utilizes a recorder in which a stylus repeatedly sweeps across the surface of an electrically-conducting recording

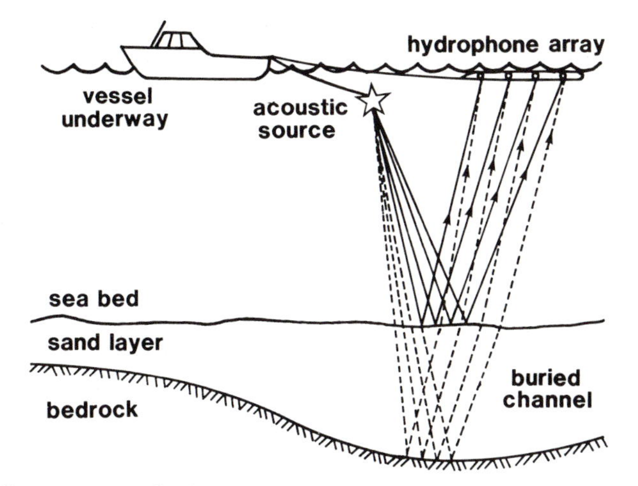

Fig. 4.35. The survey set-up for single-channel seismic reflection profiling.

paper that is continuously moving forward at a slow speed past a strip electrode in contact with the paper. A mark is burnt into the paper whenever an electrical signal is fed to the stylus and passes through the paper to the strip electrode. The seismic/acoustic source is triggered at the commencement of a stylus sweep and all seismic pulses returned during the sweep interval are recorded as a series of dark bands on the recording paper (Fig. 4.36). The triggering rate and sweep speed are variable over a wide range: for a shallow penetration survey the source may be triggered every 500 ms and the recording interval may be 0–250 ms, whereas for a deep penetration survey in deep water the source may be triggered every 8 s and the recording interval may be 2–6 s.

The analogue recording systems used in single-channel profiling are relatively cheap to operate. There are no processing costs and seismic records are produced in real time by the continuous chart recording of band-pass-filtered and amplified signals, sometimes with time variable gain (TVG). When careful consideration is given to source and hydrophone array design and deployment, good basic reflection records may be obtained from a single-channel system, but they cannot compare in quality with the type of seismic record produced by computer processing of multichannel data. Moreover, single-channel recordings cannot provide velocity information so that the conversion of reflection times into reflector depths has to utilize independent estimates of seismic velocity. Nonetheless, single-channel profiling often provides good imaging of subsurface geology and permits estimates of reflector depth and geometry that are sufficiently accurate for many purposes.

The record sections suffer from the presence of multiple reflections, especially multiples of the sea bed reflection, which may obliterate primary reflection events in the later parts of the records. Multiples are a particular problem when surveying in very shallow water, since they then occur at a short time interval after the primary events (e.g. see Fig. 4.39). Record sections are often difficult to interpret in areas of complex reflector geometry because of the presence of bow-tie effects, diffraction events, and other features of non-migrated seismic sections.

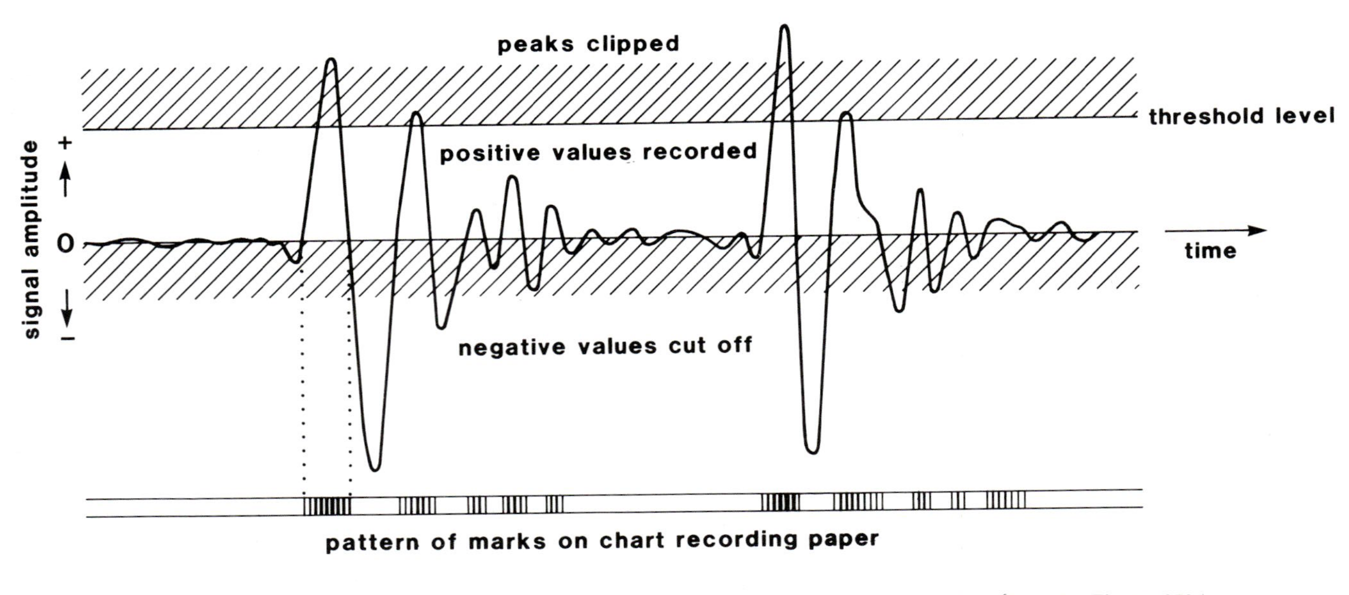

Fig. 4.36. Seismic signals and their representation on the chart recording paper of an oceanographic recorder. (From Le Tirant 1979.)

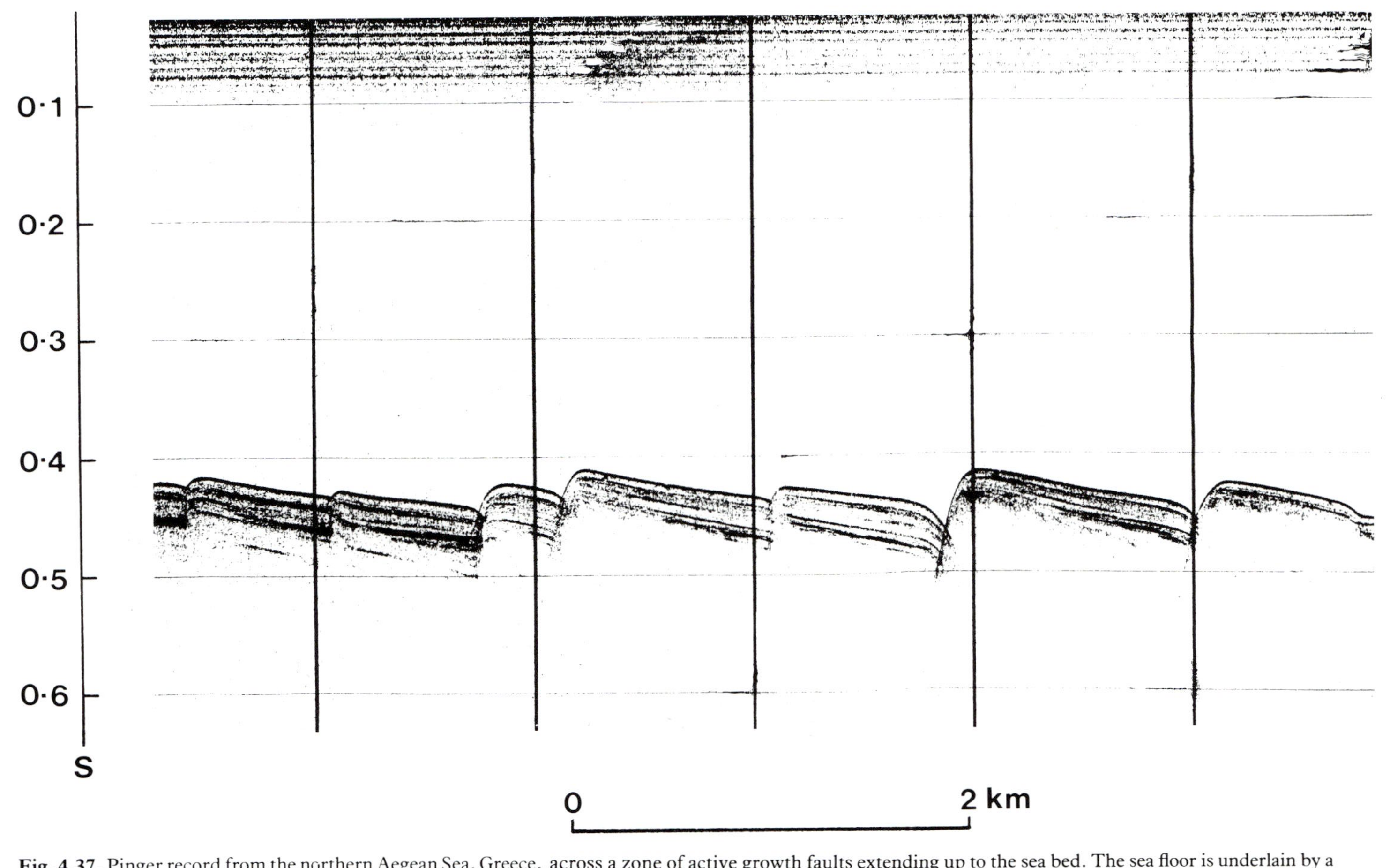

Fig. 4.37. Pinger record from the northern Aegean Sea, Greece, across a zone of active growth faults extending up to the sea bed. The sea floor is underlain by a layered sequence of Holocene muds and silts that can be traced to a depth of about 50 m. Note the diffraction patterns associated with the edges of the individual fault blocks.

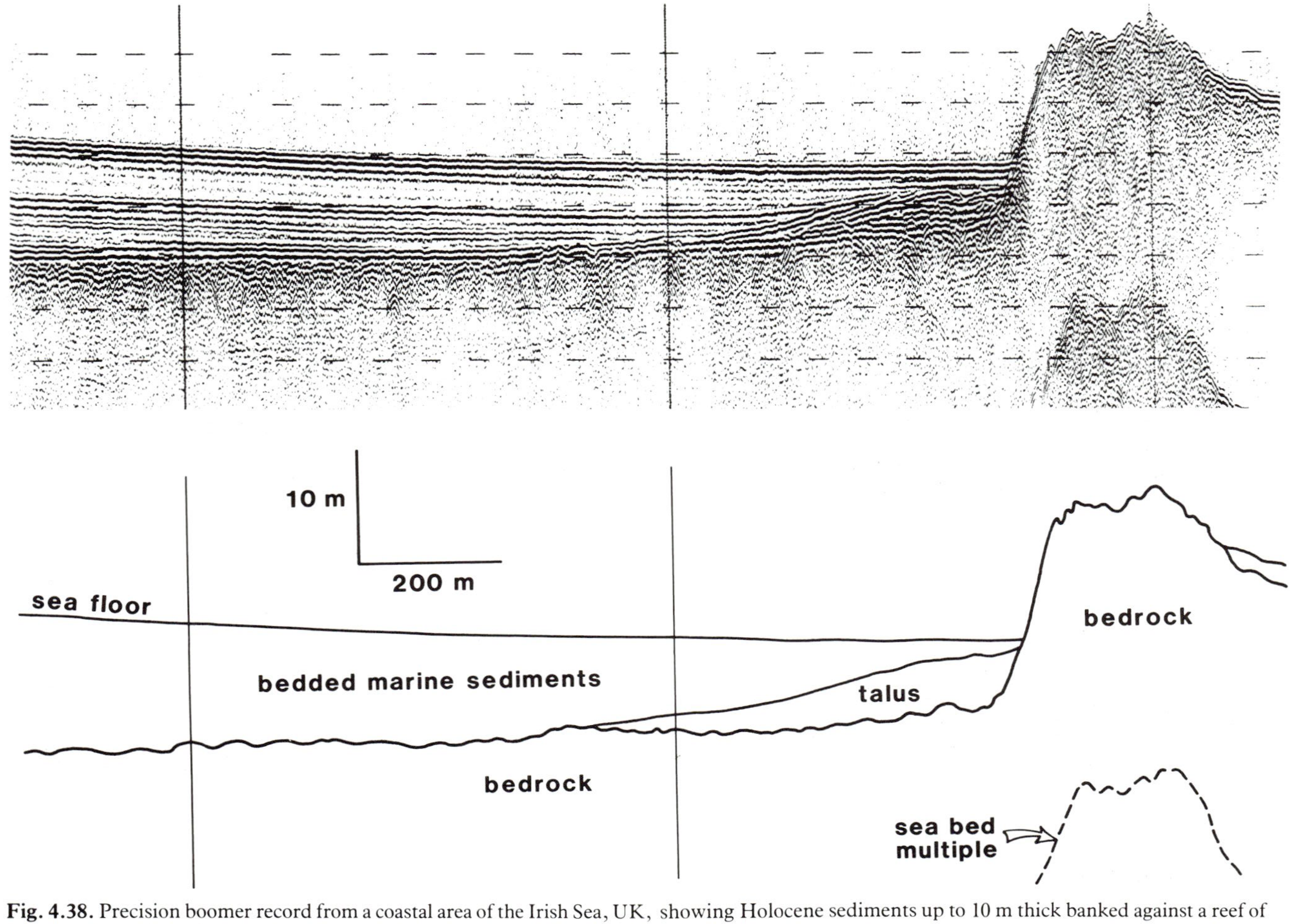

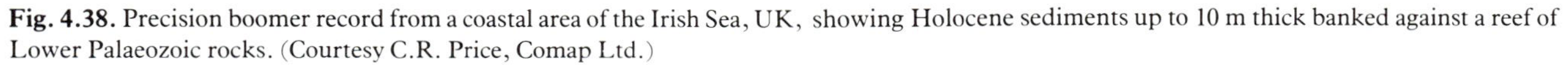

Fig. 4.38. Precision boomer record from a coastal area of the Irish Sea, UK, showing Holocene sediments up to 10 m thick banked against a reef of Lower Palaeozoic rocks. (Courtesy C.R. Price, Comap Ltd.)

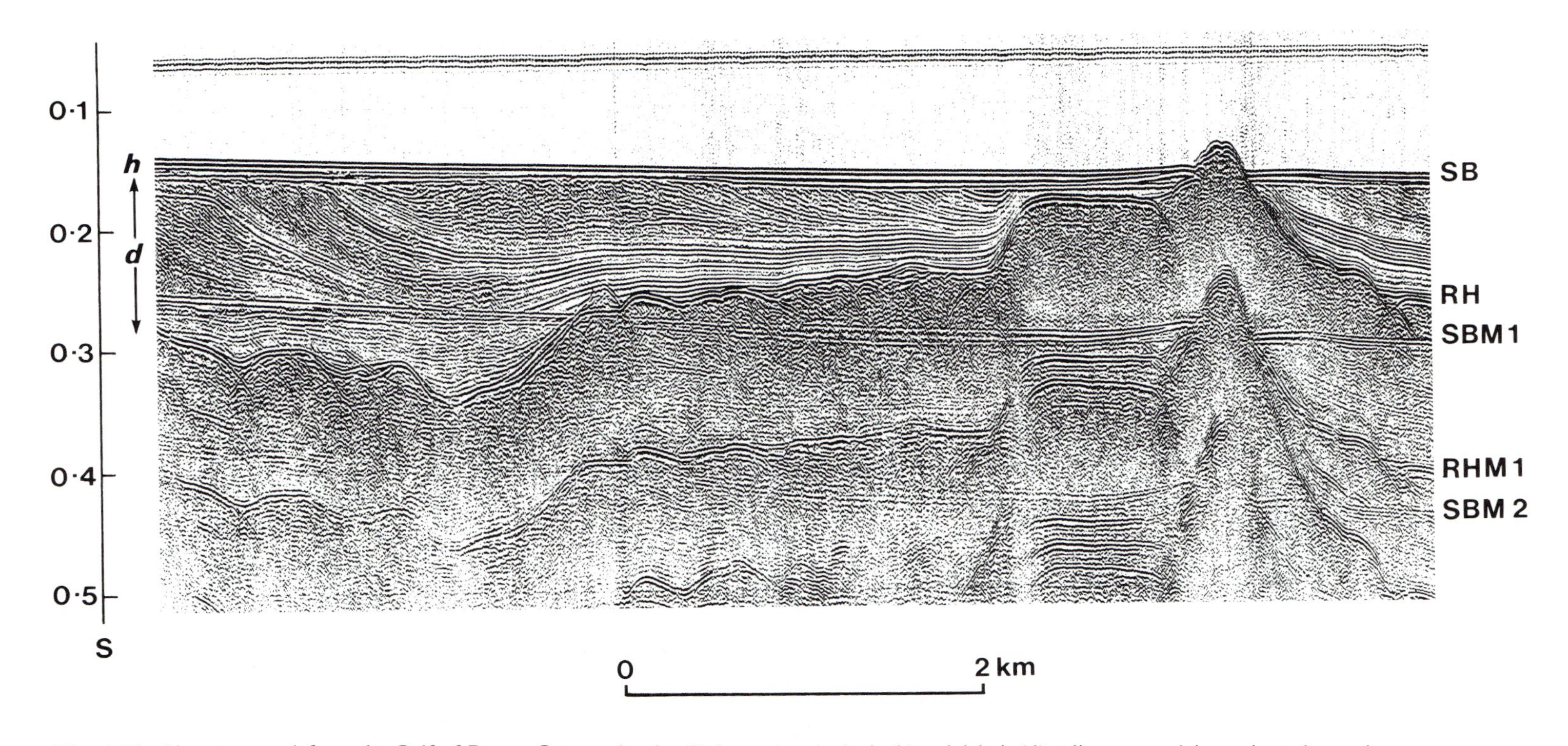

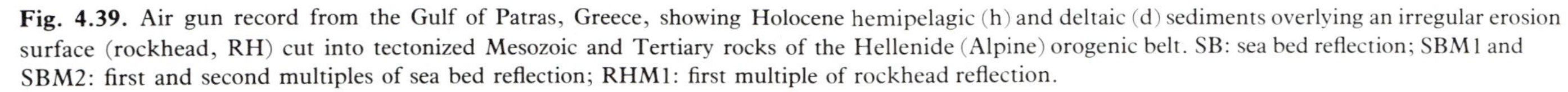

Fig. 4.39. Air gun record from the Gulf of Patras, Greece, showing Holocene hemipelagic (h) and deltaic (d) sediments overlying an irregular erosion surface (rockhead, RH) cut into tectonized Mesozoic and Tertiary rocks of the Hellenide (Alpine) orogenic belt. SB: sea bed reflection; SBM1 and SBM2: first and second multiples of sea bed reflection; RHM1: first multiple of rockhead reflection.

As discussed in Chapter 3 there is a variety of marine seismic/acoustic sources, operating at differing energy levels and characterized by different dominant frequencies. Consequently by selection of a suitable source, single-channel profiling can be applied to a wide range of offshore investigations from high-resolution surveys of near-surface sedimentary layers to surveys of deep geological structure. In general there is a trade-off between depth of penetration and degree of vertical resolution, since the higher energy sources required to transmit signals to greater depths are characterized by lower dominant frequencies and longer pulse lengths that adversely affect the resolution of the resultant seismic records.

Pingers are low energy (typically about 5 J), tunable sources that can be operated within the frequency range from 3 kHz to 12 kHz. The piezoelectric transducers used to generate the pinger signal also serve as receivers for reflected acoustic energy and, hence, a separate hydrophone streamer is not required in pinger surveying. Vertical resolution can be as good as 10–20 cm but depth penetration is limited to a few tens of metres in muddy sediments or several metres in coarse sediments, with virtually no penetration into solid rock. Pinger surveys are commonly used in offshore engineering site investigation and are of particular value in submarine pipeline route surveys. Repeated pinger surveying along a pipeline route enables monitoring of local sediment movement and facilitates location of the pipeline where it has become buried under recent sediments. A typical pinger record is shown in Fig. 4.37.

Boomer sources provide a higher energy output (typically 300–500 J) and operate at lower dominant frequencies (1–5 kHz) than pingers and therefore provide greater penetration (up to 100 m in bedrock) with good resolution (0.5–1 m). Boomer surveys are useful for mapping thick sedimentary sequences, in connection with channel dredging or sand and gravel extraction, or for high resolution surveys of shallow geological structure. A boomer record section is illustrated in Fig. 4.38.

Sparker sources can be operated over a wide range of energy levels (300–30 000 J), though the production of spark discharges of several thousand joules every few seconds requires a large power supply and a large bank of capacitors. Sparker surveying therefore represents a versatile tool for a wide range of applications, from shallow penetration surveys (100 m) with moderate resolution (2 m) to deep penetration surveys ($>$1 km) where resolution is not important. However, sparker surveying cannot match the resolution of precision boomer surveying, and sparkers do not offer as good a source signature as air guns for deeper penetration surveys.

By suitable selection of chamber size and rate of release of compressed air, air gun sources can be tailored to high resolution or deep penetration profiling applications and therefore represent the most versatile source for single-channel profiling. A reflection record obtained in a shallow water area with a small air gun (40 in^3) is shown in Fig. 4.39.

Single-channel reflection profiling systems (sometimes referred to as *sub-bottom profiling systems*) are commonly operated in conjunction with a precision echo-sounder, for high-quality bathymetric information, and/or with a sidescan

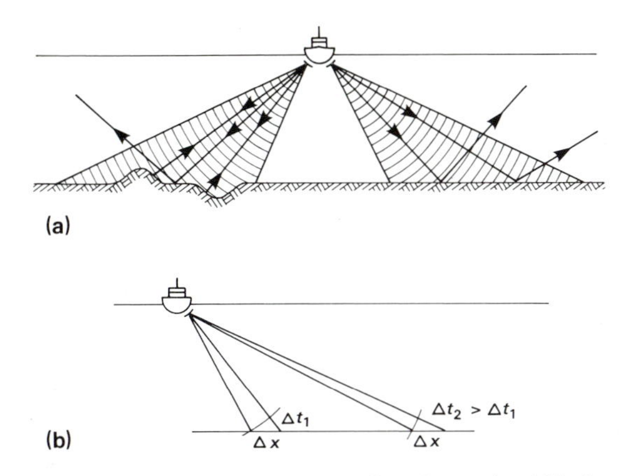

Fig. 4.40. Principles of sidescan sonar. (a) Individual reflected raypaths within the transmitted lobes, showing signal return from topographic features on the sea bed. (b) Scale distortion resulting from oblique incidence: the same widths of seafloor Δx are represented by different time intervals Δt_1 and Δt_2 at the inner and outer edges of the sonograph, respectively.

sonar system. *Sidescan sonar* is a sideway-scanning acoustic survey method in which the sea floor to one or both sides of the survey vessel is insonified by beams of high-frequency sound (30–110 kHz) transmitted by hull-mounted or fish-mounted transceiving transducers (Fig. 4.40). Sea bed features facing towards the survey vessel, such as rock outcrops or sedimentary bedforms, reflect acoustic energy back towards the transducers whilst in the case of features facing away from the vessel, or a featureless sea floor, the acoustic energy is reflected away

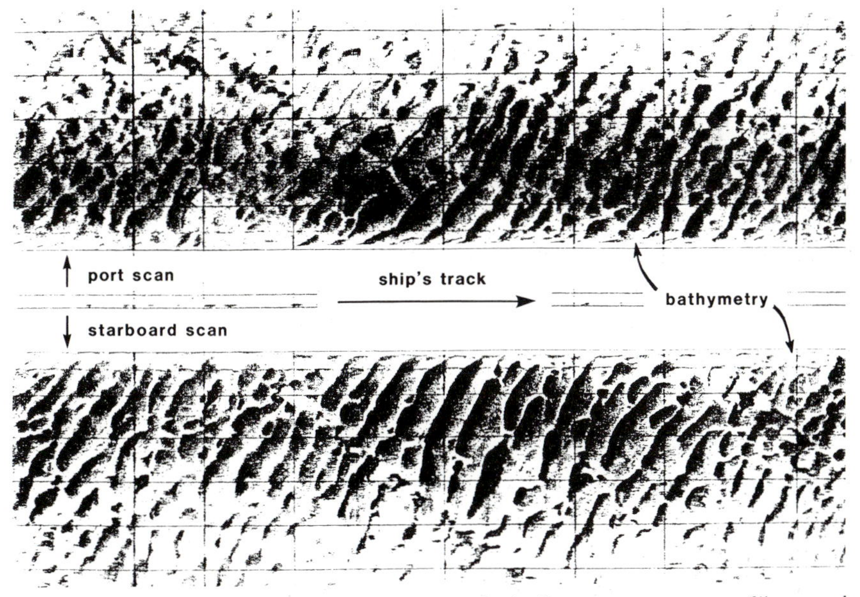

Fig. 4.41. Sonograph obtained from a dualscan survey of a pipeline route across an area of linear sand waves in the southern North Sea. The inner edges of the two swathes define the bathymetry beneath the survey vessel. (Scanning range: 100 m).

from the transducers. Signals reflected back to the transducers are fed to the same type of recorder that is used to produce seismic profiling records, and the resulting pattern of returned acoustic energy is known as a *sonograph*. The oblique insonification produces scale distortion resulting from the varying path lengths and angles of incidence of returning rays (Fig. 4.40(b)). This distortion can be automatically corrected for prior to display so that the sonograph provides an isometric plan view of sea bed features. A sonograph is shown in Fig. 4.41.

Although not strictly a seismic surveying tool, sidescan sonar provides valuable information on, for example, the configuration and orientation of sedimentary bedforms or on the pattern of rock outcrops. This information is often very useful in complementing the subsurface information derived from shallow seismic reflection surveys. Sidescan sonar is also useful for locating artifacts on the sea floor such as wrecks, cables or pipes. As with sub-bottom profiling systems, results in deep water are much improved by the use of deep-tow systems.

Further reading

Anstey, N.A. (1982) *Simple Seismics*. IHRDC, Boston.

Al-Sadi, H.N. (1980) *Seismic Exploration*. Birkhäuser Verlag, Basel.

Bally, A.W. (1983) *Seismic Expression of Structural Styles, Vol. 1 – The Layered Earth; Vol. 2 – Tectonics of Extensional Provinces*. Studies in Geology Series No. 15, American Association of Petroleum Geologists, Tulsa.

Brewer, J.A. & Oliver, J.E. (1980) Seismic reflection studies of deep crustal structure. *Ann. Rev. Earth Planet. Sci.*, **8**, 205–30.

Brown, L.F. & Fisher, W.L. (1980) *Seismic Stratigraphic Interpretation and Petroleum Exploration*. AAPG Continuing Education Course Note Series No. 16.

Dix, C.H. (1981) *Seismic Prospecting For Oil*. IHRDC, Boston.

Dobrin, M.B. (1976) *Introduction to Geophysical Prospecting*. McGraw-Hill, New York (3rd edn).

Fitch, A.A. (1976) *Seismic Reflection Interpretation*. Gebrüder Borntraeger, Berlin.

Fitch, A.A. (1981) *Developments in Geophysical Exploration Methods, Vol. 2*. Applied Science Publishers, London.

Hubral, P. & Krey, T. (1980) *Interval Velocities from Seismic Reflection Time Measurements*. Society of Exploration Geophysicists, Tulsa.

Kleyn, A.H. (1983) *Seismic Reflection Interpretation*. Applied Science Publishers, London.

Mayne, W.H. (1962) Common reflection point horizontal stacking techniques. *Geophysics*, **27**, 927–8.

Mayne, W.H. (1967) Practical considerations in the use of common reflection point technique. *Geophysics*, **32**, 225–9.

McQuillin, R., Bacon, M. & Barclay, W. (1979) *An Introduction To Seismic Interpretation*. Graham and Trotman, London.

Payton, C.E. (ed.) (1977) *Seismic Stratigraphy – Application To Hydrocarbon Exploration*. Memoir 26, American Association of Petroleum Geologists, Tulsa.

Robinson, E.A. (1983) *Migration of Geophysical Data*. IHRDC, Boston.

Robinson, E.A. (1983) *Seismic Velocity Analysis and the Convolutional Model*. IHRDC, Boston.

Sengbush, R.L. (1983) *Seismic Exploration Methods*. IHRDC, Boston.

Sheriff, R.E. (1980) *Seismic Stratigraphy*. IHRDC, Boston.

Sheriff, R.E. (1982) *Structural Interpretation of Seismic Data*. AAPG Continuing Education Course Note Series No. 23.

Sheriff, R.E. & Geldart, L.P. (1983) *Exploration Seismology, Vol. 2: Data-processing and Interpretation*. Cambridge University Press, Cambridge.

Waters, K.H. (1978) *Reflection Seismology – A Tool For Energy Resource Exploration*. Wiley, New York.

Ziolkowski, A. (1983) *Deconvolution*. IHRDC, Boston.

5

Seismic Refraction Surveying

5.1 Introduction

The seismic refraction surveying method utilizes seismic energy that returns to the surface after travelling through the ground along refracted ray paths. The method is normally used to locate refracting interfaces (refractors) separating layers of different seismic velocity, but the method is also applicable in cases where velocity varies smoothly as a function of depth or laterally.

Refraction seismology is applied to a very wide range of scientific and technical problems, from engineering site investigation surveys to large-scale experiments designed to study the structure of the entire crust or lithosphere. Refraction measurements can provide valuable velocity information for use in the interpretation of reflection surveys, and refracted arrivals recorded during land reflection surveys are used to map the weathered layer, as discussed in Chapter 4. This wide variety of applications leads to an equally wide variety of field survey methods and associated interpretation techniques.

As briefly discussed in Chapter 3, the first arrival of seismic energy at a detector offset from a seismic source always represents either a direct ray or a refracted ray. This fact allows simple refraction surveys to be performed in which attention is concentrated solely on the first arrival (or *onset*) of seismic energy, and time-distance curves of these first arrivals are interpreted to derive information on the depth to refracting interfaces. As will be seen later in the chapter, this simple approach does not always yield a full or accurate picture of the subsurface.

Refraction seismograms may, of course, contain reflection events as subsequent arrivals though generally no special attempt is made to enhance reflected arrivals in refraction surveys. Nevertheless, the relatively high reflection coefficients associated with rays incident on an interface at angles near to the critical angle often lead to strong *wide-angle reflections* which are quite commonly detected at the greater recording ranges that characterize large-scale refraction surveys. These wide-angle reflections often provide valuable additional information on subsurface structure such as, for example, indicating the presence of a low velocity layer which would not be revealed by refracted arrivals alone.

The vast majority of refraction surveying is carried out along profile lines which are normally arranged to be sufficiently long to ensure that refracted arrivals from target layers are recorded as first arrivals. To ensure that the relevant crossover distance is well exceeded, refraction profiles typically need to be between five and ten times as long as the required depth of investigation, although the actual profile length required in a particular case depends upon the distribution of velocities with depth. The requirement in refraction surveying for an increase in profile length with increase in the depth of investigation contrasts with the situation in conventional reflection surveying, where near-normal incidence reflections from deep interfaces are recorded at small offset distances. A consequence of this requirement is that large seismic sources are needed for the detection of deep refractors in order that sufficient energy is transmitted over the long range necessary for the recording of deep refracted phases as first arrivals.

In many geological situations, subsurface refractors may approximate planar surfaces over the linear extent of a refraction line. In such cases the observed travel-time curves are commonly assumed to derive from a set of planar layers and are analysed to determine depths to, and dips of, individual planar refractors. The geometry of refracted ray paths through planar layer models of the subsurface is considered below, after which, consideration is given to methods of dealing with refraction at irregular (non-planar) interfaces.

5.2 Geometry of refracted ray paths: planar interfaces

The general assumptions relating to the ray path geometries considered below are that the subsurface is composed of a series of layers separated by plane and possibly dipping interfaces, that seismic velocities are uniform within each layer, that layer velocities increase with depth, and that ray paths are restricted to a vertical plane containing the profile line (i.e. that there is no component of cross dip).

5.2.1 Two-layer case with horizontal interface

Figure 5.1 illustrated progressive positions of the wavefront associated with energy travelling directly through an upper layer and energy critically refracted in a lower layer, from a seismic source at A. Direct and refracted ray paths to a detector at D, a distance x from the source, are also shown. The layer velocities are v_1 and v_2 ($> v_1$) and the refracting interface is at a depth z.

The direct ray travels horizontally through the top of the upper layer from A to D at velocity v_1. The refracted ray travels down to the interface and back up to the surface at velocity v_1 along slant paths AB and CD that are inclined at the critical angle θ, and travels along the interface between B and C at the higher velocity v_2.

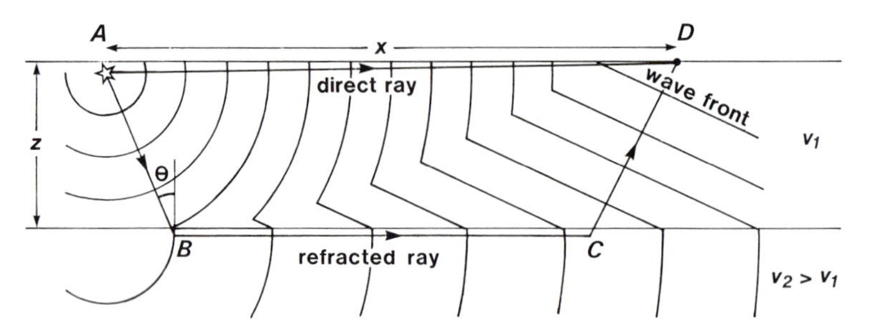

Fig. 5.1. Successive positions of the expanding wavefronts for direct and refracted waves through a two-layer model. Individual ray paths from source A to detector D are shown.

The total travel time along the refracted ray path *ABCD* is

$$t = t_{AB} + t_{BC} + t_{CD}$$

$$= \frac{z}{v_1 \cos\theta} + \frac{(x - 2z\tan\theta)}{v_2} + \frac{z}{v_1 \cos\theta}$$

Noting that $\sin\theta = v_1/v_2$ (Snell's Law) and $\cos\theta = (1 - v_1^2/v_2^2)^{1/2}$, the travel time equation may be expressed in a number of different forms, a useful general form being

$$t = \frac{x\sin\theta}{v_1} + \frac{2z\cos\theta}{v_1} \tag{5.1}$$

Alternatively

$$t = \frac{x}{v_2} + \frac{2z(v_2^2 - v_1^2)^{1/2}}{v_1 v_2} \tag{5.2}$$

Or

$$t = \frac{x}{v_2} + t_i \tag{5.3}$$

where, plotting t against x (Fig. 5.2), t_i is the intercept on the time axis of a travel-time curve or *time-distance curve* having a gradient of $1/v_2$. t_i, known as the *intercept time*, is given by

$$t_i = \frac{2z(v_2^2 - v_1^2)^{1/2}}{v_1 v_2} \qquad \text{(from (5.2))}$$

Solving for refractor depth

$$z = \frac{t_i v_1 v_2}{2(v_2^2 - v_1^2)^{1/2}}$$

Thus by analysis of the travel-time curves of direct and refracted arrivals, v_1 and v_2 can be derived (reciprocal of the gradient of the relevant travel-time curve, see Fig. 5.2) and from the intercept time t_i the refractor depth z can be determined.

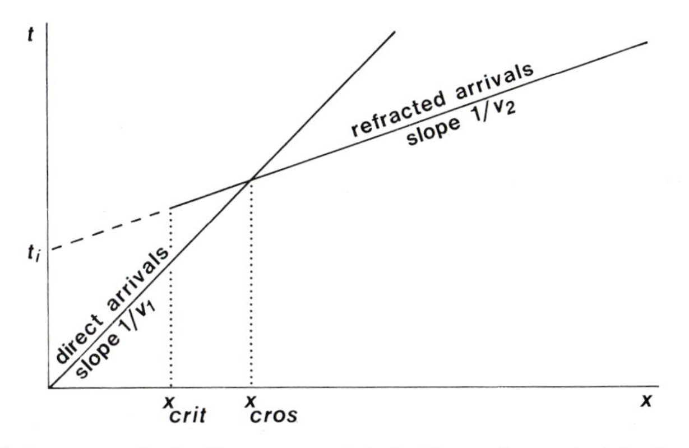

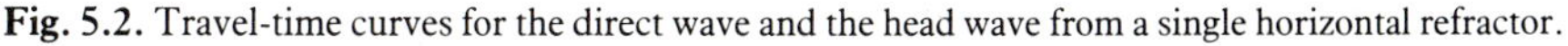

Fig. 5.2. Travel-time curves for the direct wave and the head wave from a single horizontal refractor.

At the crossover distance x_{cros} the travel times of direct and refracted rays are equal

$$\frac{x_{cros}}{v_1} = \frac{x_{cros}}{v_2} + \frac{2z(v_2^2 - v_1^2)^{1/2}}{v_1 v_2}$$

Thus, solving for x_{cros}

$$x_{cros} = 2z \left[\frac{v_2 + v_1}{v_2 - v_1} \right]^{1/2}$$

From this equation it may be seen that the crossover distance is always greater than twice the depth to the refractor.

5.2.2 *Three-layer case with horizontal interfaces*

The geometry of the ray path in the case of critical refraction at the second interface is shown in Fig. 5.3. The seismic velocities of the three layers are v_1, v_2 ($> v_1$) and v_3 ($> v_2$). The angle of incidence of the ray on the upper interface is θ_1 and on the lower interface is θ_2 (critical angle). The thicknesses of layers 1 and 2 are z_1 and z_2 respectively.

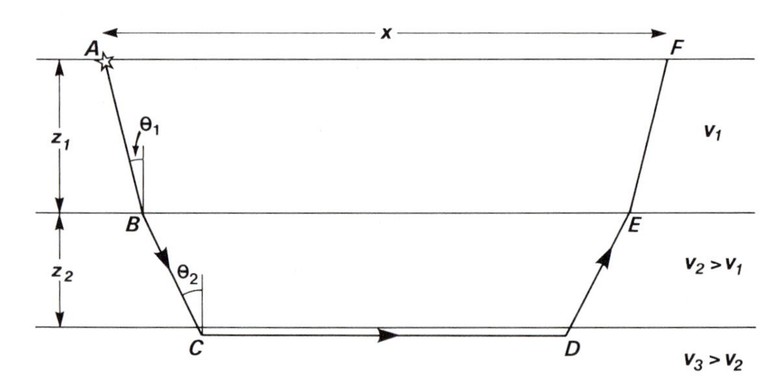

Fig. 5.3. Refracted ray path through the bottom layer of a three-layer model.

By analogy with equation (5.1) for the two-layer case, the travel time along the refracted ray path *ABCDEF* to an offset distance x, involving critical refraction at the second interface, can be written in the form

$$t = \frac{x \sin \theta_1}{v_1} + \frac{2z_1 \cos \theta_1}{v_1} + \frac{2z_2 \cos \theta_2}{v_2} \qquad (5.5)$$

where $\theta_1 = \sin^{-1}(v_1/v_3)$

and $\theta_2 = \sin^{-1}(v_2/v_3)$

or

$$t = \frac{x \sin \theta_1}{v_1} + t_{i_1} + t_{i_2} \qquad (5.6)$$

where t_{i_1} is the intercept on the time axis of the travel-time curve for rays critically refracted at the upper interface and t_{i_2} is the difference between t_{i_1} and the intercept of the curve for rays critically refracted at the lower interface (see Fig. 5.4).

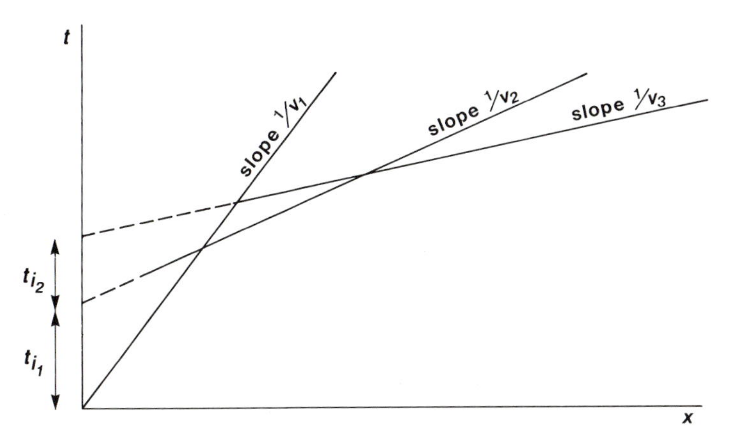

Fig. 5.4. Travel-time curves for the direct wave and the head waves from two horizontal refractors.

The interpretation of travel-time curves for a three-layer case proceeds via the initial interpretation of the top two layers. Having used the travel-time curve for rays critically refracted at the upper interface to derive z_1 and v_2, the travel-time curve for rays critically refracted at the second interface can be used to derive z_2 and v_3 using equations (5.5) and (5.6) or equations derived therefrom.

5.2.3 *Multilayer case with horizontal interfaces*

In general the travel time t_n of a ray critically refracted along the top surface of the nth layer is given by

$$t_n = \frac{x \sin \theta_1}{v_1} + \sum_{i=1}^{n-1} \frac{2z_i \cos \theta_i}{v_i} \tag{5.7}$$

where $\theta_i = \sin^{-1}(v_i/v_n)$

Equation (5.7) can be used progressively to compute layer thicknesses in a sequence of horizontal strata represented by travel-time curves of refracted arrivals.

5.2.4 *Dipping-layer case with planar interfaces*

In the case of a dipping refractor (Fig. 5.5(a)) the value of dip enters the travel time equations as an additional unknown. The reciprocal of the gradient of the travel-time curve no longer represents the refractor velocity but a quantity known as the *apparent velocity* which is higher than the refractor velocity when recording along a profile line in the updip direction from the shot point and lower when recording downdip.

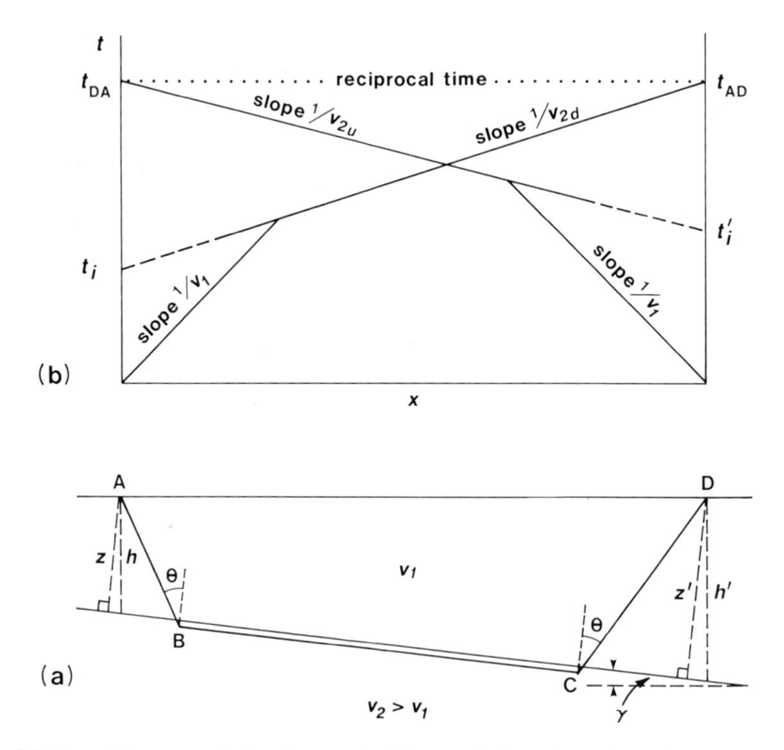

Fig. 5.5. Travel-time curves for head wave arrivals from a dipping refractor in the forward and reverse directions along a refraction profile line.

The conventional method of dealing with the possible presence of refractor dip is to *reverse* the refraction experiment by firing at each end of the profile line and recording seismic arrivals along the line from both shots. In the presence of a component of refractor dip along the profile direction, the *forward* and *reverse* travel-time curves for refracted rays will differ in their gradients and intercept times, as shown in Fig. 5.5(b).

The general form of the equation for the travel time t_n of a ray critically refracted in the nth dipping refractor (Fig. 5.6; Johnson 1976) is given by

$$t_n = \frac{x \sin \beta_1}{v_1} + \sum_{i=1}^{n-1} \frac{h_i(\cos \alpha_i + \cos \beta_i)}{v_i} \tag{5.8}$$

where h_i is the vertical thickness of the ith layer beneath the shot
v_i is the velocity of the ray in the ith layer
α_i is the angle with respect to the vertical made by the downgoing ray in the ith layer
β_i is the angle with respect to the vertical made by the upgoing ray in the ith layer
and x is the offset distance between source and detector.

Equation (5.8) is directly comparable with equation (5.7), the only differences being the replacement of θ by angles α and β that include a dip term. In the case of shooting downdip, for example (see Fig. 5.6), $\alpha_i = \theta_i - \gamma_i$ and $\beta_i = \theta_i + \gamma_i$ where γ_i is the dip of the ith layer and $\theta_i = \sin^{-1}(v_i/v_n)$ as before. Note that h is the vertical thickness rather than the perpendicular or true thickness of a layer.

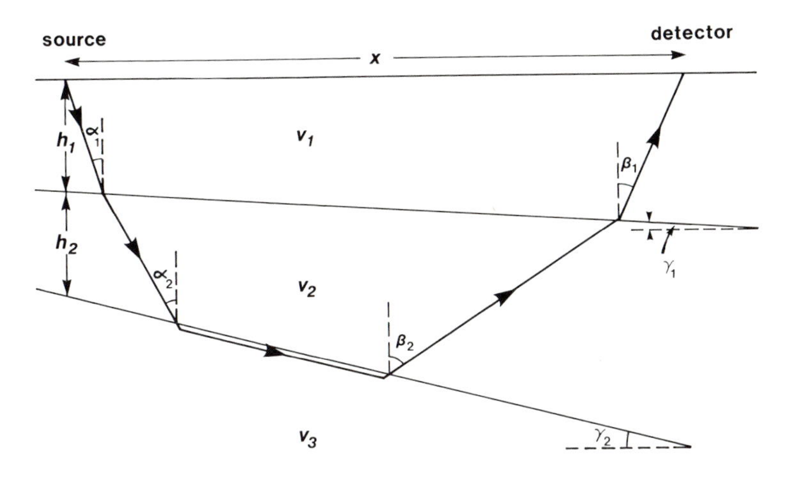

Fig. 5.6. Geometry of the refracted ray path through a multilayer, dipping model. (After Johnson 1976.)

To exemplify the use of equation (5.8) in interpreting travel-time curves, consider the two layer case illustrated in Fig. 5.5.

Shooting downdip, along the forward profile

$$t_2 = \frac{x \sin\beta}{v_1} + \frac{h(\cos\alpha + \cos\beta)}{v_1}$$

$$= \frac{x\sin(\theta+\gamma)}{v_1} + \frac{h\cos(\theta-\gamma)}{v_1} + \frac{h\cos(\theta+\gamma)}{v_1}$$

$$= \frac{x\sin(\theta+\gamma)}{v_1} + \frac{2h\cos\theta\cos\gamma}{v_1}$$

$$= \frac{x\sin(\theta+\gamma)}{v_1} + \frac{2z\cos\theta}{v_1} \qquad (5.9)$$

where z is the perpendicular distance to the interface beneath the shot.

Equation (5.9) defines a linear curve with a gradient of $\sin(\theta+\gamma)/v_1$ and an intercept time of $2z\cos\theta/v_1$.

Shooting updip, along the reverse profile

$$t_2' = \frac{x\sin(\theta-\gamma)}{v_1} + \frac{2z'\cos\theta}{v_1} \qquad (5.10)$$

where z' is the perpendicular distance to the interface beneath the second shot.

The gradients of the travel-time curves of refracted arrivals along the forward and reverse profile lines yield the downdip and updip apparent velocities v_{2_d} and

5

v_{2_u} respectively (Fig. 5.5(b)). From the forward direction

$$1/v_{2_d} = \sin(\theta+\gamma)/v_1 \qquad (5.11)$$

And from the reverse direction

$$1/v_{2_u} = \sin(\theta-\gamma)/v_1 \qquad (5.12)$$

Hence

$$\theta+\gamma = \sin^{-1}(v_1/v_{2_d})$$

$$\theta-\gamma = \sin^{-1}(v_1/v_{2_u})$$

Solving for θ and γ yields

$$\theta = \tfrac{1}{2}\,(\sin^{-1}(v_1/v_{2_d})+\sin^{-1}(v_1/v_{2_u}))$$

$$\gamma = \tfrac{1}{2}\,(\sin^{-1}(v_1/v_{2_d})-\sin^{-1}(v_1/v_{2_u}))$$

Knowing v_1, from the gradient of the direct ray travel-time curve, and θ, the true refractor velocity may be derived using Snell's Law

$$v_2 = v_1/\sin\theta$$

The perpendicular distances z and z' to the interface under the two ends of the profile are obtained from the intercept times t_i and t_i' of the travel-time curves obtained in the forward and reverse directions

$$t_i = 2z\cos\theta/v_1$$

$$\therefore z = v_1 t_i/2\cos\theta$$

And similarly

$$z' = v_1 t_i'/2\cos\theta$$

By using the computed refractor dip γ, the perpendicular depths z and z' can be converted into vertical depths h and h' using

$$h = z/\cos\gamma$$

and

$$h' = z'/\cos\gamma.$$

Note that the travel time of a seismic phase from one end of a refraction profile line to the other (i.e. from shot point to shot point) should be the same whether measured in the forward or the reverse direction. Referring to Fig. 5.5(b), this means that t_{AD} should equal t_{DA}. Establishing that there is satisfactory agreement between these *reciprocal times* is a useful means of checking that travel-time curves have been drawn correctly through a set of refracted ray arrival times derived from a reversed profile.

5.2.5 *Faulted planar interfaces*

The effect of a fault displacing a planar refractor is to offset the segments of the travel-time curve on opposite sides of the fault (see Fig. 5.7). There are thus two intercept times t_{i_1} and t_{i_2}, one associated with each of the travel-time curve segments, and the difference between these intercept times ΔT is a measure of the

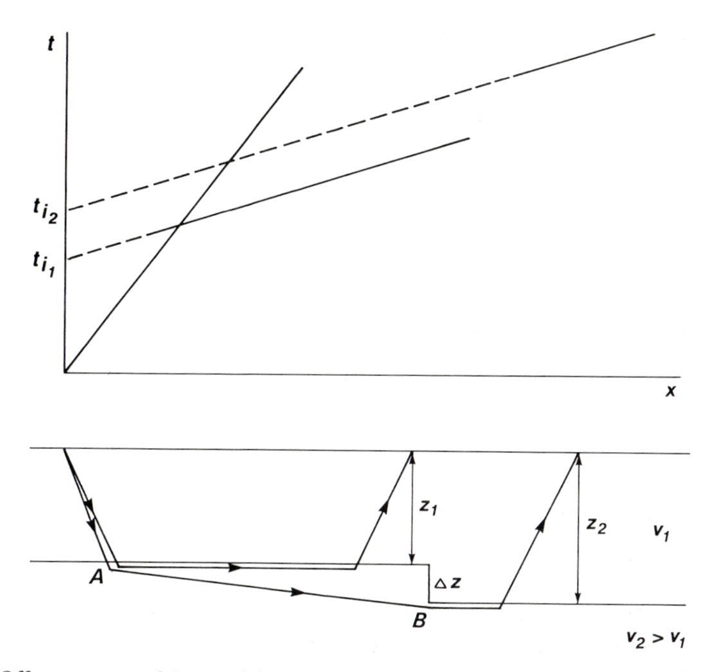

Fig. 5.7. Offset segments of the travel time curve for refracted arrivals from opposite sides of a fault.

throw of the fault. For example, in the case of the faulted horizontal refractor shown in Fig. 5.7 the throw of the fault Δz is given by

$$\Delta z = \Delta T v_1 v_2 / (v_2^2 - v_1^2)^{1/2}$$

Note that there is some error in this formulation, since the ray travelling to the downthrown side of the fault is not the critically refracted ray at A and involves diffraction at the base B of the fault step, but the error will be negligible where the fault throw is small compared with the refactor depth.

5.3 Profile geometries for studying planar layer problems

As stated above, the conventional field geometry for a refraction profile involves shooting at each end of the profile line and recording seismic arrivals along the line from both shots. As will be seen with reference to Fig. 5.5(a), only the central portion of the refractor (from B to C) is in fact sampled by refracted rays detected along the line length. Interpreted depths to the refractor under the endpoints of a profile line, using equations given above, are thus not directly measured but are inferred on the basis of the refractor geometry over the shorter length of refractor actually traversed by refracted rays. Where continuous cover of refractor geometry is required along a series of reversed profiles, therefore, individual profile lines should be arranged to overlap in order that all parts of the refractor are directly sampled by critically refracted rays.

In addition to the conventional reversed profile, illustrated schematically in Fig. 5.8(a), other methods of deriving full planar layer interpretations in the

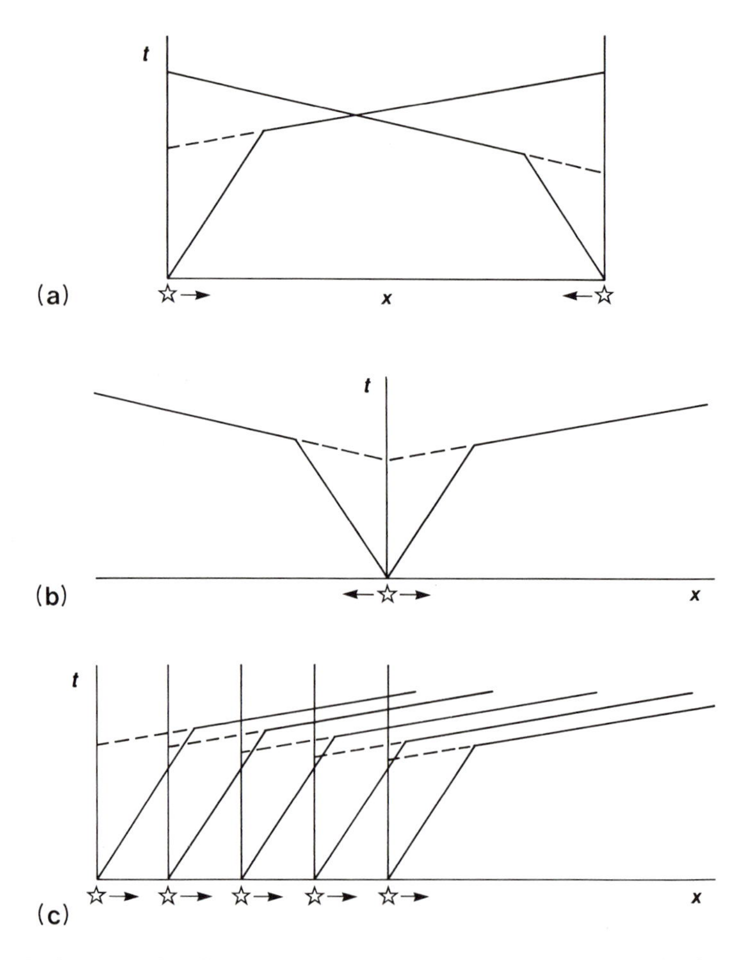

Fig. 5.8. Various types of profile geometry used in refraction surveying: (a) Conventional reversed profile with end shots; (b) Split-profile with central shot; (c) Single-ended profile with repeated shots.

presence of dip include the *split-profile* method (Johnson 1976) and the *single-ended profile method* (Cunningham 1974).

The split-profile method (Fig. 5.8(b)) involves recording outwards in both directions from a central shot point. Although the interpretation method differs in detail from that for a conventional reversed profile, it is based on the same general travel-time equation (5.8).

The single-ended profile method (Fig. 5.8(c)) was developed to derive interpretations of low velocity surface layers represented by refracted arrivals in single-ended reflection spread data, for use in the calculation of static corrections. A simplified treatment is given below.

To obtain a value of refractor dip, estimates of apparent velocity are required in both the forward and reverse directions. The repeated forward shooting of the single-ended profile method enables an apparent velocity in the forward direction to be computed from the gradient of the travel-time curves. For the method of computing the apparent velocity in the reverse direction, consider two refracted

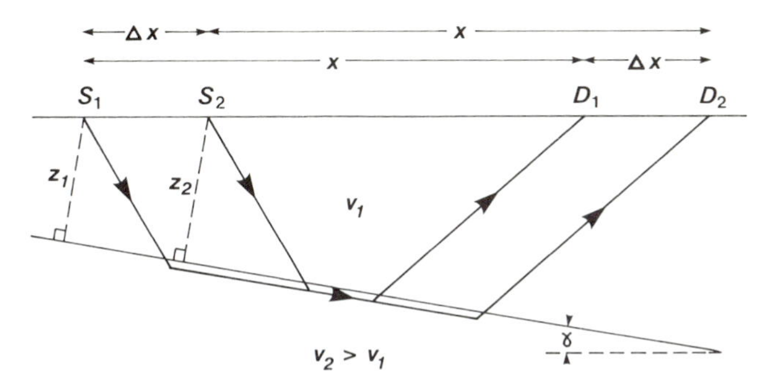

Fig. 5.9. Refraction interpretation using the single-ended profiling method. (After Cunningham 1974.)

ray paths from surface sources S_1 and S_2 to surface detectors D_1 and D_2 respectively (Fig. 5.9). The offset distance is x in both cases, the separation Δx of S_1 and S_2 being the same as that of D_1 and D_2.

Since D_1 is on the downdip side of S_1, the travel time of a refracted ray from S_1 to D_1 is given by (equation (5.9))

$$t_1 = \frac{x \sin(\theta+\gamma)}{v_1} + \frac{2z_1 \cos\theta}{v_1} \tag{5.13}$$

and from S_2 to D_2 the travel time is given by

$$t_2 = \frac{x \sin(\theta+\gamma)}{v_1} + \frac{2z_2 \cos\theta}{v_1} \tag{5.14}$$

where z_1 and z_2 are the perpendicular depths to the refractor under shot points S_1 and S_2 respectively.

Now,

$$z_2 - z_1 = \Delta x \sin\gamma$$

$$\therefore z_2 = z_1 + \Delta x \sin\gamma \tag{5.15}$$

Substituting (5.15) in (5.14) and then subtracting (5.13) from (5.14) yields

$$t_2 - t_1 = \Delta t = \frac{\Delta x}{v_1}(2\sin\gamma\cos\theta)$$

$$= \frac{\Delta x \sin(\theta+\gamma)}{v_1} - \frac{\Delta x \sin(\theta-\gamma)}{v_1} \tag{5.16}$$

Substituting equations (5.11) and (5.12) in (5.16) and rearranging terms

$$\frac{\Delta t}{\Delta x} = \frac{1}{v_{2_d}} - \frac{1}{v_{2_u}}$$

where v_{2u} and v_{2d} are the updip and downdip apparent velocities respectively. In the case considered v_{2_d} is derived from the single-ended travel-time curves, hence v_{2_u} can be calculated from the difference in travel time of refracted rays from adjacent shots recorded at the same offset distance x. With both apparent velocities calculated, interpretation proceeds by the standard methods for conventional reversed profiles discussed in Section 5.2.4.

5.4 Geometry of refracted ray paths: irregular (non-planar) interfaces

The assumption of planar refracting interfaces would often lead to unacceptable error or imprecision in the interpretation of refraction survey data. For example, a survey may be carried out to study the form of the concealed bedrock surface beneath a valley fill of alluvium of glacial drift. Clearly such a surface could not be represented adequately by a planar refractor. It is sometimes necessary, therefore, to remove the constraint that refracting interfaces be interpreted as planar and, consequently, to employ different interpretation methods.

A test of the prevailing refractor geometry is provided by the configuration of travel-time curves derived from a survey. A layered sequence of planar refractors gives rise to a travel-time graph consisting of a series of straight-line segments, each segment representing a particular refracted phase and characterized by a particular gradient and intercept time. Irregular travel-time curves are an indication of irregular refractors (or, alternatively, of lateral velocity variation within individual layers – a complication not discussed here). Methods of interpreting irregular travel-time curves, to determine the non-planar refractor geometry that gives rise to them, are based on the concept of *delay time*.

Consider a horizontal refractor separating upper and lower layers of velocity v_1 and v_2 ($> v_1$) respectively (Fig. 5.1). The travel time of a head wave arriving at an offset distance x is given (see equation (5.3)) by

$$t = x/v_2 + t_i$$

The intercept time t_i can be considered to be composed of two delay times resulting from the presence of the top layer at each end of the ray path. Referring to Fig. 5.10(a), the *delay time* (or *time-term*) a is defined as the time difference between the slant path AB through the top layer and the time that would be required for a ray to travel along the projection BC of the above path through the refractor at the refractor velocity to a position vertically below the point of emergence of the ray at the surface.

Thus
$$\begin{aligned} a &= t_{AB} - t_{BC} \\ &= \frac{AB}{v_1} - \frac{BC}{v_2} \\ &= \frac{z}{v_2 \sin\theta \cos\theta} - \frac{z}{v_2}\tan\theta \\ &= \frac{z}{v_2 \tan\theta} \\ &= z\,(v_2^2 - v_1^2)^{1/2}/v_1 v_2 \end{aligned} \tag{5.16}$$

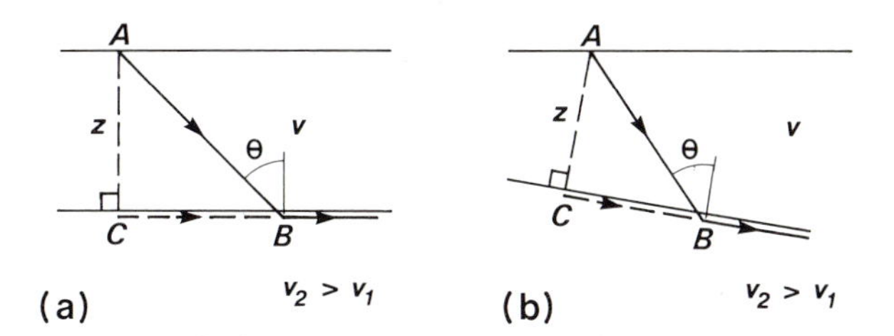

Fig. 5.10. The concept of delay time.

Solving equation (5.16) for the depth z to the refractor yields

$$z = av_1v_2/(v_2^2 - v_1^2)^{1/2} \tag{5.17}$$

Thus the delay time can be converted into a refractor depth if v_1 and v_2 are known.

The intercept time t_i in equation (5.3) can be partitioned into two delay times

$$t = x/v_2 + a_s + a_d \tag{5.18}$$

where a_s and a_d are the delay times at the shot end and detector end of the refracted ray path. Note that in this case of a horizontal refractor $a_s = a_d = \frac{1}{2}t_i = z(v_2^2 - v_1^2)^{1/2}/v_1v_2$.

In the presence of refractor dip the delay time is similarly defined except that point C is perpendicularly, not vertically, below A (see Fig. 5.10(b)), and the delay time is again related to depth by equation (5.17) where z is now the refractor depth at A measured normal to the refractor surface. Using this definition of delay time, the travel time of a ray refracted along a dipping interface (see Fig. 5.11(a)) is given by

$$t = x'/v_2 + a_s + a_d \tag{5.19}$$

where $\quad a_s = t_{AB} - t_{BC}$ and $a_d = t_{DE} - t_{DF}$

For shallow dips, x' (unknown) is closely similar to the offset distance x (known) in which case equation (5.18) can be used in place of (5.19) and methods

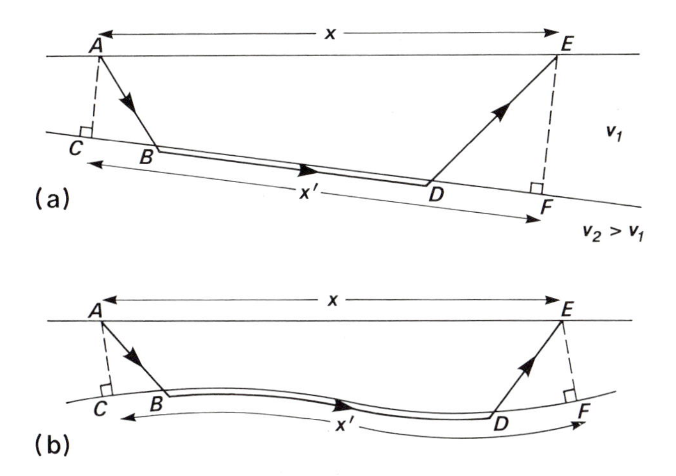

Fig. 5.11. Refracted ray paths associated with (a) a dipping and (b) an irregular refractor.

applicable to a horizontal refractor employed. This approximation is valid also in the case of an irregular refractor if the relief on the refractor is small in amplitude compared to the average refractor depth (Fig. 5.11(b)).

Delay times cannot be measured directly but occur in pairs in the travel-time equation for a refracted ray from a surface source to a surface detector. The *plus-minus method* of Hagedoorn (1959) provides a means of solving equation (5.18) to derive individual delay time values for the calculation of local depths to an irregular refractor.

Fig. 5.12(a) illustrates a two-layer ground model with an irregular refracting interface. Selected ray paths are shown associated with a reversed refraction profile line of length l between end shot points S_1 and S_2. The travel time of a refracted ray travelling from one end of the line to the other is given by

$$t_{S_1S_2} = l/v_2 + a_{S_1} + a_{S_2} \tag{5.20}$$

where a_{S_1} and a_{S_2} are the delay times at the shot points. Note that $t_{S_1S_2}$ is the reciprocal time for this reversed profile (see Fig. 5.12(b)). For rays travelling to an intermediate detector position D from each end of the line, the travel times are, for the forward ray, from shot point S_1:

$$t_{S_1D} = x/v_2 + a_{S_1} + a_D \tag{5.21}$$

for the reverse ray, from shot point S_2:

$$t_{S_2D} = (l-x)/v_2 + a_{S_2} + a_D \tag{5.22}$$

where a_D is the delay time at the detector.

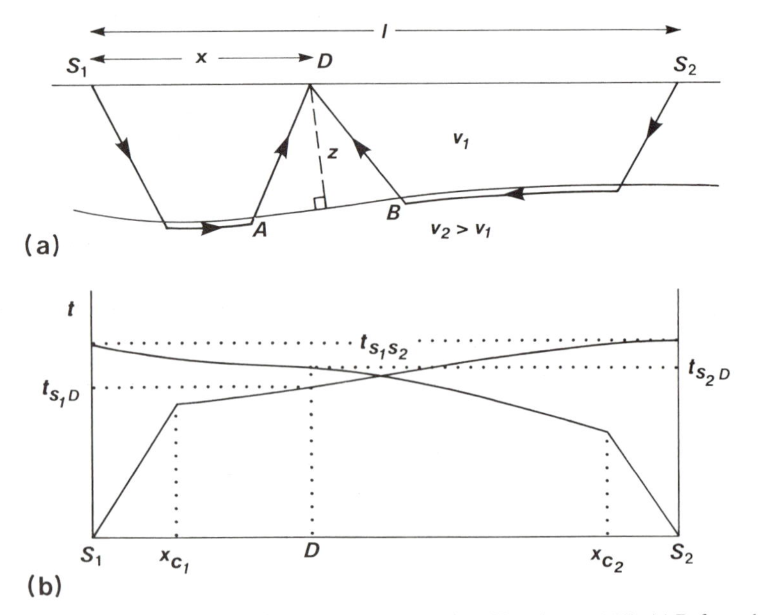

Fig. 5.12. The plus-minus method of refraction interpretation (Hagedoorn 1959). (a) Refracted ray paths from each end of a reversed seismic profile line to an intermediate detector position; (b) Travel-time curves in the forward and reverse directions.

Adding equations (5.21) and (5.22)

$$t_{S_1D}+t_{S_2D} = l/v_2+a_{S_1}+a_{S_2}+2a_D$$

Substituting equation (5.20) in the above equation yields

$$t_{S_1D}+t_{S_2D} = t_{S_1S_2}+2a_D$$

Hence

$$a_D = \tfrac{1}{2}(t_{S_1D}+t_{S_2D}-t_{S_1S_2}) \quad (5.23)$$

This delay time is the *plus* term of Hagedoorn and may be used to compute the perpendicular depth z to the underlying refractor at D using equation (5.17), once v_1 and v_2 have been determined. v_1 is computed from the slope of the direct ray travel-time curve (see Fig 5.12(b)). v_2 cannot be obtained directly from the irregular travel-time curve of refracted arrivals, but it can be estimated by means of Hagedoorn's *minus* term, obtained by taking the difference of equations (5.21) and (5.22)

$$t_{S_1D}-t_{S_2D} = 2x/v_2-l/v_2+a_{S_1}-a_{S_2}$$

This subtraction eliminates the variable (site dependent) delay time a_D from the above equation and, since the last three terms on the right hand side of the equation are constant for a particular profile line, plotting the minus term $(t_{S_1D}-t_{S_2D})$ against the offset distance x yields a graph of slope $2/v_2$ from which v_2 may be derived. Any lateral change of refractor velocity v_2 along the profile line will show up as a change of gradient in the minus term plot.

A plus term and, hence, a local refractor depth can be computed at all detector positions at which head wave arrivals are recognized from both ends of the profile line. In practice, this normally means the portion of the profile line between the crossover distances, that is, between x_{c_1} and x_{c_2} in Fig. 5.12(b).

The plus-minus method is only applicable in the case of shallow refractor dips, generally being considered valid for dips of less than 10°. With steeper dips, x' becomes significantly different from the offset distance x. Further, there is an inherent smoothing of the interpreted refractor geometry in the plus-minus method since in computing the plus term from the travel times of forward and reverse rays arriving at any detector position, the refractor is assumed to be planar between the points of emergence from the refractor of the forward and reverse rays, e.g. between A and B in Fig. 5.12(a) for rays arriving at detector D. This problem of smoothing is solved in the *generalized reciprocal method* of refraction

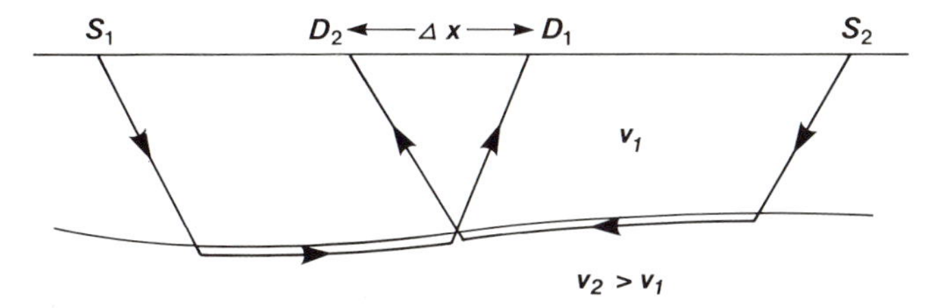

Fig. 5.13. The generalized reciprocal method of refraction interpretation (Palmer 1980.)

interpretation (Palmer 1980) by combining forward and reverse rays which, rather than arriving at the same detector, leave the refractor at approximately the same point and arrive at different detector positions separated by a distance Δx (see Fig. 5.13). The optimal value of Δx is selected on the basis of various tests associated with the method.

Where a refractor is overlain by more than one layer, equation (5.17) cannot be used directly to derive a refractor depth from a delay time (or plus term). In such a case, either the thickness of each overlying layer is computed separately using refracted arrivals from the shallower interfaces or an average overburden velocity is used in place of v_1 in equation (5.17) to achieve a depth conversion.

5.5 Construction of wavefronts and ray tracing

Given the travel-time curves in the forward and reverse directions along a profile line it is possible to reconstruct the configuration of successive wavefronts in the subsurface and thereby derive, graphically, the form of refracting interfaces. This *wavefront method* of Thornburgh (1930) represents one of the earliest refraction interpretation methods but is no longer widely used.

With the recent massive expansion in the speed and power of digital computers, and their wide availability, an increasingly important method of refraction interpretation is a modelling technique known as *ray tracing* (Červený *et al.* 1974). In this method, which is especially useful in the case of complex subsurface structures that are difficult to treat analytically, structural models are postulated and the travel times of refracted (and reflected) rays through these models are calculated by computer for comparison with observed travel times. The model is then adjusted iteratively until the calculated and observed travel

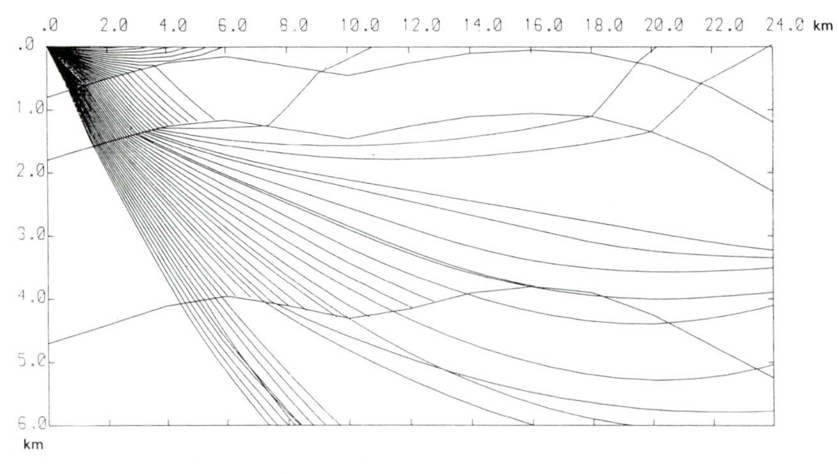

Fig. 5.14. Modelling of complex geology by ray tracing in the case of a refraction profile between two quarries in South Wales, UK. Refracted ray paths from Cornelly Quarry (located in Carboniferous Limestone) are modelled through a layered Palaeozoic sedimentary sequence overlying an irregular Precambrian basement surface at a depth of about 5 km. This model accounts for the measured travel times of refracted arrivals observed along the profile. (From Bayerly & Brooks 1980.)

times are in acceptable agreement. An example of a ray tracing interpretation is illustrated in Fig. 5.14. The ray tracing method is particularly valuable in coping with such complexities as horizontal or vertical velocity gradients within layers, highly irregular or steeply dipping refractor interfaces, and discontinuous layers.

5.6 The hidden layer problem

A *hidden layer*, or *blind layer*, is one that is undetectable by refraction surveying. In practice, there are two different types of hidden layer problem.

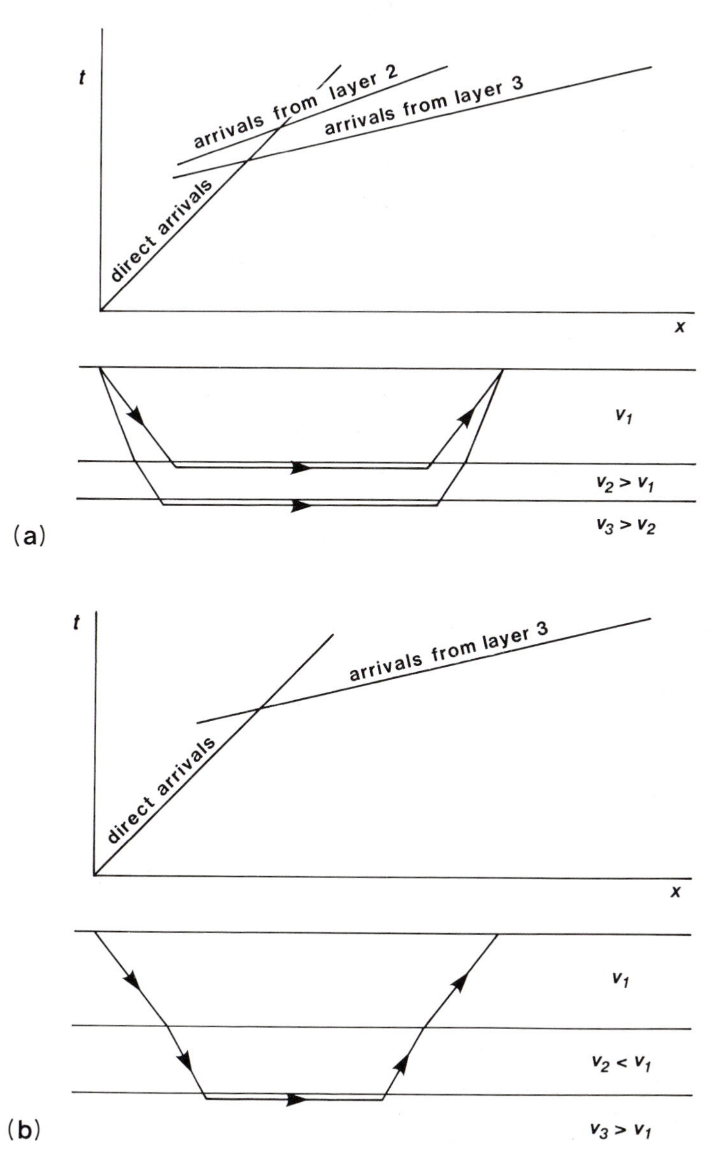

Fig. 5.15. The hidden layer problem in refraction seismology. (a) A thin layer that does not give rise to first arrivals; (b) A layer of low velocity that does not generate head waves.

Firstly, a layer may simply not give rise to first arrivals, i.e. rays travelling to deeper levels may arrive before those critically refracted at the top of the layer in question (Fig. 5.15(a)). This may result from the thinness of the layer, or from the closeness of its velocity to that of the overlying layer. In such a case, a method of survey involving recognition of only first arrivals will fail to detect the layer.

A more insidious type of hidden layer problem is associated with a low velocity layer, as illustrated in Fig. 5.15(b). Rays cannot be critically refracted at the top of such a layer and the layer will therefore not give rise to head waves. Hence, a low velocity layer cannot be detected by refraction surveying although the top of the low velocity layer gives rise to wide angle reflections that may be detected during a refraction survey.

In the presence of a low velocity layer, the interpretation of travel-time curves leads to an overestimation of the depth to underlying interfaces. Low velocity layers are a hazard in all types of refraction seismology. On a small scale, a peat layer in muds and sands above bedrock may escape detection, leading to a false estimation of foundation conditions and rockhead depths beneath a construction site; on a much larger scale, low velocity zones of regional extent are known to exist within the continental crust and may escape detection in crustal seismic experiments.

5.7 Refraction in layers of continuous velocity change

In some geological situations, velocity varies gradually as a function of depth rather than discontinuously at discrete interfaces of lithological change. In thick clastic sequences, for example, especially clay sequences, velocity increases downwards due to the progressive compaction effects associated with increasing depth of burial. A seismic ray propagating through a layer of gradual velocity change is continuously refracted to follow a curved ray path. For example, in the special case where velocity increases linearly with depth the seismic ray paths describe arcs of circles. The deepest point reached by a ray travelling on a curved path is known as its *turning point*.

In such cases of continuous velocity change with depth, the travel-time curve for refracted rays that return to the surface along curved ray paths is itself curved, and the geometrical form of the curve may be analysed to derive information on the distribution of velocity as a function of depth (see, e.g. Dobrin 1976).

Velocity increase with depth may be significant in thick surface layers of clay due to progressive compaction and dewatering, but may also be significant in buried layers. Refracted arrivals from such buried layers are not true head waves since the associated rays do not travel along the top surface of the layer but along a curved path in the layer with a turning point at some depth below the interface. Such refracted waves are referred to as *diving waves* (Červený & Ravindra 1971). Methods of interpreting refraction data in terms of diving waves are generally complex, but include ray tracing techniques. Indeed, some ray tracing programmes require velocity gradients to be introduced into all layers of an interpretation model in order to generate diving waves rather than true head waves.

5.8 Methodology of refraction profiling

Many of the basic principles of refraction surveying have been covered in the preceding sections but in this section several aspects of the design of refraction profile lines are brought together in relation to the particular objectives of a refraction survey.

5.8.1 *Field survey arrangements*

Although the same principles apply to all scales of refraction profiling, the logistical problems of implementing a profile line increase as the required line length increases. Further, the problems of surveying on land are quite different from those encountered at sea. A consequence of these logistical differences is a very wide variety of survey arrangements for the implementation of refraction profile lines and these differences are illustrated by three examples.

For a small-scale refraction survey of a construction site to locate water table or rockhead (both of which surfaces are generally good refractors), recordings out to an offset distance of about 100 m normally suffice, geophones being connected via a multicore cable to a portable 12- or 24-channel seismic recorder. A simple weight-dropping device (even a sledge hammer impacted on to a steel base plate) provides sufficient energy to traverse the short recording range. The dominant frequency of such a source exceeds 100 Hz and the required accuracy of seismic travel times is about 0.5 millisecond. Such a survey can be easily accomplished by two operators.

To carry out refraction profiling on a larger scale at sea, for example in the investigation of layered sedimentary sequences to depths of a few kilometres, refraction lines up to about 20 km long can readily be implemented using a single ship in conjunction with one or more free-floating radio-transmitting sonobuoys (Fig. 5.16). Having deployed the sonobuoys, the ship steams along the profile line repeatedly firing explosive charges or a large air gun. Seismic signals travelling back to the surface through the water layer are detected by a hydrophone suspended beneath each sonobuoy, amplified and transmitted back to the ship where they are tape recorded along with the shot-instant. An alternative method of single-ship seismic refraction surveying utilizes sea bed detectors with built-in tape recording and timing facilities. For the purposes of recovery, the detectors are 'popped-up' to surface by remotely triggering a release mechanism. Sea bed detectors provide a better signal to noise ratio than hydrophones suspended in the water column and, in deep water, recording on the sea bed allows much better definition of shallow structure. In this type of work the dominant frequency is typically in the range 10–50 Hz and travel times need to be known to about 10 milliseconds.

Finally, a large scale seismic refraction line on land to investigate deep crustal structure is typically 250–300 km long. Seismic events need to be recorded at a series of independently operated recording stations all receiving a radio-transmitted time code to provide a common time base for the recordings. Very large energy sources, such as depth charges (detonated at sea or in a lake) or large

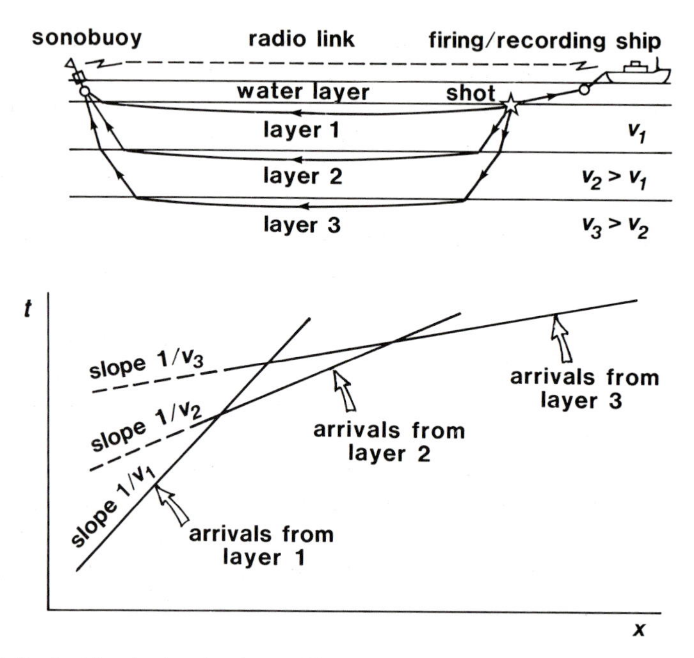

Fig. 5.16. Single-ship seismic refraction profiling.

quarry blasts, are required in order that sufficient energy is transmitted over the length of the profile line. The dominant frequency of such sources is less than 10 Hz and the required accuracy of seismic travel times is about 50 milliseconds. Such an experiment requires the active involvement of a large team of investigators.

5.8.2 *Recording scheme*

For a complete mapping of refractors beneath a profile line it is important to arrange that head wave arrivals from all refractors of interest are obtained over the same portion of line. The importance of this can be seen by reference to Fig. 5.17

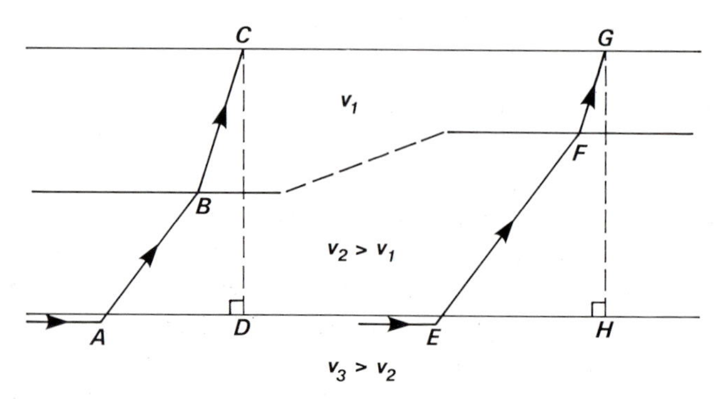

Fig. 5.17. Variation in the travel time of a head wave associated with variation in the thickness of a surface layer.

where it is shown that a change in thickness of a surface low velocity layer would cause a change in the delay time associated with arrivals from a deeper refractor and may be erroneously interpreted as a change in refractor depth. The actual geometry of the shallow refractor could be mapped by means of shorter reversed profiles along the length of the main profile, to ensure that head waves from the shallow refractor were recorded at positions where the depth to the basal refractor was required. Knowledge of the disposition of the shallow refractor derived from the shorter profiles would then allow correction of travel times of arrivals from the deeper refractor.

The general design requirement is the formulation of an overall observational scheme as illustrated in Fig. 5.18. Such a scheme might include off-end shots into individual reversed profile lines, since off-end shots extend the length of refractor traversed by recorded head waves and provide insight into the structural causes of any observed complexities in the travel-time curves. Selection of detector spacing along the individual profile lines is determined by the required detail of the refractor geometry, the sampling interval of interpretation points on the refractor being approximately equal to the detector spacing. Thus, the horizontal resolution of the method is equivalent to the detector spacing.

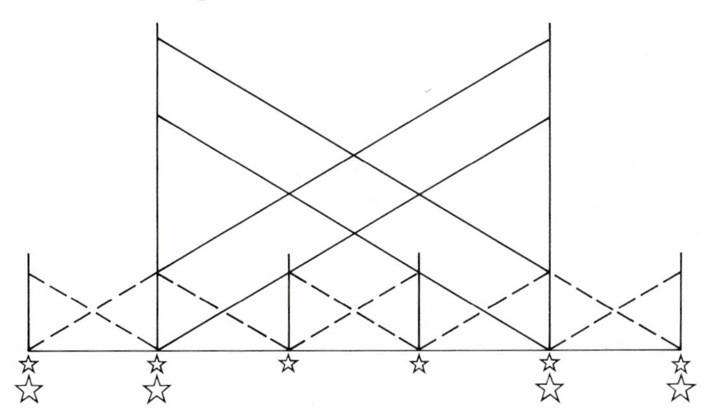

Fig. 5.18. A possible observational scheme to obtain shallow and deeper refraction coverage along a survey line. The inclined lines indicate the range of coverage from the individual shots shown.

5.8.3 *Weathering and elevation corrections*

The type of observational scheme illustrated in Fig. 5.18 is often implemented for the specific purpose of mapping the surface zone of weathering and associated low velocity across the length of a longer profile designed to investigate deeper structure. The velocity and thickness of the weathered layer are highly variable laterally and travel times of rays from underlying refractors need to be corrected for the variable delay introduced by the layer. This weathering correction is directly analogous to that applied in reflection seismology (see Section 4.8). The weathering correction is particularly important in shallow refraction surveying where the size of the correction is often a substantial percentage of the overall travel time of a refracted ray. In such cases, failure to apply an accurate weathering correction can lead to major error in interpreted depths to shallow refractors.

A weathering correction is applied by effectively replacing the weathered layer of velocity v_w with material of velocity v_1 equal to the velocity of the underlying layer. For a ray critically refracted along the top of the layer immediately underlying the weathered layer, the weathering correction is simply the sum of the delay times at the shot and detector ends of the ray path. Application of this correction replaces the refracted ray path by a direct path from shot to detector in a layer of velocity v_1. For rays from a deeper refractor a different correction is required. Referring to Fig. 5.19, this correction effectively replaces ray path *ABCD* by ray path *AD*. For a ray critically refracted in the *n*th layer the weathering correction t_w is given by

$$t_w = -(z_s+z_d)\{(v_n-v_1)^{1/2}/v_1v_n-(v_n-v_w)^{1/2}/v_wv_n\}$$

where z_s and z_d are the thicknesses of the weathered layer beneath the shot and detector respectively and v_n is the velocity in the *n*th layer.

In addition to the weathering correction, a correction is also needed to remove the effect of differences in elevation of individual shots and detectors, and an elevation correction is therefore applied to reduce travel times to a common

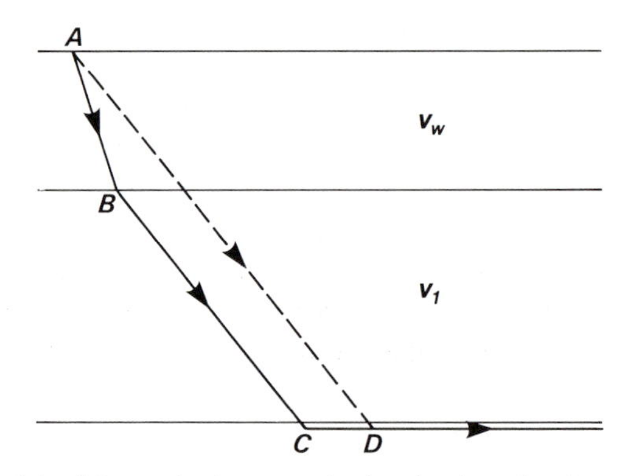

Fig. 5.19. The principle of the weathering correction in refraction seismology.

datum plane. The elevation correction t_e for rays critically refracted in the *n*th layer is given by

$$t_e = -(h_s+h_d)\{(v_n-v_1)^{1/2}/v_1v_n\}$$

where h_s and h_d are the heights above datum of the shot point and detector location respectively.

In shallow water marine refraction surveying the water layer is conventionally treated as a weathered layer and a correction applied to replace the water layer by material of velocity equal to the velocity of the sea bed.

5.8.4 *Display of refraction seismograms*

In small-scale refraction surveys the individual seismograms are conventionally plotted out in their true time relationships by a multichannel oscillographic recorder similar to that employed to display seismic traces from land reflection spreads (see Fig. 4.8). From such displays, arrival times of refracted waves may be picked and, after suitable correction, utilized to plot the time-distance curves that form the basis of refraction interpretation.

Interpretation of large-scale refraction surveys is often as much concerned with later arriving phases, such as wide-angle reflections or S-wave arrivals, as with first arrivals and it is necessary to compile the individual seismograms into an overall record section on which the various seismic phases can be correlated from seismogram to seismogram. The optimal type of display is achieved using a *reduced time* scale in which any event at time t and offset distance x is plotted at the reduced time T where

$$T = t - x/v_R$$

and v_R is a scaling factor known as the *reduction velocity*. Thus, for example, a seismic arrival from deep in the Earth's crust with an overall travel time of 30 s to an offset distance of 150 km would, with a reduction velocity of 6 km/s, have a reduced time of 5 s.

Plotting in reduced time has the effect of progressively moving seismic events forward as a function of offset and, therefore, rotating the associated time-distance curves towards the horizontal. For example, a time-distance curve with a reciprocal slope of 6 km/s on a $t-x$ graph would plot as a horizontal line on a $T-x$ graph using a reduction velocity of 6 km/s. By appropriate choice of reduction velocity, seismic arrivals from a particular refractor of interest can be arranged to plot about a horizontal datum, so that relief on the refractor will show up directly as departures of the arrivals from a horizontal line. The use of reduced time also enables the display of complete seismograms with an expanded time scale appropriate for the analysis of later arriving phases. A example of a record section from a crustal seismic experiment, plotted in reduced time, is illustrated in Fig. 5.20.

5.9 Other methods of refraction surveying

Although the vast bulk of refraction surveying is carried out along profile lines, other spatial arrangements of shots and detectors may be utilized for particular purposes. Such arrangements include fan-shooting and irregularly distributed shots and recorders as used in the time term method.

Fan-shooting (Fig. 5.21) is a convenient method of accurately delineating a subsurface zone of anomalous velocity whose approximate position and size are already known. Detectors are distributed around a segment of arc approximately centred on one or more shot points, and travel times of refracted rays are measured to each detector. Through a homogeneous medium the travel times to detectors would be linearly related to range, but any ray path which encounters an anomalous velocity zone will be subject to a time lead or time lag depending upon the velocity of the zone relative to the velocity of the surrounding medium. Localized anomalous zones capable of detection and delineation by fan-shooting include salt domes, buried valleys and backfilled mine shafts.

An irregular, areal distribution of shots and detectors (Fig. 5.22(a)) represents a completely generalized approach to refraction surveying and facilitates mapping of the three-dimensional geometry of a subsurface refractor using the *time term*

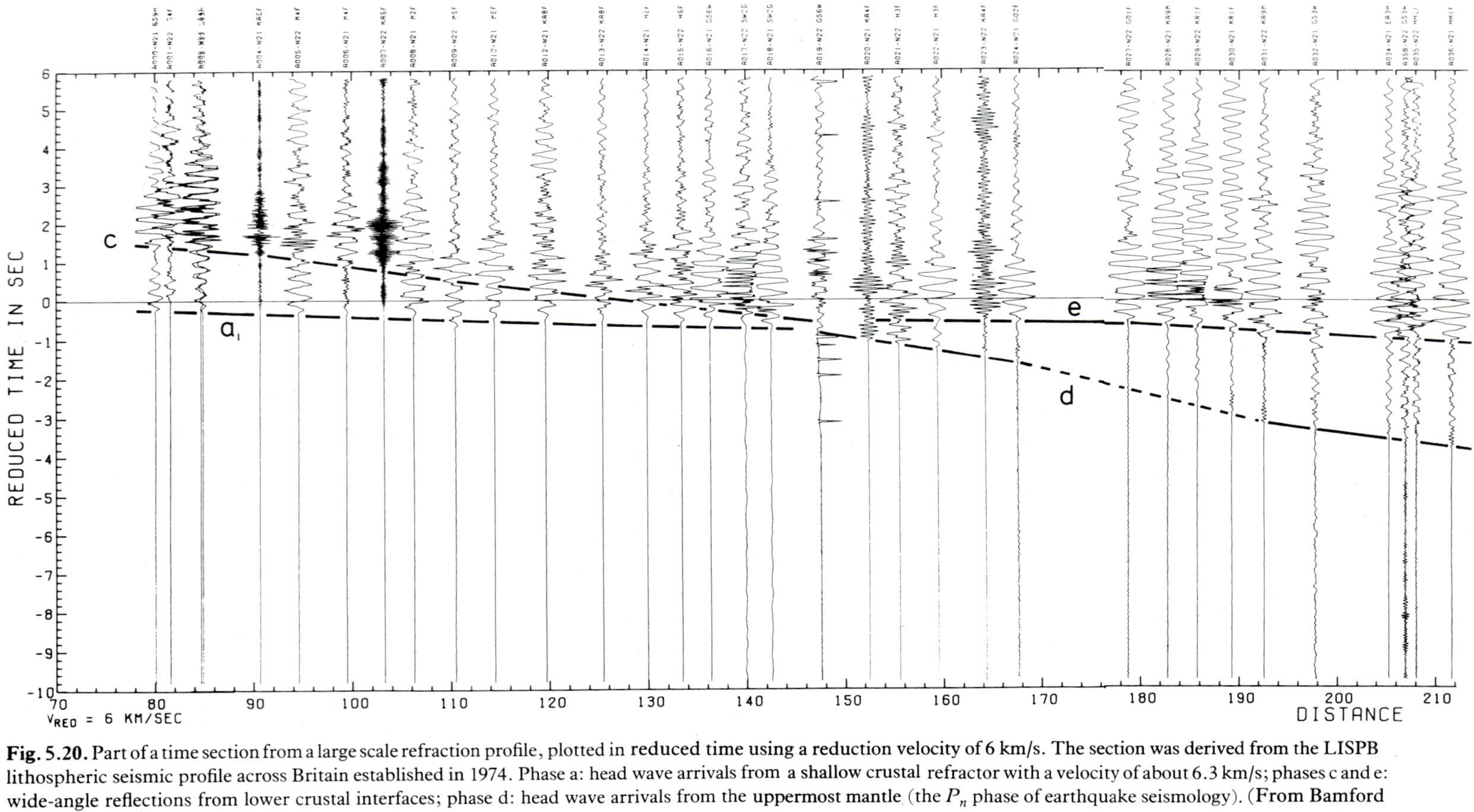

Fig. 5.20. Part of a time section from a large scale refraction profile, plotted in reduced time using a reduction velocity of 6 km/s. The section was derived from the LISPB lithospheric seismic profile across Britain established in 1974. Phase a: head wave arrivals from a shallow crustal refractor with a velocity of about 6.3 km/s; phases c and e: wide-angle reflections from lower crustal interfaces; phase d: head wave arrivals from the uppermost mantle (the P_n phase of earthquake seismology). (From Bamford *et al*. 1978.)

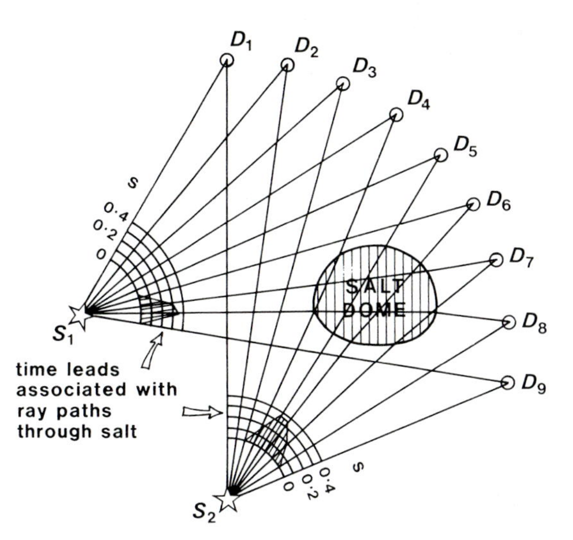

Fig. 5.21. Fan-shooting for the detection of localized zones of anomalous velocity.

method of interpretation (Willmore & Bancroft 1960; Berry & West 1966). Rather than being an intrinsic aspect of the survey design, however, an areal distribution of shot points and recording sites may result simply from an opportunistic approach to refraction surveying in which freely available sources of seismic energy such as quarry blasts are utilized to derive subsurface information from seismic recordings.

The *time term method* uses the form of the travel time equation containing delay times (equation (5.18)) and is subject to the same underlying assumptions as other interpretation methods using delay times. However, in the time term method a statistical approach is adopted to deal with a redundancy of data inherent in the method and to derive the best estimate of the interpretation parameters. Introducing an error term into the travel time equation

$$t_{ij} = x_{ij}/v + a_i + a_j + \epsilon_{ij}$$

where t_{ij} is the travel time of head waves from the ith site to the jth site
x_{ij} is the offset distance between site i and site j
a_i and a_j are the delay times (time terms)
v is the refractor velocity (assumed constant)
ϵ_{ij} is an error term associated with the measurement of t_{ij}.

If there are n sites there can be up to $n(n-1)$ observational linear equations of the above type, representing the situation of a shot and detector at each site and all sites sufficiently far apart for the observation of head waves from the underlying refractor. In practice there will be fewer observational equations than this because, normally, only a few of the sites are shot points and head wave arrivals are not recognized along every shot-detector path (Fig. 5.22(b)). There are $(n+1)$ unknowns, namely the individual delay times at the n sites and the refractor velocity v.

If the number m of observational equations equals the number of unknowns,

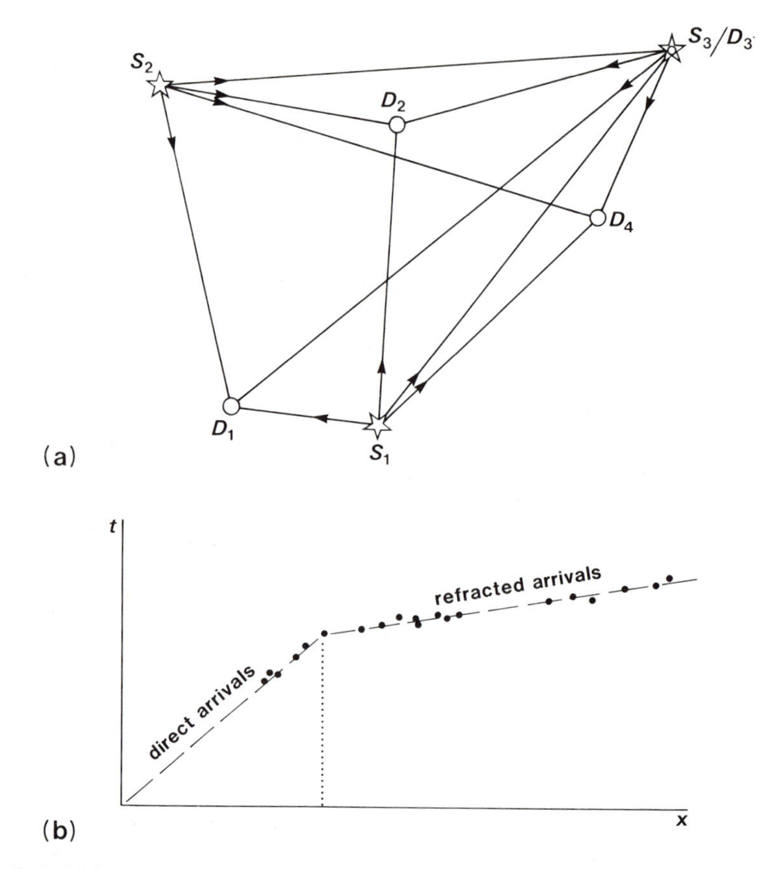

Fig. 5.22. (a) An example of the type of network of shots and detectors from which the travel times of refracted arrivals can be used in a time term analysis of the underlying refractor geometry. (b) The plot of travel time as a function of distance identifies the set of refracted arrivals that may be used in the analysis.

the equations can be solved to derive the unknown quantities, although it is necessary either that at least one shot and detector position should coincide or that the delay time should be known at one site. In fact, with the time term approach to refraction surveying it is normally arranged for m to well exceed $(n+1)$, and for several shot and detector positions to be interchanged. The resulting overdetermined set of equations is solved by deriving values for the individual delay times and refractor velocity that minimize the sum of squares of the errors ϵ_{ij}. Delay times can then be converted into local refractor depths using the same procedure as in the plus-minus method described earlier.

5.10 Two-ship seismic surveying: combined refraction and reflection surveying

Specialized methods of marine surveying involving he use of two survey vessels and multichannel recording include *expanding spread profiles* and *constant offset profiles* (Stoffa & Buhl 1979). These methods have been developed for the detailed study of the deep structure of the crust and upper mantle under continental margins and oceanic areas.

Expanding spread profiling (ESP) is designed to obtain detailed information relating a localized region of the crust. The shot-firing vessel and recording vessel travel outwards at the same speed from a central position, obtaining reflected and refracted arrivals from subsurface interfaces out to large offsets. Thus, in addition to near-normal incidence reflections such as would be recorded in a conventional common depth point (CDP) reflection survey, wide-angle reflections and refracted arrivals are also recorded from the same section of crust. The combined reflection/refraction data allow derivation of a highly-detailed velocity-depth structure for the localized region.

In constant offset profiling (COP), the shot-firing and recording vessels travel along a profile line at a fixed, wide separation. Thus, wide-angle reflections and refractions are continuously recorded along the line. This survey technique facilities the mapping of lateral changes in crustal structure over wide areas and allows continuous mapping of the types of refracting interface that do not give rise to good near-normal incidence reflections and which therefore cannot be mapped adequately using conventional reflection profiling. Such interfaces include zones of steep velocity gradient, in contrast to the first-order velocity discontinuities that constiute the best reflectors.

Further reading

Červený, V. & Ravindra, R. (1971) *Theory of Seismic Head Waves*. University of Toronto Press.

Dobrin, M.B. (1976) *Introduction to Geophysical Prospecting*. McGraw-Hill, New York (3rd edn).

Giese, P., Prodehl, C. & Stein, A. (eds.) (1976) *Explosion Seismology in Central Europe*. Springer-Verlag, Berlin.

Musgrave, A.W. (ed.) (1967) *Seismic Refraction Prospecting*. Society of Exploration Geophysicists, Tulsa.

Palmer, D. (1980) *The Generalised Reciprocal Method of Seismic Refraction Interpretation*. Society of Exploration Geophysicists, Tulsa.

Stoffa, P.L. & Buhl, P. (1979) Two-ship multichannel seismic experiments for deep crustal studies: expanded spread and constant offset profiles. *J. geophys. Res.*, **84**, 7645–60.

Willmore, P.L. & Bancroft, A.M. (1960) The time-term approach to refraction seismology. *Geophys. J.R. astron. Soc.*, **3**, 419–32.

6

Gravity Surveying

6.1 Introduction

In gravity surveying subsurface geology is investigated on the basis of variations in the Earth's gravitational field generated by differences of density between subsurface rocks. An underlying concept is the idea of a causative body, which is a rock unit of different density from its surroundings. A causative body represents a subsurface zone of anomalous mass and causes a localized perturbation in the gravitational field known as a gravity anomaly. A very wide range of geological situations give rise to zones of anomalous mass that produce significant gravity anomalies. On a small scale, buried relief on a bedrock surface, such as a buried valley, can give rise to measurable anomalies. On a larger scale, small negative anomalies are associated with salt domes, as discussed in Chapter 1. On a larger scale still, major gravity anomalies are generated by granite plutons or sedimentary basins. Interpretation of gravity anomalies allows an assessment to be made of the probable depth and shape of the causative body.

The ability to carry out gravity surveys in marine areas extends the scope of the method so that the technique may be employed in most areas of the world.

6.2 Basic theory

The basis of the gravity survey method is Newton's Law of Gravitation, which states that the force of attraction F between two masses m_1 and m_2, whose dimensions are small with respect to the distance r between them, is given by

$$F = \frac{Gm_1m_2}{r^2} \tag{6.1}$$

where G is the Gravitational Constant ($6.67 \times 10^{-11}\ \mathrm{m^3\,kg^{-1}\,s^{-2}}$).

Consider the gravitational attraction of a spherical, non-rotating, homogeneous Earth of mass M and radius R on a small mass m on its surface. It is relatively simple to show that the mass of a sphere acts as though it were concentrated at the centre of the sphere and by substitution in equation 6.1

$$F = \frac{GM}{R^2}m = mg \tag{6.2}$$

Force is related to mass by an acceleration and the term $g = GM/R^2$ is known as the gravitational acceleration or, simply, *gravity*. The weight of the mass is given by mg.

On such an Earth, gravity would be constant. However, the Earth's ellipsoidal shape, rotation, irregular surface relief and internal mass distribution cause gravity to vary over its surface.

The gravitational field is most usefully defined in terms of the *gravitational potential* U:

$$U = \frac{GM}{r} \tag{6.3}$$

Whereas the gravitational acceleration g is a vector quantity, having both magnitude and direction (vertically downwards), the gravitational potential U is a scalar, having magnitude only. The first derivative of U in any direction gives the component of gravity in that direction. Consequently a potential field approach provides computational flexibility. Equipotential surfaces can be defined on which U is constant. The sea-level surface, or *geoid*, is the most easily recognized equipotential surface, which is everywhere horizontal and orthogonal to the direction of gravity.

6.3 Units of gravity

The mean value of gravity at the Earth's surface is about 9.80 m s^{-2}. Variations in gravity caused by density variations in the subsurface are of the order of 100 μm s^{-2}. This unit of the micrometre per second per second is referred to as the *gravity unit* (gu). In gravity surveys on land an accuracy of ± 0.1 gu is readily attainable, corresponding to about one hundred millionth of the normal gravitational field. At sea the accuracy obtainable is considerably less, about ± 10 gu. The cgs unit of gravity is the *milligal* (1 mgal $= 10^{-3}$ Gal $= 10^{-3}$ cm s^{-2}), equivalent to 10 gu.

6.4 Measurement of gravity

Since gravity is an acceleration, its measurement should simply involve determinations of length and time. However, such apparently simple measurements are not easily achievable at the precision and accuracy required in gravity surveying.

The measurement of an absolute value of gravity is extremely difficult and requires complex apparatus and a lengthy period of observation. Such measurement is classically made using large pendulums or falling body techniques (see, for example, Nettleton 1976).

The measurement of relative values of gravity, i.e. the differences of gravity between locations, is simpler and is the standard procedure in gravity surveying. Absolute gravity values at survey stations may be obtained by reference to the International Gravity Standardization Network (IGSN) of 1971 (Morelli *et al.* 1971), a network of stations at which the absolute values of gravity have been determined by reference to sites of absolute gravity measurements (see Section 6.7). By using a relative reading instrument to determine the difference in gravity between an IGSN station and a field location the absolute value of gravity at that location can be determined.

Previous generations of relative reading instruments were based on small

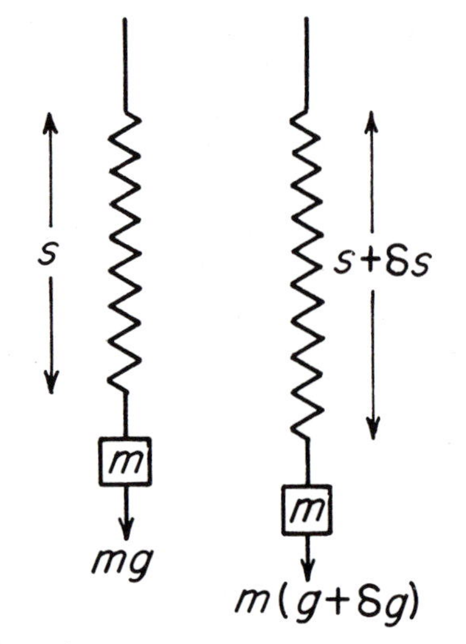

Fig. 6.1. Principle of stable gravimeter operation.

pendulums or the oscillation of torsion fibres and, although portable, took considerable time to read. Modern instruments capable of rapid gravity measurements are known as *gravity meters* or *gravimeters*.

Gravimeters are basically spring balances carrying a constant mass. Variations in the weight of the mass caused by variations in gravity cause the length of the spring to vary and give a measure of the change in gravity. In Fig. 6.1 a spring of initial length s has been stretched by an amount δs as a result of an increase in gravity δg increasing the weight of the suspended mass m. The extension of the spring is proportional to the extending force (Hooke's Law), thus

$$m\delta g = k\delta s$$

and

$$\delta s = \frac{m}{k}\delta g \qquad (6.4)$$

where k is the elastic spring constant.

δs must be measured to a precision of $1:10^8$ in instruments suitable for gravity surveying on land. Although a large mass and a weak spring would increase the ratio m/k and, hence, the sensitivity of the instrument, in practice this would make the system liable to collapse. Consequently some form of optical, mechanical or electronic amplification of the extension is in practice required.

The necessity for the spring to serve a dual function, namely to support the mass and to act as the measuring device, severely restricted the sensitivity of early gravimeters, known as stable or static gravimeters. This problem is overcome in modern meters (unstable or astatic) which employ an additional force that acts in the same sense as the extension (or contraction) of the spring and consequently amplifies the movement directly.

An example of an unstable instrument is the LaCoste and Romberg gravimeter. The meter consists of a hinged beam, carrying a mass, supported by a spring attached immediately above the hinge (Fig. 6.2). The magnitude of the moment exerted by the spring on the beam is dependent upon the extension of the spring and the sine of the angle θ. If gravity increases, the beam is depressed and the spring further extended. Although the restoring force of the spring is increased, the angle θ is decreased to θ'. By suitable design of the spring and beam geometry the magnitude of the increase of restoring moment with increasing gravity can be made as small as desired. With ordinary springs the working range of such an instrument would be very small. However, by making use of a 'zero-length' spring which is pretensioned during manufacture so that the restoring force is proportional to the physical length of the spring rather than its extension, instruments can be fashioned with a very sensitive response over a wide range. The instrument is read by restoring the beam to the horizontal by altering the vertical location of the spring attachment with a micrometer screw. Thermal effects are removed by a battery-powered thermostatting system. The range of the instrument is 50 000 gu.

The other unstable instrument in common use is the Worden-type gravimeter. The necessary instability is provided by a similar mechanical arrangement, but in this case the beam is supported by two springs. The first of these springs acts as the measuring device, while the second alters the level of the 2000 gu reading range of the instrument. In certain specialized forms of this instrument the second spring is also calibrated, so that the overall reading range is similar to that of the LaCoste and Romberg gravimeter. Thermal effects are

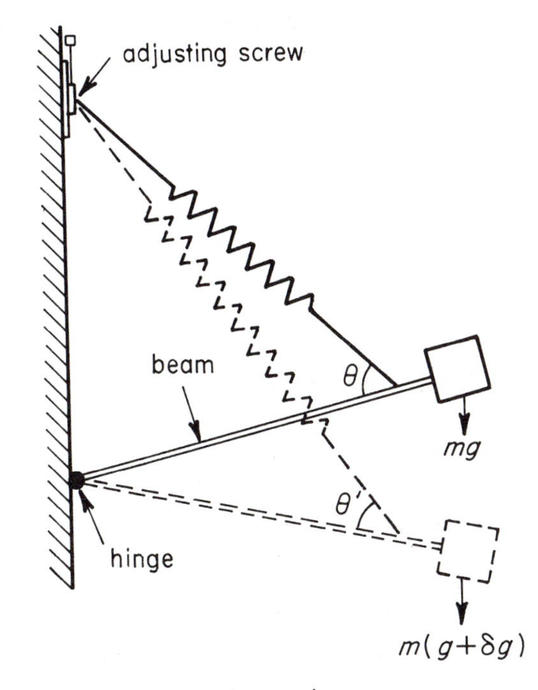

Fig. 6.2. Principle of the LaCoste and Romberg gravimeter.

normally minimized by the use of quartz components and a bimetallic beam which compensates automatically for temperature changes. Consequently no thermostatting is required and it is simply necessary to house the instrument in an evacuated flask. The restricted range of normal forms of the instrument, however, makes it unsuitable for intercontinental gravity ties or surveys in areas where gravity variation is extreme.

A shortcoming of gravimeters is the phenomenon of *drift*. This refers to a gradual change in reading with time, observable when the instrument is left at a fixed location. Drift results from the imperfect elasticity of the springs, which undergo anelastic creep with time. Drift can also result from temperature variations which, unless counteracted in some way, cause expansion or contraction of the measuring system and thus give rise to variations in measurements that are unrelated to changes in gravity. Drift is monitored by repeated meter readings at a fixed location throughout the day.

Gravity may be measured at discrete locations at sea using a remote-controlled land gravimeter, housed in a waterproof container, which is lowered over the side of the ship and, by remote operation, levelled and read on the sea bed. Measurements of comparable quality to readings on land may be obtained in this way, and the method has been used with success in relatively shallow waters. The disadvantage of the method is that the meter has to be lowered to the sea bed for each reading so that the rate of surveying is very slow. Moreover, in strong tidal currents, the survey ship needs to be anchored to keep it on station while the gravimeter is on the sea bed.

Gravity measurements may be made continuously at sea using a gravimeter modified for use on ships. Such instruments are known as shipborne, or shipboard, meters. The accuracy of measurements with a shipborne meter is considerably reduced compared to measurements on land because of the severe vertical and horizontal accelerations imposed on the shipborne meter by sea waves and the ship's motion. These external accelerations can cause variations in measured gravity of up to 10^6 gu and represent high amplitude noise from which a signal of much smaller gravity variations must be extracted. The effects of horizontal accelerations produced by waves, yawing of the ship and changes in its speed and heading can be largely eliminated by mounting the meter on a gyro-stabilized, horizontal platform, so that the meter only responds to vertical accelerations. Deviations of the platform from the horizontal produce *off-levelling errors* which are normally less than 10 gu. External vertical accelerations resulting from wave motions cannot be distinguished from gravity but their effect can be diminished by heavily damping the suspension system and by averaging the reading over an interval considerably longer than the maximum period of the wave motions (about 8 seconds). As the ship oscillates vertically above and below the plane of the mean sea surface the wave accelerations are equally negative and positive and are effectively removed by averaging over a few minutes. The operation is essentially low-pass filtering in which accelerations with periods of less than one to five minutes are rejected.

With shipborne meters employing a beam-supported sensor, such as the LaCoste and Romberg instrument, a further complication arises due to the influence of horizontal accelerations. The beam of the meter oscillates under the influence of the varying vertical accelerations caused by the ship's motions. When the beam is tilted out of the horizontal it will be further displaced by the turning force associated with any horizontal acceleration. For certain phase relationships between the vertical and horizontal components of motion of the ship, the horizontal accelerations may cause beam displacements that do not average out with time. Consider an example where the position of a meter in space describes a circular motion under the influence of sea waves (Fig. 6.3). At time t_1, as shown in Fig. 6.3, the ship is moving down, displacing the beam upwards, and the horizontal component of motion is to the right, inducing an anticlockwise torque that decreases the upward displacement of the beam. At a slightly later time t_3 the ship is moving up, displacing the beam down, and the horizontal motion is to the left, again inducing an anticlockwise torque which, now, increases the downward displacement of the beam. In such a case, the overall effect of the horizontal accelerations is to produce a systematic error in the beam position. This effect is known as *cross-coupling*, and its magnitude is dependent on the damping characteristics of the meter and the amplitude and phase relationships of the horizontal and vertical motions. It leads to an error known as the *cross-coupling error* in the measured gravity value. In general, the cross-coupling error is small or negligible in good weather conditions but can become very large in high seas. Cross-coupling errors are corrected directly from the outputs of two horizontal accelerometers mounted on the stabilized platform.

The inability to compensate fully for extraneous accelerations reduces the accuracy of shipborne measurements to 10 gu at best, the actual amount depending on prevailing sea conditions. Instrumental drift monitoring is also less precise as base ties are, of necessity, usually many days apart.

The measurement of gravity from aircraft is not at present satisfactory because of the excessive error in applying corrections. Eötvös corrections (Section 6.8.5) may be as great as 16 000 gu at a speed of 200 knots, a 1% error in velocity or

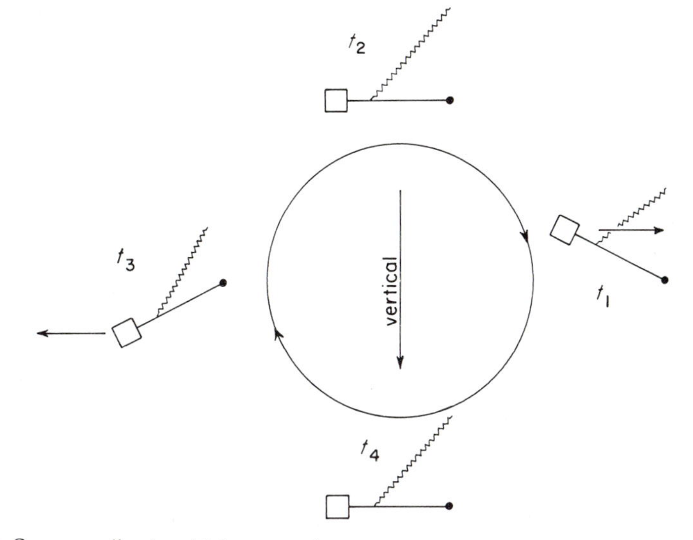

Fig. 6.3. Cross-coupling in a shipborne gravimeter.

heading producing maximum errors of 180 gu and 250 gu respectively. Vertical accelerations associated with the aircraft's motion with periods longer than the instrumental averaging time cannot readily be corrected. Such uncertainties can be overcome to a certain extent by the use of autopilots and automatic height stabilizers but the present precision of such systems is only of the order of 100 gu.

The calibration constants of gravimeters may vary with time and should be checked periodically. The most common procedure is to take readings at two or more locations where absolute or relative values of gravity are known. In calibrating Worden-type meters, these readings would be taken for several settings of the coarse adjusting screw so that the calibration constant is checked over as much of the full range of the instrument as possible. Such a procedure cannot be adopted for the LaCoste and Romberg gravimeter, where each different dial range has its own calibration constant. In this case checking can be accomplished by taking readings at different inclinations of the gravimeter on a tilt table, a task usually entrusted to the instrument's manufacturer.

6.5 Gravity anomalies

Gravimeters effectively respond only to the vertical component of the gravitational attraction of an anomalous mass. Consider the gravitational effect of an anomalous mass δg, with horizontal and vertical components δg_x and δg_z respectively, on the local gravity field g and its representation on a vector diagram (Fig. 6.4).

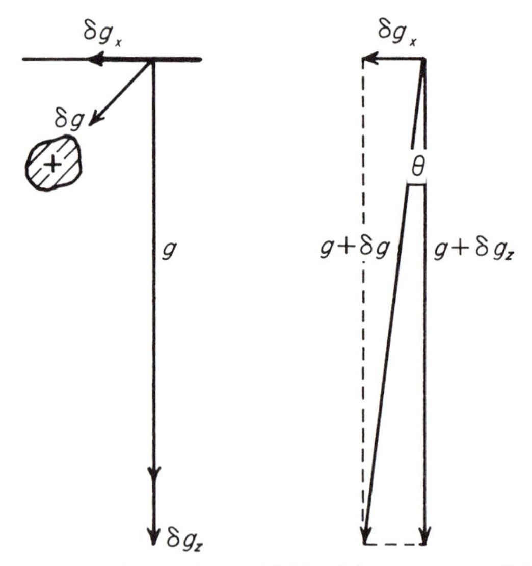

Fig. 6.4. Relationship between the gravitational field and the components of the gravity anomaly of a small mass.

Solving the rectangle of forces

$$\begin{aligned} g+\delta g &= ((g+\delta g_z)^2+\delta g_x^2)^{1/2} \\ &= (g^2+2g\delta g_z+\delta g_z^2+\delta g_x^2)^{1/2} \end{aligned}$$

Terms in δ^2 are insignificantly small and can thus be ignored. Binomial expansion of the equation then gives

$$g+\delta g \doteq g+\delta g_z$$

so that

$$\delta g \doteq \delta g_z \tag{6.5}$$

Consequently, measured perturbations in gravity effectively correspond to the vertical component of the attraction of the causative body. The local deflection of the vertical θ is given by

$$\theta = \tan^{-1} \frac{\delta g_x}{g}$$

and since $\delta g_x \ll g$, θ is usually insignificant. Very large mass anomalies such as mountain ranges can, however, produce measurable local vertical deflections.

6.6 Gravity anomalies of simple shapes

Consider the gravitational attraction of a point mass m at a distance r from the mass (Fig. 6.5).

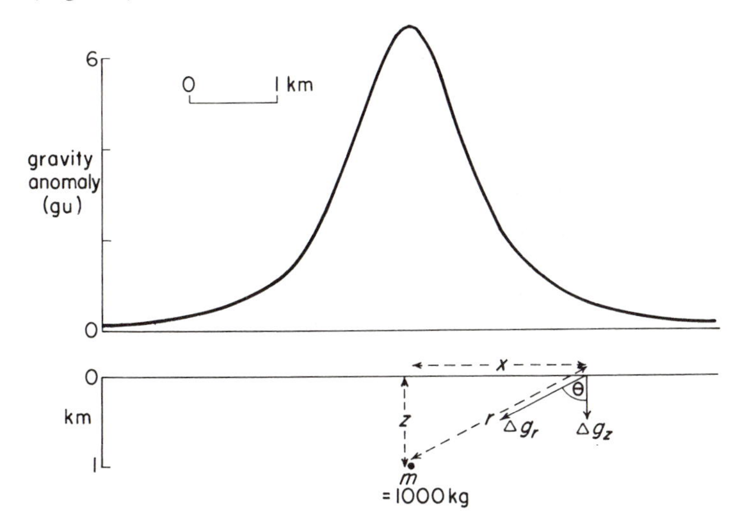

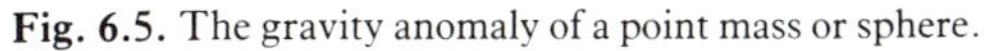
Fig. 6.5. The gravity anomaly of a point mass or sphere.

The gravitational attraction Δg_r in the direction of the mass is given by

$$\Delta g_r = \frac{Gm}{r^2} \text{ from Newton's Law.}$$

Since only the vertical component of the attraction Δg_z is measured, the gravity anomaly Δg caused by the mass is

$$\Delta g = \frac{Gm}{r^2} \cos \theta$$

or

$$\Delta g = \frac{Gmz}{r^3} \tag{6.6}$$

Since a sphere acts as though its mass were concentrated at its centre, equation (6.6) also corresponds to the gravity anomaly of a sphere whose centre lies at a depth z.

Equation (6.6) can be used to build up the gravity anomaly of many simple geometric shapes by constructing them from a suite of small elements which correspond to point masses, and then summing (integrating) the attractions of these elements to derive the anomaly of the whole body.

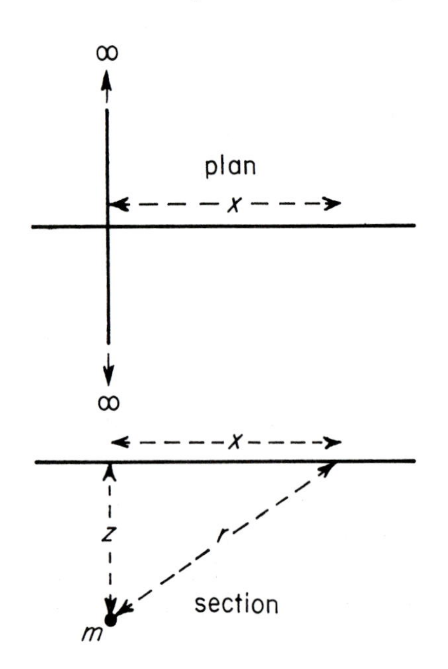

Fig. 6.6. Coordinates describing an infinite horizontal line mass.

Integration of equation (6.6) in a horizontal direction provides the equation for a line mass (Fig. 6.6) extending to infinity in this direction

$$\Delta g = \frac{2Gmz}{r^2} \tag{6.7}$$

Equation (6.7) also represents the anomaly of a horizontal cylinder, whose mass acts as though concentrated along its axis.

Integration in the second horizontal direction provides the gravity anomaly of an infinite horizontal sheet, and a further integration in the vertical direction between fixed limits provides the anomaly of an infinite horizontal slab

$$\Delta g = 2\pi G\rho t \tag{6.8}$$

where ρ is the density of the slab and t its thickness. Note that this attraction is independent of both the location of the observation point and the depth of the slab.

A similar series of integrations, this time between fixed limits, can be used to determine the anomaly of a right rectangular prism.

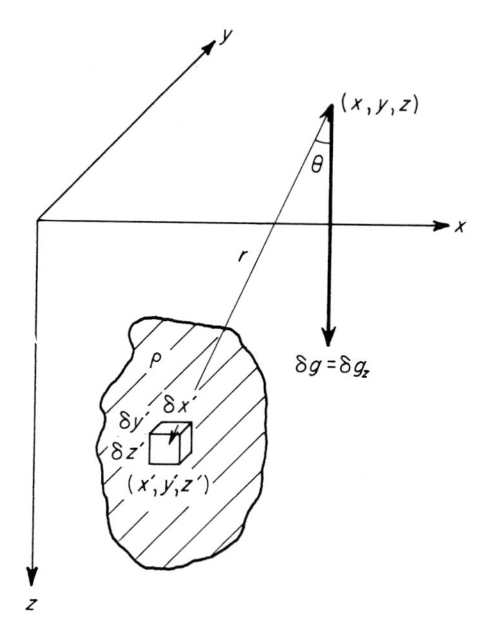

Fig. 6.7. The gravity anomaly of an element of a mass of irregular shape.

In general, the gravity anomaly of a body of *any* shape can be determined by summing the attractions of all the mass elements which make up the body. Consider a small prismatic element of such a body of density ρ, located at x', y', z', with sides of length $\delta x'$, $\delta y'$, $\delta z'$ (Fig. 6.7). The mass δm of this element, is given by

$$\delta m = \rho \delta x' \, \delta y' \, \delta z'$$

Consequently its attraction δg at a point outside the body (x, y, z), a distance r from the element, is derived from equation (6.6)

$$\delta g = G\rho \frac{(z' - z)}{r^3} \, \delta x' \, \delta y' \, \delta z'$$

The anomaly of the whole body Δg is then found by summing all such elements which make up the body

$$\Delta g = \Sigma \, \Sigma \, \Sigma \, G\rho \, \frac{(z' - z)}{r^3} \, \delta x' \, \delta y' \, \delta z' \qquad (6.9)$$

If $\delta x'$, $\delta y'$ and $\delta z'$ are allowed to approach zero, then

$$\Delta g = \int \int \int G\rho \, \frac{(z' - z)}{r^3} \, dx' \, dy' \, dz' \qquad (6.10)$$

where $\quad r = ((x' - x)^2 + (y' - y)^2 + (z' - z)^2)^{1/2}$

As shown before, the attraction of bodies of regular geometry can be determined by integrating equation (6.10) analytically. The anomalies of irregularly shaped bodies are calculated by numerical integration using equations of the form of equation (6.9).

6.7 Gravity surveying

The station spacing used in a gravity survey may vary from a few metres in the case of detailed mineral or geotechnical surveys to several kilometres in regional reconnaissance surveys. The station density should be greatest where the gravity field is changing most rapidly, as accurate measurement of gravity gradients is critical to subsequent interpretation. If absolute gravity values are required in order to interface the results with other gravity surveys, at least one easily accessible base station must be available where the absolute value of gravity is known. If the location of the nearest IGSN station is inconvenient, a gravimeter can be used to establish a local base by measuring the difference in gravity between the IGSN station and the local base. Because of instrumental drift this cannot be accomplished directly and a procedure known as *looping* is adopted. A series of alternate readings at recorded times is made at the two stations and drift

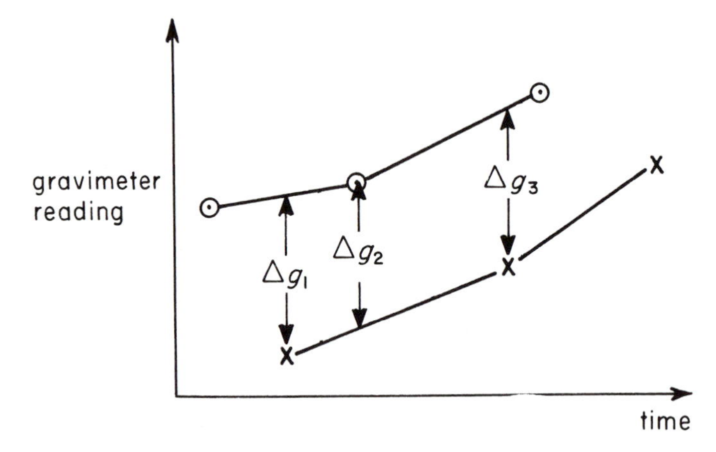

Fig. 6.8. The principle of looping. Crosses and circles represent alternate gravimeter readings taken at two base stations. The vertical separations between the drift curves for the two stations (Δg_{1-3}) provide an estimate of the gravity difference between them.

curves constructed for each (Fig. 6.8). The difference in ordinate measurements (Δg_{1-3}) for the two stations then gives a measure of the drift-corrected gravity difference.

During a gravity survey the gravimeter is read at a base station at a frequency dependent on the drift characteristics of the instrument. At each survey station, location, time, elevation/water depth and gravimeter reading are recorded.

In order to obtain a reduced gravity value accurate to ± 1 gu, the reduction procedure described in the following section indicates that the gravimeter must be read to a precision of ± 0.1 gu, the latitude of the station must be known to ± 10 m and the elevation of the station must be known to ± 10 mm. The latitude of the station must consequently be determined from maps at a scale of 1:10000 or smaller, or by the use of electronic position-fixing systems. Uncertainties in the elevations of gravity stations probably account for the greatest errors in reduced gravity values on land; at sea, water depths are easily determined with a precision depth recorder to an accuracy consistent with the gravity measurements. In well-surveyed land areas, the density of accurately determined elevations at bench marks is normally sufficiently high that gravity stations can be sited at bench marks or connected to them by levelling surveys. Reconnaissance gravity surveys of less well-mapped areas require some form of independent elevation determination. Many such areas have been surveyed using anaeroid altimeters. The accuracy of heights determined by such instruments is dependent upon the prevailing climatic conditions and is of the order of 1–5 m, leading to a relatively large uncertainty in the elevation corrections applied to the measured gravity values. The optimal equipment for surveys of this type is an inertial navigational system, which can provide elevations to ±0.5 m together with accurate locations. Such equipment is now available in a compact form suitable, for example, for mounting in helicopters, but its large cost inhibits its widespread usage.

6.8 Gravity reduction

Before the results of a gravity survey can be interpreted it is necessary to correct for all variations in the Earth's gravitational field which do not result from the differences of density in the underlying rocks. This process is known as *gravity reduction* or *reduction to the geoid*, as sea-level is usually the most convenient datum level.

6.8.1 Drift correction

Correction for instrumental drift is based on repeated readings at a base station at recorded times throughout the day. The meter reading is plotted against time (Fig. 6.9) and drift is assumed to be linear between consecutive base readings. The drift correction at time t' is d, which is subtracted from the observed value.

After drift correction the difference in gravity between an observation point and the base is found by multiplication of the difference in meter reading by the

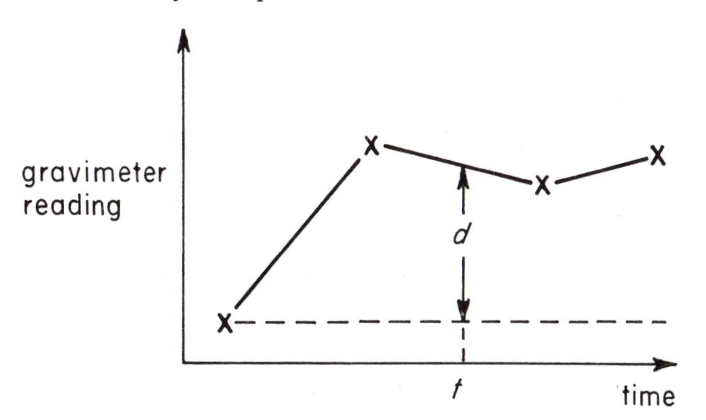

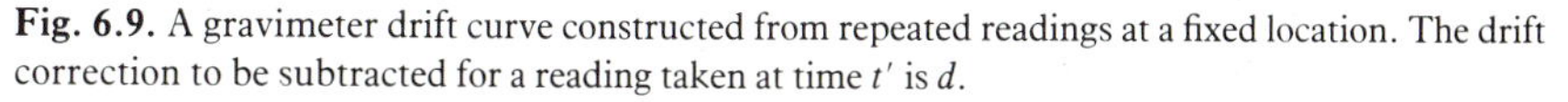
Fig. 6.9. A gravimeter drift curve constructed from repeated readings at a fixed location. The drift correction to be subtracted for a reading taken at time t' is d.

6

calibration factor of the gravimeter. Knowing this difference in gravity, the absolute gravity at the observation point g_{obs} can be computed from the known value of gravity at the base. Alternatively, readings can be related to an arbitrary datum but this practice is not desirable as the results from different surveys cannot then be tied together.

6.8.2 *Latitude correction*

Gravity varies with latitude because of the non-spherical shape of the Earth and because the angular velocity of a point on the Earth's surface decreases from a maximum at the equator to zero at the poles (Fig. 6.10(a)). The centripetal acceleration generated by this rotation has a negative radial component that consequently causes gravity to decrease from pole to equator. The true shape of the Earth is an oblate spheroid or polar flattened ellipsoid (Fig. 6.10(b)) whose difference in equatorial and polar radii is some 21 km. Consequently, points near the equator are farther from the centre of mass of the Earth than those near the poles, causing gravity to increase from the equator to the poles. The amplitude of this effect is reduced by the differing subsurface mass distributions resulting from the equatorial bulge, the mass underlying equatorial regions being greater than that underlying polar regions.

The net effect of these various factors is that gravity at the poles exceeds gravity at the equator by some 51 860 gu, with the north-south gravity gradient at latitude ϕ being 8.12 sin 2ϕ gu km^{-1}.

Clairaut's formula relates gravity to latitude on the reference spheroid according to an equation of the form

$$g_{\phi} = g_0(1+k_1 \sin^2\phi - k_2 \sin^2 2\phi) \tag{6.11}$$

where g_{ϕ} is the predicted value of gravity at latitude ϕ, g_0 is the value of gravity at the equator and k_1, k_2 are constants dependent on the shape and speed of rotation of the Earth. Equation (6.11) is, in fact, an approximation of an infinite series. The values of g_0, k_1 and k_2 in current use define the Gravity Formula 1967

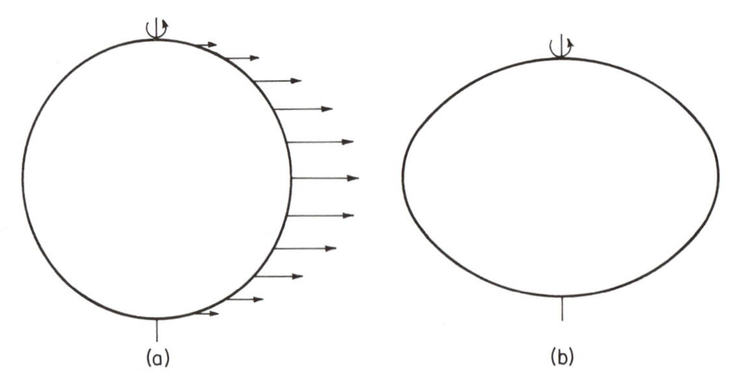

Fig. 6.10. (a) The variation in angular velocity with latitude around the Earth represented by vectors whose lengths are proportional to angular velocity. (b) An exaggerated representation of the shape of the Earth. The true shape of this oblate ellipsoid of revolution results in a difference in equatorial and polar radii of some 21 km.

(g_0 = 9780318 gu, k_1 = 0.0053024, k_2 = 0.0000059; IAG 1971). Prior to 1976 less accurate constants were employed in the International Gravity Formula (1930). Results reduced using the earlier formula must be modified before incorporation into survey data reduced using the Gravity Formula 1967 by using the relationship $g_\phi(1967)—g_\phi(1930) = (136 \sin^2\phi - 172)$ gu.

An alternative, more accurate, representation of the Gravity Formula 1967 (Mittermayer 1969), in which the constants are adjusted so as to minimize errors resulting from the truncation of the series is

$$g_\phi = 9780318.5\,(1+0.005278895 \sin^2\phi + 0.000023462 \sin^4\phi)\ \text{gu}.$$

This form, however, is less suitable if the survey results are to incorporate pre-1967 data made compatible with the Gravity Formula 1967 using the above relationship.

The value g_ϕ gives the predicted value of gravity at sea-level at any point on the Earth's surface and is subtracted from observed gravity to correct for latitude variation.

6.8.3 *Elevation corrections*

Correction for the differing elevations of gravity stations is made in three parts. The *free-air correction* (FAC) corrects for the decrease in gravity with height in free air resulting from increased distance from the centre of the Earth, according to Newton's Law. To reduce to datum an observation taken at height h (Fig. 6.11(a))

$$\text{FAC} = 3.086\,h\ \text{gu}\ (h \text{ in metres})$$

The FAC is positive for an observation point above datum to correct for the decrease in gravity with elevation.

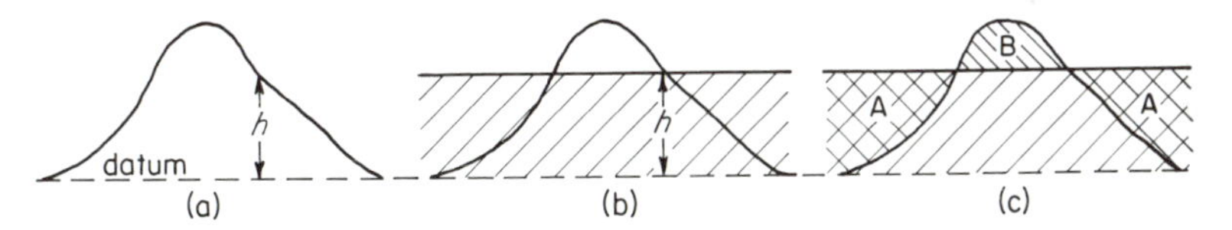

Fig. 6.11. (a) The free-air correction for an observation at a height h above datum. (b) The Bouguer correction. The shaded region corresponds to a slab of rock of thickness h extending to infinity in both horizontal directions. (c) The terrain correction.

The free-air correction accounts solely for variation in the distance of the observation point from the centre of the Earth; no account is taken of the gravitational effect of the rock present between the observation point and datum. The *Bouguer correction* (BC) removes this effect by approximating the rock layer beneath the observation point to an infinite horizontal slab with a thickness equal to the elevation of the observation above datum (Fig. 6.11(b)). If ρ is the density of the rock, from equation (6.8)

$$\text{BC} = 2\pi G\rho h = 0.4191\times 10^{-3}\,\rho h\ \text{gu}\ (h \text{ in metres}, \rho \text{ in kg m}^{-3})$$

On land the Bouguer correction must be subtracted, as the gravitational attraction of the rock between observation point and datum must be removed from the observed gravity value. The Bouguer correction of sea surface observations is positive to account for the lack of rock between surface and sea bed. The correction is equivalent to the replacement of the water layer by material of a specified rock density ρ_r. In this case

$$\mathrm{BC} = 2\pi G(\rho_r - \rho_w)z$$

where z is the water depth and ρ_w the density of water.

The free-air and Bouguer corrections are often applied together as the *combined elevation correction*.

The Bouguer correction makes the assumption that the topography around the gravity station is flat. This is rarely the case and a further correction, the *terrain correction* (TC), must be made to account for topographic relief in the vicinity of the gravity station. This correction is always positive as may be appreciated from consideration of Fig. 6.11(c). The regions designated A form part of the Bouguer correction slab although they do not consist of rock. Consequently the Bouguer correction has overcorrected for these areas and their effect must be restored by a positive terrain correction. Region B consists of rock material that has been excluded from the Bouguer correction. It exerts an upward attraction at the observation point causing gravity to decrease. Its attraction must thus be corrected by a positive terrain correction.

Classically, terrain corrections are applied using a circular graticule divided by radial and concentric lines into a large number of compartments known, after its inventor, as a Hammer chart (Fig. 6.12). The outermost zone extends to

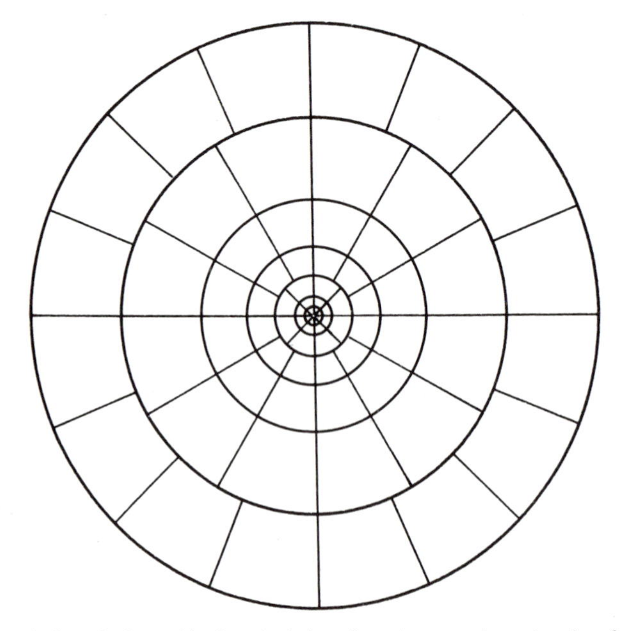

Fig. 6.12. A typical graticule used in the calculation of terrain corrections. A series of such graticules with zones varying in radius from 2 m to 21.9 km are used with topographic maps of varying scale.

almost 22 km, beyond which topographic effects are usually negligible. The graticule is laid on a topographic map with its centre on the gravity station and the average topographic elevation of each compartment is determined. The elevation of the gravity station is subtracted from these values, and the gravitational effect of each compartment is determined by reference to tables constructed using the formula for the gravitational effect of a sector of a vertical cylinder at its axis. The terrain correction is then computed by summing the gravitational contribution of all compartments. Table 6.1 shows the method of computation. Such operations are time consuming as the topography of over 130 compartments has to be averaged for each station, but terrain correction is the one operation in gravity reduction that cannot be fully automated. Labour can be reduced by averaging topography within a rectangular grid. Only a single digitization is required as the topographic effects may be calculated at any point within the grid by summing the effects of the right rectangular prisms defined by the grid squares and their elevation difference with the gravity station. This operation can effectively correct for the topography of areas distant from the gravity station and can be readily computerized. Correction for inner zones, however, must still be performed manually as any reasonable digitization scheme for a complete survey area and its environs must employ a sampling interval that is too large to provide an accurate representation of the terrain close to the station.

Table 6.1. Terrain corrections.

Zone	r_1	r_2	n	Zone	r_1	r_2	n
B	2.0	16.6	4	H	1529.4	2614.4	12
C	16.6	53.3	6	I	2614.4	4468.8	12
D	53.3	170.1	6	J	4468.8	6652.2	16
E	170.1	390.1	8	K	6652.2	9902.5	16
F	390.1	894.8	8	L	9902.5	14740.9	16
G	894.8	1529.4	12	M	14740.9	21943.3	16

$$T = 0.4191 \times 10^{-3} \frac{\rho}{n} \left(r_2 - r_1 + \sqrt{r_1^2 + z^2} - \sqrt{r_2^2 + z^2}\right)$$

T = Terrain correction of compartment (gu).
ρ = Bouguer correction density (kg m^{-3}).
n = Number of compartments in zone.
r_1 = Inner radius of zone (m).
r_2 = Outer radius of zone (m).
z = Modulus of elevation difference between observation point and mean elevation of compartment (m).

Terrain effects are low in areas of subdued topography, rarely exceeding 10 gu in flat-lying areas. In areas of rugged topography terrain effects are considerably greater, being at a maximum in steep-sided valleys, at the base or top of cliffs and at the summits of mountains.

Where terrain effects are considerably less than the desired accuracy of a survey, the terrain correction may be ignored. However, the usual necessity for this correction accounts for the bulk of time spent on gravity reduction and is thus a major contributor to the cost of a gravity survey.

6.8.4 *Tidal correction*

Gravity measured at a fixed location varies with time because of periodical variation in the gravitational effects of the Sun and Moon associated with their orbital motions, and correction must be made for this variation in a high precision survey. In spite of its much smaller mass, the gravitational attraction of the Moon is larger than that of the Sun because of its proximity. Also, these gravitational effects cause the shape of the solid Earth to vary in much the same way that the celestial attractions cause tides in the sea. These *solid Earth tides* are considerably smaller than oceanic tides and lag farther behind the lunar motion. They cause the elevation of an observation point to be altered by a few tens of millimetres and thus vary its distance from the centre of mass of the Earth. The periodic gravity variations caused by the combined effects of Sun and Moon are known as *tidal variations*. They have a maximum amplitude of some 3 gu and a minimum period of about 12 hours.

If a gravimeter with a relatively high drift rate is used, base ties are made at an interval much smaller than the minimum Earth tide period and the tidal variations are automatically removed during the drift correction. If a meter with a low drift rate is employed, base ties are normally made only at the start and end of the day so that the tidal variation has undergone a full cycle. In such a case, a separate tidal correction may need to be made. The tidal effects are predictable and published every year in the geophysical press.

6.8.5 *Eötvös correction*

The Eötvös correction (EC) is applied to gravity measurements taken on a moving vehicle such as a ship or an aircraft. Depending on the direction of travel, vehicular motion will generate a centripetal acceleration which either reinforces or opposes gravity. The correction required is

$$\text{EC} = 75.03V \sin\alpha \cos\phi + 0.04154V^2 \text{ gu}$$

where V is the speed of the vehicle in knots, α the heading and ϕ the latitude of the observation. In mid-latitudes the Eötvös correction is about +75 gu for each knot of E to W motion so that speed and heading must be accurately known.

6.8.6 *Free-air and Bouguer anomalies*

The *free-air anomaly* (FAA) and *Bouguer anomaly* (BA) may now be defined

$$\text{FAA} = g_{\text{obs}} - g_\phi + \text{FAC}\ (\pm\text{EC}) \tag{6.12}$$

$$\text{BA} = g_{\text{obs}} - g_\phi + \text{FAC} \pm \text{BC} + \text{TC}\ (\pm\text{EC}) \tag{6.13}$$

The Bouguer anomaly forms the basis for the interpretation of gravity data on land. In marine surveys Bouguer anomalies are conventionally computed for inshore and shallow water areas as the Bouguer correction removes the local gravitational effects associated with local changes in water depth. Moreover, the computation of the Bouguer anomaly in such areas allows direct comparison of gravity anomalies offshore and onshore and permits the combination of land and marine data into gravity contour maps. These may be used, for example, in tracing geological features across coastlines. The Bouguer anomaly is not appropriate for deeper water surveys, however, as in such areas the application of a Bouguer correction is an artificial device that leads to very large positive Bouguer anomaly values without significantly enhancing local gravity features of geological origin. Consequently the free-air anomaly is frequently used for interpretation in such areas. Moreover, the FAA provides a broad assessment of the degree of isostatic compensation of an area (e.g. Bott 1982).

6.9 Rock densities

Gravity anomalies result from the difference in density, or *density contrast*, between a body of rock and its surroundings. For a body of density ρ_1 embedded in material of density ρ_2 (Fig. 6.13), the density contrast $\Delta\rho$ is given by

$$\Delta\rho = \rho_1 - \rho_2$$

The sign of the density contrast determines the sign of the gravity anomaly.

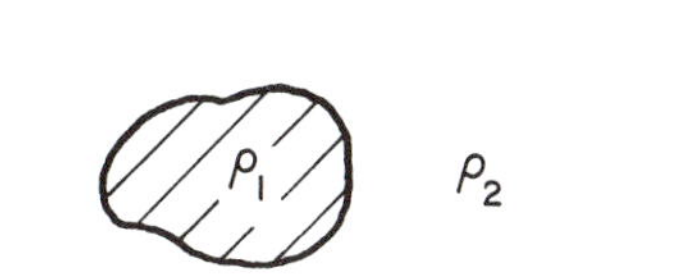

Fig. 6.13. Density contrast.

Rock densities are among the least variable of all geophysical parameters. Most common rock types have densities in the range between 1500 and 3500 kg m^{-3}. The density of a rock is dependent on both its composition and porosity.

Variation in porosity is the main cause of density variation in sedimentary rocks. Thus, in sedimentary rock sequences, density tends to increase with depth, due to compaction, and with age, due to progressive cementation.

Most igneous and metamorphic rocks have negligible porosity, and composition is the main cause of density variation. Density generally increases as acidity decreases; thus there is a progression of density increase from acid through basic to ultrabasic igneous rock types. Density ranges for common rock types are presented in Table 6.2.

6.10 Determination of rock densities

A knowledge of rock density is necessary both for application of the Bouguer correction and for the interpretation of gravity data.

Density is commonly determined by direct measurements on rock samples. A

Table 6.2. Approximate density ranges (kg m^{-3}) of some common rock types.

Alluvium (wet)	1960–2000
Clay	1630–2600
Shale	2060–2660
Sandstone, Cretaceous	2050–2350
Triassic	2250–2300
Carboniferous	2350–2550
Limestone	2600–2800
Chalk	1940–2230
Dolomite	2280–2900
Halite	2100–2400
Granite	2520–2750
Granodiorite	2670–2790
Anorthosite	2610–2750
Basalt	2700–3200
Gabbro	2850–3120
Gneiss	2610–2990
Quartzite	2600–2700
Amphibolite	2790–3140

NB. The lower end of the density range quoted in many texts is often unreasonably extended by measurements made on samples affected by physical or chemical weathering.

sample is weighed in air and in water. The difference in weights provides the volume of the sample and so the dry density can be obtained. If the rock is porous the saturation density may be calculated by following the above procedure after saturating the rock with water. The density value employed in interpretation then depends upon the location of the rock above or below the water table.

It should be stressed that the density of any particular rock type can be quite variable. Consequently it is usually necessary to measure several tens of samples of each particular rock type in order to obtain a reliable mean density and variance.

As well as these direct methods of density determination, there are several indirect (or *in situ*) methods. These usually provide a mean density of a particular rock unit which may be internally quite variable. *In situ* methods do, however, yield valuable information where sampling is hampered by lack of exposure or made impossible because the rocks concerned occur only at depth.

The measurement of gravity at different depths beneath the surface using a special borehole gravimeter or, more commonly, a standard gravimeter in a mineshaft, provides a measure of the mean density of the material between the observation levels. In Fig. 6.14 gravity has been measured at the surface and at a point underground at a depth h immediately below. If g_1 and g_2 are the values of gravity obtained at the two levels, then, applying free-air and Bouguer corrections one obtains

$$g_1 - g_2 = 3.086h - 4\pi G\rho h \qquad (6.16)$$

The Bouguer correction is double that employed on the surface as the slab of rock

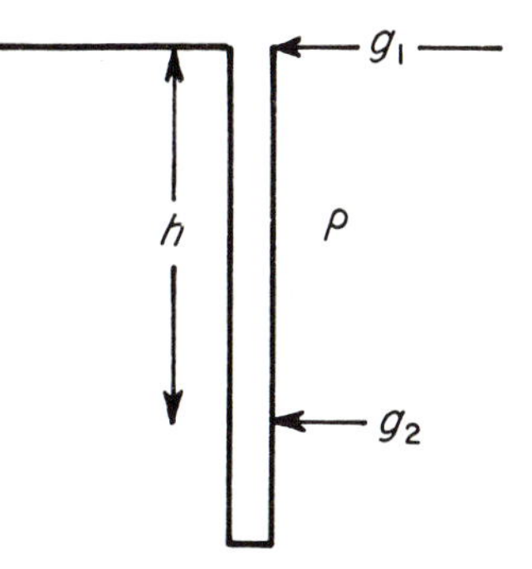

Fig. 6.14. Density determination by subsurface gravity measurements. The measured gravity difference $g_1 - g_2$ over a height difference h can be used to determine the mean density ρ of the rock separating the measurements.

between the observation levels exerts both a downward attraction at the surface location and an upward attraction at the underground location. The density ρ of the medium separating the two observations can then be found from the difference in gravity.

Density may be measured in boreholes using a density (gamma-gamma) logger. This consists of a small sonde about 0.5 m in length which carries a source of γ-rays, usually Co_{60}, at one end and a detector, generally a Geiger counter, at the other end. The sonde is spring-loaded so as to maintain intimate contact with the borehole wall. γ-rays interact with the wall-rock and are deflected back to the detector by Compton scattering. The amplitude of the returned radiation is proportional to the electron concentration in the rock, which is approximately proportional to the wall-rock density. A detailed picture is provided of the variation of density with depth through the borehole section, although only the material within about 300 mm of the sonde is sampled.

Nettleton's method of density determination involves taking gravity observations over a small isolated topographic prominence. Field data are reduced using a series of different densities for the Bouguer and terrain corrections (Fig. 6.15). The density value that yields a Bouguer anomaly with the least correlation (positive or negative) with the topography is taken to represent the density of the prominence. The method is useful in that no borehole or mineshaft is required, and a mean density of the material forming the prominence is provided. A disadvantage of the method is that isolated relief features may be formed of anomalous materials which are not representative of the area in general.

Density information is also provided by the P-wave velocity of a rock obtained in seismic surveys since there is a relationship between P-wave velocity and density. Nafe & Drake (1963) have constructed the graph reproduced in Fig. 6.16 by plotting observational data. Because of the dispersion of the data used to construct the curve, however, densities estimated from seismic velocities are probably no more accurate than about ± 100 kg m^{-3}. Other workers (e.g. Birch 1960, 1961, Christensen & Fountain 1975) have determined empirical, linear relationships between P-wave velocity and density. This is the only method available for the estimation of densities of deep strata that cannot be directly sampled.

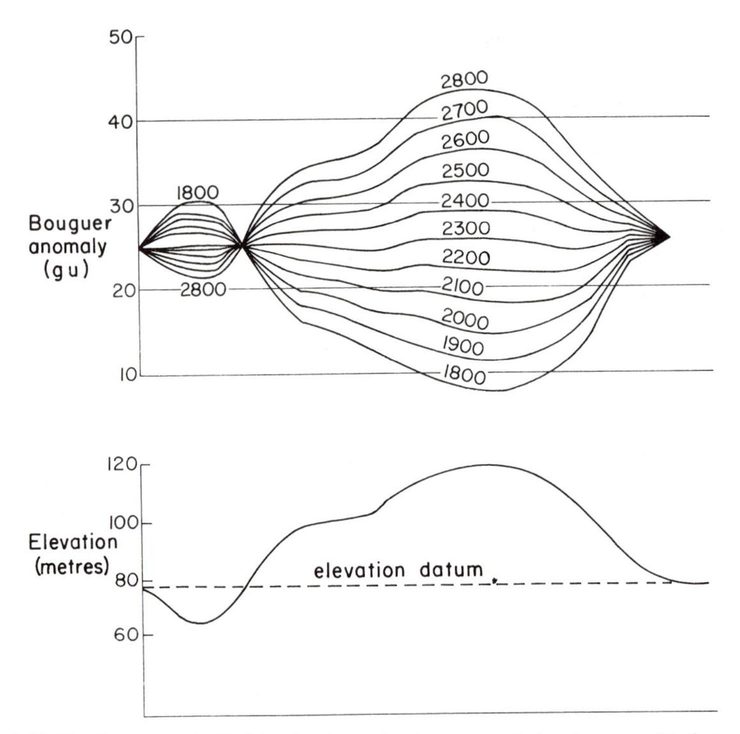

Fig. 6.15. Nettleton's method of density determination over an isolated topographic feature. Gravity reductions have been performed using densities ranging from 1800 to 2800 kg m^{-3} for both Bouguer and terrain corrections. The profile corresponding to a value of 2300 kg m^{-3} shows least correlation with topography so that this density is taken to represent the density of the feature. (Redrawn from Dobrin 1976.)

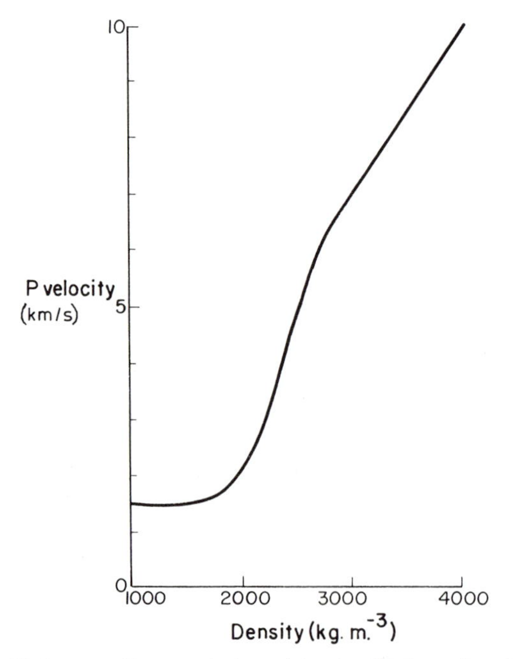

Fig 6.16. Relationship between P-wave velocity and density. (Redrawn from Nafe & Drake 1963.)

6.11 Interpretation of gravity anomalies

6.11.1 The inverse problem

The interpretation of potential field anomalies (gravity, magnetic and electrical) is inherently ambiguous. The ambiguity arises because any given anomaly could be caused by an infinite number of possible sources. For example, concentric spheres of constant mass but differing density and radius would all produce the same anomaly since their mass acts as though located at the centre of the sphere. This ambiguity represents the inverse problem of potential field interpretation, which states that although the anomaly of a given body may be calculated uniquely, there are an infinite number of bodies that could give rise to any specified anomaly. An important task in interpretation is to decrease this ambiguity by using all available external constraints on the nature and form of the anomalous body. Such constraints include geological information derived from surface outcrops, boreholes and mines, and from other, complementary, geophysical techniques.

6.11.2 Regional fields and residual anomalies

Bouguer anomaly fields are often characterized by a broad, gently varying, regional anomaly on which may be superimposed higher wavenumber local anomalies (Fig. 6.17). Usually in gravity surveying it is the local anomalies that are of prime interest and the first step in interpretation is the removal of the *regional field* to isolate the *residual anomalies*. This may be performed graphically by sketching in a linear or curvilinear field by eye. Such a method is biased by the interpreter, but this is not necessarily disadvantageous as his geological knowledge can be incorporated into the selection of the regional field. Several analytical methods of regional field analysis are available and include trend surface analysis (fitting a polynomial to the observed data) and low-pass filtering (Section 6.12). Such procedures must be used critically as fictitious residual

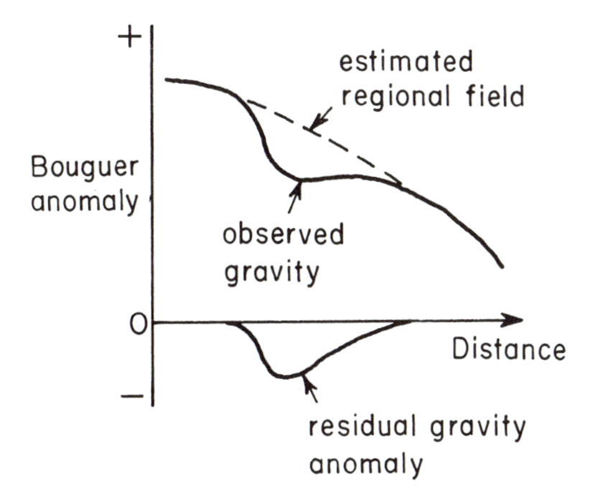

Fig. 6.17. The separation of regional and residual gravity anomalies from the observed Bouguer anomaly.

anomalies can sometimes arise when the regional field is subtracted from the observed data due to the mathematical procedures employed.

It is necessary before carrying out interpretation to differentiate between two-dimensional and three-dimensional anomalies. Two-dimensional anomalies are elongated in one horizontal direction so that the anomaly length in this direction is at least twice the anomaly width. Such anomalies may be interpreted in terms of structures which theoretically extend to infinity in the elongate direction by using profiles at right angles to the strike. Three-dimensional anomalies may have any shape and are considerably more difficult to interpret quantitatively.

Gravity interpretation proceeds via the methods of direct and indirect interpretation.

6.11.3 *Direct interpretation*

Direct interpretation provides, directly from the gravity anomalies, information on the anomalous body which is largely independent of the true shape of the body. Various methods are discussed below.

(a) *Limiting depth*

Limiting depth refers to the maximum depth at which the top of a body could lie and still produce an observed gravity anomaly. Gravity anomalies decay with the inverse square of the distance from their source so that anomalies caused by deep structures are of lower amplitude and greater extent than those caused by shallow sources. This wavenumber-amplitude relationship to depth may be quantified to compute the maximum depth (or limiting depth) at which the top of the anomalous body could be situated.

(i) Half width method. The half width of an anomaly ($x_{1/2}$) is the horizontal

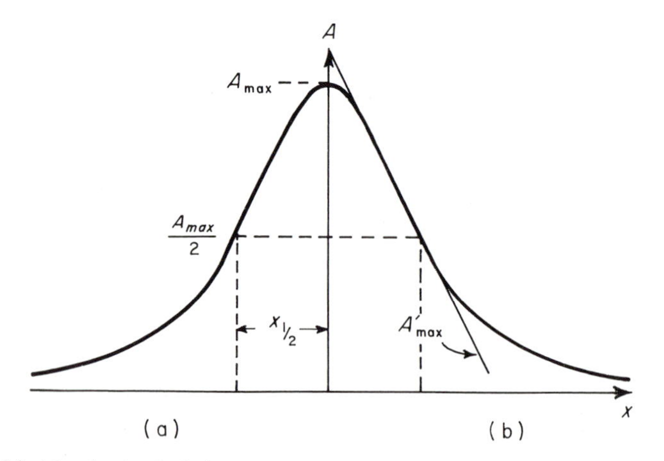

Fig. 6.18. Limiting depth calculations using (a) the half-width method and (b) the gradient-amplitude ratio.

distance from the anomaly maximum to the point at which the anomaly has reduced to half of its maximum value (Fig. 6.18(a)).

If the anomaly is three-dimensional the initial assumption is made that it is caused by a point mass. From manipulation of the point mass formula (equation (6.6)), the depth z to the point mass can be determined in terms of the half width

$$z = \frac{x_{1/2}}{(4^{1/3} - 1)^{1/2}}$$

The anomalous body will not be a point mass so z will represent an overestimate of the true depth, i.e. the top of a sphere equivalent to the point mass will lie above its centre of gravity. Thus for any three-dimensional body

$$z < \frac{x_{1/2}}{(4^{1/3} - 1)^{1/2}} \tag{6.15}$$

A similar procedure is adopted for a two-dimensional anomaly, but in this case the initial assumption is made that it is caused by a horizontal line mass (equation (6.7)). For any two-dimensional body

$$z < x_{1/2} \tag{6.16}$$

(ii) Gradient-amplitude ratio method. This method requires the computation of the maximum anomaly amplitude (A_{max}) and the maximum horizontal gravity gradient (A'_{max}) (Fig. 6.18(b)). Again the initial assumption is made that a three-dimensional anomaly is caused by a point mass and a two-dimensional anomaly by a line mass. By differentation of the relevant formulae, for any three-dimensional body

$$z < 0.86 \frac{A_{max}}{A'_{max}} \tag{6.17}$$

and for any two-dimensional body

$$z < 0.65 \frac{A_{max}}{A'_{max}} \tag{6.18}$$

(iii) Second derivative methods. There are a number of limiting depth methods based on the computation of the maximum second horizontal derivative, or maximum rate of change of gradient, of a gravity anomaly (Smith 1959). Such methods provide rather more accurate limiting depth estimates than either the half-width or gradient-amplitude ratio methods if the observed anomaly is free from noise.

(b) Excess mass

The excess mass of a body can be uniquely determined from its gravity anomaly without making any assumptions about its shape, depth or density. Excess mass refers to the difference in mass between the body and the mass of country rock that would otherwise fill the space occupied by the body. The basis of this calculation is a formula derived from Gauss' theorem, and it involves a surface integration of the residual anomalies over the area in which they occur. The survey area is divided into n grid squares of area Δa and the mean residual anomaly Δg found for each square. The excess mass M_e is then given by

$$M_e = \frac{1}{2\pi G} \sum_{i=1}^{n} \Delta g_i \Delta a_i \tag{6.19}$$

Before using this procedure it is important that the regional field is removed so that the anomaly tails to zero. The method only works well for isolated anomalies whose extremities are well defined. Gravity anomalies decay slowly with distance from source and so these tails can cover a wide area and be important contributors to the summation.

To compute the actual mass M of the body, the densities of both anomalous body (ρ_1) and country rock (ρ_2) must be known

$$M = \frac{\rho_1 M_e}{(\rho_1 - \rho_2)} \tag{6.20}$$

The method is of use in estimating the tonnage of ore bodies.

(c) *Inflection point*

The locations of inflection points on gravity profiles, i.e. positions where the horizontal gravity gradient changes most rapidly, can provide useful information on the nature of the edge of an anomalous body. Over structures with outward dipping contacts, such as granite bodies (Fig. 6.19(a)), the inflection points (arrowed) lie near the base of the anomaly. Over structures with inward dipping contacts such as sedimentary basins (Fig. 6.19(b)), the inflection points lie near the uppermost edge of the anomaly.

(d) *Approximate thickness*

If the density contrast $\Delta\rho$ of an anomalous body is known, its thickness t may be crudely estimated from its maximum gravity anomaly Δg by making use of the slab formula (equation (6.8))

$$t = \frac{\Delta g}{2\pi G \Delta\rho} \tag{6.21}$$

This thickness will always be an underestimate for a body of restricted horizontal extent. The method is commonly used in estimating the throw of a fault from the difference in the gravity fields of the upthrown and downthrown sides.

6.11.4 *Indirect interpretation*

In indirect interpretation, the causative body of a gravity anomaly is simulated by a model whose theoretical anomaly can be computed, and the shape of the model

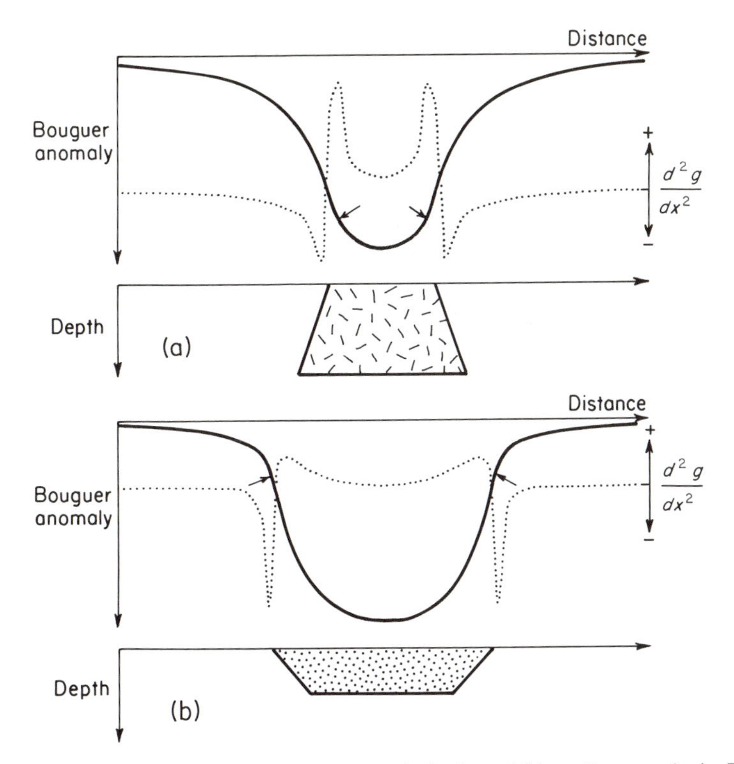

Fig. 6.19. Bouguer anomaly profiles across (a) a granite body, and (b) a sedimentary basin. The inflection points are marked with an arrow. The broken lines represent the second horizontal derivative (rate of change of gradient) of the gravity anomaly, which is at a maximum at the inflection points.

is altered until the computed anomaly closely matches the observed anomaly. Because of the inverse problem this model will not be a unique interpretation, but ambiguity can be decreased by using other constraints on the nature and form of the anomalous body.

A simple approach to indirect interpretation is the comparison of the observed anomaly with the anomaly computed for certain standard geometrical shapes whose size, position, form and density contrast are altered to improve the fit. Two-dimensional anomalies may be compared with anomalies computed for horizontal cylinders or half cylinders, and three-dimensional anomalies compared with those of spheres, vertical cylinders or right rectangular prisms. Combinations of such shapes may also be used to simulate an observed anomaly.

Fig. 6.20(a) shows a large, circular gravity anomaly situated near Darnley Bay, N.W.T., Canada. The anomaly is radially symmetrical and a profile across the anomaly (Fig. 6.20(b)) can be simulated by a model constructed from a suite of coaxial cylinders whose diameters decrease with depth so that the anomalous body has the overall form of an inverted cone. This study illustrates well the non-uniqueness of gravity interpretation. The nature of the causative body is unknown and so no information is available on its density. An alternative interpretation, again in the form of an inverted cone, but with an increased

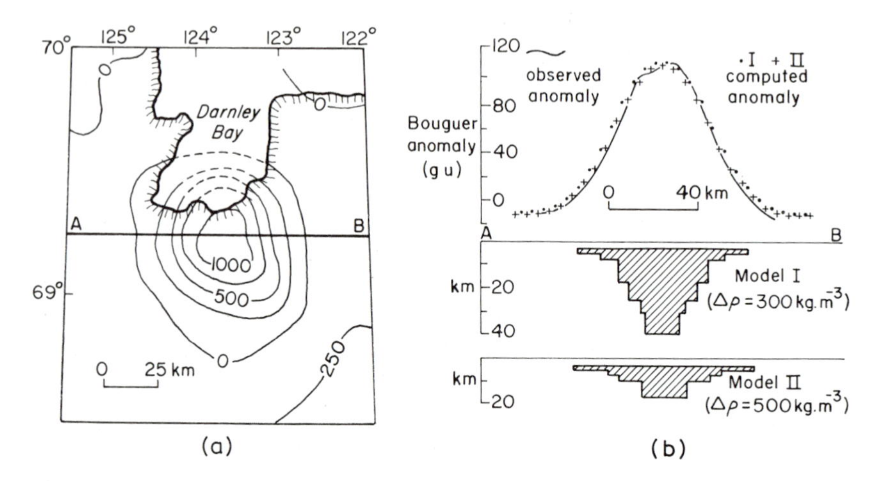

Fig. 6.20. (a) The circular gravity anomaly at Darnley Bay, N.W.T., Canada. (b) Two possible interpretations of the anomaly in terms of a model constructed from a suite of coaxial vertical cylinders. (Redrawn from Stacey 1971.)

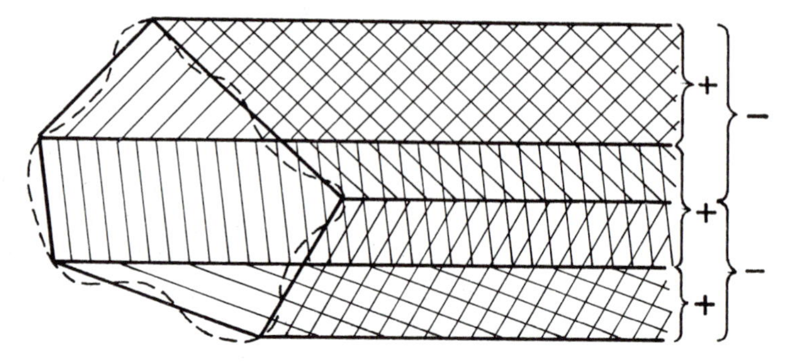

Fig. 6.21. The computation of gravity anomalies of two-dimensional bodies of irregular cross-section. The body is approximated by a polygon and the effects of semi-infinite slabs with sloping edges defined by the sides of the polygon are progressively added and subtracted until the anomaly of the polygon is obtained.

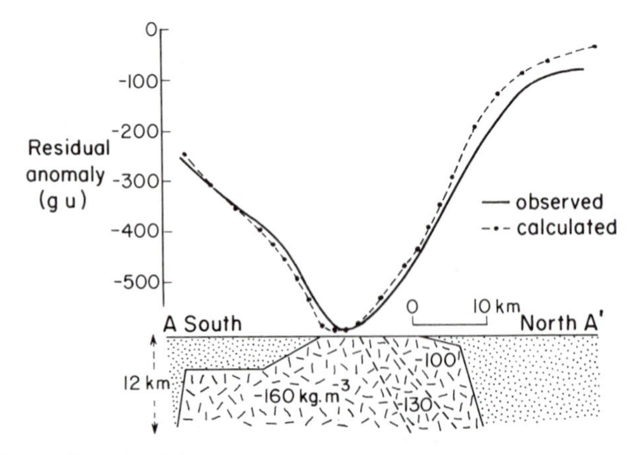

Fig. 6.22. A two-dimensional interpretation of the gravity anomaly of the Bodmin Moor granite, southwest England. See Fig. 6.22 for location. (Redrawn from Bott & Scott 1964.)

density contrast, is presented in Fig. 6.20(b). Both models provide adequate simulations of the observed anomaly, and cannot be distinguished using the information available.

The computation of anomalies over models of irregular form is accomplished by dividing the models into a series of regularly-shaped compartments and calculating the combined effect of these compartments at each observation point. At one time this operation was performed by the use of graticules but nowadays the calculations are invariably performed by computers.

The calculation of anomalies over two-dimensional bodies, which are assumed to extend to infinity in the strike direction, is considerably simpler than the calculation for three-dimensional bodies. The outline of a two-dimensional body is normally approximated by a polygon (Fig. 6.21). It is relatively simple to compute the attraction of a two-dimensional semi-infinite slab with a sloping edge at any observation point. The attraction of the polygon can then be found by proceeding around the body summing the attractions of semi-infinite prisms bounded by edges where the depth increases and subtracting the effects where the depth decreases (Talwani *et al.* 1959).

Fig. 6.22 illustrates a two-dimensional interpretation, in terms of a model of irregular geometry, of the Bodmin Moor granite of southwest England. The shape of the uppermost part of the model is controlled by the surface outcrop of granite, while the density contrasts employed are based on density measurements on rock samples. The interpretation shows unambiguously that the contacts of the granite slope outwards. Ambiguity is evident, however, in the interpretation of the gravity gradient over the northern flank of the granite. The model presented in Fig. 6.22 interprets the cause of this gradient as a northerly increase in the density of the granite; a possible alternative, however, would be a northerly thinning of a granite body of constant density contrast.

Two-dimensional methods can sometimes be extended to three-dimensional bodies by applying end-correction factors to account for the restricted extent of the causative body in the strike direction (Cady 1980). The end-correction factors are, however, only approximations and full three-dimensional modelling is preferable.

The gravity anomaly of a three-dimensional body may be calculated by dividing the body into a series of horizontal slices and approximating each slice by a polygon (Talwani & Ewing 1960). Alternatively the body may be constructed out of a suite of right rectangular prisms.

However a model calculation is performed, indirect interpretation involves four steps:

(1) Construction of a reasonable model.

(2) Computation of its gravity anomaly.

(3) Comparison of computed with observed anomaly.

(4) Alteration of model to improve correspondence of observed and calculated anomalies and return to step (2).

The process is thus iterative and the goodness of fit between observed and calculated anomalies is gradually improved. Step (4) can be performed manually for bodies of relatively simple geometry. Bodies of complex geometry in two or three dimensions are not so simply dealt with and in such cases it is advantageous to employ techniques which perform the iteration automatically.

The most flexible of such methods is *non-linear optimization* (Al-Chalabi 1972). All variables (body points, density contrasts, regional field) may be allowed to vary within defined limits. The method then attempts to minimize some function F which defines the goodness of fit, e.g.

$$F = \sum_{i=1}^{n} (\Delta g_{\text{obs}_i} - \Delta g_{\text{calc}_i})^2$$

where Δg_{obs} and Δg_{calc} are a series of n observed and calculated values.

The minimization proceeds by altering the values of the variables within their stated limits to produce a successively smaller value for F for each iteration. The technique is elegant and successful but expensive in computer time.

Other such automatic techniques involve the simulation of the observed profile by a thin layer of variable density. This *equivalent layer* is then progressively expanded so that the whole body is of a uniform, specified density contrast. The body then has the form of a series of vertical prisms in either two or three dimensions which extend either above, below or symmetrically around the original equivalent layer. Such methods are less flexible than the non-linear optimization technique in that usually only a single density contrast may be specified and the model produced must either have a flat base or top or be symmetrical about a central horizontal plane.

6.12 Elementary potential theory and potential field manipulation

Gravitational and magnetic fields are both potential fields. In general the potential at any point is defined as the work necessary to move a unit mass or pole from an infinite distance to that point through the ambient field. Potential fields obey Laplace's equation which states that the sum of the rates of change of the field gradient in three orthogonal directions is zero. In a normal Cartesian coordinate system with horizontal axes x, y and a vertical axis z, Laplace's equation is stated

$$\frac{\partial^2 A}{\partial x^2} + \frac{\partial^2 A}{\partial y^2} + \frac{\partial^2 A}{\partial z^2} = 0 \qquad (6.22)$$

where A refers to a gravitational or magnetic field and is a function of (x, y, z).

In the case of a two-dimensional field there is no variation along one of the horizontal directions so that A is a function of x and z only and equation (6.22) simplifies to

$$\frac{\partial^2 A}{\partial x^2} + \frac{\partial^2 A}{\partial z^2} = 0 \qquad (6.23)$$

Solution of this partial differential equation is easily performed by separation of variables

$$A_k(x, z) = (a \cos kx + b \sin kx)e^{kz} \qquad (6.24)$$

where a and b are constants, the positive variable k is the spatial frequency or wavenumber, A_k is the potential field amplitude corresponding to that wavenumber and z is the level of observation. Equation (6.24) shows that a potential field can be represented in terms of sine and cosine waves whose amplitude is controlled exponentially by the level of observation.

Consider the simplest possible case where the two-dimensional anomaly measured at the surface $A(x, 0)$ is a sine wave

$$A(x, 0) = A_0 \sin kx \qquad (6.25)$$

where A_0 is a constant and k the wavenumber of the sine wave. Equation (6.24) enables the general form of the equation to be stated for any value of z

$$A(x, z) = (A_0 \sin kx)e^{kz} \qquad (6.26)$$

The field at a height h above the surface can then be determined by substitution in equation (6.26)

$$A(x, -h) = (A_0 \sin kx)e^{-kh} \qquad (6.27)$$

and the field at depth d below the surface

$$A(x, d) = (A_0 \sin kx)e^{kd} \qquad (6.28)$$

The sign of h and d is important as the z-axis is normally defined as positive downwards.

Equation (6.26) is an over-simplification in that a potential field is never a function of a single sine wave. Invariably such a field is composed of a range of wavenumbers. However, the technique is still valid as long as the field can be expressed in terms of all its component wavenumbers, a task easily performed by use of the Fourier transform (Section 2.3). If, then, instead of the terms $(a \cos kx + b \sin kx)$ in equation (6.24) or $(A_0 \sin kx)$ in equation (6.26), the full Fourier spectrum, derived by Fourier transformation of the field into the wavenumber domain, is substituted, the results of equations (6.27) and (6.28) remain valid.

These latter equations show that the field measured at the surface can be used to predict the field at any level above or below the plane of observation. This is the basis of the upward and downward field continuation methods in which the potential field above or below the original plane of measurement is calculated in order to accentuate the effects of deep or shallow structures respectively.

Upward continuation methods are employed in gravity interpretation to determine the form of regional gravity variation over a survey area, since the regional field is assumed to originate from relatively deep-seated structures. Fig. 6.23(a) is a Bouguer anomaly map of the Saguenay area in Quebec, Canada, and Fig. 6.23(b) represents the field continued upward to an elevation of 16 km. Comparison of the two figures clearly illustrates how the high wavenumber components of the observed field have been effectively removed by the continuation process. The upward continued field must result from relatively deep structures and consequently represents a valid regional field for the area. Upward continuation is also useful in the interpretation of magnetic anomaly fields (see

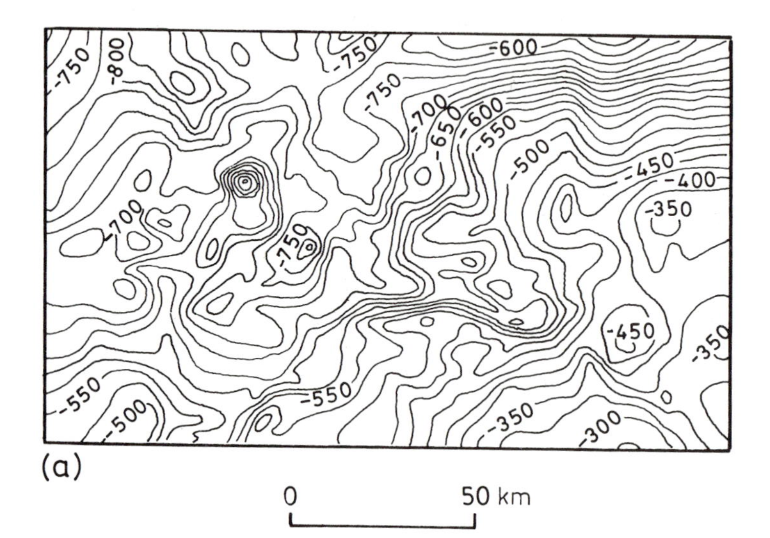

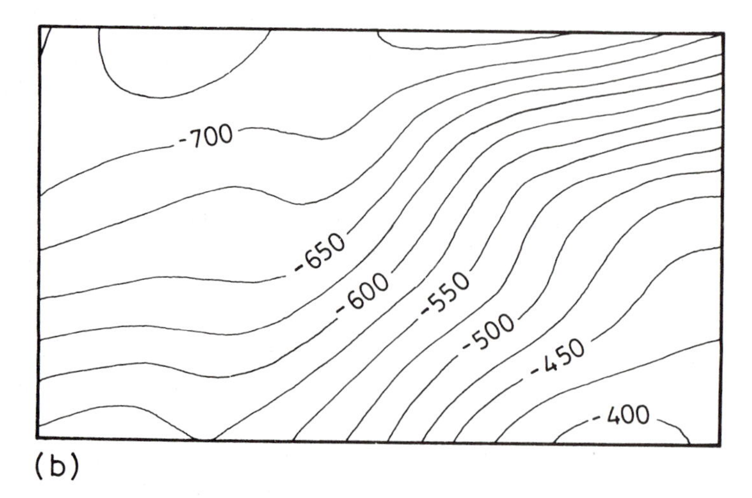

Fig. 6.23. (a) Observed Bouguer anomalies (gu) over the Saguenay area, Quebec, Canada. (b) The gravity field continued upward to an elevation of 16 km. (Redrawn from Duncan & Garland 1977.)

Chapter 7) over areas containing many near-surface magnetic sources such as dykes and other intrusions. Upward continuation attenuates the high wavenumber anomalies associated with such features and enhances, relatively, the anomalies of the deeper-seated sources.

Downward continuation of potential fields is of more restricted application. The téchnique may be used in the resolution of the separate anomalies caused by adjacent structures whose effects overlap at the level of observation. On downward continuation, high wavenumber components are relatively enhanced and the anomalies show extreme fluctuations if the field is continued to a depth greater than that of its causative structure. The level at which these fluctuations commence provides an estimate of the limiting depth of the anomalous body. The effectiveness of this method is diminished if the potential field is contaminated with noise, as the noise is accentuated on downward continuation.

The selective enhancement of the low or high wavenumber components of potential fields may be achieved in a different but analogous manner by the application of *wavenumber filters*. Gravitational and magnetic fields may be processed and analysed in a similar fashion to seismic data, replacing frequency by wavenumber. Such processing is more complex than the equivalent seismic filtering as potential field data are generally arranged in two horizontal dimensions, i.e. contour maps, rather than a single dimension. However, it is possible to devise two-dimensional filters for the selective removal of high or low wavenumber components from the observed anomalies. The consequence of the application of such techniques is similar to upward or downward continuation in that shallow structures are mainly responsible for the high wavenumber components of anomalies and deep structures for the low wavenumbers. However, it is not possible fully to isolate local or regional anomalies by wavenumber filtering because the wavenumber spectra of deep and shallow sources overlap.

Other manipulations of potential fields may be accomplished by the use of more complex filter operators (e.g. Gunn 1975). Vertical or horizontal derivatives of any order may be computed from the observed field. Such computations are not widely employed, but second horizontal derivative maps are occasionally used for interpretation as they accentuate anomalies associated with shallow bodies.

6.13 Applications of gravity surveying

Gravity studies are extensively used in the investigation of large and medium scale geological structures. Very early marine surveys, performed from submarines, indicated the existence of large positive and negative gravity anomalies associated with island arcs and oceanic trenches respectively; subsequent work has demonstrated their lateral continuity and has shown that most of the major features of the Earth's surface can be delineated by gravity surveys. On a smaller scale gravity anomalies reveal the form of igneous intrusions, such as granitic plutons and anorthositic massifs, and have consequently provided constraints on their mechanisms of emplacement, composition and origin. Similarly the gravity method has been extensively used in the location of sedimentary basins, and their interpreted structure has provided important constraints on theories of their mode of origin.

Gravity studies can be used to reveal the state of isostatic compensation of large structures, i.e. to determine if their near-surface mass distributions are balanced by complementary mass distributions at depth. Although most of the major relief features of the Earth's surface have been demonstrated to be in broad isostatic equilibrium, gravity surveys cannot uniquely define the form of the compensation.

The gravity method was once extensively used by the petroleum industry for the location of possible hydrocarbon traps. In the 1930s the method proved considerably more successful than seismic techniques and in the Gulf Coast region of the USA salt domes were once located by gravity surveying at a rate of one per week. Although the method has the obvious application of being capable of delineating sedimentary basins, the vastly improved efficiency and technology of seismic surveying have led to the demise of gravity surveying as a primary exploration tool. However, it is probable that specialized borehole gravimeters and surface gravity surveys will be increasingly used in the analysis of the interwell regions of hydrocarbon reservoirs. Gravity surveying is not used extensively in mineral exploration, but has an important application in the determination of the tonnage of a previously located ore deposit.

Modern technology has produced gravimeters capable of measuring gravity changes as small as a microgal (= 10^{-8} m s^{-1}). Microgravimetric techniques find occasional geotechnical applications in the location of subsurface voids (see Section 10.4) and can also be used, for example, to study the temporal movement of ground water through a region. Another important recent development in gravity surveying is the design of a portable instrument capable of measuring absolute gravity with high precision. Although the cost of such instruments is enormous it is possible that they will be used in the future to investigate large scale mass movements in the Earth's interior.

Gravitational studies, both of the type described in this chapter and satellite observations, are important in geodesy, the study of the shape of the Earth. Gravity surveying also has military significance, since the trajectory of a missile is affected by gravity variation along its flight path.

Further reading

Baranov, W. (1975) *Potential Fields and Their Transformations in Applied Geophysics*. Gebrüder Borntraeger, Berlin.

Bott, M.H.P. (1973) Inverse methods in the interpretation of magnetic and gravity anomalies. *In:* Alder, B., Fernbach, S. & Bolt, B.A. (eds.), *Methods in Computational Physics*, **13**, 133–62.

Dehlinger, P. (1978) *Marine Gravity*. Elsevier, Amsterdam.

Garland, G.D. (1965) *The Earth's Shape and Gravity*. Pergamon, Oxford.

Grant, F.S. & West, G.F. (1965) *Interpretation Theory in Applied Geophysics*. McGraw-Hill, New York.

LaCoste, L.J.B. (1967) Measurement of gravity at sea and in the air. *Rev. Geophys.*, **5**, 477–526.

LaCoste, L.J.B., Ford, J., Bowles, R. & Archer, K. (1982) Gravity measurements in an airplane using state-of-the-art navigation and altimetry. *Geophysics*, **47**, 832–7.

Nettleton, L.L. (1976) *Gravity and Magnetics in Oil Exploration*. McGraw-Hill, New York.

Ramsey, A.S. (1964) *An Introduction to the Theory of Newtonian Attraction*. Cambridge University Press, Cambridge.

7

Magnetic Surveying

7.1 Introduction

The aim of a magnetic survey is to investigate subsurface geology on the basis of anomalies in the Earth's magnetic field resulting from the magnetic properties of the underlying rocks. Although most rock-forming minerals are effectively non-magnetic, certain rock types contain sufficient magnetic minerals to produce significant magnetic anomalies. Similarly, man-made ferrous objects also generate magnetic anomalies. Magnetic surveying thus has a broad range of applications, from small-scale engineering or archaeological surveys to detect buried metallic objects, to large-scale surveys carried out to investigate regional geological structure.

Magnetic surveys can be performed on land, at sea and in the air. Consequently the technique is widely employed, and the speed of operation of airborne surveys makes the method very attractive in the search for types of ore deposit that contain magnetic minerals.

7.2 Basic concepts

Within the vicinity of a bar magnet a magnetic flux is developed which flows from one end of the magnet to the other (Fig. 7.1). This flux can be mapped from the directions assumed by a small compass needle suspended within it. The points within the magnet where the flux converges are known as the *poles* of the magnet. A freely-suspended bar magnet similarly aligns in the flux of the Earth's magnetic field. The pole of the magnet which tends to point in the direction of the Earth's north pole is called the north-seeking or positive pole, and this is balanced by a south-seeking or negative pole of identical strength at the opposite end of the magnet.

The force F between two magnetic poles of strengths m_1 and m_2 separated by a distance r is given by

$$F = \frac{\mu_0 m_1 m_2}{4\pi\mu_R r^2} \qquad (7.1)$$

where μ_0 and μ_R are constants corresponding to the *magnetic permeability of vacuum* and the *relative magnetic permeability* of the medium separating the poles (see later). The force is attractive if the poles are of different sign and repulsive if they are of like sign.

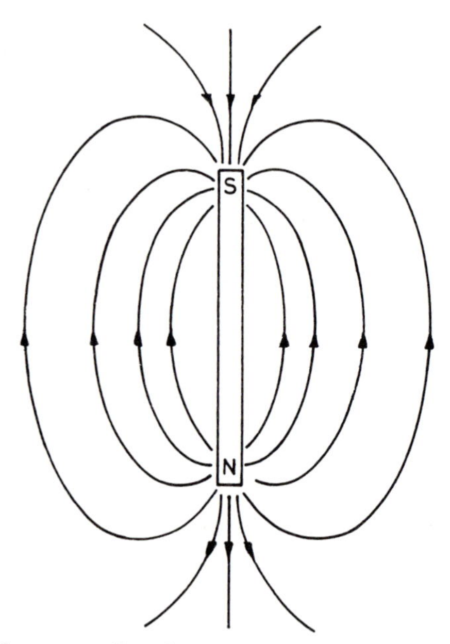

Fig. 7.1. The magnetic flux surrounding a bar magnet.

The *magnetic field B* due to a pole of strength *m* at a distance *r* from the pole is defined as the force exerted on a unit positive pole at that point

$$B = \frac{\mu_0 m}{4\pi\mu_R r^2} \tag{7.2}$$

Magnetic fields can be defined in terms of *magnetic potentials* in a similar manner to gravitational fields. For a single pole of strength *m*, the magnetic potential *V* at a distance *r* from the pole is given by

$$V = \frac{\mu_0 m}{4\pi\mu_R r} \tag{7.3}$$

The magnetic field component in any direction is then given by the partial derivative of the potential in that direction.

In the SI system of units, magnetic parameters are defined in terms of the flow of electrical current (see, for example, Reilly 1972). If a current is passed through a coil consisting of several turns of wire, a *magnetic flux* flows through and around the coil annulus which arises from a *magnetizing force H*. The magnitude of *H* is proportional to the number of turns in the coil and the strength of the current, and inversely proportional to the length of the wire, so that *H* is expressed in A m^{-1}. The density of the magnetic flux, measured over an area perpendicular to the direction of flow, is known as the *magnetic induction* or *magnetic field B* of the coil. *B* is proportional to *H* and the constant of proportionality μ is known as the *magnetic permeability*. Lenz's law of induction relates the rate of change of magnetic flux in a circuit to the voltage developed within it so that *B* is expressed in volt s m^{-2} (Weber (Wb) m^{-2}). The unit of the Wb m^{-2} is designated the *tesla* (T). Permeability is consequently expressed in Wb A^{-1} m^{-1} or Henry (H) m^{-1}. The cgs unit of magnetic field strength is the *Gauss* (G), numerically equivalent to 10^{-4}T.

The tesla is too large a unit in which to express the small magnetic anomalies caused by rocks, and a subunit, the *nanotesla* (nT), is employed (1nT = 10^{-9}T). The cgs system employs the numerically equivalent *gamma* (γ), equal to 10^{-5}G.

Common magnets exhibit a pair of poles and are therefore referred to as dipoles. The *magnetic moment* M of a dipole with poles of strength m a distance l apart is given by

$$M = ml \qquad (7.4)$$

The magnetic moment of a current-carrying coil is proportional to the number of turns in the coil, its cross-sectional area and the magnitude of the current so that magnetic moment is expressed in A m^2.

When a material is placed in a magnetic field it may acquire a magnetization in the direction of the field which is lost when the material is removed from the field. This phenomenon is referred to as *induced magnetization* or *magnetic polarization*, and results from the alignment of elementary dipoles (see below) within the material in the direction of the field. As a result of this alignment the material has magnetic poles distributed over its surface which correspond to the ends of the dipoles (Fig. 7.2). The intensity of induced magnetization J_i of a material is

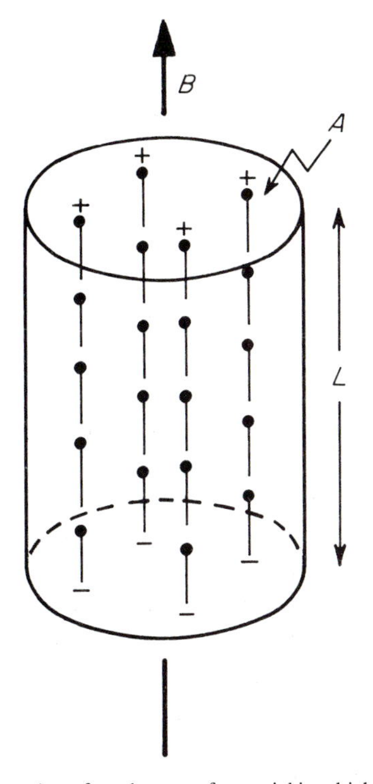

Fig. 7.2. Schematic representation of an element of material in which elementary dipoles align in the direction of an external field B to produce an overall induced magnetization.

defined as the dipole moment per unit volume of material:

$$\mathcal{J}_i = \frac{M}{LA} \tag{7.5}$$

where M is the magnetic moment of a sample of length L and cross-sectional area A. $\mathcal{J}_i$ is consequently expressed in A m^{-1}. In the cgs system intensity of magnetization is expressed in emu cm^{-3} (emu = electromagnetic unit), where 1 emu cm^{-3} = 1000 A m^{-1}.

The induced intensity of magnetization is proportional to the strength of the magnetizing force H of the inducing field:

$$\mathcal{J}_i = kH \tag{7.6}$$

where k is the *magnetic susceptibility* of the material. Since $\mathcal{J}_i$ and H are both measured in A m^{-1}, susceptibility is dimensionless in the SI system. In the cgs system susceptibility is similarly dimensionless, but a consequence of rationalizing the SI system is that SI susceptibility values are a factor 4π greater than corresponding cgs values.

In a vacuum the magnetic field strength B and magnetizing force H are related by $B = \mu_0 H$ where μ_0 is the permeability of vacuum ($4\pi \times 10^{-7}$ H m^{-1}). Air and water have very similar permeabilities to μ_0 and so this relationship can be taken to represent the Earth's magnetic field when it is undisturbed by magnetic materials. When a magnetic material is placed in this field, the resulting magnetization gives rise to an additional magnetic field in the region occupied by the material whose strength is given by $\mu_0 \mathcal{J}_i$. Within the body the total magnetic field, or magnetic induction, B is given by

$$B = \mu_0 H + \mu_0 \mathcal{J}_i$$

Substituting equation (7.6)

$$B = \mu_0 H + \mu_0 kH = (1+k)\mu_0 H = \mu_R \mu_0 H$$

where μ_R is a dimensionless constant known as the *relative magnetic permeability*. The magnetic permeability μ is thus equal to the product of the relative permeability and the permeability of vacuum, and has the same dimensions as μ_0. μ_R for air and water is thus close to unity.

All substances are magnetic at an atomic scale. Each atom acts as a dipole due to both the spin of its electrons and the orbital path of the electrons around the nucleus. Quantum theory allows two electrons to exist in the same state (or electron shell) provided that their spins are in opposite directions. Two such electrons are called paired electrons and their spin magnetic moments cancel. In *diamagnetic* materials all electron shells are full and no unpaired electrons exist. When placed in a magnetic field the orbital paths of the electrons rotate so as to produce a magnetic field in opposition to the applied field. Consequently the susceptibility of diamagnetic substances is weak and negative. In *paramagnetic* substances the electron shells are incomplete so that a magnetic field results from the spin of their unpaired electrons. When placed in an external magnetic field the dipoles corresponding to the unpaired electron spins rotate to produce a field in the same sense as the applied field so that the susceptibility is positive. This is still, however, a relatively weak effect.

In small grains of certain paramagnetic substances whose atoms contain several unpaired electrons, the dipoles associated with the spins of the unpaired electrons are magnetically coupled between adjacent atoms. The grain is then said to constitute a single *magnetic domain*. Depending on the degree of overlap of the electron orbits, this coupling may be either parallel or antiparallel. In *ferromagnetic* materials the dipoles are parallel (Fig. 7.3), giving rise to a very strong spontaneous magnetization which can exist even in the absence of an external magnetic field, and a very high susceptibility. Ferromagnetic substances include iron, cobalt, and nickel, and rarely occur naturally in the Earth's crust. In *antiferromagnetic* materials such as haematite, the dipole coupling is antiparallel with equal numbers of dipoles in each direction. The magnetic fields of the dipoles are self-cancelling so that there is no external magnetic effect. However, defects in the crystal lattice structure of an antiferromagnetic material may give rise to a small net magnetization, called *parasitic antiferromagnetism*. In *ferrimagnetic* materials such as magnetite, the dipole coupling is similarly antiparallel, but the numbers of dipoles in each direction are unequal. Consequently ferrimagnetic materials can exhibit a strong spontaneous magnetization and a high susceptibility. Virtually all the minerals responsible for the magnetic properties of common rock types (Section 7.3) fall into this category.

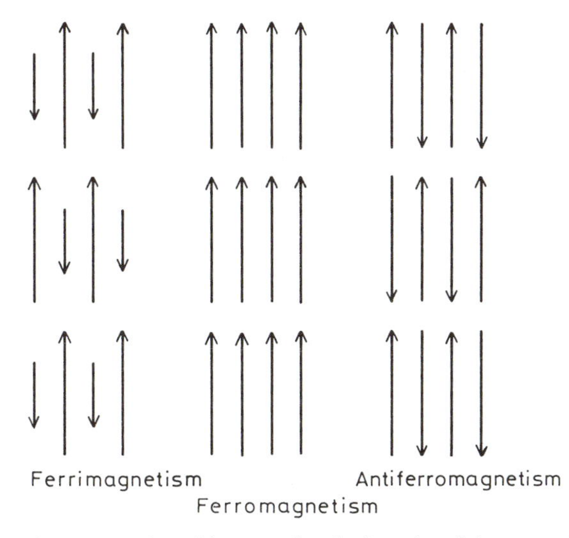

Fig. 7.3. Schematic representation of the strength and orientation of elementary dipoles within ferrimagnetic, ferromagnetic and antiferromagnetic domains.

The strength of the magnetization of ferromagnetic and ferrimagnetic substances decreases with temperature and disappears at the *Curie temperature*. Above this temperature interatomic distances are increased to separations which preclude electron coupling, and the material behaves as an ordinary paramagnetic substance.

In larger grains, the total magnetic energy is decreased if the magnetization of each grain subdivides into individual volume elements (magnetic domains) with diameters of the order of a micrometre, within which there is parallel coupling of dipoles. In the absence of any external magnetic field the domains become oriented in such a way as to reduce the magnetic forces between adjacent domains. The boundary between two domains, the *Bloch wall*, is a narrow zone in which the dipoles cant over from one domain direction to the other.

When a multidomain grain is placed in a weak external magnetic field, the Bloch wall unrolls and causes a growth of those domains magnetized in the direction of the field at the expense of domains magnetized in other directions. This induced magnetization is lost when the applied field is removed as the domain walls rotate back to their original configuration. When stronger fields are applied domain walls unroll irreversibly across small imperfections in the grain so that those domains magnetized in the direction of the field are permanently enlarged. The inherited magnetization remaining after removal of the applied field is known as *remanent*, or *permanent*, *magnetization* J_r. The application of even stronger magnetic fields causes all possible domain wall movements to occur and the material is then said to be *magnetically saturated*.

Primary remanent magnetization may be acquired either as an igneous rock solidifies and cools through the Curie temperature of its magnetic minerals (thermoremanent magnetization, TRM) or as the magnetic particles of a sediment align within the Earth's field during sedimentation (detrital remanent magnetization, DRM). Secondary remanent magnetizations may be impressed later in the history of a rock as magnetic minerals recrystallize or grow during diagenesis or metamorphism (chemical remanent magnetization, CRM). Remanent magnetization may develop slowly in a rock standing in an ambient magnetic field as the domain magnetizations relax into the direction of the field (viscous remanent magnetization, VRM).

Any rock containing magnetic minerals may possess both induced and remanent magnetizations J_i and J_r. These may be in different directions and may differ significantly in magnitude. The magnetic effects of such a rock arise from the resultant J of the two magnetization vectors (Fig. 7.4). The magnitude of J controls the amplitude of the magnetic anomaly and the orientation of J determines its shape.

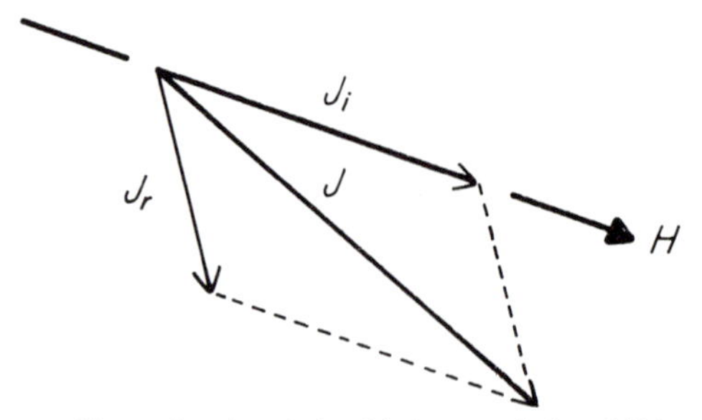

Fig. 7.4. Vector diagram illustrating the relationship between induced (J_i), remanent (J_r) and total (J) magnetization components.

7.3 Rock magnetism

Most common rock-forming minerals exhibit a very low magnetic susceptibility and rocks owe their magnetic character to the generally small proportion of magnetic minerals that they contain. There are only two geochemical groups which provide such minerals. The iron-titanium-oxygen group possesses a solid solution series of magnetic minerals from magnetite (Fe_3O_4) to ulvöspinel (Fe_2TiO_4). The other common iron oxide, haematite (Fe_2O_3) is antiferromagnetic and thus does not give rise to magnetic anomalies (see Section 10.2.4) unless a parasitic antiferromagnetism is developed. The iron-sulphur group provides the magnetic mineral pyrrhotite (FeS_{1+x}, $0 < x < 0.15$) whose magnetic susceptibility is dependent upon the actual composition.

By far the most common magnetic mineral is magnetite, which has a Curie temperature of 578°C. Although the size, shape and dispersion of the magnetite grains within a rock affect its magnetic character, it is reasonable to classify the magnetic behaviour of rocks according to their overall magnetite content. Histograms illustrating the susceptibilities of common rock types are presented in Fig. 7.5.

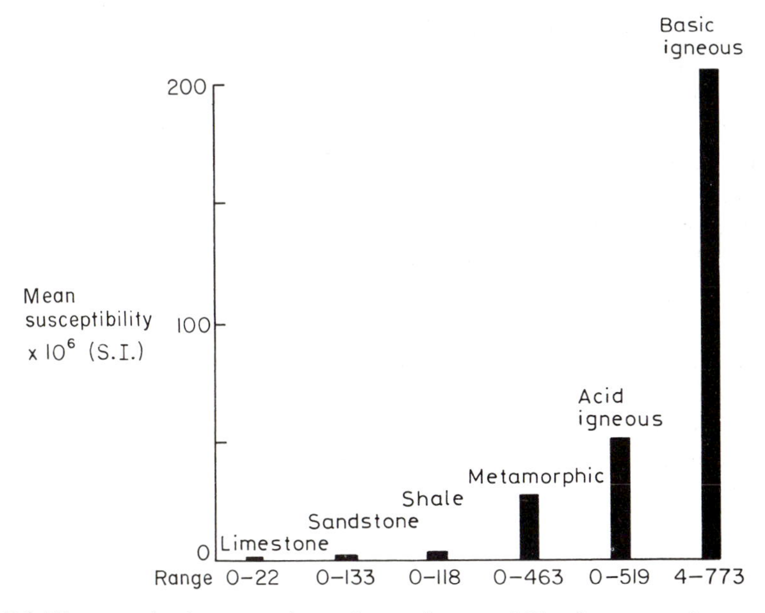

Fig. 7.5. Histogram showing mean values and ranges in susceptibility of common rock types. (Redrawn from Dobrin 1976.)

Basic igneous rocks are usually highly magnetic due to their relatively high magnetite content. The proportion of magnetite in igneous rocks tends to decrease with increasing acidity so that acid igneous rocks, although variable in their magnetic behaviour, are usually less magnetic than basic rocks. Metamorphic rocks are also variable in their magnetic character. If the partial pressure of oxygen is relatively low, magnetite becomes resorbed and the iron and oxygen are incorporated into other mineral phases as the grade of metamorphism increases. Relatively high oxygen partial pressure can, however, result in the formation of magnetite as an accessory mineral in metamorphic reactions.

In general the magnetite content and, hence, the susceptibility of rocks is extremely variable and there can be considerable overlap between different lithologies. It is not usually possible to identify with certainty the causative lithology of any anomaly from magnetic information alone. However, sedimentary rocks are effectively non-magnetic unless they contain a significant amount of magnetite in the heavy mineral fraction. Where magnetic anomalies are observed over sediment-covered areas the anomalies are generally caused by an underlying igneous or metamorphic basement, or by intrusions into the sediments.

Common causes of magnetic anomalies include dykes, faulted, folded or truncated sills and lava flows, massive basic intrusions, metamorphic basement rocks and magnetite ore bodies. Magnetic anomalies range in amplitude from a few tens of nT over deep metamorphic basement to several hundred nT over basic intrusions and may reach an amplitude of several thousand nT over magnetite ores.

7.4 The geomagnetic field

Magnetic anomalies caused by rocks are localized effects superimposed on the normal magnetic field of the Earth (geomagnetic field). Consequently, knowledge of the behaviour of the geomagnetic field is necessary in both the reduction of magnetic data to a suitable datum and in the interpretation of the resulting anomalies. The geomagnetic field is geometrically more complex than the gravity field of the Earth and exhibits irregular variation in both orientation and magnitude with latitude, longitude and time.

At any point on the Earth's surface a freely suspended magnetic needle will

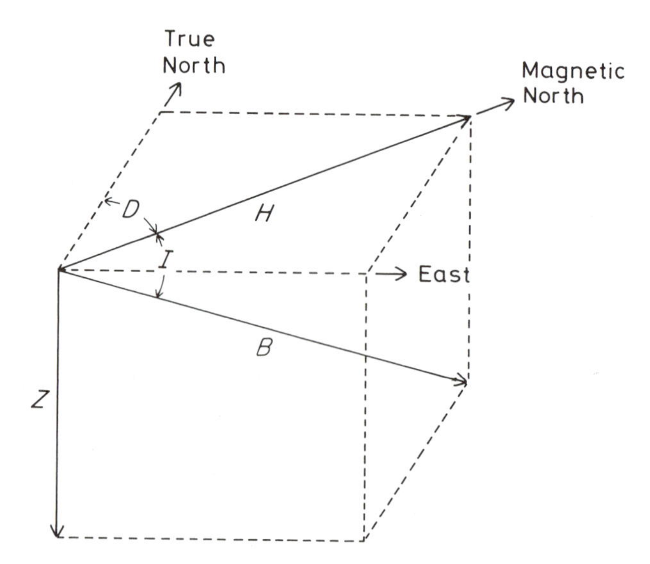

Fig. 7.6. The geomagnetic elements.

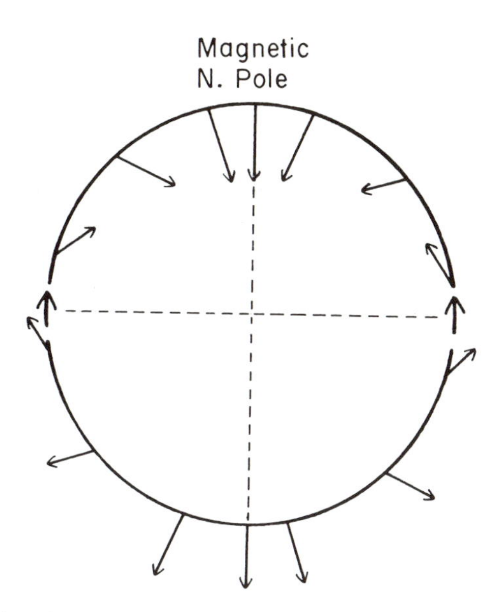

Fig. 7.7. The variation of the inclination of the total magnetic field with latitude based on a simple dipole approximation of the geomagnetic field. (After Sharma 1976.)

assume a position in space in the direction of the ambient geomagnetic field. This will generally be at an angle to both the vertical and geographic north. In order to describe the magnetic field vector, use is made of descriptors known as the geomagnetic elements (Fig. 7.6). The *total field vector* B has a vertical component Z, and a horizontal component H in the direction of magnetic north. The dip of B is the *inclination* I of the field and the horizontal angle between geographic and magnetic north is the *declination* D. B varies in strength from about 25 000 nT in equatorial regions to about 70 000 nT at the poles.

In the northern hemisphere the magnetic field generally dips downward towards the north and becomes vertical at the north magnetic pole (Fig. 7.7). In the southern hemisphere the dip is generally upwards towards the north. The line of zero inclination approximates the geographic equator, and is known as the magnetic equator.

About 90% of the Earth's field can be represented by the field of a theoretical magnetic dipole at the centre of the Earth inclined at about 11.5° to the axis of rotation. The magnetic moment of this fictitious *geocentric dipole* can be calculated from the observed field. If this dipole field is subtracted from the observed magnetic field, the residual field can then be approximated by the effects of a second, smaller, dipole. The process can be continued by fitting dipoles of ever decreasing moment until the observed geomagnetic field is simulated to any required degree of accuracy. The effects of each fictitious dipole contribute to a function known as an harmonic and the technique of successive approximations of the observed field is known as spherical harmonic analysis – the equivalent of Fourier analysis in spherical polar coordinates. The method has been used to compute the formula of the International Geomagnetic Reference Field (IGRF) which defines the theoretical undisturbed magnetic field at any point on the

Earth's surface. In magnetic surveying, the IGRF is used to remove from the magnetic data those magnetic variations attributable to this theoretical field. The formula is considerably more complex than the equivalent Gravity Formula used for latitude correction (see Section 6.8.2) as a large number of harmonics is employed (Barraclough & Malin 1971; Peddie 1983).

The geomagnetic field cannot in fact result from permanent magnetism in the Earth's deep interior. The required dipolar magnetic moments are far greater than is considered realistic and the prevailing high temperatures are far in excess of the Curie temperature of any known magnetic material. The cause of the geomagnetic field is attributed to a dynamo action produced by the circulation of charged particles in coupled convective cells within the outer, fluid, part of the Earth's core (Elsasser 1958). The exchange of dominance between such cells is believed to produce the periodic changes in polarity of the geomagnetic field revealed by palaeomagnetic studies. The circulation patterns within the core are not fixed and change slowly with time. This is reflected in a slow, progressive, temporal change in all the geomagnetic elements known as *secular variation*. Such variation is predictable and a well-known example is the gradual rotation of the north magnetic pole around the geographic pole.

Magnetic effects of external origin cause the geomagnetic field to vary on a daily basis to produce *diurnal variations*. Under normal conditions (Q or quiet days) the diurnal variation is smooth and regular and has an amplitude of about 20–80 nT, being at a maximum in polar regions. Such variation results from the magnetic field induced by the flow of charged particles within the ionosphere towards the magnetic poles, as both the circulation patterns and diurnal variations vary in sympathy with the tidal effects of the Sun and Moon.

Some days (D or disturbed days) are distinguished by far less regular diurnal variations and involve large, short-term disturbances in the geomagnetic field, with amplitudes of up to 1000 nT, known as magnetic storms. Such days are usually associated with intense solar activity and result from the arrival in the ionosphere of charged solar particles. Magnetic surveying should be discontinued during such storms because of the impossibility of correcting the data collected for the rapid and high-amplitude changes in the magnetic field.

7.5 Magnetic anomalies

All magnetic anomalies caused by rocks are superimposed on the geomagnetic field in the same way that gravity anomalies are superimposed on the Earth's gravitational field. The magnetic case is more complex, however, as the geomagnetic field varies not only in amplitude, but also in direction, whereas the gravitational field is everywhere, by definition, vertical.

Describing the normal geomagnetic field by a vector diagram (Fig. 7.8(a)), the geomagnetic elements are related

$$B^2 = H^2 + Z^2 \tag{7.7}$$

A magnetic anomaly is now superimposed on the Earth's field causing a change ΔB in the strength of the total field vector B. Let the anomaly produce a

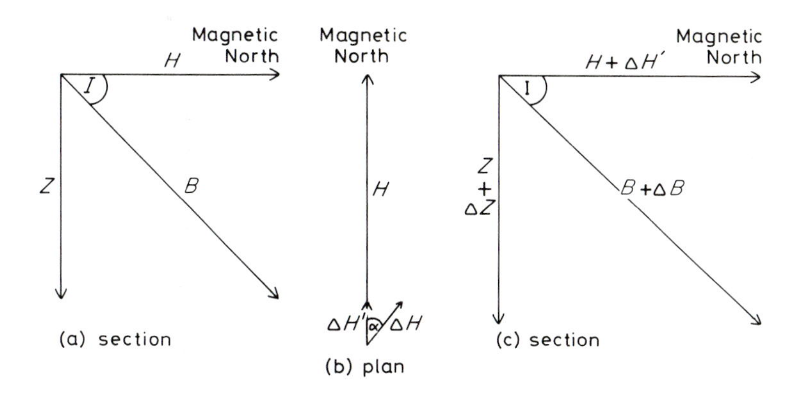

Fig. 7.8. Vector representation of the geomagnetic field with and without a superimposed magnetic anomaly.

vertical component ΔZ and a horizontal component ΔH at an angle α to H (Fig. 7.8(b)). Only that part of ΔH in the direction of H, namely $\Delta H'$, will contribute to the anomaly

$$\Delta H' = \Delta H \cos \alpha \tag{7.8}$$

Using a similar vector diagram to include the magnetic anomaly (Fig. 7.8(c))

$$(B+\Delta B)^2 = (H+\Delta H')^2+(Z+\Delta Z)^2$$

If this equation is expanded, the equality of equation (7.7) substituted and the insignificant terms in Δ^2 ignored, the equation reduces to

$$\Delta B = \Delta Z(Z/B)+\Delta H'(H/B)$$

Substituting equation (7.8) and angular descriptions of geomagnetic element ratios

$$\Delta B = \Delta Z \sin I+\Delta H \cos I \cos \alpha \tag{7.9}$$

where I is the inclination of the geomagnetic field.

This approach can be used to calculate the magnetic anomaly caused by a small isolated magnetic pole of strength m, defined as the effect of this pole on a unit positive pole at the observation point. The pole is situated at depth z, a horizontal distance x and radial distance r from the observation point (Fig. 7.9). The force of repulsion ΔB_r on the unit positive pole in the direction r is given by substitution in equation (7.1), with $\mu_R = 1$

$$\Delta B_r = \frac{Cm}{r^2} \quad \text{where } C = \frac{\mu_0}{4\pi}$$

If it is assumed that the profile lies in the direction of magnetic north so that the horizontal component of the anomaly lies in this direction, the horizontal

7

(ΔH) and vertical (ΔZ) components of this force can be computed by resolving in the relevant directions

$$\Delta H = \frac{Cm}{r^2} \cos \theta = \frac{Cmx}{r^3} \tag{7.10}$$

$$\Delta Z = \frac{-Cm}{r^2} \sin \theta = \frac{-Cmz}{r^3} \tag{7.11}$$

The vertical field anomaly is negative as, by convention, the z-axis is positive downwards. Plots of the form of these anomalies are shown in Fig. 7.9. The horizontal field anomaly is a positive/negative couplet and the vertical field anomaly is centred over the pole.

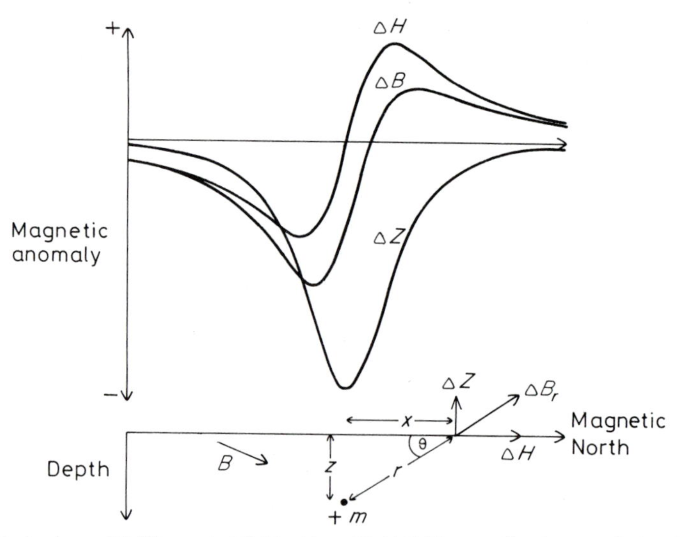

Fig. 7.9. The horizontal (ΔH), vertical (ΔZ) and total field (ΔB) anomalies due to an isolated positive pole.

The total field anomaly ΔB is then obtained by substituting the expressions of equations (7.10) and (7.11) in equation (7.9) where $\alpha = 0$. If the profile were not in the direction of magnetic north, the angle α would represent the angle between magnetic north and the profile direction.

7.6 Magnetic surveying instruments

Since the early 1900s an assortment of surveying instruments has been designed that are capable of measuring the geomagnetic elements Z, H and B. Most modern survey instruments, however, are designed to measure B only. The precision normally required is $\pm$ 1 nT which is approximately one part in 5×10^5 of the background field, a considerably lower requirement of precision than is necessary for gravity measurements (see Chapter 6).

In early magnetic surveys the geomagnetic elements were measured using *magnetic variometers*. There were several types, including the torsion head magnetometer and the Schmidt vertical balance, but all consisted essentially of bar magnets suspended in the Earth's field. Such devices required accurate levelling and a stable platform for measurement so that readings were time-consuming and limited to sites on land.

Since the 1940s, a new generation of instruments has been developed which provides virtually instantaneous readings and requires only coarse orientation so that magnetic measurements can be taken on land, at sea and in the air.

The first such device to be developed was the *fluxgate magnetometer*, which

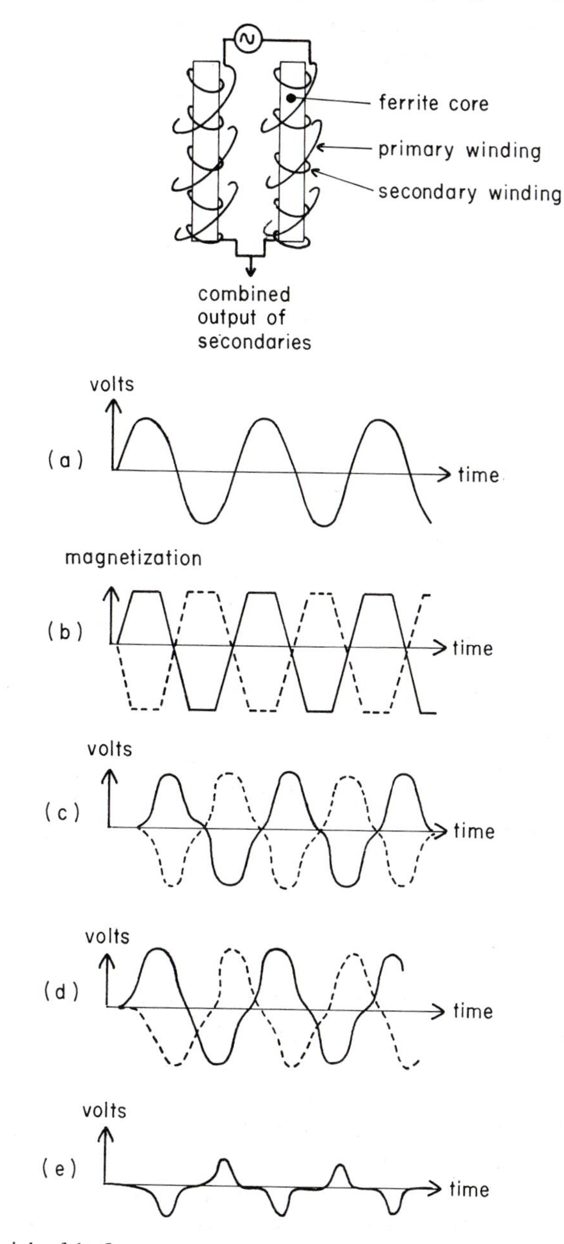

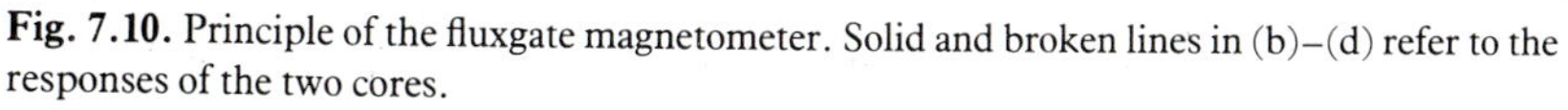
Fig. 7.10. Principle of the fluxgate magnetometer. Solid and broken lines in (b)–(d) refer to the responses of the two cores.

found early application during the second world war in the detection of submarines from the air. The instrument employs two identical ferromagnetic cores of such high permeability that the geomagnetic field can induce a magnetization that is a substantial proportion of their saturation value (see Section 7.2). Identical primary and secondary coils are wound in opposite directions around the cores (Fig. 7.10). An alternating current of 50–1000 Hz is passed through the primary coils (Fig. 7.10(a)), generating an alternating magnetic field. In the absence of any external magnetic field, the cores are driven to saturation near the peak of each half-cycle of the current (Fig. 7.10(b)). The alternating magnetic field in the cores induces an alternating voltage in the secondary coils which is at a maximum when the field is changing most rapidly (Fig. 7.10(c)). Since the coils are wound in opposite directions the voltage in the coils is equal and of opposite sign so that their combined output is zero. In the presence of an external magnetic field, such as the Earth's field, which has a component parallel to the axis of the cores, saturation occurs earlier for the core whose primary field is reinforced by the external field and later for the core opposed by the external field. The induced voltages are now out of phase as the cores reach saturation at different times (Fig. 7.10(d)). Consequently the combined output of the secondary coils is no longer zero but consists of a series of voltage pulses (Fig. 7.10(e)), the magnitude of which can be shown to be proportional to the external field component.

The instrument can be used to measure Z or H by aligning the cores in these directions, but the required accuracy of orientation is some eleven seconds of arc to achieve a reading accuracy of $\pm$ 1 nT. Such accuracy is difficult to obtain on the ground and impossible when the instrument is mobile. The total geomagnetic field can, however, be measured to $\pm$ 1 nT with far less precise orientation as the field changes much more slowly as a function of orientation about the total field direction. Airborne versions of the instrument employ orienting mechanisms of various types to maintain the axis of the instrument in the direction of the geomagnetic field. This is accomplished by making use of the feedback signal generated by additional sensors whenever the instrument moves out of orientation to drive servo-motors which realign the cores into the desired direction.

The fluxgate magnetometer is a continuous-reading instrument and is relatively insensitive to magnetic field gradients along the length of the cores. The instrument may be temperature sensitive, requiring correction.

The most commonly used magnetometer for both survey work and observatory monitoring is currently the *nuclear precession* or *proton magnetometer*. The sensing device of the proton magnetometer is a canister containing a liquid rich in hydrogen atoms, such as kerosene or water, surrounded by a coil (Fig. 7.11(a)). The hydrogen nuclei (protons) act as small dipoles and normally align parallel to the ambient geomagnetic field B_e (Fig. 7.11(b)). A current is passed through the coil to generate a magnetic field B_p fifty to a hundred times larger than the geomagnetic field, and in a different direction, causing the protons to realign in this new direction (Fig. 7.11(c)). The current to the coil is then switched off so that the polarizing field is rapidly removed. The protons return to their original alignment with B_e by spiralling, or precessing, in phase around this direction (Fig. 7.11(d)) with a period of about 0.5 ms, taking some 1–3 seconds to achieve their original orientation. The frequency f of this precession is given by

$$f = \frac{\gamma_p B_e}{2\pi}$$

where γ_p is the gyromagnetic ratio of the proton, an accurately known constant. Consequently measurement of f, about 2 kHz, provides a very accurate measurement of the strength of the total geomagnetic field. f is determined by measurement of the alternating voltage of the same frequency induced to flow in the coil by the precessing protons.

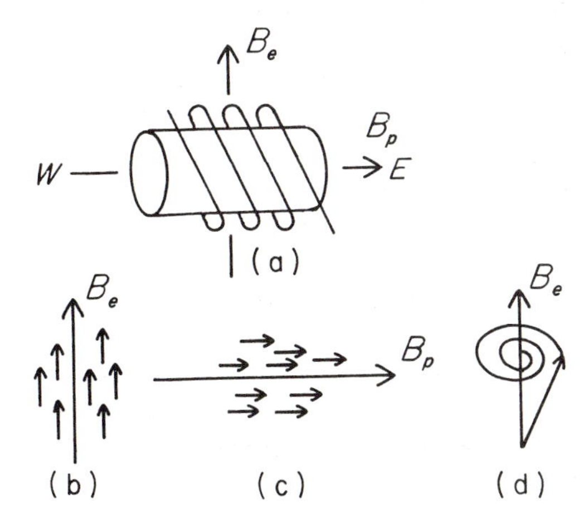

Fig. 7.11. Principle of the proton magnetometer.

Field instruments provide absolute readings of the total magnetic field accurate to ± 1 nT although much greater precision can be attained if necessary. The sensor does not have to be accurately oriented, although it should ideally lie at an appreciable angle to the total field vector. Consequently, readings may be taken by sensors towed behind ships or aircraft without the necessity of orienting mechanisms. Aeromagnetic surveying with proton magnetometers may suffer from the slight disadvantage that readings are not continuous due to the finite cycle period. Small anomalies may be missed since an aircraft travels a significant distance between the discrete measurements, which may be spaced at intervals of a few seconds. This problem has been largely obviated by modern instruments with recycling periods of the order of a second. The proton magnetometer is sensitive to acute magnetic gradients which may cause protons in different parts of the sensor to precess at different rates with a consequent adverse effect on signal strength.

The sensing elements of fluxgate or proton magnetometers can be used in pairs to measure either horizontal or vertical magnetic field gradients. *Magnetic gradiometers* are differential magnetometers in which the spacing between the sensors is fixed and small with respect to the distance of the causative body whose magnetic field gradient is to be measured. Magnetic gradients can be measured, albeit less conveniently, with a magnetometer by taking two successive measurements at close vertical or horizontal spacings. Magnetic gradiometers are employed in surveys of shallow magnetic features as the gradient anomalies tend to resolve complex anomalies into their individual components, which can be used in the determination of the location, shape and depth of the causative bodies. The method has the further advantages that regional and temporal variations in the geomagnetic field are automatically removed.

Since modern magnetic instruments require no precise levelling a magnetic survey on land invariably proceeds more rapidly than a gravity survey.

7.7 Ground magnetic surveys

Ground magnetic surveys are usually performed over relatively small areas on a previously defined target. Consequently, station spacing is commonly of the order of 10–100 m, although smaller spacings may be employed where magnetic gradients are high. Readings should not be taken in the vicinity of metallic objects such as railway lines, cars, roads, fencing, houses etc. which might perturb the local magnetic field. For similar reasons, operators of magnetometers should not carry metallic objects.

Base station readings are not necessary for monitoring instrumental drift as fluxgate and proton magnetometers do not drift, but may be used to monitor diurnal variations (see Section 7.9).

7.8 Aeromagnetic and marine surveys

The vast majority of magnetic surveys are carried out in the air, with the sensor towed in a housing known as a 'bird' to remove the instrument from the magnetic effects of the aircraft or fixed in a 'stinger' in the tail of the aircraft, in which case inboard coil installations compensate for the aircraft's magnetic field.

Aeromagnetic surveying is rapid and cost effective, typically costing some 40% less per line kilometre than a ground survey. Vast areas can be surveyed rapidly without the cost of sending a field party into the survey area and data can be obtained from areas inaccessible to ground survey.

The most difficult problem in airborne surveys is position fixing. Where available, electronic positioning systems are employed. Without these it is necessary to use aerial photography. Terrain photographs are taken simultaneously with the magnetic readings so that the location can subsequently be determined by reference to topographic maps.

Marine magnetic surveying techniques are similar to those of airborne surveying. The sensor is towed in a 'fish' at least 2½ ship's lengths behind the vessel to remove its magnetic effects. Marine surveying is obviously slower than aeromagnetic surveying, but is frequently carried out in conjunction with several other geophysical methods, such as gravity surveying and continuous seismic profiling, which cannot be employed in the air.

7.9 Reduction of magnetic observations

The reduction of magnetic data is necessary to remove all causes of magnetic variation from the observations other than those arising from the magnetic effects of the subsurface.

7.9.1 *Diurnal variation correction*

The effects of diurnal variation may be removed in several ways. On land a method similar to gravimeter drift monitoring may be employed in which the magnetometer is read at a fixed base station periodically throughout the day. The differences observed in base readings are then distributed among the readings at stations occupied during the day according to the time of observation. It should be remembered that base readings taken during a gravity survey are made to correct for both the drift of the gravimeter and tidal effects: magnetometers do not drift and base readings are taken solely to correct for temporal variation in the measured field. Such a procedure is inefficient as the instrument has to be returned periodically to a base location and is not practical in marine or airborne surveys. These problems may be overcome by use of a base magnetometer, a continuous-reading instrument which records magnetic variations at a fixed location within or close to the survey area. This method is preferable on land as the survey proceeds faster and the diurnal variations are fully charted. Where the survey is of regional extent the records of a magnetic observatory may be used. Such observatories continuously record changes in all the geomagnetic elements. However, diurnal variations differ quite markedly from place to place and so the observatory used should not be more than about 100 km from the survey area.

Diurnal variation during an aeromagnetic survey may alternatively be assessed by arranging numerous cross-over points in the survey plan (Fig. 7.12).

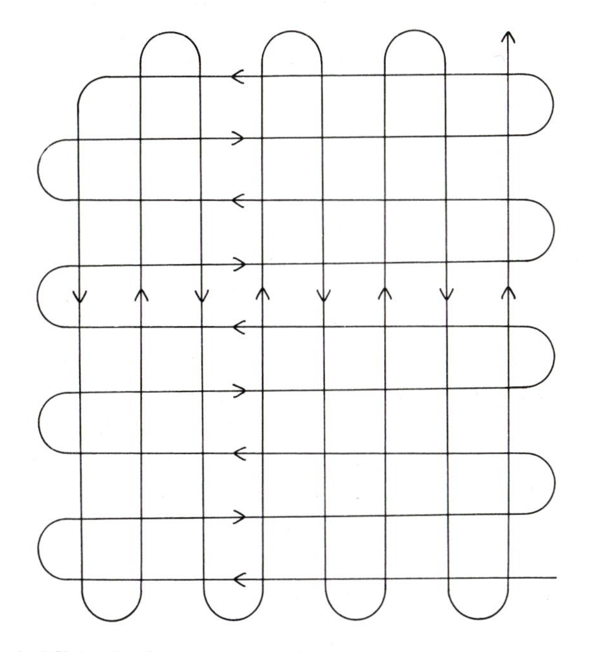

Fig. 7.12. A typical flight plan for an aeromagnetic survey.

Analysis of the differences in readings at each cross-over, representing the field change over a series of different time periods, allows the whole survey to be corrected for diurnal variation by a process of network adjustment, without the necessity of a base instrument.

Diurnal variations, however recorded, must be examined carefully. If large, high frequency variations are apparent, resulting from a magnetic storm, the survey results should be discarded.

7.9.2 *Geomagnetic correction*

The magnetic equivalent of the latitude correction in gravity surveying is the *geomagnetic correction* which removes the effect of a geomagnetic reference field from the survey data. The most rigorous method of geomagnetic correction is the use of the IGRF (Section 7.4), which expresses the undisturbed geomagnetic field in terms of a large number of harmonics and includes temporal terms to correct for secular variation. The complexity of the IGRF requires the calculation of corrections by computer. It must be realized, however, that the IGRF is imperfect as the harmonics employed are based on observations at relatively few, scattered, magnetic observatories. Consequently the IGRF in areas remote from observatories can be substantially in error.

Over the area of a magnetic survey the geomagnetic reference field may be approximated by a uniform gradient defined in terms of latitudinal and longitudinal gradient components. For example, the geomagnetic field over the British Isles is approximated by the following gradient components: 2.13 nT/km N; 0.26 nT/km W. For any survey area the relevant gradient values may be assessed from magnetic maps covering a much larger region.

The appropriate regional gradients may also be obtained by employing a single dipole approximation of the Earth's field and using the well-known equations for the magnetic field of a dipole to derive local field gradients:

$$Z = \frac{\mu_0}{4\pi}\frac{2M}{R^3}\cos\theta, \; H = \frac{\mu_0}{4\pi}\frac{M}{R^3}\sin\theta \tag{7.12}$$

$$\frac{\partial Z}{\partial \theta} = -2H, \; \frac{\partial H}{\partial \theta} = \frac{Z}{2} \tag{7.13}$$

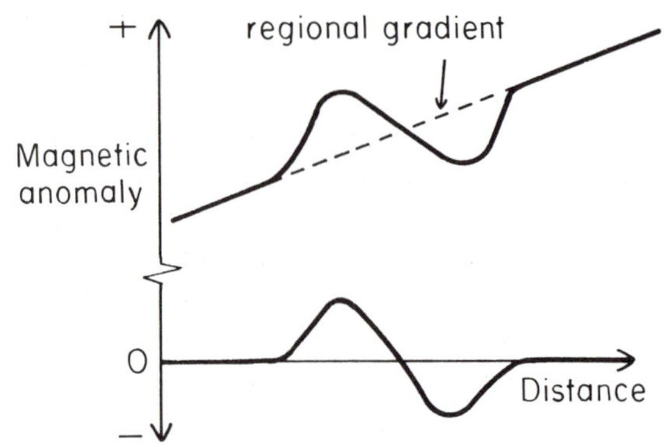

Fig. 7.13. The removal of a regional gradient from a magnetic field by trend analysis. The regional field is approximated by a linear trend.

where Z and H are the vertical and horizontal field components, θ the colatitude in radians, R the radius of the Earth, M the magnetic moment of the Earth and $\partial Z/\partial\theta$ and $\partial H/\partial\theta$ the rate of change of Z and H with colatitude.

An alternative method of removing the regional field over a relatively small survey area is by use of trend analysis. A trend line (for profile data) or trend surface (for areal data) is fitted to the observations using the least squares criterion, and subsequently subtracted from the observed data to leave the local anomalies as positive and negative residuals (Fig. 7.13).

7.9.3 *Elevation and terrain corrections*

The vertical gradient of the geomagnetic field is only some 0.03 nT m^{-1} at the poles and -0.015 nT m^{-1} at the equator, so an *elevation correction* is not usually applied. The influence of topography can be significant in ground magnetic surveys but is not completely predictable as it depends upon the magnetic properties of the topographic features. Therefore, in magnetic surveying *terrain corrections* are rarely applied.

Having applied diurnal and geomagnetic corrections, all remaining magnetic field variations should be caused solely by spatial variations in the magnetic properties of the subsurface and are referred to as magnetic anomalies.

7.10 Interpretation of magnetic anomalies

7.10.1 *Introduction*

The interpretation of magnetic anomalies is similar in its procedures and limitations to gravity interpretation as both techniques utilize natural potential fields based on inverse square laws of attraction. There are several differences, however, which increase the complexity of magnetic interpretation.

Whereas the gravity anomaly of a causative body is entirely positive or negative, depending on whether the body is more or less dense than its surroundings, the magnetic anomaly of a finite body invariably contains positive and negative elements arising from the dipolar nature of magnetism (Fig. 7.14). Moreover, whereas density is a scalar, intensity of magnetization is a vector, and the direction of magnetization in a body closely controls the shape of its magnetic anomaly. Thus bodies of identical shape can give rise to very different magnetic anomalies. For the above reasons magnetic anomalies are often much less closely related to the shape of the causative body than are gravity anomalies.

The intensity of magnetization of a rock is largely dependent upon the amount, size, shape and distribution of its contained ferrimagnetic minerals and these represent only a small proportion of its constituents. By contrast, density is a bulk property. Intensity of magnetization can vary by a factor of 10^6 between different rock types, and is thus considerably more variable than density, where the range is commonly 1500–3500 kg m^{-3}.

Magnetic anomalies are independent of the distance units employed. For example, the same magnitude anomaly is produced by, say, a 3 m cube (on a

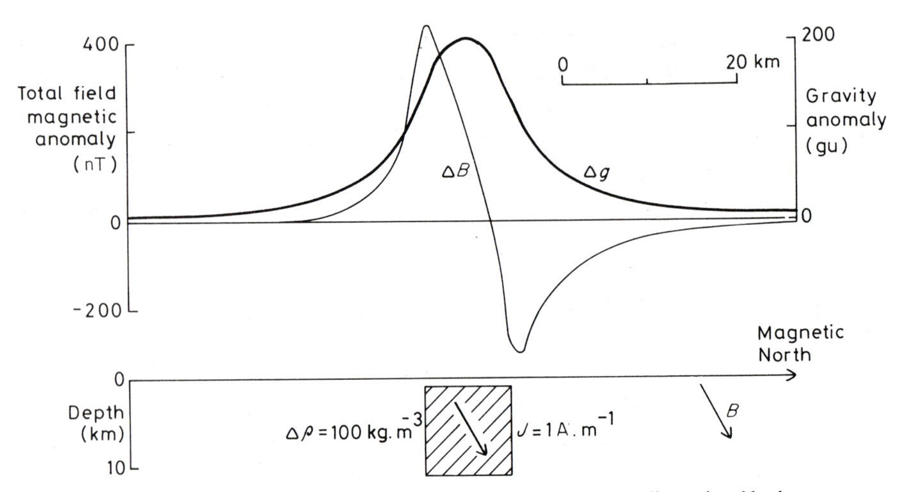

Fig. 7.14. Gravity (Δg) and magnetic (ΔB) anomalies over the same two-dimensional body.

metre scale) as a 3 km cube (on a kilometre scale) with the same magnetic properties. The same is not true of gravity anomalies.

The problem of ambiguity in magnetic interpretation is the same as for gravity, i.e. the same inverse problem is encountered. Thus, just as with gravity, all external controls on the nature and form of the causative body must be

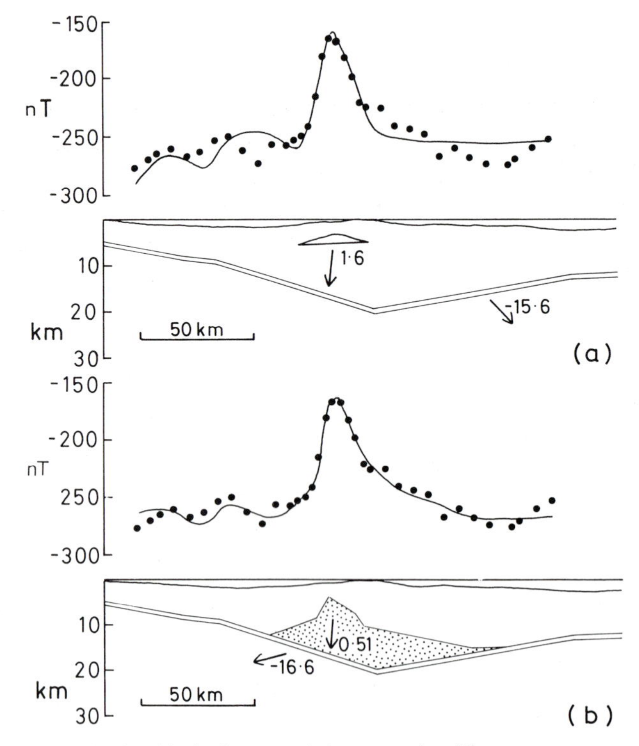

Fig. 7.15. An example of ambiguity in magnetic interpretation. The arrows correspond to the direction of the magnetization vector, whose magnitude is given in A m^{-1}. (Redrawn from Westbrook 1975.)

employed to reduce the ambiguity. An example of this problem is illustrated in Fig. 7.15, which shows two possible interpretations of a magnetic profile across the Barbados Ridge in the eastern Caribbean. In both cases the regional variations are attributed to the variation in depth of a 1 km thick oceanic crustal layer 2. The high amplitude central anomaly, however, can be explained either by the presence of a detached sliver of oceanic crust (Fig. 7.15(a)) or a rise of metamorphosed sediments at depth (Fig. 7.15(b)).

Much qualitative information may be derived from a magnetic contour map. This applies especially to aeromagnetic maps which often provide major clues as to the geology and structure of a broad region from an assessment of the shapes and trends of anomalies. Sediment-covered areas with relatively deep basement are typically represented by smooth magnetic contours reflecting basement structures and magnetization contrasts. Igneous and metamorphic terrains generate far more complex magnetic anomalies, and the effects of deep geological features may be obscured by high wavenumber anomalies of near-surface origin. In most types of terrain an aeromagnetic map can be a useful aid to reconnaissance geological mapping.

In carrying out quantitative interpretation of magnetic anomalies, both direct and indirect methods may be employed, but the former are much more limited than for gravity interpretation and no equivalent general equations exist for total field anomalies.

7.10.2 Direct interpretation

Limiting depth is the most important parameter derived by direct interpretation, and this may be deduced from magnetic anomalies by making use of their property of decaying rapidly with distance from source. Magnetic anomalies caused by shallow structures are more dominated by high wavenumber components than those resulting from deeper sources. This effect may be quantified by computing the power spectrum of the anomaly as it can be shown, for certain

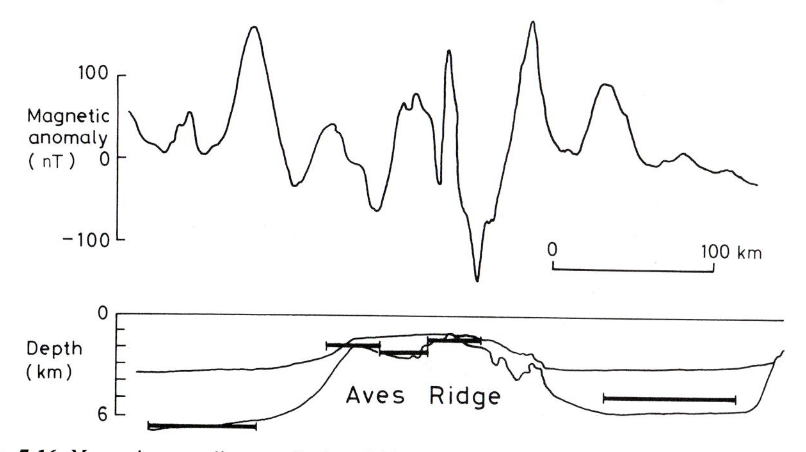

Fig. 7.16. Magnetic anomalies over the Aves Ridge, eastern Caribbean. Lower diagram illustrates bathymetry and basement/sediment interface. Horizontal bars indicate depth estimates of the magnetic basement derived by spectral analysis of the magnetic data.

types of source body, that the log-power spectrum has a linear gradient whose magnitude is dependent upon the depth of the source (Spector & Grant 1970). Such techniques of spectral analysis provide rapid depth estimates from regularly-spaced digital field data: no geomagnetic or diurnal corrections are necessary as these remove only low wavenumber components and do not affect the depth estimates which are controlled by the high wavenumber components of the observed field. Fig. 7.16 shows a magnetic profile across the Aves Ridge in the eastern Caribbean. In this region the configuration of the sediment/basement interface is reasonably well-known from both seismic reflection and refraction surveys. The magnetic anomalies clearly show their most rapid fluctuation over areas of relatively shallow basement, and this observation is quantified by the power spectral depth estimates (horizontal bars) which show excellent correlation with the known basement relief.

7.10.3 *Indirect interpretation*

Indirect interpretation of magnetic anomalies is similar to gravity interpretation in that an attempt is made to match the observed anomaly with that calculated for a model by iterative adjustments to the model. Simple magnetic anomalies may be simulated by a single dipole. Such an approximation to the magnetization of a real geological body is often valid for highly magnetic ore bodies whose direction of magnetization tends to align with their long dimension (Fig. 7.17). In such cases the anomaly is calculated by summing the effects of both poles at the observation points, employing equations (7.10), (7.11) and (7.9). More complicated magnetic bodies, however, require a different approach.

The magnetic anomaly of most regularly-shaped bodies can be calculated by building up the bodies from a series of dipoles parallel to the magnetization

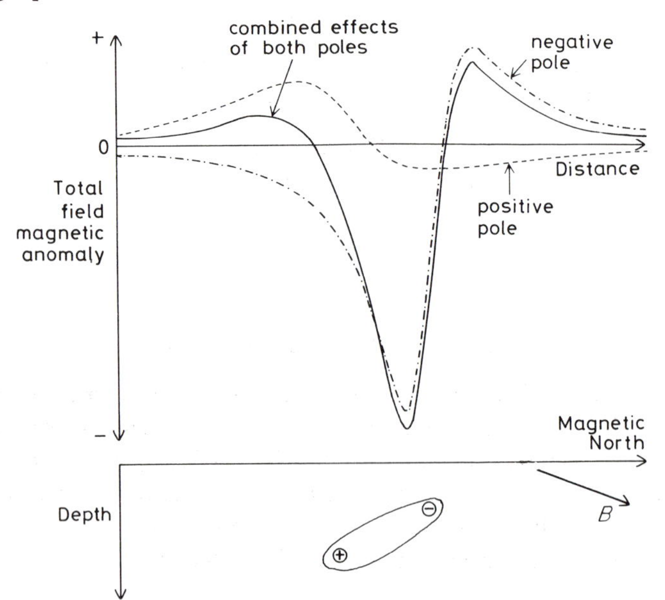

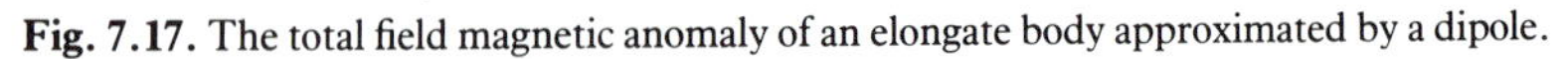
Fig. 7.17. The total field magnetic anomaly of an elongate body approximated by a dipole.

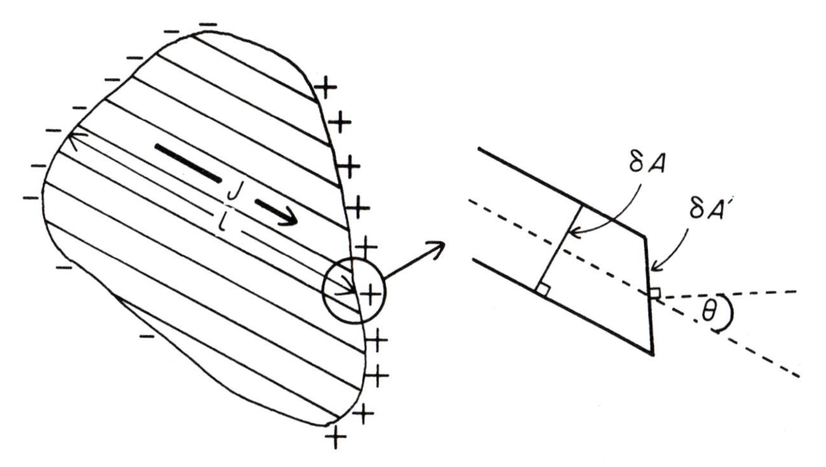

Fig. 7.18. The representation of the magnetic effects of an irregularly shaped body in terms of a number of elements parallel to the magnetization direction. Inset shows in detail the end of one such element.

direction (Fig. 7.18). The poles of the magnets are negative on the surface of the body where the magnetization vector enters the body and positive where it leaves the body. Thus any uniformly-magnetized body can be represented by a set of magnetic poles distributed over its surface. Consider one of these elementary magnets of length l and cross-sectional area δA in a body with intensity of magnetization J and magnetic moment M. From equation (7.5).

$$M = J\delta A l \tag{7.14}$$

If the pole strength of the magnet is m, from equation (7.4), $m = M/l$, and substituting in equation (7.14)

$$m = J\delta A \tag{7.15}$$

If $\delta A'$ is the area of the end of the magnet and θ the angle between the magnetization vector and a direction normal to the end face

$$\delta A = \delta A' \cos\theta$$

Substituting in equation (7.15)

$$m = J\delta A' \cos\theta$$

thus

$$\text{the pole strength per unit area} = J\cos\theta \tag{7.16}$$

A consequence of the distribution of an equal number of positive and negative poles over the surface of a magnetic body is that an infinite horizontal layer produces no magnetic anomaly since the effects of the poles on the upper and lower surfaces are self-cancelling. Consequently, magnetic anomalies are not produced by continuous sills or lava flows. Where, however, the horizontal structure is truncated, the vertical edge will produce a magnetic anomaly (Fig. 7.19).

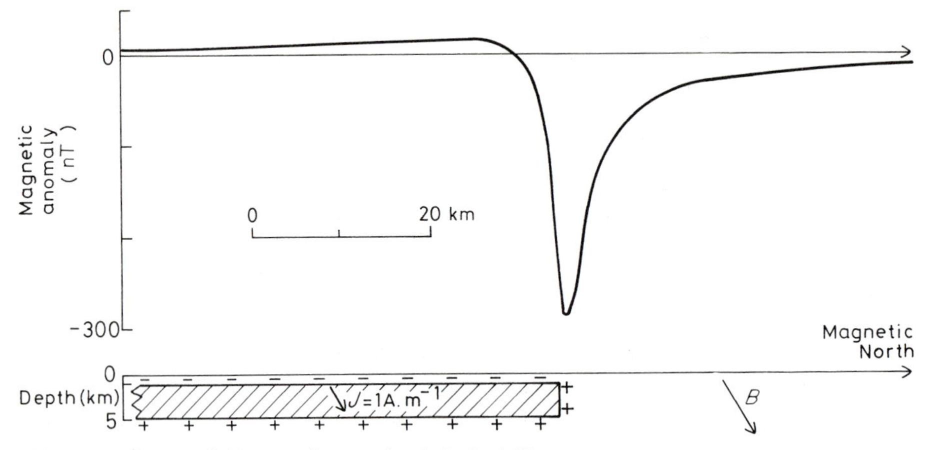

Fig. 7.19. The total field magnetic anomaly of a faulted sill.

The magnetic anomaly of a body of regular shape is calculated by determining the pole distribution over the surface of the body using equation (7.16). Each small element of the surface is then considered and its vertical and horizontal component anomalies are calculated at each observation point using equations (7.10) and (7.11). The effects of all such elements are summed (integrated) to produce the vertical and horizontal anomalies for the whole body and the total field anomaly is calculated using equation (7.9). The integration can be performed analytically for bodies of regular shape, while irregularly-shaped bodies may be split into regular shapes and the integration performed numerically.

In two-dimensional modelling the cross-sectional form of the body is approximated by a polygonal outline and a similar approach adopted to gravity interpretation by making use of the magnetic anomaly of a semi-infinite slab with a sloping edge (Talwani 1965) (Fig. 7.20). Examples of this technique have been presented in Fig. 7.15. An important difference from gravity interpretation is the increased stringency with which the two-dimensional approximation should be made. It can be shown that two-dimensional magnetic interpretation is much more sensitive to errors associated with variation along strike than is the case with gravity interpretation: the length-width ratio of a magnetic anomaly should be at least 10 : 1 for a two-dimensional approximation to be valid, in contrast to gravity interpretation where a 2 : 1 length–width ratio is sufficient to validate two-dimensional interpretation.

Three-dimensional modelling of magnetic anomalies is complex. Probably the most convenient methods are to approximate the causative body by a cluster of right rectangular prisms or by a series of horizontal slices of polygonal outline.

Because of the dipolar nature of magnetic anomalies, trial and error methods of indirect interpretation are difficult to perform manually since anomaly shape is not closely related to the geometry of the causative body. Consequently the automatic methods of interpretation described in Section 6.11.4 are widely employed.

The continuation and filtering operations used in gravity interpretation and described in Section 6.12 are equally applicable to magnetic fields. A further processing operation that may be applied to magnetic anomalies is known as

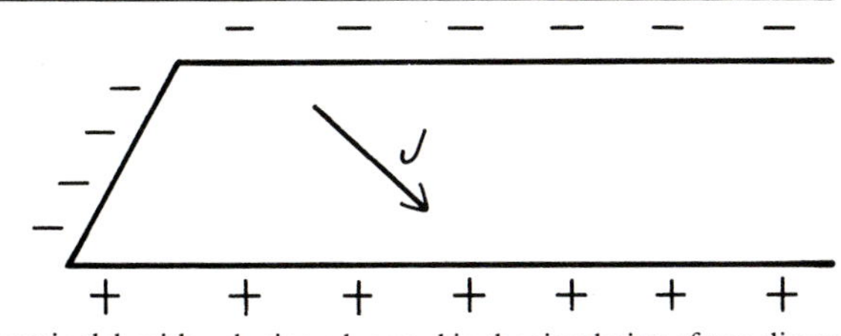

Fig. 7.20. The magnetic slab with a sloping edge used in the simulation of two-dimensional bodies of irregular cross-section.

reduction to the pole, and involves the conversion of the anomalies into their equivalent form at the north magnetic pole (Baranov & Naudy 1964). This process usually simplifies the magnetic anomalies as the ambient field is then vertical and bodies with magnetizations which are solely induced produce anomalies that are axisymmetric. The existence of remanent magnetization, however, commonly prevents reduction to the pole from producing the desired simplification in the resultant pattern of magnetic anomalies.

7.11 Potential field transformations

The formulae for the gravitational potential caused by a point mass and the magnetic potential due to an isolated pole were presented in equations (6.3) and (7.3). A consequence of the similar laws of attraction governing gravitating and magnetic bodies is that these two equations have the variable of inverse distance ($1/r$) in common. Elimination of this term between the two formulae provides a relationship between the gravitational and magnetic potentials known as *Poisson's equation.* In reality the relationship is more complex than implied by equations (6.3) and (7.3) as isolated magnetic poles do not exist. However, the validity of the relationship between the two potential fields remains. Since gravity or magnetic fields can be determined by differentiation of the relevant potential in the required direction, Poisson's equation provides a method of transforming magnetic fields into gravitational fields and *vice versa* for bodies in which the ratio of intensity of magnetization to density remains constant. Such transformed fields are known as *pseudogravitational* and *pseudomagnetic* fields (Garland 1951).

One application of this technique is the transformation of magnetic anomalies into pseudogravity anomalies for the purposes of indirect interpretation, as the latter are significantly easier to interpret than their magnetic counterpart. The method is even more powerful when the pseudofield is compared with a corresponding measured field. For example the comparison of gravity anomalies with the pseudogravity anomalies derived from magnetic anomalies over the same area can show whether the same geological bodies are the cause of the two types of anomaly. Performing the transformation for different orientations of the magnetization vector provides an estimate of the true vector orientation since this will produce a pseudogravity field which most closely approximates the observed gravity field. The relative amplitudes of these two fields then provide a measure of the ratio of intensity of magnetization to density. These potential field transformations provide an elegant means of comparing gravity and magnetic

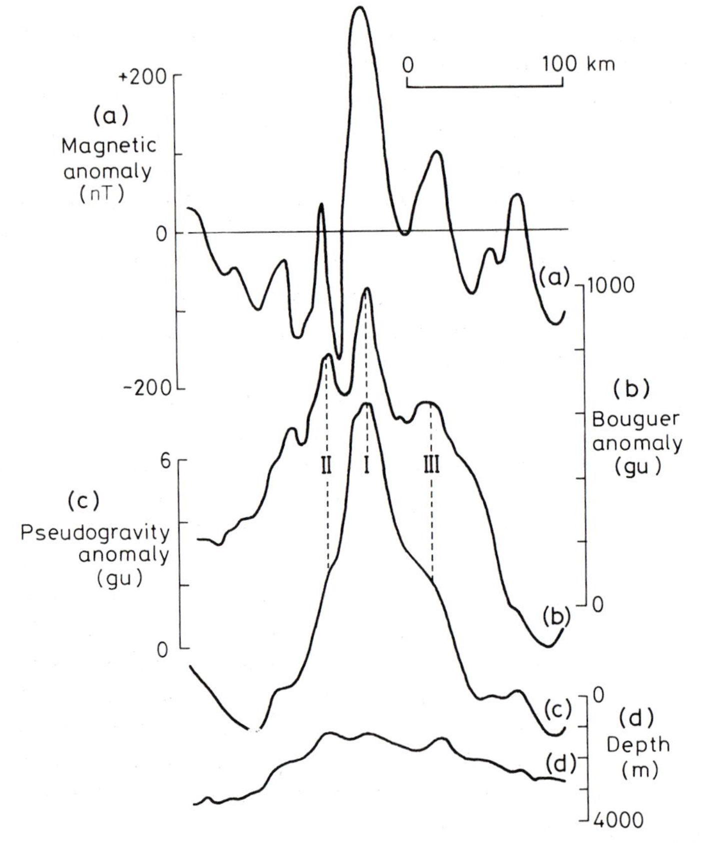

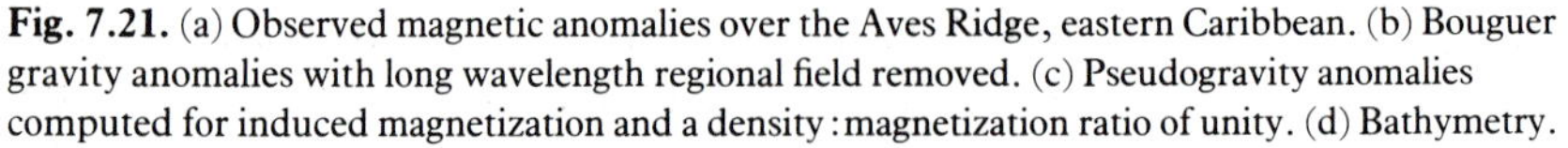

Fig. 7.21. (a) Observed magnetic anomalies over the Aves Ridge, eastern Caribbean. (b) Bouguer gravity anomalies with long wavelength regional field removed. (c) Pseudogravity anomalies computed for induced magnetization and a density : magnetization ratio of unity. (d) Bathymetry.

anomalies over the same area and sometimes allow greater information to be derived about their causative bodies than would be possible if the techniques were treated in isolation.

Figs. 7.21(a) and (b) show magnetic and residual gravity anomaly profiles across the Aves Ridge, a submarine prominence in the eastern Caribbean which runs parallel to the island arc of the Lesser Antilles. The pseudogravity profile calculated from the magnetic profile assuming induced magnetization is presented in Fig. 7.21(c). It is readily apparent that the main pseudogravity peak correlates with peak I on the gravity profile and that peaks II and III correlate with much weaker features on the pseudofield profile. The data thus suggest that the density features responsible for the gravity maxima are also magnetic, with the causative body of the central peak having a significantly greater susceptibility than the flanking bodies.

7.12 Applications of magnetic surveying

Magnetic surveying is one of the most widely used geophysical methods in terms of line length surveyed, and is a rapid and cost-effective technique.

Magnetic surveys are used extensively in the search for mineral deposits, a task accomplished rapidly and economically by airborne methods. Deposits of magnetite, ilmenite, pyrrhotite, and, to a lesser extent, haematite can give rise to major magnetic anomalies. Manganese and chromium ores may also, but not invariably, produce detectable anomalies. Other ore minerals, themselves non-magnetic, may be associated with magnetic rocks: for example, the occurrence of gold in intrusive igneous rocks, or diamonds in kimberlite pipes. The magnetic anomalies associated with base-metal sulphides and asbestos may result from the presence of small quantities of magnetic minerals within the deposit.

A major use of magnetic surveying is as an aid to geological mapping. Over extensive regions with a thick sedimentary cover, structural features may be revealed if magnetic horizons such as ferruginous sandstones and shales, tuffs and lava flows are present within the sedimentary sequence. In the absence of magnetic sediments, magnetic survey data can provide information on the nature and form of the crystalline basement. Both cases are applicable to petroleum exploration in the location of structural traps within sediments or features of basement topography which might influence the overlying sedimentary sequence. The magnetic method may also be used to assist a programme of reconnaissance geological mapping based on widely-spaced grid samples, since aeromagnetic anomalies can be employed to delineate geological boundaries between sampling points. On a smaller scale, ground magnetic surveying is often of use to the engineering geologist and hydrogeologist in locating buried features such as subsurface contacts in the basement, faults and dykes. It may also be applied to archaeological and geotechnical investigations for the location of buried, man-made, ferrous objects.

In academic studies the magnetic method can be used to provide information on geological structures at all scales. A significant contribution in this field was the discovery of linear magnetic anomalies within the ocean basins, which led directly to the theory of sea floor spreading (see Section 10.5).

Further reading

Baranov, W. (1975) *Potential Fields and Their Transformation in Applied Geophysics*. Gebrüder Borntraeger, Berlin.

Bott, M.H.P. (1973) Inverse methods in the interpretation of magnetic and gravity anomalies. *In:* Alder B., Fernbach, S. & Bolt, B.A. (eds.) *Methods in Computational Physics*, **13**, 133–62.

Garland, G.D. (1951) Combined analysis of gravity and magnetic anomalies. *Geophysics*, **16**, 51–62.

Grant, F.S. & West, G.F. (1965) *Interpretation Theory in Applied Geophysics*. McGraw-Hill, New York.

Gunn, P.J. (1975) Linear transformations of gravity and magnetic fields. *Geophys. Prosp.*, **23**, 300–12.

Kanasewich, E.R. & Agarwal, R.G. (1970) Analysis of combined gravity and magnetic fields in wave number domain. *J. Geophys. Res.*, **75**, 5702–12.

Nettleton, L.L. (1971) *Elementary Gravity and Magnetics for Geologists and Seismologists*. Society of Exploration Geophysicists, Tulsa. Monograph Series No. 1.

Nettleton, L.L. (1976) *Gravity and Magnetics in Oil Exploration*. McGraw-Hill, New York.

Stacey, F.D. & Banerjee, S.K. (1974) *The Physical Principles of Rock Magnetism*. Elsevier, Amsterdam.

Sharma, P. (1976) *Geophysical Methods in Geology*. Elsevier, Amsterdam.

Tarling, D.H. (1983) *Palaeomagnetism*. Chapman & Hall, London.

Vacquier, V., Steenland, N.C., Henderson, R.G. & Zeitz, I. (1951) Interpretation of aeromagnetic maps. *Geol. Soc. Am. Mem.*, **47**.

8

Electrical Surveying

8.1 Introduction

There are many methods of electrical surveying. Some make use of naturally-occurring fields within the Earth while others require the introduction of artificially-generated currents into the ground. The resistivity method is used in the study of horizontal and vertical discontinuities in the electrical properties of the ground, and also in the detection of three-dimensional bodies of anomalous electrical conductivity. It is routinely used in engineering and hydrogeological investigations to investigate the shallow subsurface geology. The induced polarization method makes use of the capacitative action of the subsurface to locate zones where conductive minerals are disseminated within their host rocks. The self potential method makes use of natural currents flowing in the ground that are generated by electrochemical processes to locate shallow bodies of anomalous conductivity.

Electrical methods utilize direct currents or low frequency alternating currents to investigate the electrical properties of the subsurface, in contrast to the electromagnetic methods discussed in the next chapter that use alternating electromagnetic fields of higher frequency to this end.

8.2 Resistivity method

8.2.1 Introduction

In the resistivity method, artificially-generated electric currents are introduced into the ground and the resulting potential differences are measured at the surface. Deviations from the pattern of potential differences expected from homogeneous ground provide information on the form and electrical properties of subsurface inhomogeneities.

8.2.2 Resistivities of rocks and minerals

The *resistivity* of a material is defined as the resistance in ohms between the opposite faces of a unit cube of the material. For a conducting cylinder of resistance δR, length δL and cross-sectional area δA (Fig. 8.1) the resistivity ρ is given by

$$\rho = \frac{\delta R \delta A}{\delta L} \tag{8.1}$$

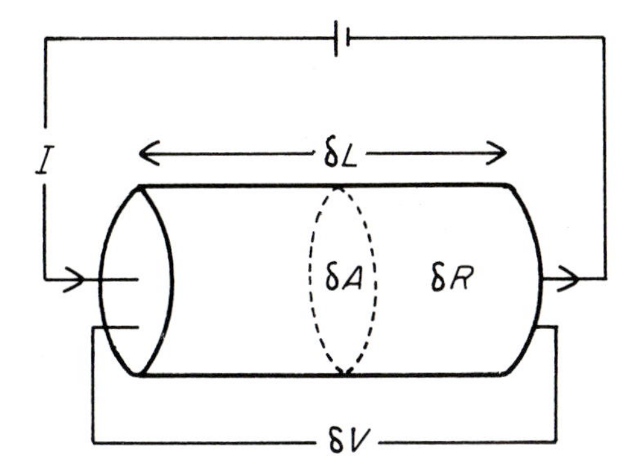

Fig. 8.1. The parameters used in defining resistivity.

The SI unit of resistivity is the ohm-metre (ohm m) and the reciprocal of resistivity is termed *conductivity* (units: siemens (S) per metre; 1 S m^{-1} = 1 ohm^{-1} m^{-1}).

Resistivity is one of the most variable of physical properties. Certain minerals such as native metals and graphite conduct electricity via the passage of electrons. Most rock-forming minerals are, however, insulators, and electrical current is carried through a rock mainly by the passage of ions in pore waters. Thus most rocks conduct electricity by electrolytic rather than electronic processes. It follows that porosity is the major control of the resistivity of rocks, and that resistivity generally increases as porosity decreases. However, even crystalline rocks with negligible intergranular porosity are conductive along cracks and fissures. Fig. 8.2 shows the range of resistivities expected for common rock types. It is apparent that there is considerable overlap between different rock types and, consequently, identification of a rock type is not possible solely on the basis of resistivity data.

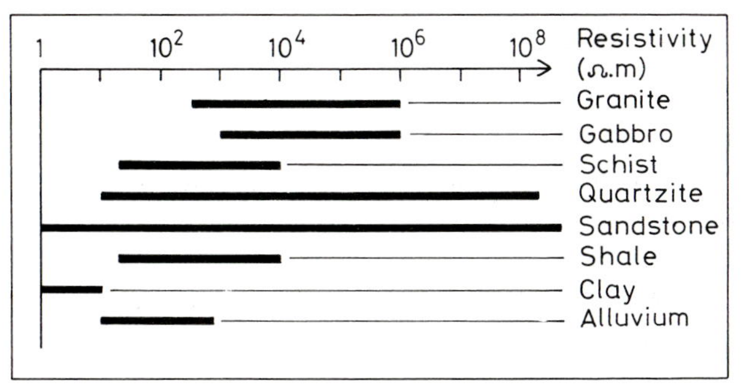

Fig. 8.2. The approximate range of resistivity values of common rock types.

Strictly, equation (8.1) refers to electronic conduction but it may still be used to describe the effective resistivity of a rock, i.e. the resistivity of the rock and its pore water. The effective resistivity can also be expressed in terms of the resistivity and volume of the pore water present according to an empirical formula given by Archie

$$\rho = a\phi^{-b}f^{-c}\rho_w \tag{8.2}$$

where ϕ is the porosity, f the fraction of pores containing water of resistivity ρ_w and a, b and c are empirical constants. ρ_w can vary considerably according to the quantities and conductivities of dissolved materials.

8.2.3 *Current flow in the ground*

Consider the element of homogeneous material shown in Fig. 8.1. A current I is passed through the cylinder causing a potential drop $-\delta V$ beteen the ends of the element.

Ohm's law relates the current, potential difference and resistance such that $-\delta V = \delta RI$, and from equation (8.1) $\rho\delta R = \delta L/\delta A$. Substituting

$$\frac{\delta V}{\delta L} = -\frac{\rho I}{\delta A} = -\rho i \tag{8.3}$$

$\delta V/\delta L$ represents the potential gradient through the element in volt m^{-1} and i the current density in amp m^{-2}. In general the current density in any direction within a material is given by the negative partial derivative of the potential in that direction divided by the resistivity.

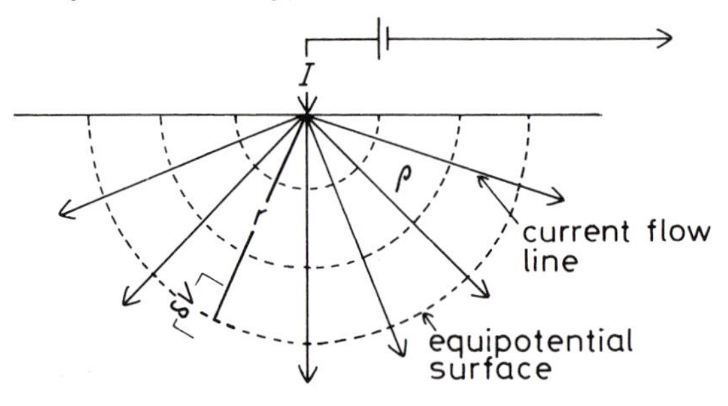

Fig. 8.3. Current flow from a single surface electrode.

Now consider a single current electrode on the surface of a medium of uniform resistivity ρ (Fig. 8.3). The circuit is completed by a current sink at a large distance from the electrode. Current flows radially away from the electrode so that the current distribution is uniform over hemispherical shells centred on the source. At a distance r from the electrode the shell has a surface area of $2\pi r^2$, so the current density i is given by

$$i = \frac{I}{2\pi r^2} \tag{8.4}$$

From equation (8.3), the potential gradient associated with this current density is

$$\frac{\partial V}{\partial r} = -\rho i = -\frac{\rho I}{2\pi r^2} \tag{8.5}$$

The potential V_r at distance r is then obtained by integration

$$V_r = \int \partial V = -\int \frac{\rho I \partial r}{2\pi r^2} = \frac{\rho I}{2\pi r} \tag{8.6}$$

The constant of integration is zero since $V_r = 0$ when $r = \infty$.

Equation (8.6) allows the calculation of the potential at any point on or below the surface of a homogeneous half space. The hemispherical shells in Fig. 8.3 mark surfaces of constant voltage and are termed *equipotential surfaces*.

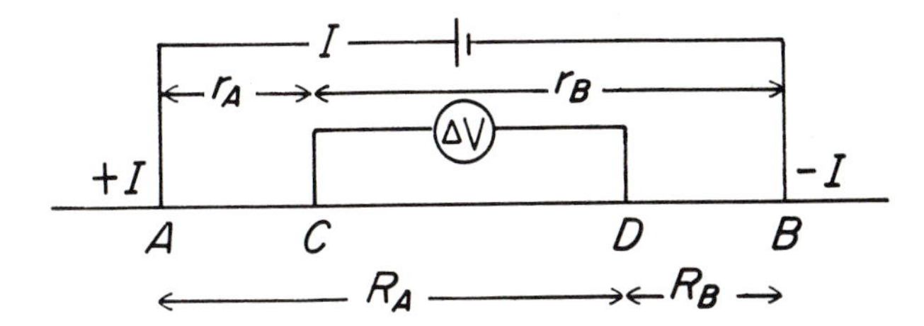

Fig. 8.4. The generalized form of the electrode configuration used in resistivity measurements.

Now consider the case where the current sink is a finite distance from the source (Fig. 8.4). The potential V_C at an internal electrode C is the sum of the potential contributions V_A and V_B from the current source at A and the sink at B

$$V_C = V_A + V_B$$

From equation (8.6)

$$V_C = \frac{\rho I}{2\pi}\left(\frac{1}{r_A} - \frac{1}{r_B}\right) \tag{8.7}$$

Similarly

$$V_D = \frac{\rho I}{2\pi}\left(\frac{1}{R_A} - \frac{1}{R_B}\right) \tag{8.8}$$

Absolute potentials are difficult to monitor so the potential difference ΔV between electrodes C and D is measured

$$\Delta V = V_C - V_D = \frac{\rho I}{2\pi}\left\{\left(\frac{1}{r_A} - \frac{1}{r_B}\right) - \left(\frac{1}{R_A} - \frac{1}{R_B}\right)\right\}$$

Thus

$$\rho = \frac{2\pi\Delta V}{\mathrm{I}\left\{\left(\frac{1}{r_A} - \frac{1}{r_B}\right) - \left(\frac{1}{R_A} - \frac{1}{R_B}\right)\right\}} \tag{8.9}$$

Where the ground is uniform, the resistivity calculated from equation (8.9) should be constant and independent of both electrode spacing and surface location. When subsurface inhomogeneities exist, however, the resistivity will vary with the relative positions of the electrodes. Any computed value is then known as the *apparent resistivity* ρ_a and will be a function of the form of the inhomogeneity. Equation (8.9) is the basic equation for calculating the apparent resistivity for any electrode configuration.

In homogeneous ground the depth of current penetration increases as the separation of the current electrodes is increased, and Fig. 8.5 shows the proportion of current flowing beneath a given depth Z as the ratio of electrode separation L to depth increases. When $L = Z$ about 30% of the current flows below

Z and when $L = 2Z$ about 50% of the current flows below Z. The current electrode separation must be chosen so that the ground is energized to the required depth, and should be at least equal to this depth. This places practical limits on the depths of penetration attainable by normal resistivity methods due to the difficulty in laying long lengths of cable and the generation of sufficient power. Depths of penetration of about 1 km are the limit for normal equipment.

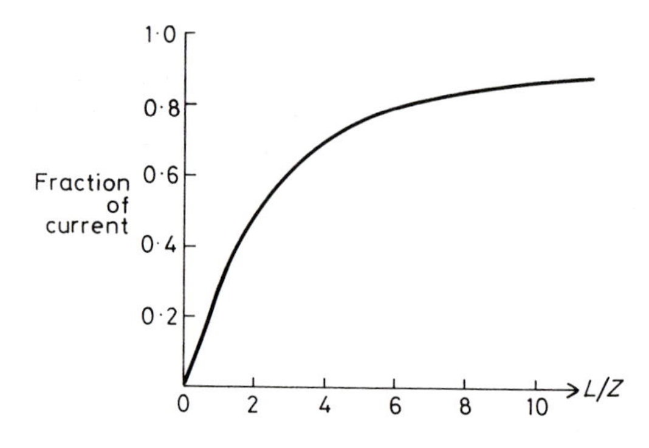

Fig. 8.5. The fraction of current penetrating below a depth Z for a current electrode separation L. (After Telford *et al.* 1976.)

Two main types of procedure are employed in resistivity surveys.

Vertical electrical sounding (VES), also known as 'electrical drilling' or 'expanding probe', is used mainly in the study of horizontal or near-horizontal interfaces. The current and potential electrodes are maintained at the same relative spacing and the whole spread is progressively expanded about a fixed central point. Consequently, readings are taken as the current reaches progressively greater depths. The technique is extensively used in geotechnical surveys to determine overburden thickness and also in hydrogeology to define horizontal zones of porous strata.

Constant separation traversing (CST), also known as 'electrical profiling', is used to determine lateral variations of resistivity. The current and potential electrodes are maintained at a fixed separation and progressively moved along a profile. This method is employed in mineral prospecting to locate faults or shear zones and to detect localized bodies of anomalous conductivity. It is also used in geotechnical surveys to determine variations in bedrock depth and the presence of steep discontinuities. Results from a series of CST traverses with a fixed electrode spacing can be employed in the production of resistivity contour maps.

8.2.4 Electrode spreads

Many configurations of electrodes have been designed (Habberjam 1979) and although several are occasionally employed in specialized surveys only two are in common use. The *Wenner configuration* is the simpler in that current and potential electrodes are maintained at an equal spacing a (Fig. 8.6). Substitution of this condition into equation (8.9) yields

$$\rho_a = 2\pi a \frac{\Delta V}{I} \tag{8.10}$$

During VES the spacing a is gradually increased about a fixed central point and in CST the whole spread is moved along a profile with a fixed value of a. The efficiency of performing vertical electrical sounding can be greatly increased by making use of a multicore cable to which a number of electrodes are permanently attached at standard separations (Barker 1981). A sounding can then be rapidly accomplished by switching between different sets of four electrodes. Such a system has the additional advantage that, by measuring ground resistances at two electrode array positions, the effects of near-surface lateral resistivity variations can be substantially reduced.

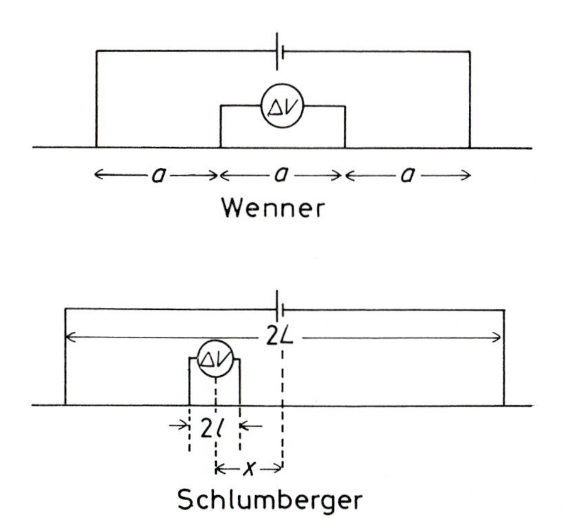

Fig. 8.6. The Wenner and Schlumberger electrode configurations.

In surveying with the Wenner configuration all four electrodes need to be moved between successive readings. This labour is partially overcome by the use of the *Schlumberger configuration* (Fig. 8.6) in which the inner, potential electrodes have a spacing $2l$ which is a small proportion of that of the outer, current electrodes ($2L$). In CST surveys with the Schlumberger configuration several lateral movements of the potential electrodes may be accommodated without the necessity of moving the current electrodes. In VES surveys the potential electrodes remain fixed and the current electrodes are expanded symmetrically about the centre of the spread. With very large values of L it may, however, be necessary to increase l also in order to maintain a measurable potential.

For the Schlumberger configuration

$$\rho_a = \frac{\pi}{2l} \frac{(L^2-x^2)^2}{(L^2+x^2)} \frac{\Delta V}{I} \tag{8.11}$$

where x is the separation of the mid-points of the potential and current electrodes. When used symmetrically, $x = 0$ so

$$\rho_a = \frac{\pi L^2}{2l} \frac{\Delta V}{I} \tag{8.12}$$

8.2.5 *Resistivity surveying equipment*

Resistivity survey instruments are designed to measure the resistance of the ground, i.e. the ratio $(\Delta V/I)$ in equations (8.10), (8.11) and (8.12), to a very high accuracy. They must be capable of reading to the very low levels of resistance commonly encountered in resistivity surveying. Apparent resistivity values are computed from the resistance measurements using the formula relevant to the electrode configuration in use.

Most modern resistivity meters employ low-frequency alternating current rather than direct current for two main reasons. Firstly, if direct current were employed there would eventually be a build up of anions around the negative electrode and cations around the positive electrode, i.e. electrolytic polarization would occur, and this would inhibit the arrival of further ions at the electrodes. Periodic reversal of the current prevents such an accumulation of ions and thus overcomes electrolytic polarization. Secondly, the use of alternating current overcomes the effects of telluric currents (see Chapter 9), which are natural electric currents in the ground that flow parallel to the Earth's surface and cause regional potential gradients. The use of alternating current nullifies their effects since at each current reversal the telluric currents alternately increase or decrease the measured potential difference by equal amounts. Summing the results over several cycles thus removes telluric effects (Fig. 8.7). The frequency of the alternating current used in resistivity surveying depends upon the required depth of penetration (see equation (9.2)). For penetration of the order of 10 m, a frequency of 100 Hz is suitable, and this is decreased to less than 10 Hz for depths of investigation of about 100 m. For very deep ground penetration direct currents must be used, and more complex measures adopted to overcome electrolytic polarization and telluric current effects.

Resistivity meters are designed to measure potential differences when no current is flowing. Such a null method is used to overcome the effects of contact

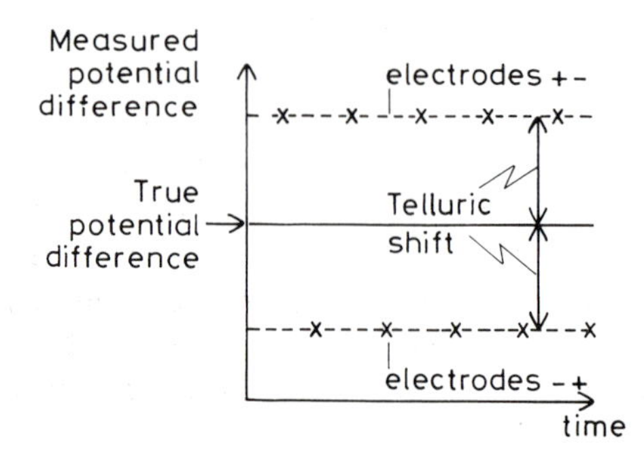

Fig. 8.7. The use of alternating current to remove the effects of telluric currents during a resistivity measurement. Summing the measured potential difference over several cycles provides the true potential difference.

resistance of the electrodes with the ground. The potential between the potential electrodes is balanced by the potential tapped from a variable resistance. No current then flows in the resistivity circuit so that contact resistance will not register, and the variable resistance reading represents the true resistance of the ground (equal to the ratio $\Delta V/I$ in the relevant equations).

Previous generations of resistivity meters required the nulling of a displayed voltage by manual manipulation of a resistor bank. Modern instruments are available with microprocessor-controlled electronic circuitry which accomplishes this operation internally and, moreover, performs checks on the circuitry before display of the result.

8.2.6 *Interpretation of resistivity data*

Electrical surveys are among the most difficult of all the geophysical methods to interpret quantitatively because of the complex theoretical basis of the technique. In resistivity interpretation mathematical analysis is most highly developed for VES, less well for CST over two-dimensional structures and least well for CST over three-dimensional bodies. The resistivity method utilizes a potential field and consequently suffers from similar ambiguity problems to the gravitational and magnetic methods.

Since a potential field is involved, the apparent resistivity signature of any structure should be computed by solution of Laplace's equation (Section 6.12) and insertion of the boundary conditions for the particular structure under consideration, or by integrating it directly. In practice such solutions are invariably complex. Consequently a simplified approach is initially adopted here in which electric fields are assumed to act in a manner similar to light. It should be remembered, however, that such an optical analogue is not strictly valid in all cases.

8.2.7 *Vertical electrical sounding interpretation*

Consider a Wenner electrode spread above a single horizontal interface between media with resistivities ρ_1 (upper) and ρ_2 (lower) with $\rho_1 > \rho_2$ (Fig. 8.8). On passing through the interface the current flow lines are deflected towards the interface in a fashion similar to refracted seismic waves (Chapter 3) since the less-resistive lower layer provides a more attractive path for the current. When the electrode separation is small, most of the current flows in the upper layer with the consequence that the apparent resistivity tends towards ρ_1. As the electrode separation is gradually increased more and more current flows within the lower layer and the apparent resistivity then approaches ρ_2. A similar situation obtains when $\rho_2 > \rho_1$, although in this case the apparent resistivity approaches ρ_2 more gradually as the more resistive lower layer is a less attractive path for the current.

Where three horizontal layers are present the apparent resistivity curves are more complex (Fig. 8.9). Although the apparent resistivity approaches ρ_1 and ρ_3 for small and large electrode spacings, the presence of the intermediate layer causes a deflection of the apparent resistivity curve at intermediate spacings. If the resistivity of the intermediate layer is greater or less than the resistivities of the

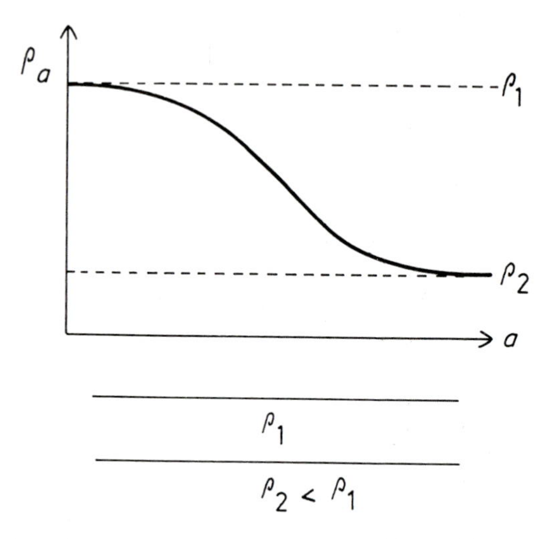

Fig. 8.8. The variation of apparent resistivity ρ_a with electrode separation a over a single horizontal interface between media with resistivities ρ_1 and ρ_2.

upper and lower layers the apparent resistivity curve is either bell-shaped or basin-shaped (Fig. 8.9(a)). A middle layer with a resistivity intermediate between ρ_1 and ρ_3 produces apparent resistivity curves characterized by a progressive increase or decrease in resistivity as a function of electrode spacing (Fig. 8.9(b)). The presence of four or more layers further increases the complexity of apparent resistivity curves.

Simple examination of the way in which apparent resistivity varies with electrode spacing may thus provide estimates of the resistivities of the upper and lowest layers and indicate the relative resistivities of any intermediate layers. In order to compute layer thicknesses it is necessary to be able to calculate the apparent resistivity of a layered structure. The first computation of this type was performed by Hummel in the 1930s using an optical analogue to calculate the apparent resistivity signature of a simple two-layered model.

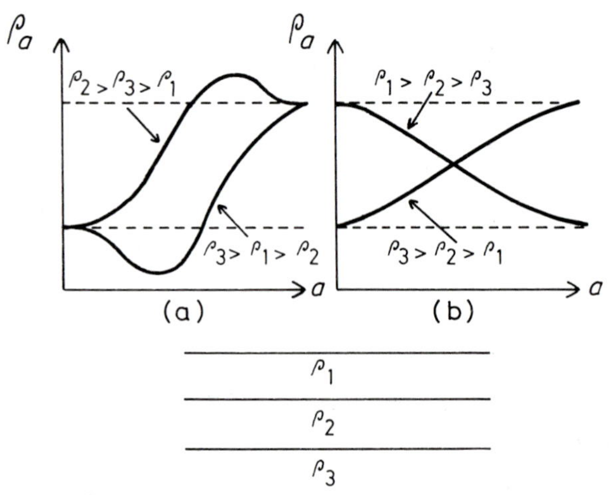

Fig. 8.9. The variation of apparent resistivity ρ_a with electrode separation a over three horizontal layers.

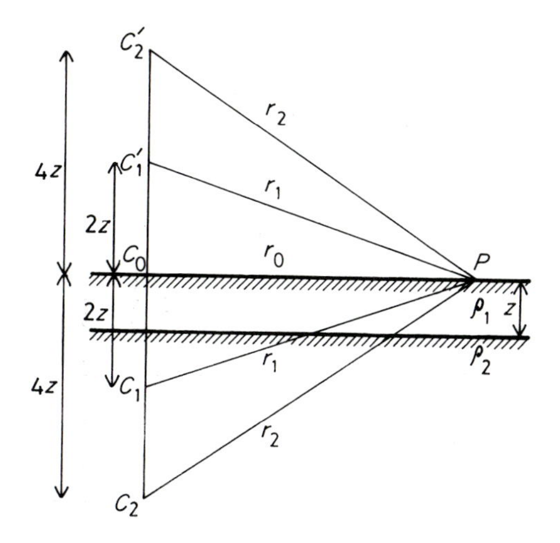

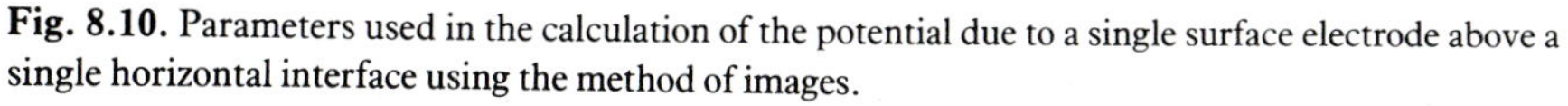

Fig. 8.10. Parameters used in the calculation of the potential due to a single surface electrode above a single horizontal interface using the method of images.

Referring to Fig. 8.10, current I is introduced into the ground at point C_0 above a single interface at depth z between an upper medium 1 of resistivity ρ_1 and a lower medium 2 of resistivity ρ_2. The two parallel interfaces between media 1 and 2 and between medium 1 and the air produce an infinite series of images of the source, located above and below the surface. Thus C_1 is the image of C_0 in the medium 1/2 interface at depth $2z$, C_1' is the image of C_1 in the medium 1/air interface at height $2z$, C_2 is the image of C_1' in the medium 1/2 interface at depth $4z$, etc. Each image in the medium 1/2 interface is reduced in intensity by a factor k, the reflection coefficient of the interface. (There is no reduction in intensity of images in the medium 1/air interface, as its reflection coefficient is unity.) A consequence of the progressive reduction in intensity is that only a few images have to be considered in arriving at a reasonable estimate of the potential at point P.

Table 8.1 summarizes this argument.

Table 8.1.

Source	Intensity	Depth/height	Distance
C_0	I	0	r_0
C_1	kI	$2z$	r_1
C_1'	kI	$2z$	r_1
C_2	k^2I	$4z$	r_2
C_2'	k^2I	$4z$	r_2
.	.	.	.
.	.	.	.

The potential V_p at point P is the sum of the contributions of all sources. Employing equation (8.6)

$$V_p = \frac{I\rho_1}{2\pi r_0} + \frac{2kI\rho_1}{2\pi r_1} + \frac{2k^2I\rho_1}{2\pi r_2} + \cdots + \frac{2k^iI\rho_1}{2\pi r_i} + \cdots$$

Thus
$$V_p = \frac{I\rho_1}{2\pi}\left(\frac{1}{r_0} + 2\sum_{n=1}^{\infty}\frac{k^n}{r_n}\right) \tag{8.13}$$

where
$$r_n = (r_0^2 + (2nz)^2)^{1/2}$$

The first term in the brackets of equation (8.13) refers to the normal potential attaining if the subsurface were homogeneous, and the second term to the disturbing potential caused by the interface. The series is convergent as the dimming factor, or reflection coefficient, k is less than unity ($k = (\rho_2 - \rho_1)/(\rho_2 + \rho_1)$, cf. Section 3.6.1).

Knowledge of the potential resulting at a single point from a single current electrode allows the computation of the potential difference ΔV between two electrodes, resulting from two current electrodes, by the addition and subtraction of their contribution to the potential at these points. For the Wenner system with spacing a

$$\Delta V = \frac{I\rho_1}{2\pi a}(1 + 4F) \tag{8.14}$$

where
$$F = \sum_{n=1}^{\infty} k^n \left[(1 + 4n^2z^2/a^2)^{-1/2} - (4 + 4n^2z^2/a^2)^{-1/2}\right] \tag{8.15}$$

Relating this to the apparent resistivity ρ_a measured by the Wenner system (equation (8.10))

$$\rho_a = \rho_1(1 + 4F) \tag{8.16}$$

Consequently the apparent resistivity can be computed for a range of electrode spacings.

Similar computations can be performed for multilayer structures, although the calculations are more easily executed using recurrence formulae and filtering techniques designed for this purpose (see later). Field data can then be compared

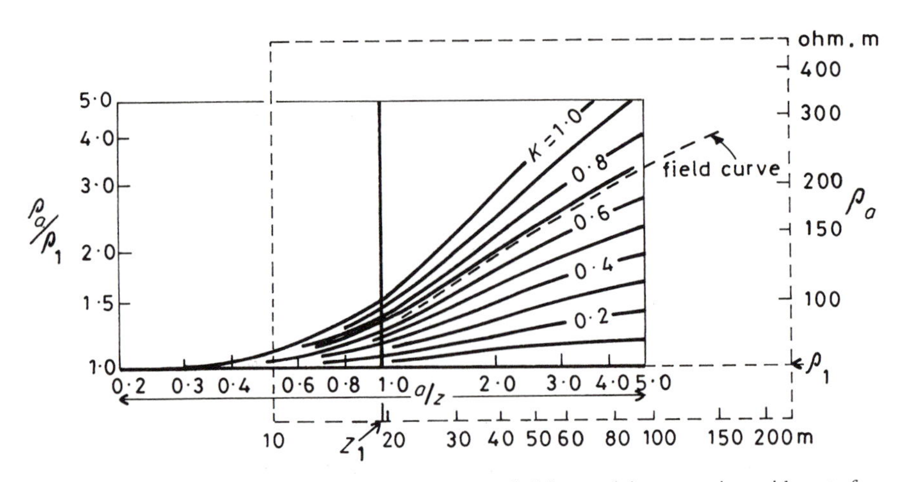

Fig. 8.11. The interpretation of a two-layer apparent resistivity graph by comparison with a set of master curves. The upper layer resistivity ρ_1 is 68 Ω. m and its thickness z_1 is 19.5 m. (After Griffiths & King 1981.)

with graphs (master curves) representing the calculated effects of layered models derived by such methods, a technique known as *curve matching*. Figure 8.11 shows an interpretation using a set of master curves for vertical electrical sounding with a Wenner spread over two horizontal layers. The master curves are prepared in dimensionless form for a number of values of the reflection coefficient k by dividing the calculated apparent resistivity values ρ_a by the upper layer resistivity ρ_1 (the latter derived from the field curve at electrode spacings approaching zero), and by dividing the electrode spacings a by the upper layer thickness z_1. The curves are plotted on logarithmic paper, which has the effect of producing a more regular appearance as the fluctuations of resistivity then tend to be of similar wavelength over the entire length of the curves. The field curve to be interpreted is plotted on transparent logarithmic paper with the same modulus as the master curves. It is then shifted over the master curves, keeping the coordinate axes parallel, until a reasonable match is obtained with one of the master curves or with an interpolated curve. The point at which $\rho_a/\rho_1 = a/z = 1$ on the master sheet gives the true values of ρ_1 and z_1 on the relevant axes. ρ_2 is obtained from the k-value of the best-fitting curve.

Curve matching is simple for the two-layer case since only a single sheet of master curves is required. When three layers are present much larger sets of curves are required to represent the increased number of possible combinations of resistivities and layer thicknesses. Curve matching is simplified if the master curves are arranged according to curve type (Fig. 8.9), and sets of master curves are arranged according to curve type (Fig. 8.9), and sets of master curves for both Wenner and Schlumberger electrode configurations are available (Orrellana & Mooney 1966, 1972). The number of master curves required for full interpretation of a four-layer field curve is prohibitively large although limited sets have been published.

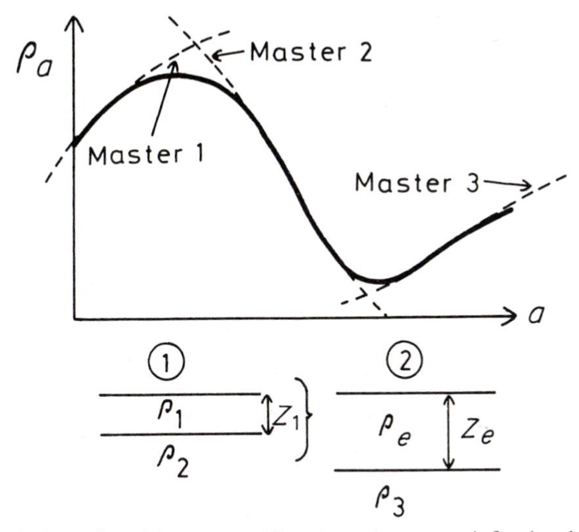

Fig. 8.12. The technique of partial curve matching. A two-layer curve is fitted to the early part of the graph and the resistivities ρ_1 and ρ_2 and thickness z_1 of the upper layer determined. ρ_1, ρ_2 and z_1 are combined into a single equivalent layer of resistivity ρ_e and thickness z_e which then forms the upper layer in the interpretation of the next segment of the graph with a second two-layer curve.

The interpretation of resistivity curves over multilayered structures may alternatively be performed by *partial curve matching* (Bhattacharya & Patra 1968). The method involves the matching of successive portions of the field curve by a set of two-layer curves. After each segment is fitted the interpreted resistivities and layer thickness are combined by use of auxilliary curves into a single layer with an equivalent thickness z_e and resistivity ρ_e. This equivalent layer then forms the upper layer in the interpretation of the next segment of the field curve with another two-layer curve (Fig. 8.12). Similar techniques are available in which successive use is made of three-layer master curves.

The curve matching methods described above are not now widely used because of the general availability of the more sophisticated interpretational techniques described below. Curve matching methods might still be used, however, to obtain interpretations in the field in the absence of computing facilities, or to derive an approximate model that is to be used as a starting point for one of the more complex routines.

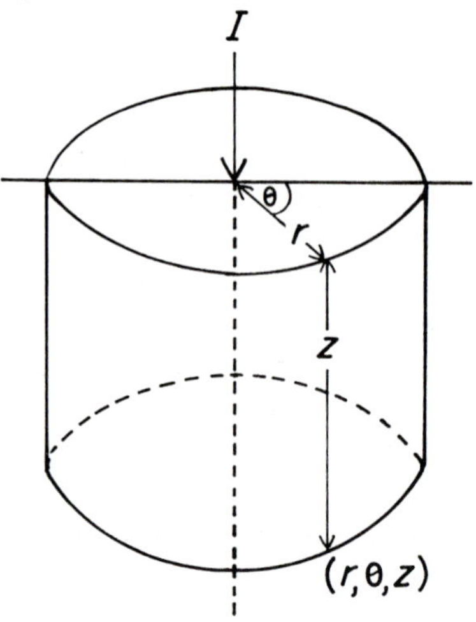

Fig. 8.13. Cylindrical polar coordinates.

Equation (8.13) represents the potential at the surface resulting from a single point of current injection over two horizontal layers as predicted by the method of images. In general, however, the potential arising from any number of horizontal layers is derived by solution of Laplace's equation (Section 6.12). The equation in this case is normally represented in cylindrical coordinates as electrical fields have cylindrical symmetry with respect to the vertical line through the current source (Fig. 8.13). The solution and application of the relevant boundary conditions are complex (e.g. Koefoed 1979), but show that the potential V at the surface over a series of horizontal layers, the uppermost of which has a resistivity ρ_1, at a distance r from a current source of strength I is given by

$$V = \frac{\rho_1 I}{2\pi} \int_0^\infty K(\lambda)\, J_0(\lambda r) d\lambda \qquad (8.17)$$

λ is the variable of integration. $J_0(\lambda r)$ is a specialized function known as a

Bessel function of order zero whose behaviour is known completely. $K(\lambda)$ is known as a kernel function and is controlled by the thicknesses and resistivities of the underlying layers. The kernel function can be built up relatively simply for any number of layers using *recurrence relationships* (Koefoed 1979) which progressively add the effects of successive layers in the sequence. A useful additional parameter is the resistivity transform $T(\lambda)$ defined by

$$T_i(\lambda) = \rho_i K_i(\lambda) \tag{8.18}$$

where $T_i(\lambda)$ is the resistivity transform of the ith layer which has a resistivity ρ_i and a kernel function $K_i(\lambda)$. $T(\lambda)$ can similarly be constructed using recurrence relationships.

By methods analogous to those used to construct equation (8.16), a relationship between the apparent resistivity and resistivity transform can be derived. For example, this relationship for a Wenner spread with electrode spacing a is

$$\rho_a = 2a \int_0^\infty T(\lambda)[\mathcal{J}_0(\lambda a) - \mathcal{J}_0(2\lambda a)]d\lambda \tag{8.19}$$

The resistivity transform function has the dimensions of resistivity and the variable λ has the dimensions of inverse length. It has been found that if $T(\lambda)$ is plotted as a function of λ^{-1} the relationship is similar to the variation of apparent resistivity with electrode spacing for the same sequence of horizontal layers. Indeed only a simple filtering operation is required to transform the $T(\lambda)$:λ^{-1} relationship (resistivity transform) into the ρ_a: a relationship (apparent resistivity function). Such a filter is known as an indirect filter. The inverse operation, i.e. the determination of the resistivity transform from the apparent resistivity function, can be performed using a direct filter.

Apparent resistivity curves over multilayered models can be computed relatively easily by determining the resistivity transform from the layer parameters using a recurrence relationship and then filtering the transform to derive the apparent resistivity function. Such a technique is considerably more efficient than the method used in the derivation of equation (8.13).

This method leads to a form of interpretation similar to the indirect interpretation of gravity and magnetic anomalies, in which field data are compared with data calculated for a model whose parameters are varied in order to simulate the field observations. This comparison can be made either between observed and calculated apparent resistivity profiles or the equivalent resistivity transforms, the latter method requiring the derivation of the resistivity transform from the field resistivity data by direct filtering. Such techniques lend themselves well to automatic iterative processes of interpretation in which a computer performs the adjustments necessary to a layered model derived by an approximate interpretation method in order to improve the correspondence between observed and calculated functions.

In addition to this indirect modelling there are also a number of direct methods of interpreting resistivity data which derive the layer parameters directly from the field profiles. Such methods usually involve the following steps:

(1) Determination of the resistivity transform of the field data by direct filtering.
(2) Determination of the parameters of the upper layer by fitting the early part of the resistivity transform curve with a synthetic two-layer curve.
(3) Subtraction of the effects of the upper layer by reducing all observations to the base of the previously determined layer by the use of a *reduction equation* (the inverse of a recurrence relationship).

Steps (2) and (3) are then repeated so that the parameters of successively deeper layers are determined. Such methods suffer from the drawback that errors increase with depth so that any error made early in the interpretation becomes magnified. The direct interpretation methods consequently employ various techniques to suppress such error magnification.

The indirect and direct methods described above have now largely superseded curve-matching techniques and provide considerably more accurate interpretations.

Interpretation of VES data suffers from non-uniqueness arising from problems known as *equivalence* and *suppression*. The problem of equivalence is illustrated by the fact that identical bell-shaped or basin-shaped resistivity curves (Fig. 8.9(a)) can be obtained for different layered models. Identical bell-shaped curves are obtained if the product of the thickness z and resistivity ρ, known as the transverse resistance, of the middle layer remains constant. For basin-shaped curves the equivalence function of the middle layer is z/ρ, known as the longitudinal conductance. The problem of suppression applies to resistivity curves in which apparent resistivity progressively increases or decreases as a function of electrode spacing (Fig. 8.9(b)). In such cases the addition of an extra intermediate layer causes a slight horizontal shift of the curve without altering its overall shape. In the interpretation of relatively noisy field data such an intermediate layer may not be detected.

It is the conventional practice in VES interpretation to make the assumption that layers are horizontal and isotropic. Deviations from these assumptions result in errors in the final interpretation.

The assumption of isotropy can be incorrect for individual layers. For example, in sediments such as clay or shale the resistivity perpendicular to the layering is usually greater than in the direction of the layering. Anisotropy cannot be detected in subsurface layers during vertical electrical sounding and normally results in too large a thickness being assigned to the layers. Other anisotropic effects are depth dependent, e.g. the reduction with depth of the degree of weathering, and the increase with depth of both compaction of sediments and salinity of pore fluids. The presence of a vertical contact, such as a fault, gives rise to lateral inhomogeneity which can greatly affect the interpretation of an electrical sounding in its vicinity.

If the layers are dipping, the basic theory discussed above is invalid. Using the optical analogue, the number of images produced by a dipping interface is finite, the images being arranged around a circle (Fig. 8.14). Because the intensity of the images progressively decreases, only the first few need to be considered in deriving a reasonable estimate of the resulting potential. Consequently the effect

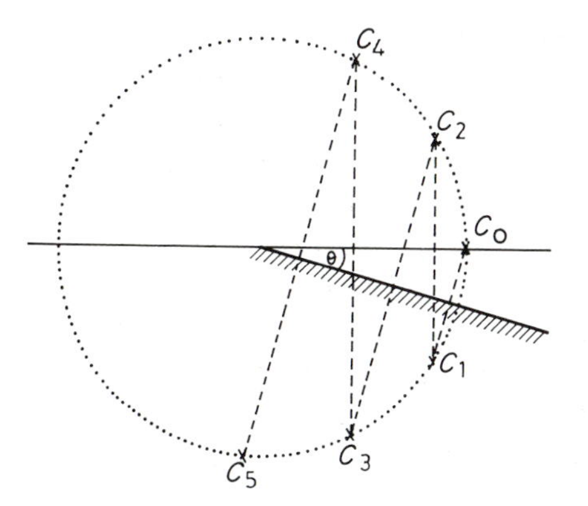

Fig. 8.14. Apparent current sources caused by a dipping interface. The sources C_1–C_5 are successive images of the primary source C_0 in the interface and the surface. The sources lie on a circle centred on the outcrop of the interface, and their number is dependent upon the magnitude of the dip of the interface θ.

of dip can probably be ignored for inclinations up to about 20°, which provide a sufficient number of images.

Topography can influence electrical surveys as current flow lines tend to follow the ground surface. Equipotential surfaces are thus distorted and anomalous readings can result.

8.2.8 Constant separation traversing interpretation

Constant separation traverses are obtained by moving an electrode spread with fixed electrode separation along a traverse line, the array of electrodes being aligned either in the direction of the traverse (longitudinal traverse) or at right angles to it (transverse traverse). The former technique is more efficient as only a single electrode has to be moved from one end of the spread to the other, and the electrodes reconnected, between adjacent readings.

Fig. 8.15 shows a transverse traverse across a single vertical contact between two media of resistivities ρ_1 and ρ_2. The apparent resistivity curve varies smoothly from ρ_1 to ρ_2 across the contact.

A longitudinal traverse over a similar structure shows the same variation from ρ_1 to ρ_2 at its extremities, but the intermediate parts of the curve exhibit a number of cusps (Fig. 8.16), which correspond to locations where successive electrodes cross the contact. There will be four cusps on a Wenner profile but two on a Schlumberger profile where only the potential electrodes are mobile.

Fig. 8.17 shows the results of transverse and longitudinal traversing across a series of faulted strata in Illinois, USA. Both sets of results illustrate well the strong resistivity contrasts between the relatively conductive sandstone and relatively resistive limestone.

A vertical discontinuity distorts the direction of current flow and thus the overall distribution of potential in its vicinity. The potential distribution at the surface can be determined by an optical analogue in which the discontinuity is

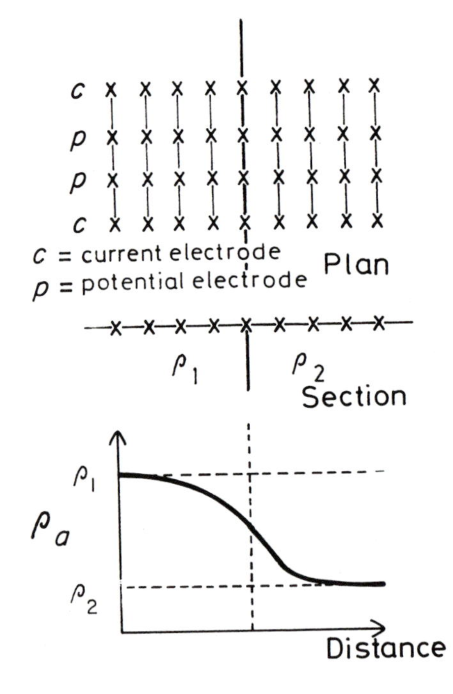

Fig. 8.15. A transverse traverse across a single vertical interface.

compared with a semi-transparent mirror which both reflects and transmits light. Referring to Fig. 8.18, current I is introduced at point C on the surface of a medium of resistivity ρ_1 in the vicinity of a vertical contact with a second medium of resistivity ρ_2.

In the optical analogue, a point P on the same side of the mirror as the source

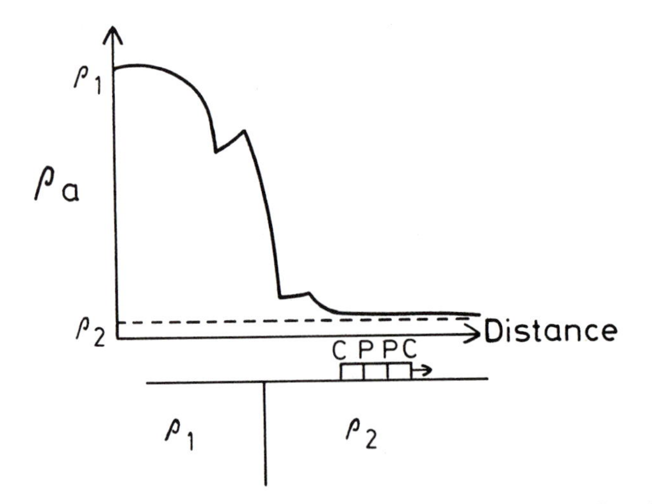

Fig. 8.16. A longitudinal traverse across a single vertical interface employing a configuration in which all four electrodes are mobile. (Redrawn from Parasnis 1973.)

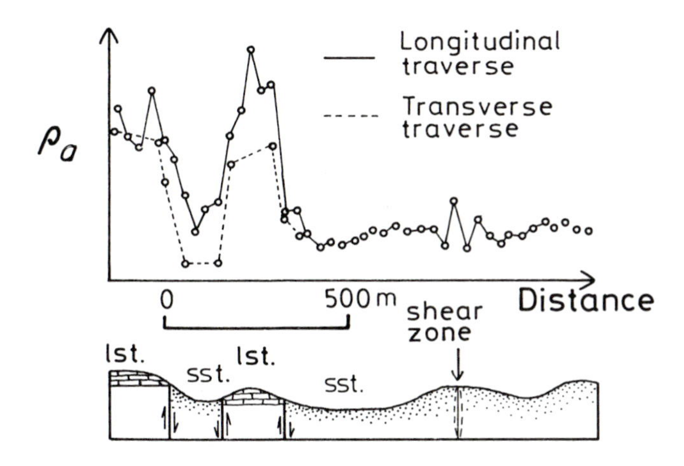

Fig. 8.17. Longitudinal and transverse traverses across a series of faulted strata in Illinois, USA. (Redrawn from Hubbert 1934.)

would receive light directly and via a single reflection. In the latter case the light would appear to originate from the image of C in the mirror C' and would be decreased in intensity with respect to the source by a factor corresponding to the reflection coefficient. Similarly, both the electric source and its image contribute to the potential V_p at P, the latter being decreased in intensity by a factor k, the reflection coefficient. From equation (8.6)

$$V_p = \frac{I\rho_1}{2\pi}\left(\frac{1}{r_1}+\frac{k}{r_2}\right) \tag{8.20}$$

For a point P' on the other side of the interface from the source, the optical analogue indicates that light would be received only after transmission through the mirror, resulting in a reduction in intensity by a factor corresponding to the transmission coefficient. Similarly, the only contributor to the potential $V_{p'}$ at P' is the current source reduced in intensity by the factor $(1-k)$. From equation (8.6)

$$V_{p'} = \frac{I(1-k)\rho_2}{2\pi r_3} \tag{8.21}$$

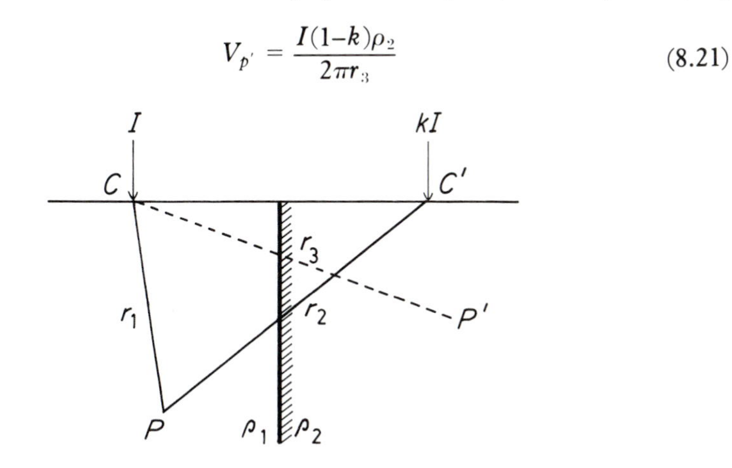

Fig. 8.18. Parameters used in the calculation of the potential due to a single surface current electrode on either side of a single vertical interface.

Equations (8.20) and (8.21) may be used to calculate the measured potential difference for any electrode spread between two points in the vicinity of the interface and thus to construct the form of an apparent resistivity profile produced by longitudinal constant separation traversing. In fact, five separate equations are required, corresponding to the five possible configurations of a four-electrode spread with respect to the discontinuity. The method can also be used to construct apparent resistivity profiles for constant separation traversing over a number of adjacent discontinuities. Albums of master curves are available for single and double vertical contacts (Logn 1954).

Three-dimensional resistivity anomalies may be obtained by contouring apparent resistivity values from a number of CST lines. The detection of a three-dimensional body is usually only possible when its top is close to the surface, and traverses must be made directly over the body or very near to its edges if its anomaly is to be registered.

Three-dimensional anomalies may be interpreted by laboratory modelling. For example, metal cylinders, blocks or sheets may be immersed in water whose resistivity is altered by adding various salts and the model moved beneath a set of stationary electrodes. The shape of the model can then be varied until a reasonable approximation to the field curves is obtained.

The mathematical analysis of apparent resistivity variations over bodies of regular or irregular form is complex but equations are available for simple shapes such as spheres or hemispheres (Fig. 8.19), and it is also possible to compute the resistivity response of two-dimensional bodies with an irregular cross-section (Dey & Morrison 1979).

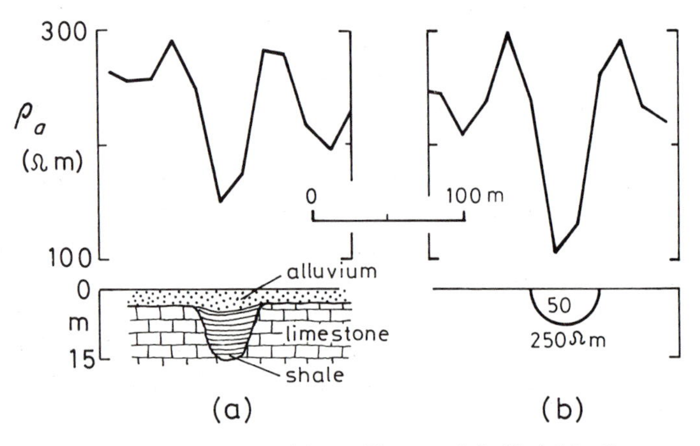

Fig. 8.19. (a) The observed Wenner resistivity profile over a shale-filled sink of known geometry in Kansas, USA. (b) The theoretical profile for a buried hemisphere. (Redrawn from Cook & Van Nostrand 1954.)

Three-dimensional anomalies may also be obtained by an extension of the CST technique known as the *mise-á-la-masse method*. This is employed when part of a conductive body, for example, an ore body, has been located either at outcrop or by drilling. One current electrode is sited within the body, the other being placed a large distance away on the surface (Fig. 8.20). A pair of potential

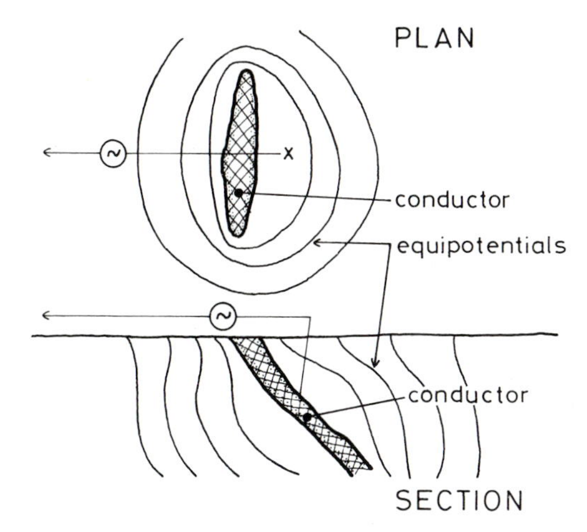

Fig. 8.20. The mise-à-la-masse method.

electrodes is then moved over the surface mapping equipotential lines (lines joining the electrodes when the indicated potential difference is zero). The method provides much more information on the extent, dip, strike and continuity of the body than the normal CST techniques.

8.2.9 Limitations of the resistivity method

Resistivity surveying is an efficient method for delineating shallow layered sequences or vertical discontinuities involving changes of resistivity. It does, however, suffer from a number of limitations:

(1) Interpretations are ambiguous. Consequently, independent geophysical and geological controls are necessary to discriminate between valid alternative interpretations of the resistivity data.

(2) Interpretation is limited to simple structural configurations. Any deviations from these simple situations may be impossible to interpret.

(3) Topography and the effects of near-surface resistivity variations can mask the effects of deeper variations.

(4) The depth of penetration of the method is limited by the maximum electrical power that can be introduced into the ground and by the practical difficulties of laying out long lengths of cable. The practical depth limit for most surveys is about 1 km.

8.3 Induced polarization (IP) method

8.3.1 Principles

If, when using a standard four-electrode resistivity spread in a DC mode, the current is abruptly switched off, the voltage between the potential electrodes does not drop to zero immediately. After a large initial decrease the voltage suffers a gradual decay and can take many seconds to reach a zero value (Fig. 8.21). A similar phenomenon is observed as the current is switched on. After an initial sudden voltage increase, the voltage increases gradually over a discrete time interval to a steady state value. The ground thus acts as a capacitor and stores electrical charge, i.e. becomes electrically polarized.

If, instead of using a DC source for the measurement of resistivity, a variable low frequency AC source is used, it is found that the measured apparent resistivity of the subsurface decreases as the frequency is increased. This is because the capacitance of the ground inhibits the passage of direct currents but transmits alternating currents with increasing efficiency as the frequency rises.

The capacitative property of the ground causes both the transient decay of a residual voltage and the variation of apparent resistivity as a function of frequency. The two effects are representations of the same phenomenon in the time and frequency domains, and are linked by Fourier transformation (see Chapter 2). These two manifestations of the capacitance property of the ground provide two different survey methods for the investigation of the effect.

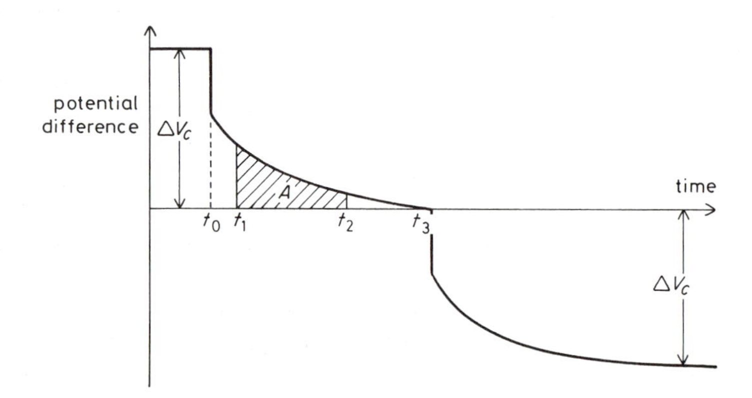

Fig. 8.21. The phenomenon of induced polarization. At time t_0 the current is switched off and the measured potential difference, after an initial large drop from the steady state value ΔV_c, decays gradually to zero. A similar sequence occurs when the current is switched on at time t_3. A represents the area under the decay curve for the time increment t_1–t_2.

The measurement of a decaying voltage over a certain time interval is known as *Time Domain* IP surveying. Measurement of apparent resistivity at two or more low AC frequencies is known as *Frequency Domain* IP surveying.

8.3.2 *Mechanisms of induced polarization*

Laboratory experiments indicate that electrical energy is stored in rocks mainly by electrochemical processes. This is achieved in two ways.

The passage of current through a rock as a result of an impressed voltage is accomplished mainly by electrolytic flow in the pore fluid. Most of the rock-forming minerals have a net negative charge on their interface with the pore fluid and attract positive ions onto this surface (Fig. 8.22(a)). The concentration of positive ions extends about 100 μm into the pore fluid, and if this distance is of the same order as the diameter of the pore throats, the movement of ions in the fluid resulting from the impressed voltage is inhibited. Negative and positive ions thus build up on either side of the blockage and, on removal of the impressed voltage, return to their original locations over a finite period of time causing a gradually decaying voltage.

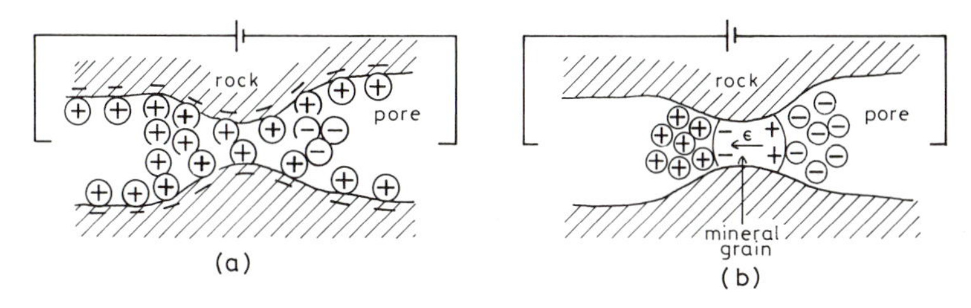

Fig. 8.22. Mechanisms of induced polarization (a) membrane polarization (b) electrode polarization.

This effect is known as *membrane polarization* or *electrolytic polarization*. It is most pronounced in the presence of clay minerals where the pores are particularly small, but the effect decreases with increasing salinity of the pore fluid.

When metallic minerals are present in a rock, an alternative, electronic, path is available for current flow. Fig. 8.22(b) shows a rock in which a metallic mineral grain blocks a pore. When a voltage is applied to either side of the pore space, positive and negative charges are impressed on opposite sides of the grain. Negative and positive ions then accumulate on either side of the grain which are attempting either to release electrons to the grain or to accept electrons conducted through the grain. The rate at which the electrons are conducted is slower than the rate of electron exchange with the ions. Consequently ions accumulate on either side of the grain and cause a build up of charge. When the impressed voltage is removed the ions slowly diffuse back to their original locations and cause a transitory decaying voltage.

This effect is known as *electrode polarization* or *overvoltage*. All minerals which are good conductors (e.g. metallic sulphides and oxides, graphite) contribute to this effect. The magnitude of the electrode polarization effect depends upon both the magnitude of the impressed voltage and the mineral concentration. It is most pronounced when the mineral is disseminated throughout the host rock as the surface area available for ionic-electronic interchange is then at a maximum. The effect decreases with increasing porosity as more alternative paths become available for the more efficient ionic conduction.

In prospecting for metallic ores, interest is obviously in the electrode polarization (overvoltage) effect. Membrane polarization, however, is indistinguishable from this effect during IP measurements. Membrane polarization consequently reduces the effectiveness of IP surveys and causes geological 'noise' which may be equivalent to the overvoltage effect of a rock with up to 2% metallic minerals.

8.3.3 Induced polarization measurements

Time domain IP measurements involve the monitoring of the decaying voltage after the current is switched off. The most commonly measured parameter is the *chargeability* M, defined as the area A beneath the decay curve over a certain time interval (t_1–t_2) normalized by the steady-state potential difference ΔV_c (Fig. 8.21).

$$M = \frac{A}{\Delta V_c} = \frac{1}{\Delta V_c} \int_{t_1}^{t_2} V(t)\,dt \tag{8.22}$$

Chargeability is measured over a specified time interval shortly after the polarizing current is cut off (Fig. 8.21). The area A is determined within the measuring apparatus by analogue integration. Different minerals are distinguished by characteristic chargeabilities, e.g. pyrite has $M = 13.4$ ms over an interval of 1 s, and magnetite 2.2 ms over the same interval. Figure 8.21 also shows that current polarity is reversed between successive measurements in order to destroy any remanent polarization.

Frequency domain techniques involve the measurement of apparent resistivity at two or more AC frequencies. Figure 8.23 shows the relationship between apparent resistivity and log current frequency. Three distinct regions are apparent: region 1 is in low frequencies where resistivity is independent of frequency; region 2 is the Warberg region where resistivity is a linear function of log frequency; region 3 is the region of electromagnetic induction (Chapter 9) where current flow is by induction rather than simple conduction. Since the relationship illustrated in Fig. 8.23 varies with rock type and mineral concentration, IP measurements are usually made at frequencies at, or below, 10 Hz to remain in the non-inductive regions.

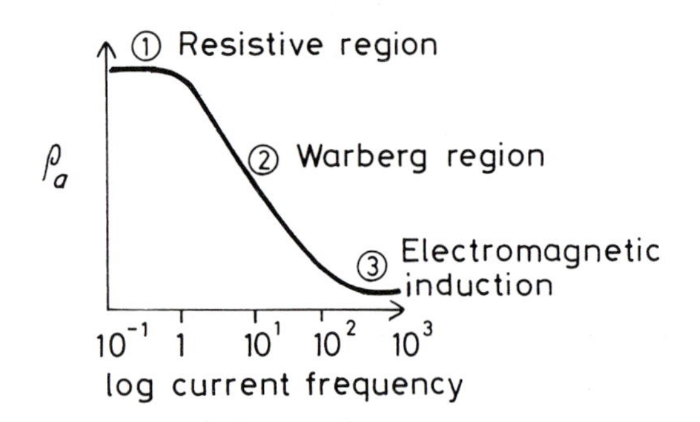

Fig. 8.23. The relationship between apparent resistivity and measuring current frequency.

Two measurements are commonly made. The *percentage frequency effect* PFE is defined as

$$\text{PFE} = 100(\rho_{0.1} - \rho_{10})/\rho_{10} \tag{8.23}$$

where $\rho_{0.1}$ and ρ_{10} are apparent resistivities at measuring frequencies of 0.1 and 10 Hz. The *metal factor* MF is defined as

$$\text{MF} = 2\pi 10^5(\rho_{0.1} - \rho_{10})/\rho_{0.1}\rho_{10} \tag{8.24}$$

This factor normalizes the PFE with respect to the lower frequency resistivity and consequently removes, to a certain extent, the variation of the IP effect with the effective resistivity of the host rock.

8.3.4 *Field operations*

IP equipment is similar to resistivity apparatus but is rather more bulky and elaborate. Theoretically, any standard electrode spread may be employed but in practice the double-dipole, pole-dipole and Schlumberger configurations (Fig. 8.24) are the most effective. Electrode spacings may vary from 3–300 m with the larger spacings used in reconnaissance surveys. To reduce the labour of moving current electrodes and generator, several pairs of current electrodes may be used, all connected via a switching device to the generator. Traverses are made over the area of interest plotting the IP reading at the mid-point of the electrode array (marked by crosses in Fig. 8.24).

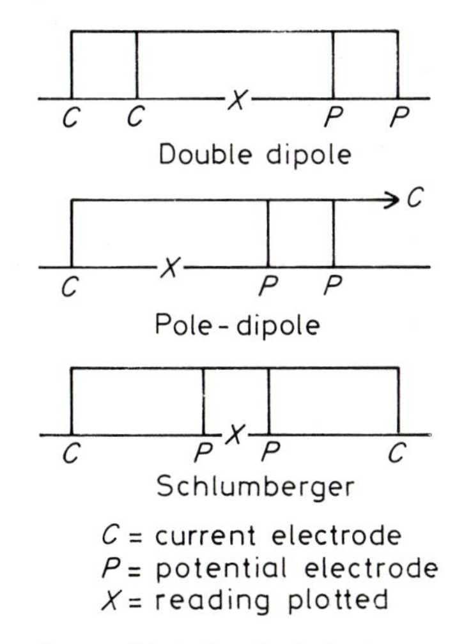

Fig. 8.24. Electrode configurations used in induced polarization measurements.

Noise in an IP survey can result from several phenomena. Telluric currents cause similar anomalous effects to those encountered in resistivity measurements. Noise also results from the general IP effect of barren rocks caused by membrane polarization. Noise generated by the measuring equipment results from electromagnetic coupling between adjacent wires. Such effects are common when alternating current is used since currents can be induced to flow in adjacent conductors. Consequently, cables should be at least 10 m apart and if they must cross they should do so at right angles to minimize electromagnetic induction effects.

8.3.5 *Interpretation of induced polarization data*

Quantitative interpretation is considerably more complex than for the resistivity method. The IP response has been computed analytically for simple features such as spheres, ellipsoids, dykes, vertical contacts and horizontal layers, enabling indirect interpretation (numerical modelling) techniques to be used.

Laboratory modelling can also be employed in indirect interpretation to simulate an observed IP anomaly. For example, apparent resistivities may be measured for various shapes and resistivities of a gelatin-copper sulphate body immersed in water.

Much IP interpretation is, however, only qualitative. Simple parameters of the anomalies, such as sharpness, symmetry, amplitude and spatial distribution may be used to estimate the location, lateral extent, dip and depth of the anomalous zone.

The IP method suffers from the same disadvantages as resistivity surveying (see Section 8.2.9). Further, the sources of significant IP anomalies are often not of economic importance, e.g. water-filled shear zones and graphite-bearing sediments can both generate strong IP effects. Field operations are slow and the method is consequently far more expensive than most other ground geophysical techniques, survey costs being comparable with those of a gravity investigation.

In spite of these drawbacks, the IP method is extensively used in base metal exploration as it has a high success rate in locating low-grade ore deposits such as disseminated sulphides. These have a strong IP effect but are non-conducting and therefore are not readily detectable by the electromagnetic methods discussed in Chapter 9. IP is by far the most effective geophysical method that can be used in the search for such targets.

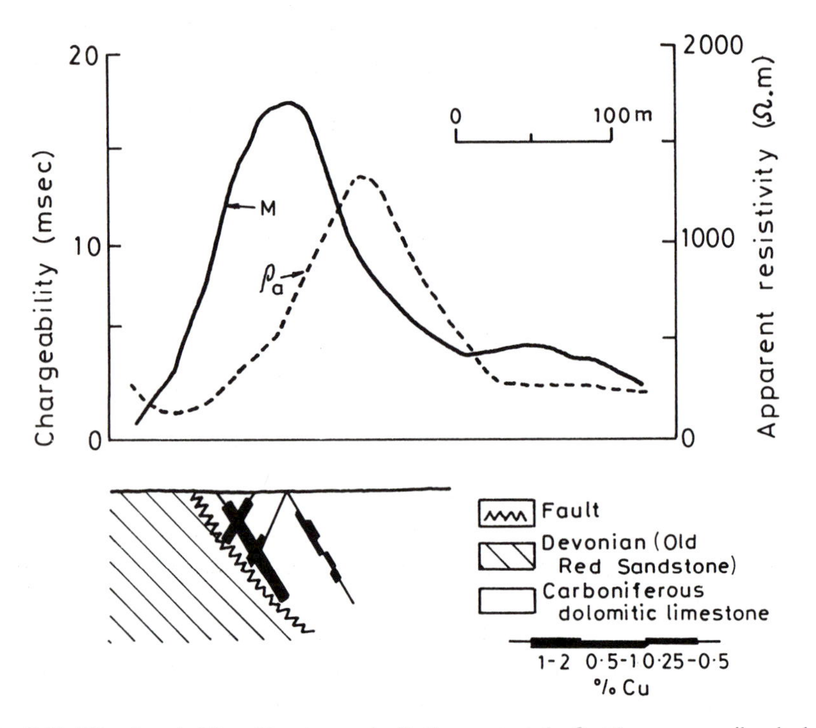

Fig. 8.25. Time domain IP profile using a pole-dipole array over the Gortdrum copper-silver body, Ireland. (Redrawn from Seigel 1967.)

Fig. 8.25 shows the chargeability profile for a time domain IP survey using a pole-dipole array across the Gortdrum copper-silver orebody in Ireland. Although the deposit is of low grade, containing less than 2% conducting minerals, the chargeability anomaly is well-defined and centred over the orebody. In contrast, the corresponding apparent resistivity profile reflects the large resistivity contrast between the Old Red Sandstone and dolomitic limestone but gives no indication of the presence of the mineralization.

8.4 Self Potential (SP) Method

8.4.1 Introduction

The self potential (or spontaneous polarization) method is based on the surface measurement of natural potential differences resulting from electrochemical reactions in the subsurface. Typical SP anomalies may have an amplitude of several hundred millivolts with respect to barren ground. They invariably exhibit a central negative anomaly and are stable over long periods of time. They are usually associated with deposits of metallic sulphides, magnetite or graphite.

8.4.2 Mechanism of self potential

Field studies indicate that for a self potential anomaly to occur its causative body must lie partially in a zone of oxidation. A widely accepted mechanism of self potential (Sato & Mooney 1960; for a more recent analysis see Kilty 1984) requires the causative body to straddle the water table (Fig. 8.26). Below the

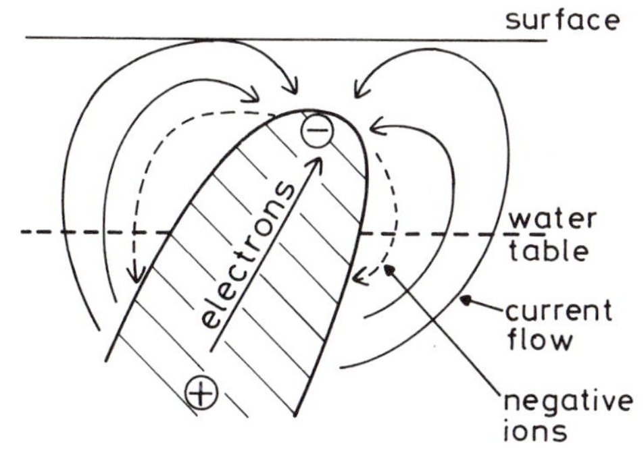

Fig. 8.26. The mechanism of self potential anomalies. (Redrawn from Sato & Mooney 1960.)

water table electrolytes in the pore fluids undergo oxidation and release electrons which are conducted upwards through the ore body. At the top of the body the released electrons cause reduction of the electrolytes. A circuit thus exists in which current is carried electrolytically in the pore fluids and electronically in the body so that the top of the body acts as a negative terminal. This explains the negative SP anomalies that are invariably observed and, also, their stability as the ore body itself undergoes no chemical reactions and merely serves to transport electrons from depth. As a result of the subsurface currents, potential differences are produced at the surface.

8.4.3 *Self potential equipment and survey procedure*

Field equipment consists simply of a pair of electrodes connected via a high impedance millivoltmeter. The electrodes must be non-polarizing as simple metal spikes would generate their own SP effects. Non-polarizing electrodes consist of a metal immersed in a saturated solution of its own salt, such as copper in copper sulphate. The salt is contained in a porous pot which allows slow leakage of the solution into the ground.

Station spacing is generally less than 30 m. Traverses may be performed by leapfrogging successive electrodes or, more commonly, by fixing one electrode in barren ground and moving the other over the survey area.

8.4.4 *Interpretation of self potential anomalies*

The interpretation of SP anomalies is similar to magnetic interpretation because dipole fields are involved in both cases. It is thus possible to calculate the potential distributions around polarized bodies of simple shape such as spheres and ellipsoids by making assumptions about the distribution of charge over their surfaces.

Most interpretation, however, is qualitative. The anomaly minimum is assumed to occur directly over the anomalous body, although it may be displaced downhill in areas of steep topography. The anomaly half-width provides a rough estimate of depth. The symmetry or asymmetry of the anomaly provides information on the attitude of the body, the steep slope and positive tail of the anomaly lying on the downdip side.

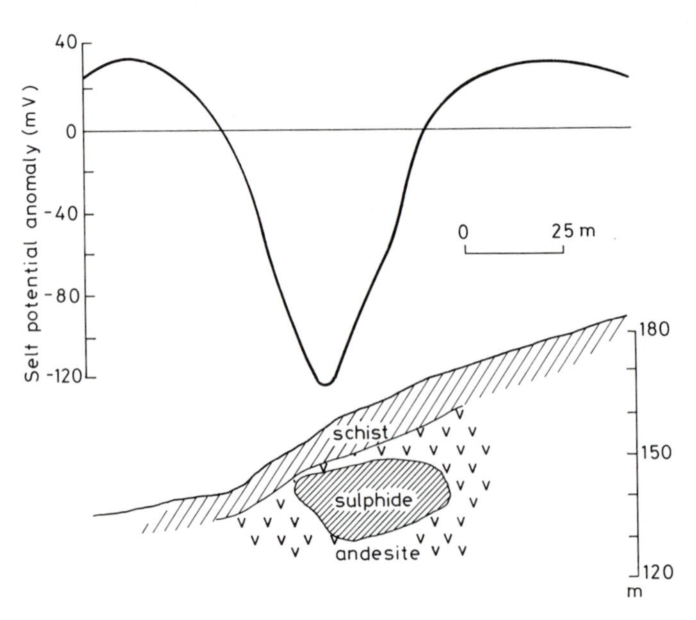

Fig. 8.27. The SP anomaly over a sulphide ore body atSariyer, Turkey. (Redrawn from Yüngül 1954.)

The type of overburden can have a pronounced effect on the presence or absence of SP anomalies. Sand has little effect but a clay cover can mask the SP anomaly of an underlying body.

The SP method is only of minor importance in exploration. This is because quantitative interpretation is difficult and the depth of penetration is limited to about 30 m. It is, however, a rapid and cheap method requiring only simple field equipment. Consequently it can be useful in rapid ground reconnaissance for base metal deposits when used in conjunction with magnetic, electromagnetic and geochemical techniques.

Fig. 8.27 shows the SP profile over a sulphide orebody in Turkey which contains copper concentrations of up to 14%. The SP anomaly is negative and has an amplitude of some 140 mV. The steep topography has displaced the anomaly minimum downhill from the true location of the orebody.

Further reading

Bertin, J. (1976) *Experimental and Theoretical Aspects of Induced Polarisation*, Vols. 1 and 2. Gebrüder Borntraeger, Berlin.

Griffiths, D.H. & King, R.F. (1981) *Applied Geophysics for Geologists and Engineers*. Pergamon, Oxford.

Habberjam, G.M. (1979) *Apparent Resistivity and the Use of Square Array Techniques*. Gebrüder Borntraeger, Berlin.

Keller, G.V. & Frischnecht, F.C. (1966) *Electrical Methods in Geophysical Prospecting*. Pergamon, Oxford.

Koefoed, O. (1968) *The Application of the Kernel Function in Interpreting Resistivity Measurements*. Gebrüder Borntraeger, Berlin.

Koefoed, O. (1979) *Geosounding Principles 1–Resistivity Sounding Measurements*. Elsevier, Amsterdam.

Kunetz, G. (1966) *Principles of Direct Current Resistivity Prospecting*. Gebrüder Borntraeger, Berlin.

Marshall, D.J. & Madden, T.R. (1959) Induced polarisation: a study of its causes. *Geophysics*, **24**, 790–816.

Parasnis, D.S. (1973) *Mining Geophysics*. Elsevier, Amsterdam.

Parasnis, D.S. (1979) *Principles of Applied Geophysics*. Chapman & Hall, London.

Parkhomenko, E.I. (1967) *Electrical Properties of Rocks*. Plenum, New York.

Sato, M. & Mooney, H.M. (1960) The electrochemical mechanism of sulphide self potentials. *Geophysics*, **25**, 226–49.

Sumner, J.S. (1976) *Principles of Induced Polarisation for Geophysical Exploration*. Elsevier, Amsterdam.

Telford, W.M., Geldart, L.P., Sheriff, R.E. & Keys, D.A. (1976) *Applied Geophysics*. Cambridge Univ. Press, Cambridge.

9

Electromagnetic Surveying

9.1 Introduction

Electromagnetic (EM) surveying methods make use of the response of the ground to the propagation of electromagnetic fields, which are composed of an alternating electric intensity and magnetizing force. Primary electromagnetic fields may be generated by passing alternating current through a small coil made up of many turns of wire or through a large loop of wire. The response of the ground is the generation of secondary electromagnetic fields and the resultant fields may be detected by the alternating currents that they induce to flow in a receiver coil by the process of electromagnetic induction.

The primary electromagnetic field travels from the transmitter coil to the receiver coil via paths both above and below the surface. Where the subsurface is homogeneous there is no difference between the fields propagated above the surface and through the ground other than a slight reduction in amplitude of the latter with respect to the former. However, in the presence of a conducting body the magnetic component of the electromagnetic field penetrating the ground induces alternating currents, or eddy currents, to flow in the conductor (Fig. 9.1). The eddy currents generate their own secondary electromagnetic field which travels to the receiver. The receiver then responds to the resultant of the arriving primary and secondary fields so that the response differs in both phase and amplitude from the response to the primary field alone. These differences between the transmitted and received electromagnetic fields reveal the presence of the conductor and provide information on its geometry and electrical properties.

The induction of current flow results from the magnetic component of the electromagnetic field. Consequently there is no need for physical contact of either transmitter or receiver with the ground. Surface EM surveys can thus proceed

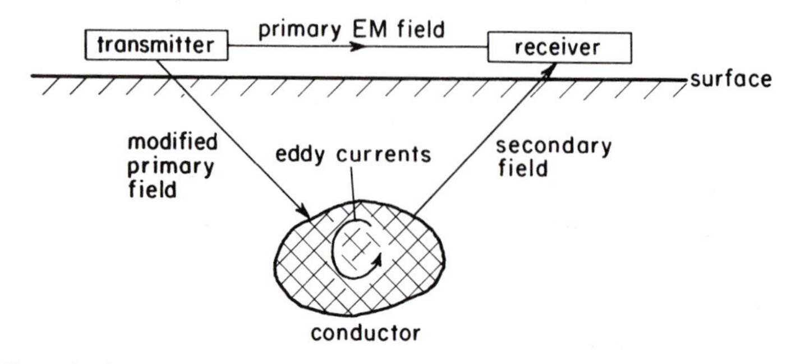

Fig. 9.1. General principle of electromagnetic surveying.

much more rapidly than electrical surveys, where ground contact is required. More importantly, both transmitter and receiver can be mounted in aircraft or towed behind them. Airborne EM methods are widely used in prospecting for conductive ore bodies (see Section 10.2).

All anomalous bodies with high electrical conductivity (see Section 8.2.2) produce strong secondary electromagnetic fields. Some ore bodies containing minerals that are themselves insulators may produce secondary fields if sufficient quantities of an accessory mineral with a high conductivity are present. For example, electromagnetic anomalies observed over certain sulphide ores are due to the presence of the conducting mineral pyrrhotite distributed throughout the ore body.

9.2 Depth of penetration of electromagnetic fields

The depth of penetration of an electromagnetic field depends upon its frequency and the electrical conductivity of the medium through which it is propagating. Electromagnetic fields are attenuated during their passage through the ground, their amplitude decreasing exponentially with depth. The depth of penetration d can be defined as the depth at which the amplitude of the field A_d is decreased by a factor e^{-1} compared with its surface amplitude A_0

$$A_d = A_0 e^{-1} \tag{9.1}$$

In this case

$$d = 503.8\,(\sigma f)^{-1/2} \tag{9.2}$$

where d is in metres, the conductivity of the ground σ is in S m^{-1} and the frequency f of the field is in Hz.

The depth of penetration thus increases as both the frequency of the electromagnetic field and the conductivity of the ground decrease. Consequently, the frequency used in an EM survey can be tuned to a desired depth range in any particular medium. For example, in relatively dry glacial clays with a conductivity of 5×10^{-4} S m^{-1}, d is about 225 m at a frequency of 10 kHz.

Equation (9.2) represents a theoretical relationship. Practically, an effective depth of penetration z_e can be defined which represents the maximum depth at which a conductor may lie and still produce a recognizable electromagnetic anomaly

$$z_e \doteq 100(\sigma f)^{-1/2} \tag{9.3}$$

The relationship is approximate as the penetration depends upon such factors as the nature and magnitude of the effects of near-surface variations in conductivity, the geometry of the subsurface conductor and instrumental noise. The frequency dependence of depth penetration places constraints on the EM method. Normally, very low frequencies are difficult to generate and measure and the maximum penetration is of the order of 500 m.

9.3 Detection of electromagnetic fields

Electromagnetic fields may be mapped in a number of ways, the simplest of which employs a small search coil consisting of several hundred turns of copper wire wound on a circular or rectangular frame typically between 0.5 m and 1 m across. The ends of the coil are connected via an amplifier to earphones. The amplitude of the alternating voltage induced in the coil by an electromagnetic field is proportional to the component of the field perpendicular to the plane of the coil. Consequently, the strength of the signal in the earphones is at a maximum when the plane of the coil is at right angles to the direction of the arriving field. Since the ear is more sensitive to sound minima than maxima, the coil is usually turned until a null position is reached. The plane of the coil then lies in the direction of the arriving field.

9.4 Tilt-angle methods

When only a primary electromagnetic field H_p is present at a receiver coil, a null reading is obtained when the plane of the coil lies parallel to the field direction. There are an infinite number of such null positions as the coil is rotated about a horizontal axis in the direction of the field (Fig. 9.2).

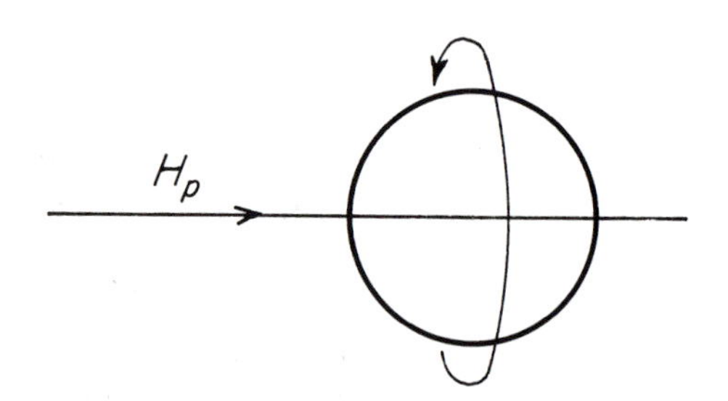

Fig. 9.2. The rotation of a search coil about an axis corresponding to the direction of arriving electromagnetic radiation H_p producing an infinite number of null positions.

In many EM systems the induced secondary field H_s lies in a vertical plane. Since the primary and secondary fields are both alternating, the total field vector describes an ellipse in the vertical plane with time (Fig. 9.3). The resultant field is then said to be *elliptically polarized* in the vertical plane. In this case there is only one null position of the search coil, namely where the plane of the coil coincides with the plane of polarization.

For good conductors it can be shown that the direction of the major axis of the ellipse of polarization corresponds reasonably accurately to that of the resultant of the primary and secondary electromagnetic field directions. The angular deviation of this axis from the horizontal is known as the *tilt-angle* θ of the resultant field (Fig. 9.3). There are a number of EM techniques (known as *tilt-angle* or *dip-angle* methods) which simply measure spatial variations in this angle. The primary field may be generated by a fixed transmitter, which usually consists of a large horizontal or vertical coil, or by a small mobile transmitter. Traverses are

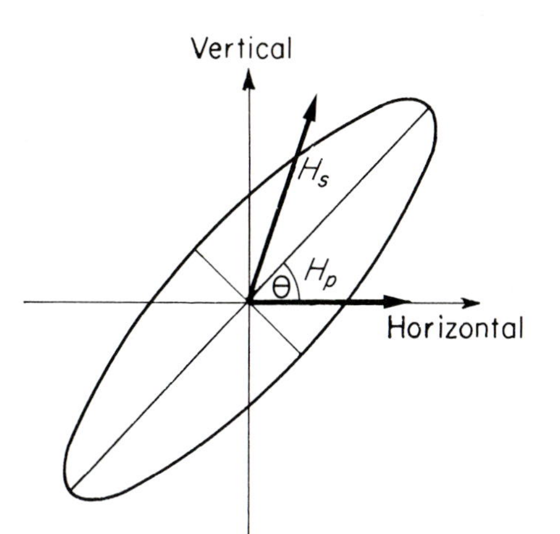

Fig. 9.3. The polarization ellipse and tilt-angle θ. H_p and H_s represent the primary and secondary electromagnetic fields.

made across the survey area normal to the geological strike. At each station the search coil is rotated about three orthogonal axes until a null signal is obtained so that the plane of the coil lies in the plane of the polarization ellipse. The tilt-angle may then be determined by rotating the coil about a horizontal axis at right angles to this plane until a further minimum is encountered.

9.4.1 Tilt-angle methods employing local transmitters

In the case of a fixed, vertical transmitter coil, the primary field is horizontal. Eddy currents within a subsurface conductor then induce a magnetic field whose lines of force describe concentric circles around the eddy current source, which is assumed to lie along its upper edge (Fig. 9.4(a)). On the side of the body nearest the transmitter the resultant field dips upwards. The tilt decreases towards the body and dips downwards on the side of the body remote from the transmitter. The body is located directly below the crossover point where the tilt-angle is zero, as here both primary and secondary fields are horizontal. When the fixed transmitter is horizontal the primary field is vertical (Fig. 9.4(b)) and the body is located where the tilt is at a minimum. An example of the use of tilt-angle methods (vertical transmitter) in the location of a massive sulphide body is presented in Fig. 9.5.

If the conductor is near the surface both the amplitude and gradients of the tilt-angle profile are large. These quantities decrease as the depth to the conductor increases and may consequently be used to derive semi-quantitative estimates of the conductor depth. A vertical conductor would provide a symmetrical tilt-angle profile with equal gradients on either side of the body. As the inclination of the conductor decreases the gradients on either side become progressively less similar. The asymmetry of the tilt-angle profile can thus be used to obtain an estimate of the dip of the conductor.

Tilt-angle methods employing fixed transmitters have been largely superseded by survey arrangements in which both transmitter and receiver are mobile and which can provide much more quantitative information on subsurface con-

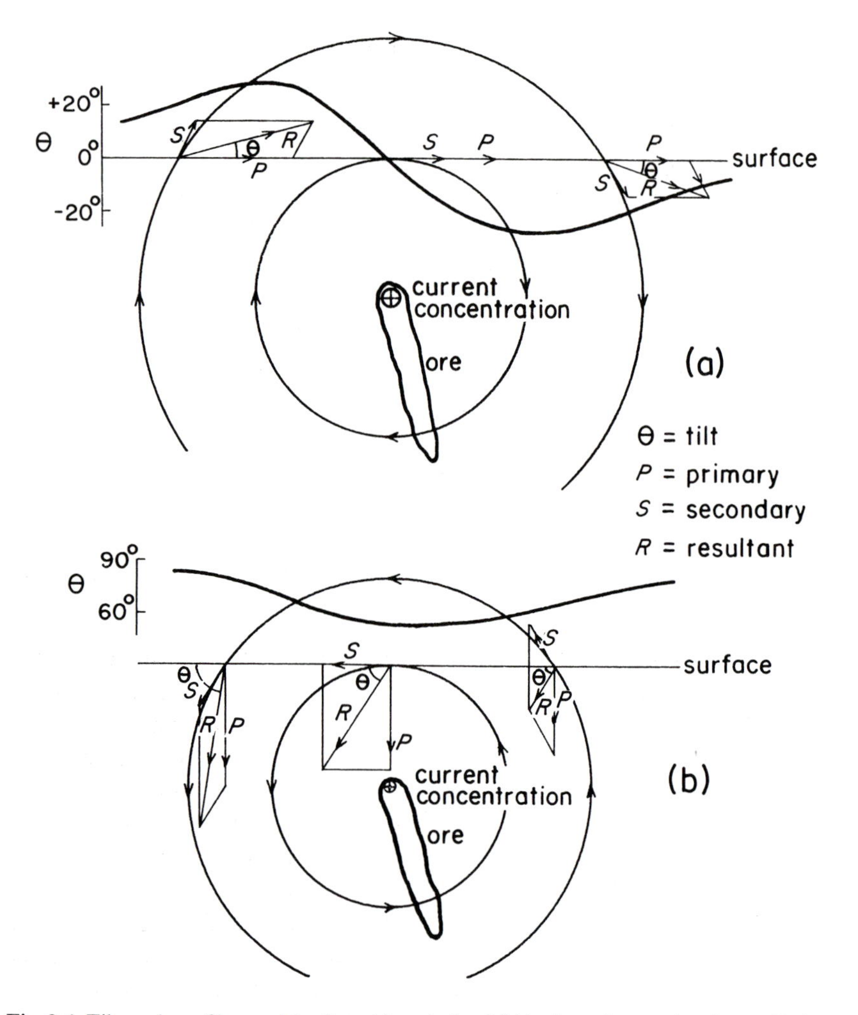

Fig. 9.4. Tilt-angle profiles resulting from (a) vertical and (b) horizontal transmitter loops. (Redrawn from Parasnis 1973.)

ductors. However, two tilt-angle methods still in common use are the Very Low Frequency (VLF) and Audio Frequency MAGnetic Field (AFMAG) methods, neither of which requires the erection of a special transmitter.

9.4.2 *The VLF Method*

The source utilized by the VLF method is electromagnetic radiation generated in the low frequency band of 15–25 kHz by the powerful radio transmitters used in long-range communications and navigational systems. Several stations using this frequency range are available around the world and transmit continuously either an unmodulated carrier wave or a wave with superimposed morse code. Such signals may be used for surveying up to distances of several thousand kilometres from the transmitter.

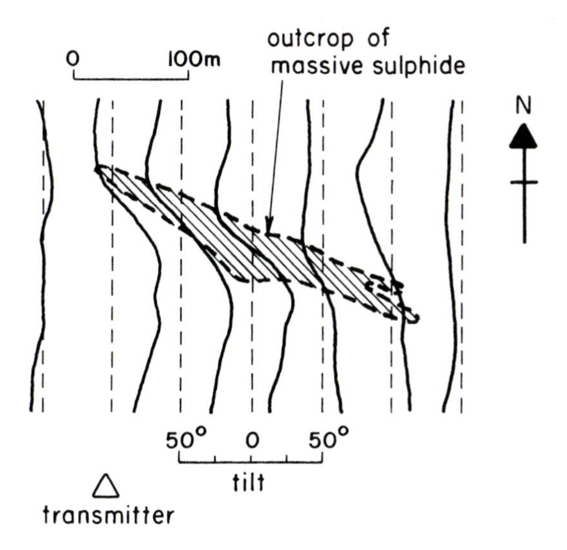

Fig. 9.5. Example of tilt-angle survey using a vertical loop transmitter. (Redrawn from Parasnis 1973.)

At large distances from source the electromagnetic field is essentially planar and horizontal (Fig. 9.6). The electric component E lies in a vertical plane and the magnetic component H lies at right angles to the direction of propagation in a horizontal plane. A conductor that strikes in the direction of the transmitter is cut by the magnetic vector and the induced eddy currents produce a secondary electromagnetic field. Conductors striking at right angles to the direction of propagation are not cut effectively by the magnetic vector.

The VLF receiver is a small hand-held device incorporating two orthogonal aerials which can be tuned to the particular frequencies of the transmitters. The direction of a transmitter is found by rotating the horizontal instrument around a vertical axis until a null position is found. Traverses are then performed over the survey area at right angles to this direction. The instrument is rotated about a horizontal axis orthogonal to the traverse and the tilt recorded at the null position.

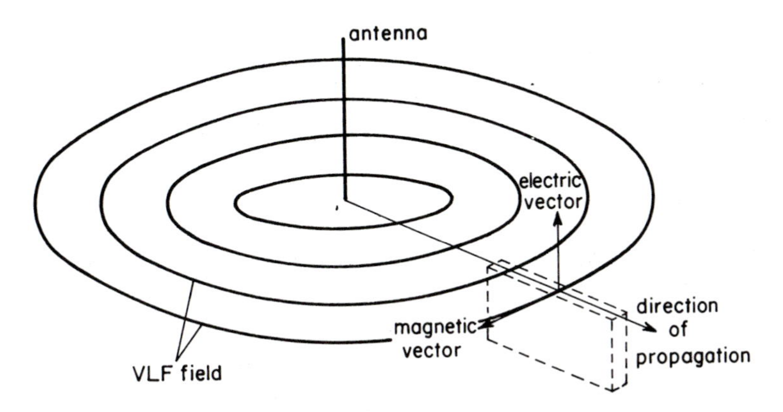

Fig. 9.6. Principle of VLF method. Dashed line shows a tabular conductor striking towards the antenna which is cut by the magnetic vector of the electromagnetic field.

Profiles are similar in form to Fig. 9.4(a), with the conductor lying beneath locations of zero tilt.

The VLF method has the advantages that the field equipment is small and light, being conveniently operated by one person, and that there is no need to install a transmitter. However, for a particular survey area, there may be no suitable transmitter providing a magnetic vector across the geological strike. A further disadvantage is that the depth of penetration is somewhat less than that attainable by tilt-angle methods using a local transmitter. The VLF method can be used in airborne EM surveying.

9.4.3 *The AFMAG method*

The AFMAG method can similarly be used on land or in the air. The source in this case is the natural electromagnetic fields generated by thunderstorms and known as *sferics*. Sferics propagate around the Earth between the ground surface and the ionosphere. This space constitutes an efficient electromagnetic waveguide and the low attenuation means that thunderstorms anywhere in the world make an effective contribution to the field at any given point. The field also penetrates the subsurface where, in the absence of electrically-conducting bodies, it is practically horizontal. The sferic sources are random so that the signal is generally quite broadband between 1 and 1000 Hz.

The AFMAG receiver differs from conventional tilt-angle coils since random variations in the direction and intensity of the primary field make the identification of minima impossible with a single coil. The receiver consists of two orthogonal coils each inclined at 45 degrees to the horizontal (Fig. 9.7). In the absence of a secondary field the components of the horizontal primary field perpendicular to the coils are equal and their subtracted output is zero (Fig. 9.7(a)). The presence of a conductor gives rise to a secondary field which causes deflection of the resultant field from the horizontal (Fig. 9.7(b)). The field components orthogonal to the two coils are then unequal, so that the combined output is no longer zero and the presence of a conductor is indicated. The output provides a measure of the tilt.

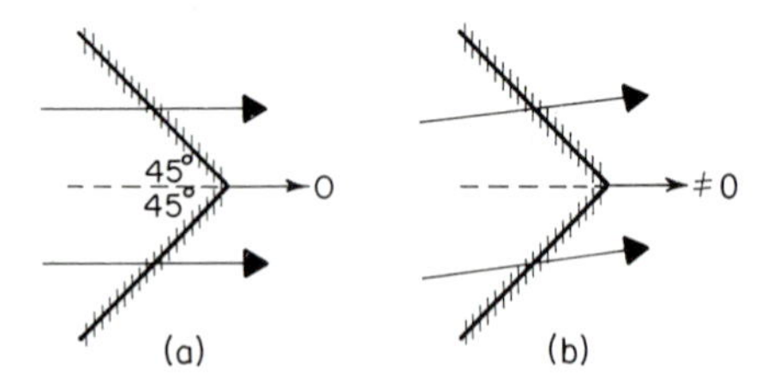

Fig. 9.7. Principle of AFMAG receiver (a) conductor absent, (b) conductor present.

On land both the azimuths and tilts of the resultant electromagnetic field can be determined by rotating the coils about a vertical axis until a maximum signal is obtained. These are conventionally plotted as dip vectors. In the air, azimuths cannot be determined as the coils are attached to the aircraft so that their orientation is controlled by the flight direction. Consequently, only perturbations from the horizontal are monitored along the flight lines. The output signal is normally fed into an amplifier tuned to two frequencies of about 140 and 500 Hz. Comparison of the amplitude of the signals at the two frequencies provides an indication of the conductivity of the anomalous structure as it can be shown that the ratio of low-frequency response to high-frequency response is greater than unity for a good conductor and less than unity for a poor conductor.

The AFMAG method has the advantage that the frequency range of the natural electromagnetic fields used extends to an order of magnitude lower than can be produced artificially so that depths of investigation of several hundred metres are feasible.

9.5 Phase measuring systems

Tilt-angle methods such as VLF and AFMAG are widely used since the equipment is simple, relatively cheap and the technique is rapid to employ. However, they provide little quantitative information on the conductor. More sophisticated EM surveying systems measure the phase and amplitude relationships between primary, secondary and resultant electromagnetic fields.

An alternating electromagnetic field can be represented by a sine wave with a wavelength of 2π (360°) (Fig. 9.8(a)). When one such wave lags behind another the waves are said to be out of phase. The phase difference can be represented by a phase angle θ corresponding to the angular separation of the waveforms. The phase relationships of electromagnetic waves can be represented on special vector diagrams in which vector length is proportional to field amplitude and the angle measured counterclockwise from the primary vector to the secondary vector represents the angular phase lag of the secondary field behind the primary.

The primary field P travels directly from transmitter to receiver above the ground and suffers no modification other than a small reduction in amplitude caused by geometric spreading. As the primary field penetrates the ground it is reduced in amplitude to a greater extent but remains in phase with the surface

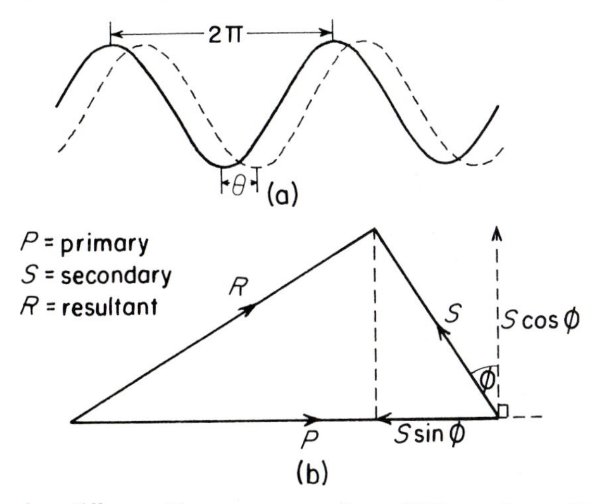

Fig. 9.8. (a) The phase difference θ between two waveforms. (b) Vector diagram illustrating the phase and amplitude relationships between primary, secondary and resultant electromagnetic fields.

primary. The primary field induces an alternating voltage in a subsurface conductor with the same frequency as the primary but with a phase lag of $\pi/2$ (90°) according to the laws of electromagnetic induction. This may be represented on the vector diagram (Fig. 9.8(b)) by a vector $\pi/2$ counterclockwise to P.

The electrical properties of the conductor cause a further phase lag ϕ,

$$\phi = \tan^{-1}(2\pi f L/r) \tag{9.4}$$

where f is the frequency of the electromagnetic field, L the inductance of the conductor (its tendency to oppose a change in the applied field) and r the resistance of the conductor. For a good conductor ϕ will approach $\pi/2$ while for a poor conductor ϕ will be almost zero.

The net effect is that the secondary field S produced by the conductor lags behind the primary with a phase angle of $(\pi/2 + \phi)$. The resultant field R can now be constructed (Fig. 9.8(b)).

The projection of S on the horizontal (primary field) axis is $S \sin \phi$ and is an angle π out of phase with P. It is known as the *in-phase* or *real component* of S. The vertical projection is $S \cos \phi$, $\pi/2$ out of phase with P, and is known as the *out-of-phase*, *imaginary* or *quadrature* component.

Modern instruments are capable of splitting the secondary electromagnetic field into its real (Re) and imaginary (Im) components. The larger the ratio Re/Im, the better the conductor. Some systems, mainly airborne, simply measure the phase angle ϕ.

Classical phase-measuring systems employed a fixed source, usually a very large loop of wire laid on the ground. These systems include the *Two-frame*, *Compensator* and *Turam* systems. They are still in use but are more cumbersome than modern systems in which both transmitter and receiver are mobile.

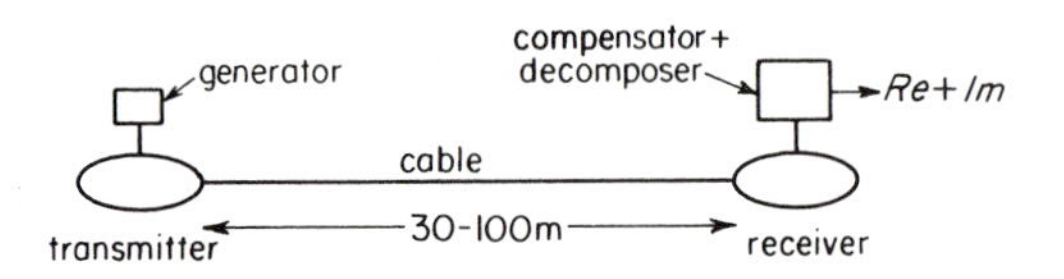

Fig. 9.9. Mobile transmitter-receiver EM field equipment.

A typical field set is shown in Fig. 9.9. The transmitter and receiver coils are about one metre in diameter and are usually carried horizontally, although different orientations may be used. The coils are linked by a cable which carries a reference signal and also allows the coil separation to be accurately maintained at, normally, between 30 m and 100 m. The transmitter is powered by a portable AC generator. Output from the receiver coil passes through a compensator and decomposer (see below). The equipment is first read on barren ground and the compensator adjusted to produce zero output. By this means, the primary field is compensated so that the system subsequently responds only to secondary fields. Consequently, such EM methods reveal the presence of bodies of anomalous conductivity without providing information on absolute conductivity values. Over the survey area the decomposer splits the secondary field into real and imaginary components which are usually displayed as a percentage of the primary field whose magnitude is relayed via the interconnecting cable. Traverses are generally made perpendicular to geological strike and readings plotted at the mid-point of the system. The maximum detection depth is about half the transmitter-receiver separation.

Fieldwork is simple and requires a crew of only two or three operators. The spacing and orientation of the coils is critical as a small percentage error in spacing can produce appreciable error in phase measurement. The coils must also be kept accurately horizontal and coplanar as small relative tilts can produce substantial errors. The required accuracy of spacing and orientation is difficult to maintain with large spacings and over uneven terrain.

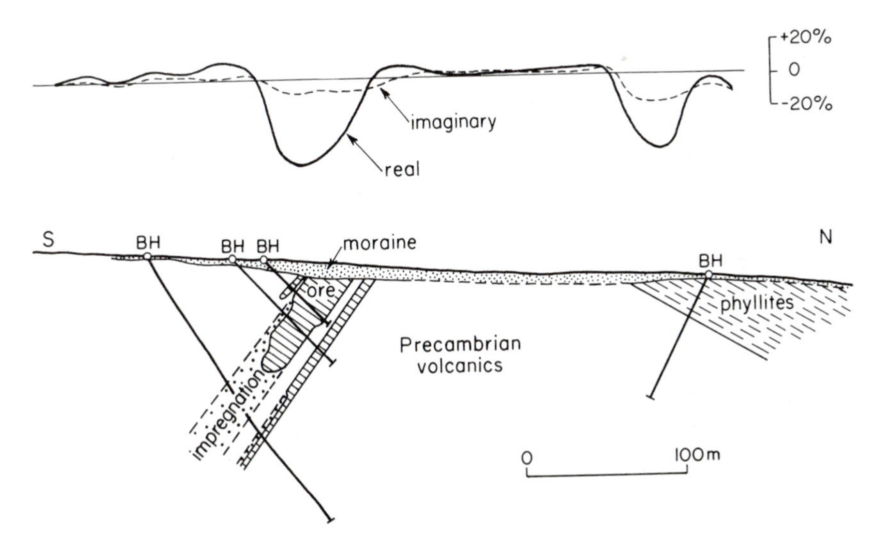

Fig. 9.10. Mobile transmitter-receiver profile, employing horizontal coplanar coils with a separation of 60 m and an operating frequency of 3.6 kHz, in the Kankberg area, north Sweden. Real and imaginary components are expressed as a percentage of the primary field. (Redrawn from Parasnis 1973.)

Fig. 9.10 shows a mobile transmitter-receiver EM profile across a sheet-like conductor in the Kankberg area of northern Sweden. A consequence of the coplanar horizontal coil system employed is that conducting bodies produce negative anomalies in both real and imaginary components with maximum amplitudes immediately above the conductor. The asymmetry of the anomalies is diagnostic of the inclination of the body, with the maximum gradient lying on the downdip side. In this case the large ratio of real to imaginary components over the ore body indicates the presence of a very good conductor, while a lesser ratio is observed over a sequence of graphite-bearing phyllites to the north.

A significant problem with many of the EM survey techniques is that a small secondary field must be measured in the presence of a much larger primary field. This problem may be overcome by using a primary field which is not continuous but consists of a series of pulses between which no primary is generated. The secondary field induced by the primary is then measured only when the primary is inactive. The better the conductivity of the body, the longer do eddy currents flow in it and the longer is the duration of the secondary field. Techniques using a pulsed primary field are known as *transient-field methods*. An example of these methods will be given later (Section 9.6.1).

9.6 Airborne electromagnetic surveying

Airborne EM techniques are widely used because of their speed and cost-effectiveness and a large number of systems are available.

There is a broad division into *passive systems*, where only the receiver is airborne, and *active systems*, where both transmitter and receiver are mobile. Passive systems include airborne versions of the VLF and AFMAG methods. Independent transmitter methods can also be used with an airborne receiver, but are not very attractive as prior ground access to the survey area is required.

Active systems are more commonly used as surveys can be performed in areas where ground access is difficult and provide more information than the passive tilt-angle methods. They are, basically, ground mobile transmitter-receiver systems lifted into the air and interfaced with a continuous recording device. Certain specialized methods, described later, have been adopted to overcome the specific difficulties encountered in airborne work. Active systems comprise two main types, *fixed separation* and *quadrature*.

9.6.1 Fixed separation systems

In fixed separation systems the transmitter and receiver are maintained at a fixed separation, and real and imaginary components are monitored as in ground surveys. The coils are generally arranged to be vertical and either coplanar or coaxial. Accurate maintenance of separation and height is essential, and this is usually accomplished by mounting the transmitter and receiver either on the wings of an aircraft or on a beam carried beneath a helicopter. Compensating methods have to be employed to correct for minute changes in the relative positions of transmitter and receiver resulting from such factors as flexure of the wing mountings, vibration and temperature changes. Since only a small transmitter-receiver separation is used to generate and detect an electromagnetic field over a relatively large distance, such minute changes in separation would cause significant distortion of the signal. Fixed-wing systems are generally flown at a ground clearance of 100–200 m, while helicopters can survey at elevations as low as 20 m.

Greater depth of penetration can be achieved by the use of two planes flying in tandem (Fig. 9.11), the rear plane carrying the transmitter and the forward plane towing the receiver mounted in a bird. Although the aircraft have to fly at a strictly regulated speed, altitude and separation, the use of a rotating primary field compensates for relative rotation of the receiver and transmitter. The rotating primary field is generated by a transmitter consisting of two orthogonal coils in the plane perpendicular to the flight direction. The coils are powered by the same AC source with the current to one coil shifted $\pi/2$ (90°) out of phase with

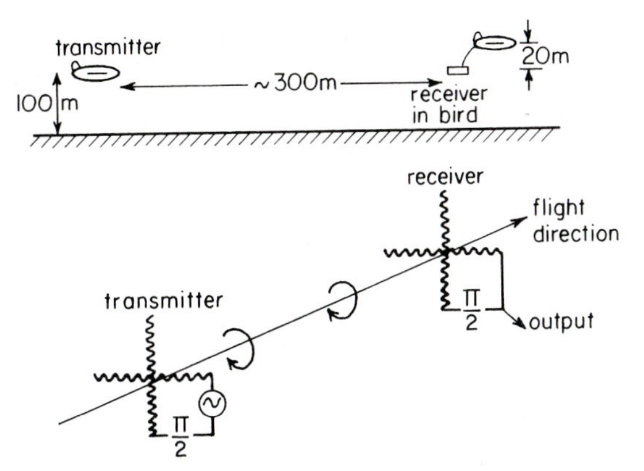

Fig. 9.11. The two-plane, rotary field, EM system.

respect to the other. The resulting field rotates about the flight line and is detected by a receiver with a similar coil configuration which passes the signals through a phase-shift network so that the output over a barren area is zero. The presence of a conductor is then indicated by non-zero output and the measured secondary field decomposed into real and imaginary components. Although penetration is increased and orientation errors minimized, the method is relatively expensive and the interpretation of data is complicated by the complex coil system.

Airborne transient field methods may be employed to overcome the problem of measuring a relatively small secondary field in the presence of a large primary

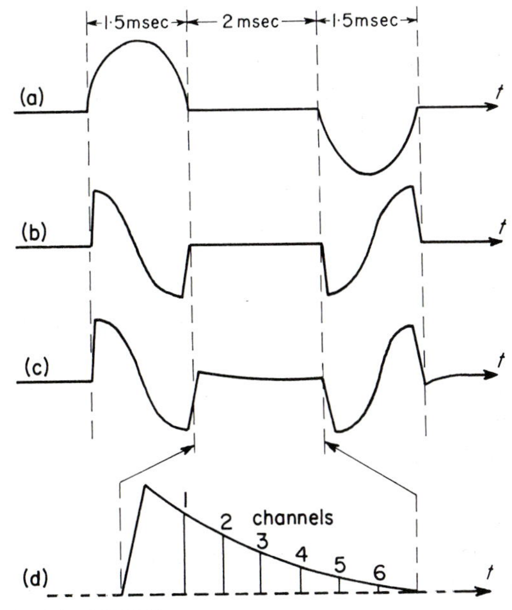

Fig. 9.12. Principle of the INPUT® system.
(a) Primary field.
(b) Receiver response to primary alone.
(c) Receiver response in the presence of a secondary field.
(d) Enlargement of the receiver signal during primary field cut-off. The amplitude of the decaying induced voltage is here sampled on six channels.

field. One such system is INPUT® (INduced PUlsed Transient). This system transmits a transient primary field in the form of a half sine wave by generating pulses of current in a transmitter coil strung horizontally about a large aircraft. The pulses have a duration of some 1.5 ms and are spaced about 2 ms apart (Fig. 9.12). The transient primary field induces transient currents within a subsurface conductor. When the primary field is cut off the induced currents decay exponentially. The receiver only becomes active in the absence of the primary field so the secondary field generated by the eddy currents induces a continuously decaying voltage in the receiver during the 2 ms primary cut-off. The duration of the voltage is directly related to the conductivity of the conductor. The decay curve is sampled at several points and the signals displayed on a strip chart. The signal amplitude in successive sampling channels is, to a certain extent, diagnostic of the type of conductor present. Poor conductors produce a rapidly decaying voltage and only register on those channels sampling the voltage shortly after primary cut-off. Good conductors appear on all channels.

INPUT® is more expensive than other airborne EM methods but provides greater depth penetration, possibly in excess of 100 m, because the secondary signal can be monitored more accurately in the absence of the primary field. It also provides a direct indication of the type of conductor present from the duration of the induced secondary field.

As well as being employed in the location of conducting ore bodies, airborne EM surveys can also be used as an aid to geological mapping. In humid and sub-tropical areas a weathered surface layer develops whose thickness and conductivity depend upon the local rock type. Fig. 9.13 shows an INPUT® profile across part of the Itapicuru Greenstone Belt in Brazil, with sampling times increasing from 0.3 ms at channel 1 to 2.1 ms at channel 6. The transient response over mafic volcanic rocks and Mesozoic sediments is developed in all six channels, indicating that their weathered layer is highly conductive, while the response over graywacke is only apparent in channels 1 to 4, indicating a comparatively less conductive layer.

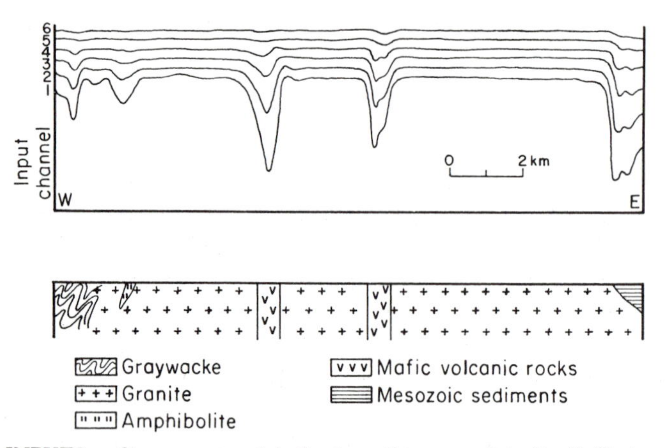

Fig. 9.13. INPUT® profile across part of the Itapicuru Greenstone Belt, Brazil. (Redrawn from Palacky 1981.)

9.6.2 *Quadrature systems*

Quadrature systems were the first airborne EM methods devised. The transmitter is usually a large aerial slung between the tail and wingtips of a fixed wing aircraft and a nominally-horizontal receiver is towed behind the aircraft on a cable some 150 m long.

In quadrature systems the orientation and height of the receiver cannot be rigorously controlled as the receiver 'bird' oscillates in the slipstream. Consequently, the measurement of real and imaginary components is not possible as the strength of the field varies irregularly with movement of the receiver coil. However, the phase difference between the primary field and the resultant field caused by a conductor is independent of variation in the receiver orientation. A disadvantage of the method is that a given phase shift ϕ may be caused by either a good or a poor conductor (Fig. 9.14). This problem is overcome by measuring the phase shift at two different primary frequencies, usually of the order of 400 and 2300 Hz. It can be shown that if the ratio of low frequency to high frequency response exceeds unity, a good conductor is present.

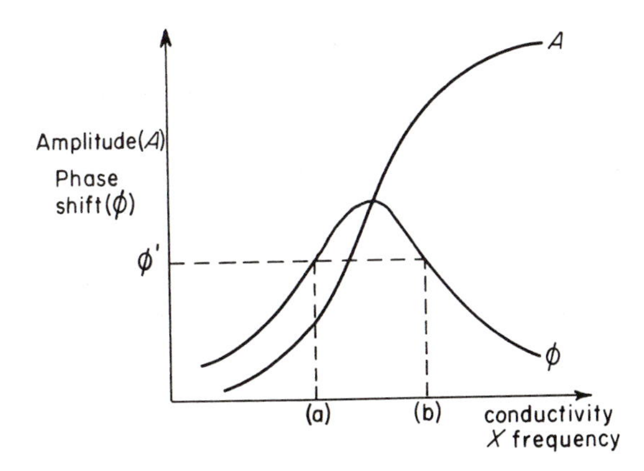

Fig. 9.14. The relationship between the phase/amplitude of a secondary electromagnetic field and the product of conductivity and frequency. A given phase shift ϕ' could result from a poor conductor (a) or a good conductor (b).

Fig. 9.15 shows a contour map of real component anomalies (in ppm of the primary field) over the Skellefteå orefield, northern Sweden. A fixed separation system was used, with vertical, coplanar coils mounted perpendicular to the flight direction on the wingtips of a small aircraft. Only contours above the noise level of some 100 ppm are presented. The pair of continuous anomaly belts in the southwest, with amplitudes exceeding 1000 ppm, corresponds to graphitic shales, which serve as guiding horizons in this orefield. The belt to the north of these is not continuous, and although in part related to sulphide ores, also results from a power cable. In the northern part of the area the three distinct anomaly centres all correspond to strong sulphide mineralization.

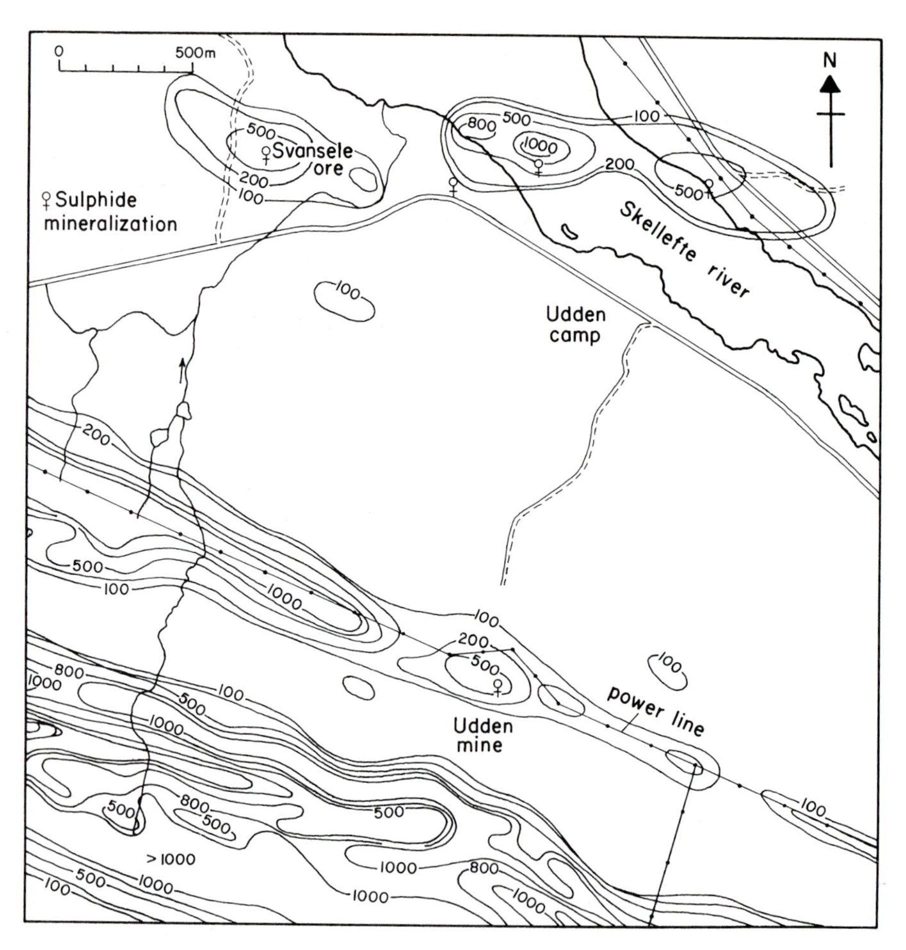

Fig. 9.15. Contour map of real component anomalies over part of the Skellefteå orefield, northern Sweden, obtained using an airborne system with vertical coplanar coils. Mean ground clearance 30 m, operating frequency 3.5 kHz. Contours in ppm of the primary field. (Redrawn from Parasnis 1973.)

9.7 Interpretation of electromagnetic data

As with other types of geophysical data an indirect approach can be adopted in the interpretation of electromagnetic anomalies. The observed electromagnetic response is compared with the theoretical response, for the type of equipment used, to conductors of various shapes and conductivities. Theoretical computations of this type are quite complex and limited to simple geometric shapes such as spheres, cylinders, thin sheets and horizontal layers.

If the causative body is of complex geometry and variable conductivity, laboratory modelling may be used. Because of the complexity of theoretical computations, this technique is used far more extensively in electromagnetic interpretation than in other types of geophysical interpretation. For example, to model a massive sulphide body in a well-conducting host rock, an aluminium model immersed in salt water may be used.

Master curves are available for simple interpretation of moving source-receiver data in cases where it may be assumed that the conductor has a simple geometric form. Fig. 9.16 shows such a set of curves for a simple sheet-like dipping conductor of thickness t and depth d where the distance between horizontal, coplanar coils is a. The point corresponding to the maximum real and imaginary values, expressed as a percentage of the primary field, is plotted on the curves. From the curves coinciding with this point, the corresonding λ/a and d/a values are determined. The latter ratio is readily converted into conductor depth. λ corresponds to $10^7(\sigma f t)^{-1}$ where σ is the conductivity of the sheet and f the frequency of the field. Since a and f are known, the product σt can be determined. By performing measurements at more than one frequency, σ and t can be computed separately.

Much electromagnetic interpretation is, however, only qualitative, particularly for airborne data. Contour maps of real or imaginary components provide information on the length and conductivity of conductors while the asymmetry of the profiles provides an estimate of the inclination of sheet-like bodies.

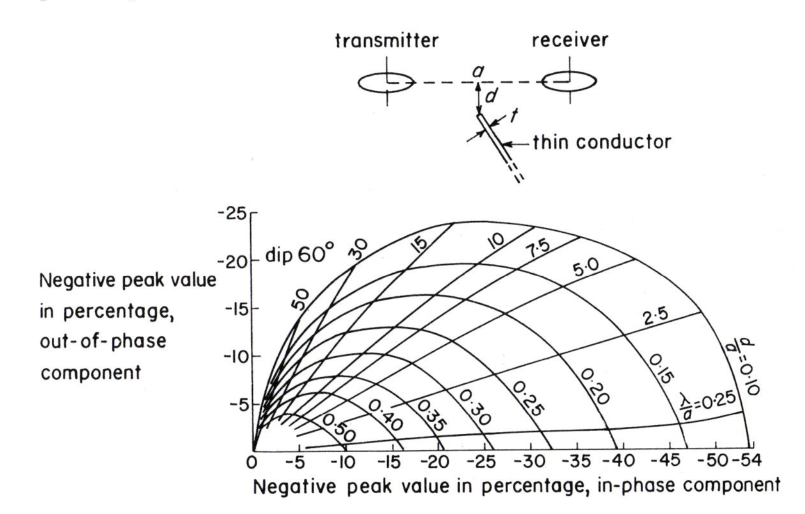

Fig. 9.16. Example of a vector diagram used in estimating the parameters of a thin dipping conductor from the peak real and imaginary component values. (Redrawn from Nair *et al.* 1968.)

9.8 Limitations of the electromagnetic method

The electromagnetic method is a versatile and efficient survey technique, but it suffers from several drawbacks. As well as being caused by economic sources with a high conductivity such as ore bodies, electromagnetic anomalies can also result from non-economic sources such as graphite, water-filled shear zones, bodies of water and man-made sources. Superficial layers with a high conductivity such as wet clays and graphite-bearing rocks may screen the effects of deeper conductors. Penetration is not very great, being limited by the frequency range that can be generated and detected. Unless natural fields are used, maximum penetration in ground surveys is limited to about 500 m, and is only about 50 m in airborne work. Finally, the quantitative interpretation of electromagnetic anomalies is complex.

9.9 Telluric and magnetotelluric field methods

9.9.1 *Introduction*

Within and around the Earth there exist large scale, low frequency, natural magnetic fields known as *magnetotelluric fields*. These induce natural alternating electric fields to flow within the Earth, known as *telluric currents*. Both of these natural fields can be used in prospecting.

Magnetotelluric fields are believed to result from the flow of charged particles in the ionosphere, as fluctuations in the fields correlate with diurnal variations in the geomagnetic field caused by solar emissions. Magnetotelluric fields penetrate the ground and there induce telluric currents to flow. The fields are of variable frequency, ranging from 10^{-5} Hz up to the audio range, and overlap the frequency range utilized in the AFMAG method (Section 9.4.3).

9.9.2 *Surveying with telluric currents*

Telluric currents flow within the Earth in large circular patterns that stay fixed with respect to the Sun. They normally flow in sheets parallel to the surface and extend to depths of several kilometres in the low frequencies. The telluric method is, in fact, the only electrical technique capable of penetrating to the depths of interest to the oil industry. Although variable in both their direction and intensity, telluric currents cause a mean potential gradient at the Earth's surface of about 10 mV km^{-1}.

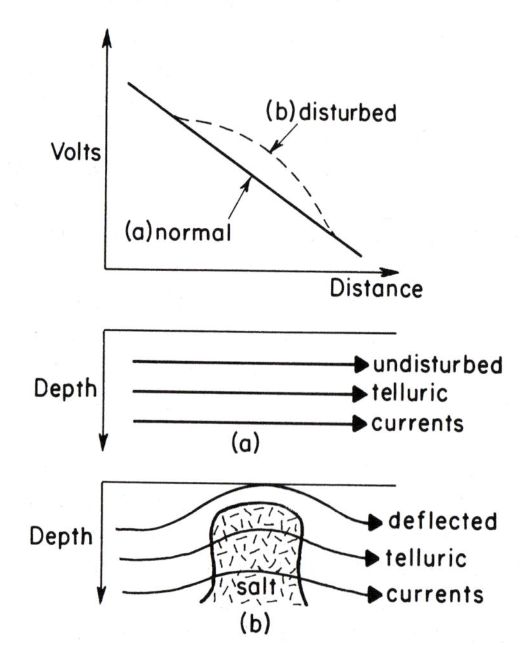

Fig. 9.17. The instantaneous potential gradient associated with telluric currents. (a) normal, undisturbed gradient, (b) disturbed gradient resulting from deflection of current flow by a salt dome.

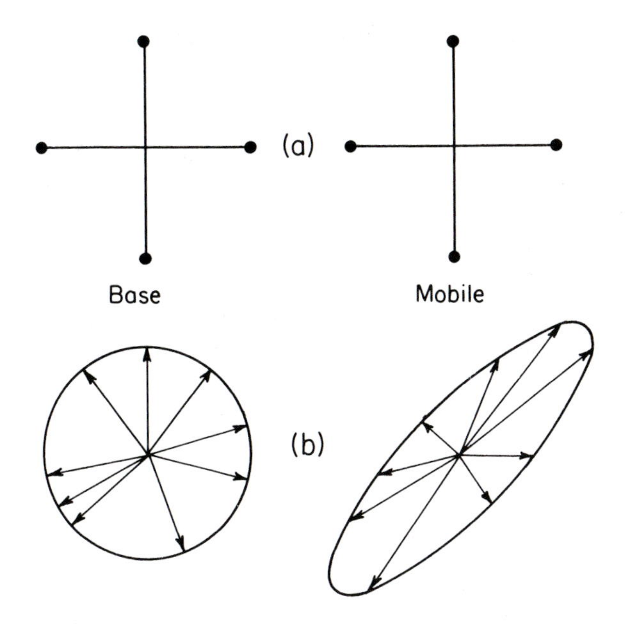

Fig. 9.18. (a) Base and mobile potential electrode sets used in telluric surveys. (b) The figure traced by the horizontal component of the telluric field over an undisturbed area (circle) and in the presence of a subsurface conductor (ellipse) after correction for temporal variations in telluric current intensity.

Telluric currents are used in prospecting by measuring the potential differences they cause between points at the surface. Obviously no current electrodes are required and potential differences are monitored using non-polarizing electrodes or plates made of a chemically inert substance such as lead. Electrode spacing is typically 300–600 m in oil exploration and 30 m or less in mineral surveys. The potential electrodes are connected to an amplifier which drives a strip chart recorder or tape recorder.

If the electrical conductivity of the subsurface were uniform the potential gradient at the surface would be constant (Fig. 9.17(a)). Zones of differing conductivity deflect the current flow from the horizontal and cause distortion of the potential gradients measured at the surface. Fig. 9.17(b) shows the distortion of current flow lines caused by a salt dome which, since it is a poor conductor, deflects the current lines into the overlying layers. Similar effects may be produced by anticlinal structures. Interpretation of anomalous potential gradients measured at the surface permits the location of subsurface zones of distinctive conductivity.

Telluric potential gradients are measured using orthogonal electrode pairs (Fig. 9.18(a)). In practice, the survey technique is complicated by temporal variation in direction and intensity of the telluric currents. To overcome this problem, one orthogonal electrode pair is read at a fixed base located on nearby barren ground and another moved over the survey area. At each observation point the potential differences between the pairs of electrodes at the base and at the mobile station are recorded simultaneously over a period of about 10 minutes. From the magnitude of the two horizontal components of the electrical field it is simple to find the variation in direction and magnitude of the resultant field at the two locations over the recording interval. The assumption is made that the ground is uniform beneath the base electrodes so that the conductivity is the same in all directions. The resultant electrical field should also be constant in all directions and would describe a circle with time (Fig. 9.18(b)). To correct for variations in intensity of the telluric currents, a function is determined which, when applied to the base electrode results, constrains the resultant electric vector to describe a circle of unit radius. The same function is then applied to the mobile electrode data. Over an anomalous structure the conductivity of the ground is not the same in all directions and the magnitude of the corrected resultant electric field varies with direction. The resultant field vector traces an ellipse whose major axis lies in the direction of maximum conductivity. The relative disturbance at this point is conveniently measured by the ratio of the area of the ellipse to the area of the corresponding base circle. The results of a survey of this type over the Haynesville Salt Dome, Texas, USA are presented in Fig. 1.4. The solid circles represent locations where ellipse areas relative to a unit base circle have been computed. Contours of these values outline the known location of the dome with reasonable accuracy.

The telluric method is applicable to oil exploration as it is capable of detecting salt domes and anticlinal structures, both of which constitute potential hydrocarbon traps. As such, the method has been used in Europe, North Africa and the USSR. It is not widely used in the USA where oil traps tend to be too small in area to cause a significant distortion of telluric current flow. The telluric method can also be adapted to mineral exploration.

9.9.3 *Magnetotelluric surveying*

Prospecting using magnetotelluric fields is more complex than the telluric method as both the electric and magnetic fields must be measured. The technique does, however, provide more information on subsurface structure. The method is, for example, used in investigations of the crust and upper mantle (e.g. Hutton *et al.* 1980).

Telluric currents are monitored as before, although no base station is required. The magnetotelluric field is measured by its inductive effect on a coil about a metre in diameter or by use of a sensitive fluxgate magnetometer. Two orthogonal components are measured at each station.

The depth z to which a magnetotelluric field penetrates is dependent on its frequency f and the resistivity ρ of the substrate, according to equations of the form of (9.2) and (9.3), i.e.

$$z = k(\rho/f)^{1/2} \tag{9.5}$$

where k is a constant. Consequently, depth penetration increases as frequency decreases. It can be shown that the amplitudes of the electric and magnetic fields, E and B, are related

$$\rho_a = \frac{0.2}{f}\left(\frac{E}{B}\right)^2 \tag{9.6}$$

where f is in Hz, E in mV km^{-1} and B in nT. The apparent resistivity ρ_a thus varies inversely with frequency. The calculation of ρ_a for a number of decreasing frequencies thus provides resistivity information at progressively increasing depths and is essentially a form of vertical electrical sounding (Section 8.2.3).

Interpretation of magnetotelluric data is most reliable in the case of horizontal layering. Master curves of apparent resistivity against period are available for two and three horizontal layers, vertical contacts and dykes, and interpretation may proceed in a similar manner to curve-matching techniques in the resistivity method (Section 8.2.7). Routines are now available, however, which allow the modelling of two-dimensional structures.

Further reading

Boissonas, E. & Leonardon, E.G. (1948) Geophysical exploration by telluric currents with special reference to a survey of the Haynesville Salt Dome, Wood County, Texas. *Geophysics*, **13**, 387–403.

Cagniard, L. (1953) Basic theory of the magnetotelluric method of geophysical prospecting. *Geophysics*, **18**, 605–35.

Dobrin, M.B. (1976) *Introduction to Geophysical Prospecting*. McGraw-Hill, New York (3rd edn).

Jewell, T.R. & Ward, S.H. (1963) The influence of conductivity inhomogeneities upon audiofrequency magnetic fields. *Geophysics*, **28**, 201–21.

Keller, G.V. & Frischnecht, F.C. (1966) *Electrical Methods in Geophysical Prospecting*. Pergamon, Oxford.

Parasnis, D.S. (1973) *Mining Geophysics*. Elsevier, Amsterdam.

Parasnis, D.S. (1979) *Principles of Applied Geophysics*. Chapman & Hall, London.

Pemberton, R.H. (1962) Airborne EM in review. *Geophysics*, **27**, 691–713.

Telford, W.M., Geldart, L.P., Sheriff, R.E. & Keys, D.A. (1976) *Applied Geophysics*. Cambridge Univ. Press, Cambridge.

Wait, J.R. (1982) *Geo-Electromagnetism*. Academic Press, New York.

10

Major Fields of Application of Geophysical Exploration

In this chapter the major fields of application of geophysical exploration are described, and are illustrated by a series of case histories. These show how the various surveying methods typically complement one another, and how geophysical data may be integrated with the known geology to solve a very wide variety of problems related to the subsurface.

10.1 Geophysics in the search for hydrocarbons

Hydrocarbons are normally found in association with thick sedimentary sequences in major sedimentary basins. There are several pre-requisites for the accumulation of oil or gas in commercial quantities: a suitable source rock, reservoir rock and cap rock; a sediment burial history conducive to the conversion of original organic matter contained within the sedimentary pile into hydrocarbons; and a suitable trap to allow the accumulation of the oil or gas and prevent its upward escape to the surface. There are many types of trap, including tectonic structures such as anticlines or tilted fault blocks, structures associated with halokinesis such as tilted strata on the flanks of salt domes, and stratigraphic traps such as local sand bodies surrounded by clay envelopes, or local reef developments in limestone sequences.

Water, oil and gas can accumulate in the pore spaces of reservoir rocks in these trapping environments with a disposition that is determined by their relative specific gravities: gas at the top, oil in the middle and water at the base. Although seismic reflection surveys can sometimes directly detect the boundaries of acoustic impedance between different fluid layers in a reservoir rock (Section 4.11.2), geophysical exploration for hydrocarbons normally employs an indirect approach, searching for the traps, such as anticlinal closures, within which the oil or gas may be present.

Exploration is usually carried out in several phases. In cases where the subsurface geology is completely unknown, such as unexplored areas of continental shelf or areas where thin cover sequences overlie major unconformities beneath which the deeper geology cannot be predicted, the initial reconaissance may involve gravity and/or aeromagnetic surveying. Gravity surveying is capable of identifying areas of thick sediments by virtue of their relatively low densities and the large-scale negative Bouguer anomalies with which they are consequently associated. Aeromagnetic surveying can be used to estimate variations of depth to an igneous or metamorphic basement underlying a sedimentary sequence and, hence, to determine indirectly the areas of main sediment accumulation. Once a prospective sedimentary basin environment has been identified, further geophysical surveying is normally carried out using seismic methods, especially reflection profiling.

The initial round of seismic exploration normally involves speculative surveys along widely-spaced profile lines covering large areas. In this way the major structural or stratigraphic elements of the regional geology are delineated, so enabling the planning of detailed, follow-up reflection surveys in more restricted areas containing the main prospective targets. Where good geological mapping of known sedimentary sequences exists, the need for expenditure on initial speculative seismic surveys is often much reduced and effort can be concentrated from an early stage on the seismic investigation of areas of particular interest.

Detailed reflection surveys involve closely-spaced profile lines and a high density of profile intersection points in order that reflections can be correlated reliably from profile to profile and used to define the prevailing structure. Initial seismic interpretation is likely to involve structural mapping, using time-structure and/or isochron maps (Section 4.11.1) in the search for the structural closures that may contain oil or gas. Any closures that are identified may need further delineation by a second round of detailed seismic surveying before the geophysicist is sufficiently confident to select the location of an exploration borehole from a time-structure map. In cases of complex structure or where fine structural detail is required, seismic interpretation is based on migrated records (Section 4.10).

Exploration boreholes are normally sited on seismic profile lines so that the borehole logs can be correlated directly with the local seismic section. This facilitates precise geological identification of specific seismic reflectors. Particularly in offshore areas where drilling is highly expensive (typically 10–20 times the cost of drilling on land), and where the best quality seismic reflection data are generally obtained, seismic stratigraphy (Section 4.11.2) is often employed to obtain more insight into prevailing sedimentary lithologies and palaeoenvironments. Seismic stratigraphy provides additional criteria on which to select areas for detailed study, for example, the recognition of local deltaic or reef facies developments, with an associated high reservoir potential, in a broader sedimentary sequence.

The contribution of seismic surveying to the development of hydrocarbon reserves does not end with the discovery of an oil or gas field. Refinement of the seismic interpretation, possibly using information from additional seismic lines, will optimize the location of production boreholes. In addition, seismic modelling (Section 4.11.3) of amplitude variations and other aspects of reflection character displayed on seismic sections across the producing structure can be used to obtain detailed information on the geometry of the reservoir and on internal lithological variations that may affect the hydrocarbon yield.

The above general account of exploration strategy is illustrated with reference to the North Sea, which occupies the continental shelf of north-west Europe between Britain and the continental mainland (Fig. 10.1). The North Sea contains two major hydrocarbon provinces: the southern North Sea gas province and the northern North Sea oil and gas province, the latter representing one of the world's largest oil fields.

There is a long history of geophysical surveying in the North Sea. Early gravity surveys were carried out in the southern North Sea by German investigators in the 1930s. During the 1950s, US workers established seismic refraction lines in the central North Sea and recorded the presence of sedimentary sequences over 4 km thick. Subsequent gravity surveys by Dutch workers revealed large negative Bouguer anomalies over southern and northern sedimentary basins containing thick post-Carboniferous sedimentary successions. The basins are separated by an east-west positive Bouguer anomaly coincident with an intervening structural high in the mid-North Sea area (Fig. 10.1) which has only a thin Mesozoic cover. During the 1950s and 60s, aeromagnetic coverage of the North

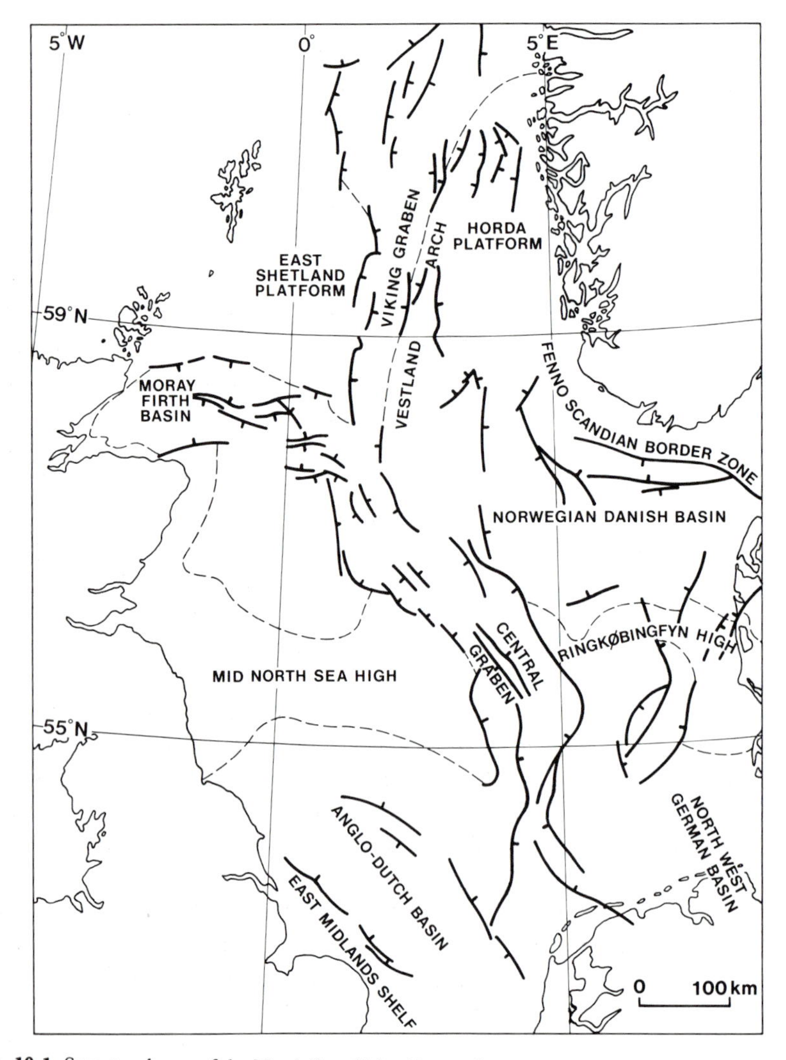

Fig. 10.1. Structural map of the North Sea. (After Day *et al.* 1981.)

Sea was obtained and was used to deduce the pre-Permian structure on the basis of estimated depths to the magnetic basement. Aeromagnetic interpretation suggested that a much greater thickness of Palaeozoic sediments underlay the northern North Sea than the southern North Sea basin.

Seismic reflection surveying activity began in the late 1950s and increased in intensity through the 1960s and 1970s to such an extent that by the early 1980s several million line-kilometres of multichannel profiling had been obtained throughout the North Sea area. Early reflection records showed good marker horizons that could be traced over wide areas, notably, the base of the Tertiary, the base of the Upper Cretaceous and the top and base of the Permian. Widespread evidence of halokinesis was obtained from the southern North Sea area. In the northern North Sea a regional structural pattern was identified, involving block faulting and tilting, associated with rifting, throughout most of the Mesozoic and broad downwarping during the later Mesozoic and succeeding Cenozoic, together with major structural inversions.

The offshore geophysics, together with the associated drilling of exploration and production wells, has provided a detailed picture of the geology of the North Sea area (Woodland 1975; Illing & Hobson 1981).

The southern North Sea area occupies the site of a major sedimentary basin of Permian age that extends eastwards into the Netherlands and northern Germany. The basin was bordered by the London-Brabant massif to the south and the Mid-North Sea structural high to the north. In this basin, the deposition of thick basal sands, mainly aeolian, was followed by several cycles of evaporite deposition. The earliest commercial finds showed that the southern North Sea represented an offshore extension of the Gröningen natural gas province of the Netherlands. Upper Carboniferous coal measures beneath the Permian basin provide the source of gas and the main reservoir is the lower Permian Rötliegendes sandstone. The overlying Zechstein evaporites provide a cap rock.

Further north, persistent rift faulting along the median zone of the North Sea throughout the Mesozoic led to the accumulation of thick marine sediments in the Central and Viking grabens (Fig. 10.1). Later regional downwarping resulted in their burial beneath often very thick Late Cretaceous and Cenozoic sequences deposited in a major sedimentary basin whose main depocentre overlay the earlier zone of rift faulting. Several major oil fields, some with gas production, exploit a variety of structural and stratigraphic traps associated with the earlier graben tectonics and the later regional basin subsidence. The main source rocks are Upper Jurassic clays, and reservoir rocks range in age from Devonian to Tertiary. Estuarine and deltaic sands of Jurassic and early Tertiary age represent the main reservoir horizons. The long history of subsidence, and its detailed cataloguing by means of a wealth of subsurface information derived from geophysical surveys and boreholes, have made this part of the North Sea a classic area for the study of sedimentary basin formation and evolution and for the investigation of causal mechanisms of basin subsidence (see, e.g. Wood & Barton 1983).

The type of structural trap characteristic of the southern North Sea gas province is illustrated in Fig. 10.2 which represents a seismic section across the North Viking gas field. The gas is trapped in the core of a NW-SE trending

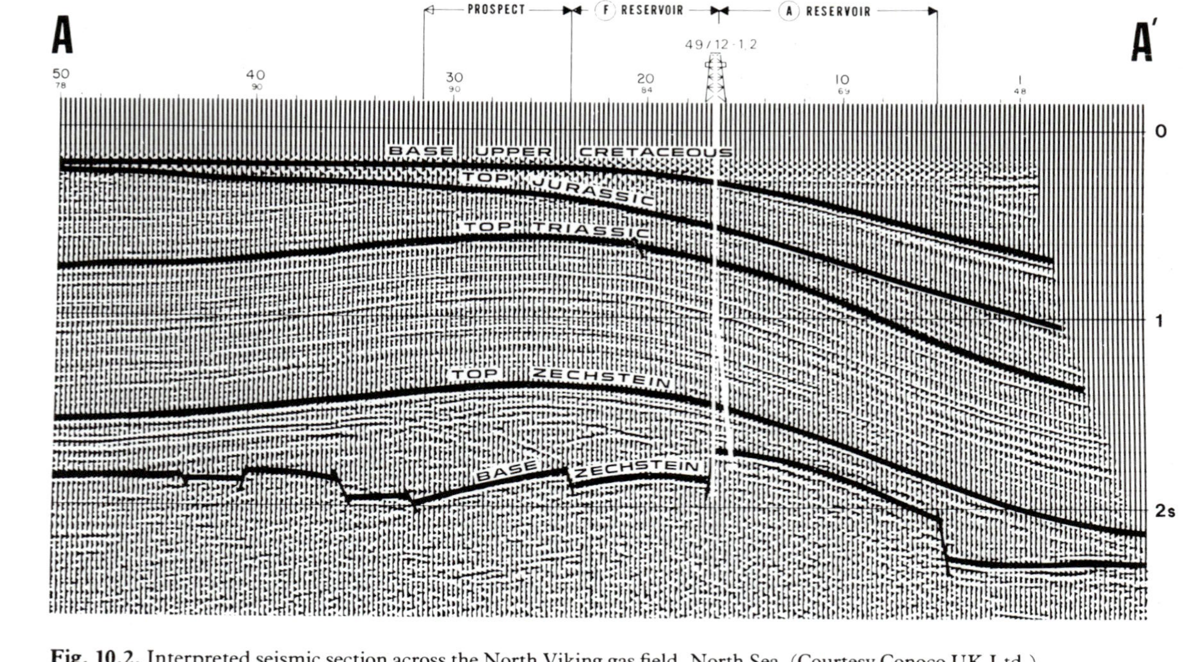

Fig. 10.2. Interpreted seismic section across the North Viking gas field, North Sea. (Courtesy Conoco UK Ltd.)

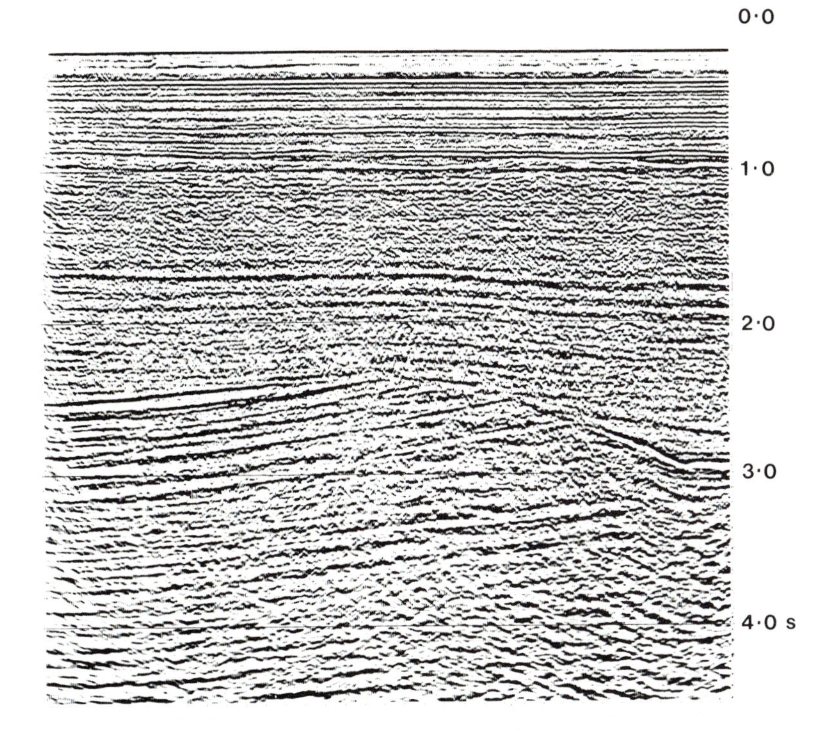

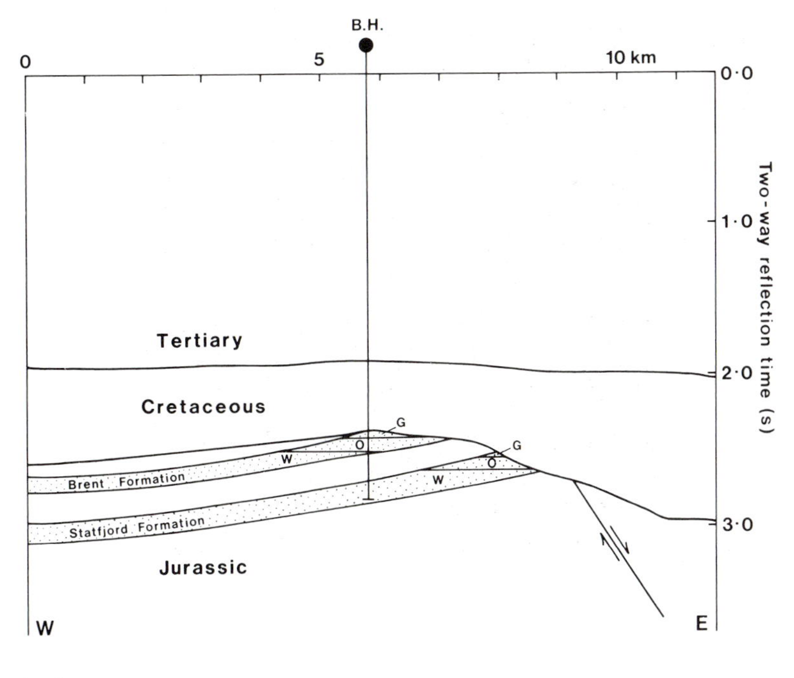

Fig. 10.3. (a) Seismic section (courtesy Shell UK Ltd.) and (b) line interpretation across the Brent oilfield, North Sea. G: gas, O: oil, W: water.

anticlinal structure that is extensively faulted at the level of the Lower Permian. A typical combined structural/stratigraphic trap in the northern North Sea province is represented by the Brent oilfield structure and Fig. 10.3 illustrates a seismic section across the field. A tilted fault block containing Upper Palaeozoic, Triassic and Jurassic strata is overlain unconformably by Upper Jurassic, Cretaceous and Tertiary sediments. Two Jurassic sands in the tilted fault block constitute the main reservoirs, the oil and gas being trapped beneath a capping of unconformably overlying shales of Upper Jurassic and Cretaceous age.

10.2 Geophysics in mineral exploration

Geophysical methods are extensively used in the search for economically valuable mineral deposits. Many materials fall into this category, including the bulk minerals sand, gravel and limestone, but in this section consideration is restricted to selected ore deposits which are of major importance to the metalliferous mining industry, namely massive sulphides, disseminated sulphides and iron ores. These deposits differ significantly from their host rocks in their physical properties and consequently give rise to geophysical anomalies of various types.

The initial aim of a geophysical survey for ore deposits is to locate mineralized areas of potential interest. For this purpose the airborne magnetic and electromagnetic techniques are eminently suitable since large areas can be surveyed rapidly at relatively low cost. Airborne measurements, however, especially electromagnetic, are limited in their depth of penetration and may not detect deeply buried ore bodies.

Once possible target areas are determined, further information on causative bodies within the anomalous zones is obtained by ground surveys which enable the prospector to determine whether the anomalous bodies are of economic importance. Ground verification surveys frequently involve the use of several different survey techniques. If ore bodies are present, the geophysical data will provide information on their depth, extent and attitude and consequently control the location of exploratory boreholes or trenches.

It is customary to refer to the 'returns-ratio' for geophysical surveys, defined as the ratio of the estimated value of the ore to the cost of the geophysical work. That geophysical surveying is of major importance in mineral exploration is shown by the fact that for many ore deposits this ratio is several hundred to one.

Several examples of the use of geophysical methods in the location of ores have been given in previous chapters dealing with individual survey techniques. In the case studies described below, stress is placed on the importance of integrating the results from several geophysical methods to derive the maximum information about the ore deposits concerned.

10.2.1 Massive sulphide ores

Massive sulphide ore bodies are usually considered to be a single mass with a cross-sectional area of at least 100 m^2 comprising 50% or more of metallic sulphides. Such ore has a minimum density of 3800 kg m^{-3}. It may contain the magnetic minerals pyrrhotite and magnetite, and if these are present in reasonable quantity the ore will produce large magnetic anomalies. The electrical conductivity of massive sulphides is normally very high, in the range 10^2–10^4 S.m^{-1}. Consequently, the geophysical methods applicable to the search for such ores are those responding to very dense, highly magnetic and conductive materials.

Airborne prospecting techniques for massive sulphides usually exploit the property of high conductivity, and extensive use is made of electromagnetic methods. The survey aircraft usually also carries a magnetometer to provide additional information at little extra cost, as the coincidence of electromagnetic and magnetic anomalies is highly indicative of massive sulphides.

Subsequent ground surveys similarly employ electrical and electromagnetic methods. Self potential methods (Section 8.4) are cheap and effective if the correct subsurface conditions exist and the ore body lies at a depth of less than about 30 m. Standard moving source-receiver EM methods (Section 9.5) are extensively used, although in rugged or forested terrain the AFMAG method (Section 9.4.3) may be more cost-effective as no heavy equipment is required and there is no need to cut tracks for survey lines. If it is required to establish the relationship of the conducting body to its host rock, resistivity rather than EM methods are employed as they provide estimates of absolute conductivity rather than simply revealing bodies with a high relative conductivity.

Gravity surveying is essentially a secondary ground exploration tool because of the high cost of obtaining gravity coverage over large areas and ambiguities in interpretation. It does, however, provide accurate estimates of ore tonnage on the basis of the total mass anomaly (Section 6.11.3) once the location of the ore body has been established.

Although electrical and electromagnetic methods are the major exploration techniques, they suffer from the drawback that anomalies may result from non-economic sources such as graphite or water-filled shear zones. However, by use of a combination of electrical, magnetic and gravity methods it is possible to eliminate most non-economic sources.

An example of an integrated geophysical study of a massive sulphide ore body in Quebec, Canada has been described by White (1966). Ninety-five percent of the area prospected is covered by glacial deposits about 15 m thick. All exposures of bedrock are high-grade metamorphic rocks of Precambrian age, although younger volcanic rocks are known to occur in neighbouring areas. The airborne survey revealed an EM anomaly 1.2–1.6 km in length with a ratio of real to imaginary secondary field components exceeding unity, indicating the presence of a good conductor (Section 9.5). The eastern part of the causative body was later found to consist of massive pyrrhotite in a host of andesite, rhyolite and silicified tuff, while the western part contained up to 20% sulphides but was mainly graphitic in nature.

The airborne survey was followed by a sequence of ground surveys. A series of EM traverses (Fig. 10.4) was made with a coplanar horizontal coil system operating at a frequency of 1600 Hz with a source-receiver separation of 61 m. The results enabled the subsurface conductor to be accurately located, and the

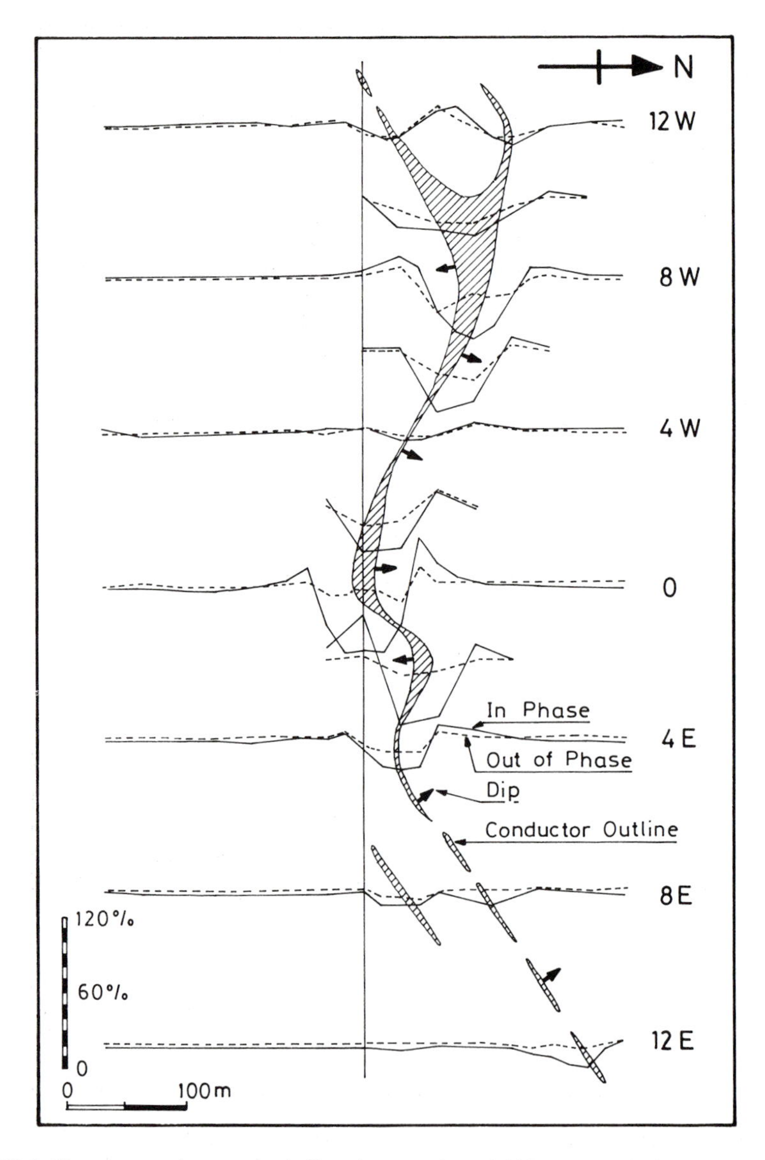

Fig. 10.4. Electromagnetic anomaly profiles over a massive sulphide ore body in Quebec, Canada. Anomalies are expressed as a percentage of the primary field. The shaded area represents the interpreted locations of the ore body. (Redrawn from White 1966.)

asymmetry of the profiles allowed estimates to be made of the conductor dip.

Profiles of the vertical field magnetic anomalies were made along the same traverses (Fig. 10.5). The strong correlation between electromagnetic and magnetic anomalies suggested that a high proportion of pyrrhotite was present, a conclusion in accord with the composition of other known ore bodies in the region. The change in character of the anomalies between traverses 4W and 8W indicated a change in nature of the conductor from sulphides to graphitic sediments as the conductor was at approximately the same depth beneath both profiles. The decrease in anomaly amplitude towards the east resulted from an

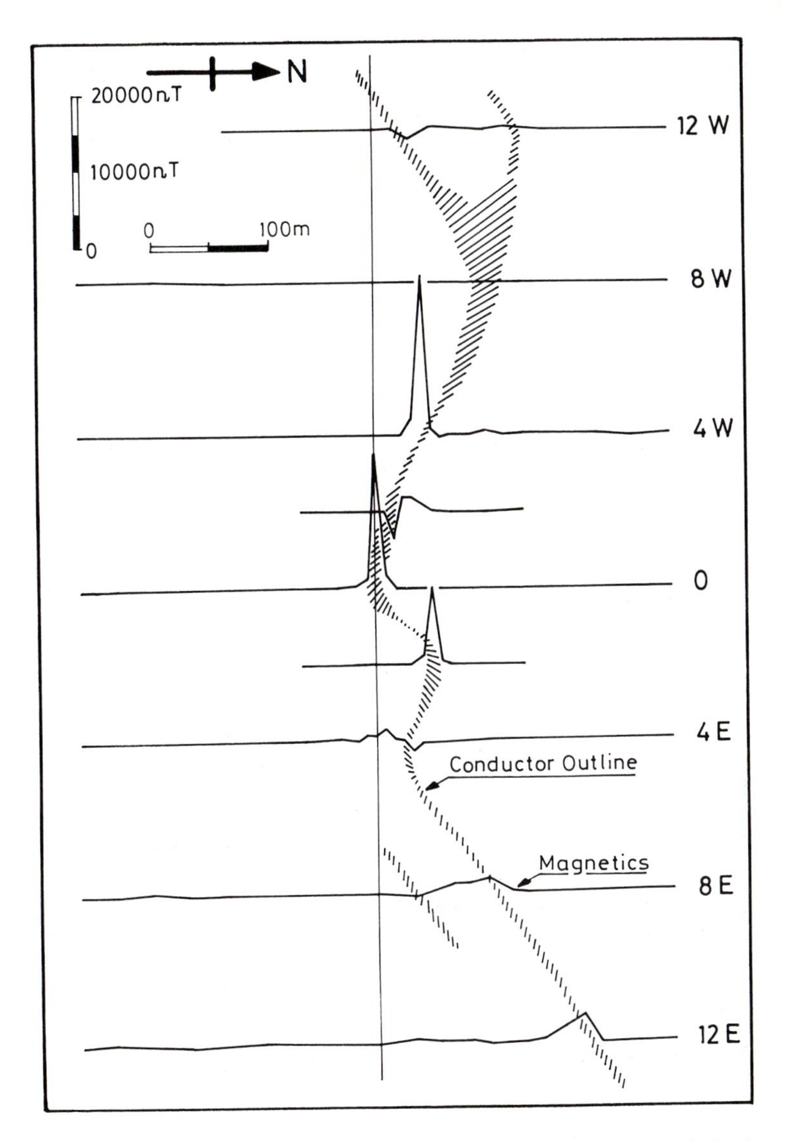

Fig. 10.5. Vertical field magnetic anomaly profiles over a massive sulphide ore body in Quebec, Canada. The shaded area represents the location of the ore body inferred from the EM survey (Fig. 10.4). (Redrawn from White 1966.)

increased overburden thickness, but indicated that pyrrhotite-bearing ore continued at least to traverse 12 E.

Gravity observations over three of the central traverses are presented in Fig. 10.6. The only significant anomaly occurs on traverse 0, where the amplitude is 3–4 gu. It was later found that this resulted from a sulphide body about 12 m thick, illustrating that gravity surveys are only of use in this type of application where other geophysical methods have indicated a body of reasonable thickness.

The geophysical data were subsequently employed to control the location of several boreholes which allowed the nature of the conductor to be determined in the most cost-effective manner.

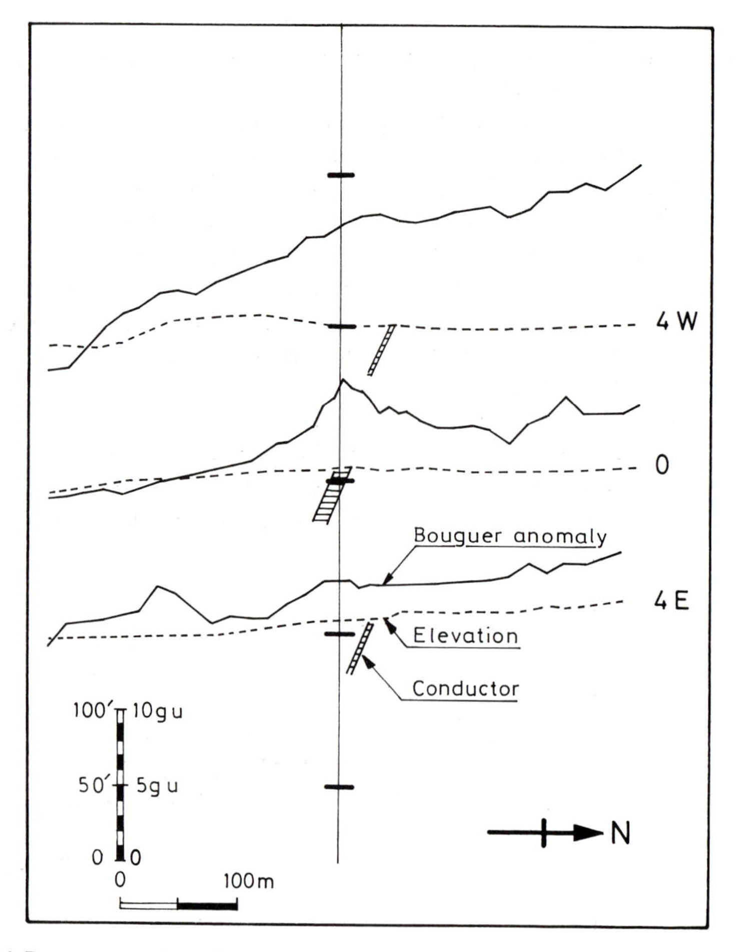

Fig. 10.6. Bouguer anomaly profiles over a massive sulphide ore body in Quebec, Canada. The conductor location is inferred from the EM survey (Fig. 10.4). (Redrawn from White 1966.)

10.2.2 Disseminated sulphide ores

Disseminated sulphide deposits are defined as those bodies in which sulphides are scattered as specks and veinlets throughout the host rock and constitute not more than 20% of the total volume. A disseminated ore body contains metallic sulphides (usually of copper and/or molybdenum) at a mineable depth and must normally exceed 100 m^2 in horizontal section to be profitable.

The density distribution in such bodies is complex since, although the metallic sulphides themselves are very dense, the density of the host can be highly variable. Consequently the gravity method is not applicable to the direct search for such ores. Similarly, their magnetic susceptibility is normally low so that magnetic surveying cannot be relied upon to provide a direct indication of a disseminated sulphide ore.

The electrical and electromagnetic methods appear to be the most suitable survey techniques. However, the conductivity of a disseminated sulphide ore body is highly variable because of the irregular dispersion of the sulphides throughout the host. Consequently, diagnostic resistivity and EM anomalies are unlikely to be encountered.

Since electrical conduction through the metallic sulphides is electronic, but electrolytic through the host, the conditions in disseminated sulphide ore bodies exist to produce strong induced polarization anomalies (Section 8.3) so the IP method is the most likely to detect such bodies. However, the physical properties of economically important sulphides such as chalcopyrite ores are not greatly different from zones of disseminated uneconomic minerals such as pyrite. Hence the economic importance of a deposit cannot be judged solely from its IP response and further geological and geochemical surveying need to be executed prior to any costly drilling programme.

The results of an IP survey over a low grade copper-silver body in Ireland are shown in Fig. 8.25. A further example of a time domain IP traverse over a copper porphyry body in British Columbia, Canada is shown in Fig. 10.7. Both IP and resistivity traverses were made at three different electrode spacings of a pole-

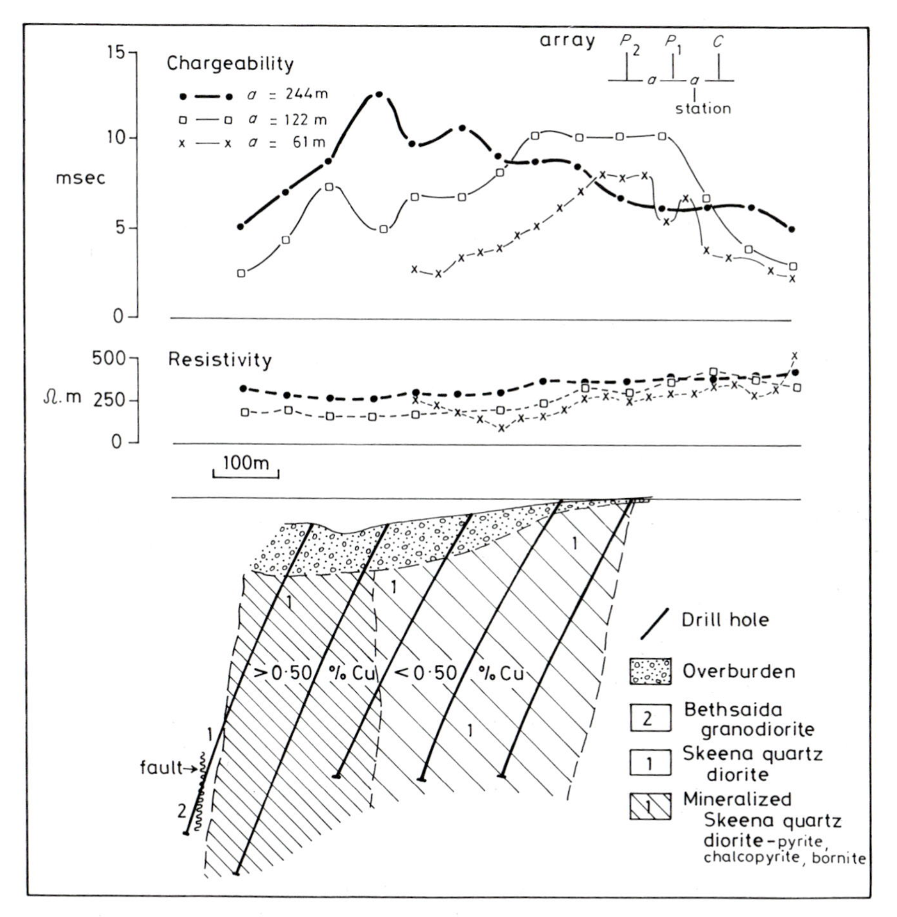

Fig. 10.7. Time domain induced polarization and resistivity profiles over a copper porphyry body in British Columbia, Canada. (Redrawn from Seigel 1967.)

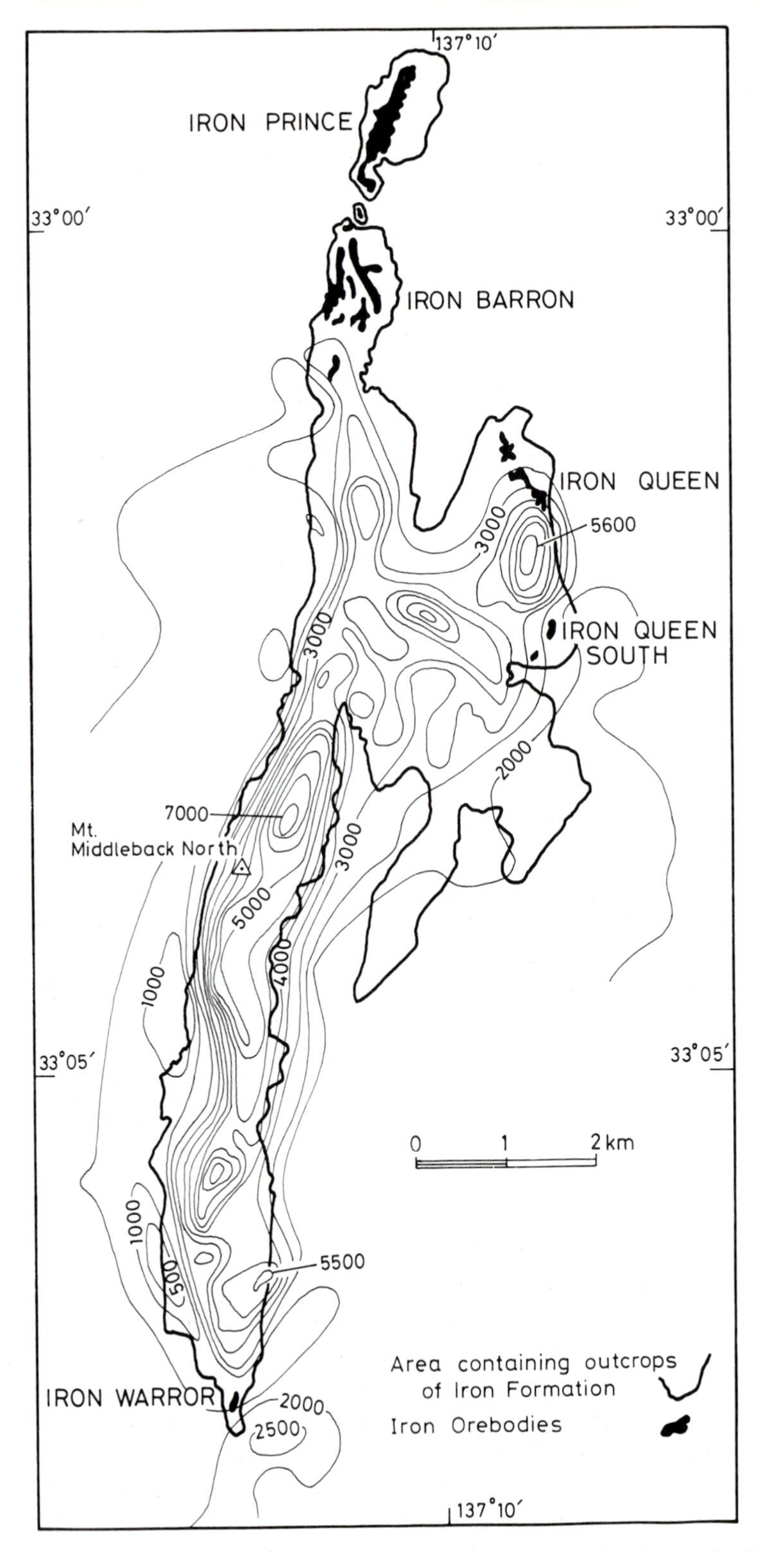

Fig. 10.8. Aeromagnetic anomalies over the Northern Middleback Range, South Australia. Contour interval 500 nT. (Redrawn from Webb 1966.)

dipole array. The resistivity results show no variation over the body, but the IP (chargeability) profiles clearly show the presence of the mineralization, allow its limits to be determined and provide estimates of the depth to its upper surface.

10.2.3 Iron ores

There are many different types of iron ore, the most abundant being sedimentary iron ores which account for about 90% of world iron production.

The density of iron ores is quite variable, for although the iron minerals themselves have a high density, the bulk density of the ore is significantly reduced by the presence of low density minerals and the development of porosity resulting from the leaching of soluble constituents. Consequently, gravitational methods are not always suitable for the location of ores of this type.

The most widely exploited physical property of iron ores in geophysical exploration is their magnetic susceptibility. However, the ratio of magnetite to haematite must be high for the ore to produce significant magnetic anomalies, as haematite is commonly non-magnetic (Section 7.2). This is illustrated in Fig. 10.8, which shows total field aeromagnetic anomalies over the Northern Middle-back Range, South Australia. The principal iron mineral is haematite and it is apparent that there is no correlation between the magnetic anomalies and the ore bodies.

Fig. 10.9 presents results from a high-level (450 m ground clearance) aero-magnetic survey of part of the Eyre Peninsula of South Australia. The elongate east-west anomaly was subsequently surveyed at a lower level (91 m) to provide greater definition and a decision was made to execute ground surveys. Both gravity and magnetic surveys were performed and the results are presented in Fig. 10.10. Electromagnetic and resistivity surveys were also attempted, but were found to be ineffective due to the presence of saline surface water which masked all other anomalies. The magnetic and gravity profiles exhibit coincident highs, although the latter are somewhat masked by a regional field increasing from south to north. Drilling was subsequently performed on the anomalously high regions and revealed the presence of a magnetite-bearing ore body at shallow depth with an iron content of about 30%.

10.3 Geophysics in hydrogeology

Many geophysical methods find application in locating and defining subsurface water resources. They provide rapidly collected information on the geological structure and prevailing lithologies of a region without the large cost of an extensive drilling programme. The geophysical survey results determine the location of the minimum number of exploratory boreholes required for both essential aquifer tests and control of the geophysical interpretation.

The magnetic method is rarely used in this context, but may find occasional application in the location of faults and shear zones which could affect the pattern of ground water flow. The gravity method is widely used in regional recon-naissance surveys to delineate the form and extent of porous sedimentary deposits

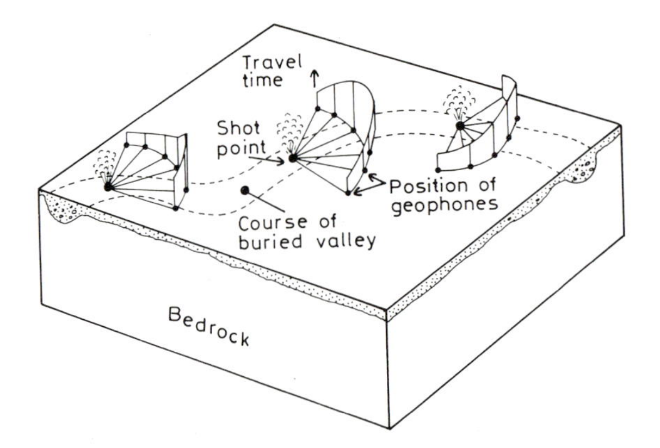

Fig. 10.11. The fan-shooting technique adapted to the location of an alluvium-filled channel.

Seismic refraction surveys also provide important controls on the interpretation of the gravity anomalies associated with certain aquifers.

The most widely used geophysical methods in hydrogeology are the electrical techniques. Resistivity surveys are routinely employed in groundwater exploration to locate zones of relatively high conductivity corresponding to saturated strata at depths down to about 400 m. As well as providing structural and lithological information, resistivity surveys may also provide indications of groundwater quality. The method provides adequate depth penetration and quantitative results.

Most constituent minerals of sedimentary rocks are insulating and the passage of electricity thus takes place mainly by ionic conduction in the pore waters. The resistivity of the rock is thus controlled by the volume of water present and will decrease as the salinity of the water increases (equation (8.2)). Consequently, in an homogeneous aquifer, it is possible to distinguish fresh from saline groundwater and even to trace the subsurface flow of contaminated groundwater resulting from pollution if the polluted water has a distinctive resistivity.

The resistivity method has been used by Bugg & Lloyd (1976) for the quantitative delineation of fresh water lenses in the Cayman Islands of the northern Caribbean. In many small oceanic islands such lenses constitute the main source of potable water.

Because of their relatively low density, fresh water lenses rest on top of the saline water that penetrates the substrate of the islands from the sea. Since the densities ρ_w and ρ_s of fresh and salt water are known, under static conditions where there is an immiscible contact, the thickness z_w of fresh water beneath mean sea-level can be predicted from the height z_s of the fresh water table above this datum (Fig. 10.12), according to the Ghyben-Herzberg relationship

$$z_w = \frac{\rho_w}{(\rho_s - \rho_w)} z_s = k z_s$$

where k is a constant. For $\rho_s = 1026 \text{ kg m}^{-3}$ and $\rho_w = 1000 \text{ kg m}^{-3}$, $k = 38$.

Fresh water lenses are subject to the dynamic effects of tides so that significant saline transition zones develop along their lower boundaries. Consequently, a modified form of density relationship must be employed with the constant k in the Ghyben-Herzberg relationship decreased to about 25. Theoretically, the base of the freshwater lens could be determined by measurement of the height of the water table. Practically though, due to the tidal effects, this would require simultaneous measurement at a large number of boreholes. Thus, in the Cayman Islands study a different approach was adopted, employing geophysical methods.

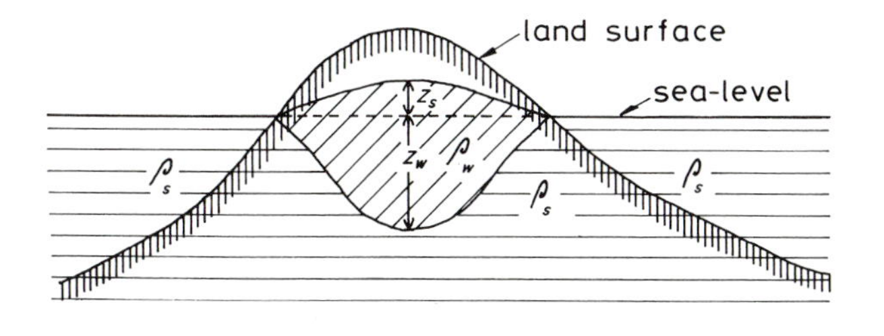

Fig. 10.12. The Ghyben-Herzberg relationship between fresh and saline water.

On the island of Grand Cayman (Fig. 10.13) three fresh water lenses had previously been discovered although their size was unknown. Resistivity methods were used to solve this problem, and their application to the study of the Central Lens is considered below. The island is structurally simple and underlain by a suite of limestones.

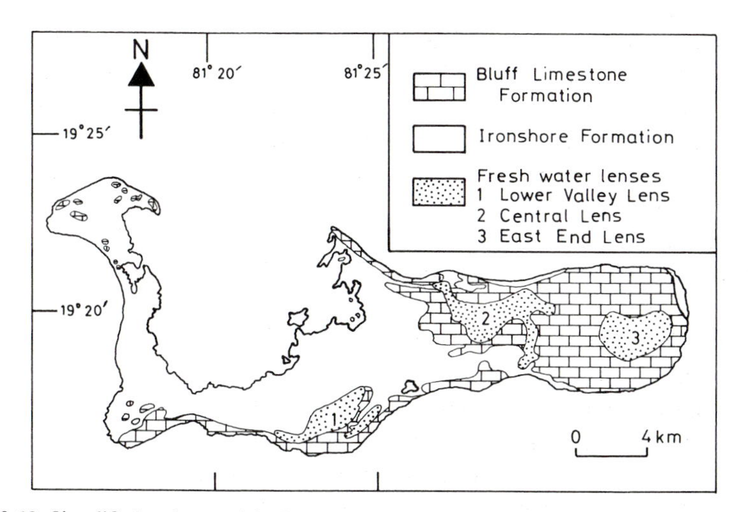

Fig. 10.13. Simplified geology and fresh water lenses of Grand Cayman. (Redrawn from Bugg & Lloyd 1976.)

A test borehole was sunk in the lens and fluid conductivity measurements were made at 0.5 m intervals (Fig. 10.14). In this environment the dominant anion is chloride. The World Health Organization potable limit for chloride is 600 ppm so for this particular study a concentration of 500 ppm was taken to define fresh water. It was found that this constraint further decreased the Ghyben-Herzberg constant k to about 20. Use of a value of 40 delineated the base of the transition zone.

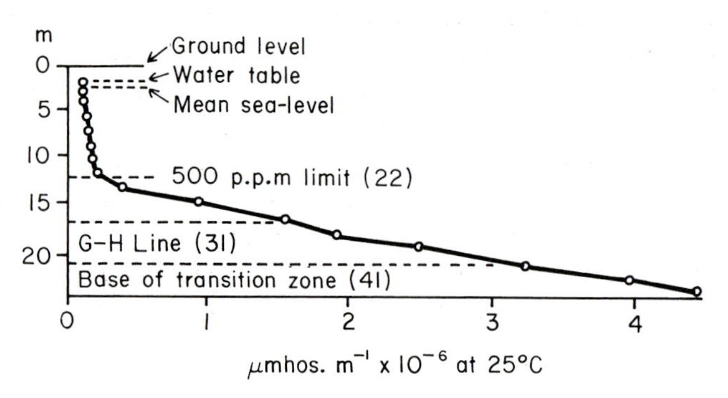

Fig. 10.14. Conductivity profile of a borehole in the Central Lens, Grand Cayman. Numbers in brackets refer to the constant k in the Ghyben-Herzberg relationship. (Redrawn from Bugg & Lloyd 1976.)

The resistivity survey consisted of a number of vertical electrical soundings using a Wenner configuration with electrode spacings ranging from 0.3 m to 58.5 m. Interpretation was initially performed by partial curve matching with the geoelectric section so derived subsequently improved using an iterative technique (Section 8.2.7). The first VES was performed in the vicinity of the test borehole. The apparent resistivity curve, interpreted geoelectric section and salinity profile,

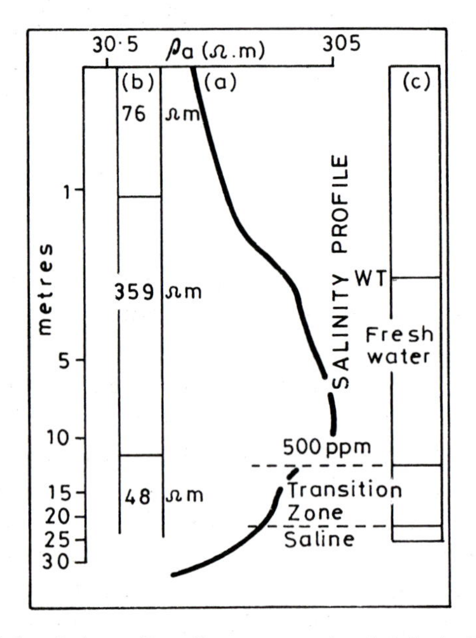

Fig. 10.15. (a) Vertical electrical sounding adjacent to a test borehole in the Central Lens, Grand Cayman. (b) Layered model interpretation of the VES. (c) Interpreted salinity profile. (Redrawn from Bugg & Lloyd 1976.)

derived from the conductivity curve of Fig. 10.14, are presented in Fig. 10.15. It is readily seen that the large resistivity contrast between fresh and saline water causes a distinct decrease of apparent resistivity with depth so that the base of the freshwater lens is easily defined, and deflections of the curve below this point enable the base of the transition zone to be determined. Resistivity variations corresponding to the water table, however, are only obvious in certain cases since measurements at small electrode spacings were influenced by the presence of infilled karstic fissures at the top of the limestone aquifer.

The VES near the borehole allowed accurate interpretations of the other profiles to be made. Additional information on the water table and, hence, the thickness of the fresh water layer from the Ghyben-Herzberg relationship, was provided by the water level observed in many dug pits. All this information was integrated to provide an accurate map of the base of the fresh water lens (Fig. 10.16).

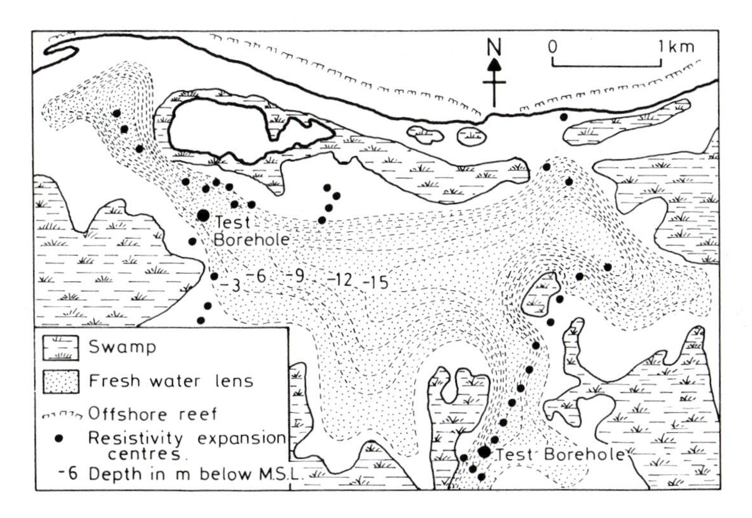

Fig. 10.16. Configuration of base of Central Lens, Grand Cayman. (Redrawn from Bugg & Lloyd 1976.)

Surface resistivity techniques thus provide a rapid and cheap method of mapping the base of fresh water lenses and substantially reduce the cost of drilling investigations.

Van Overmeeren (1975) has described the use of gravity and seismic techniques in a hydrogeological investigation near Taltal, Chile. The region is extremely arid and ground water supply and storage are consequently controlled by deep geological features. The area of interest is a broad flat valley, filled with alluvial deposits, bordered by hills of granodioritic composition.

The gravity survey was used to obtain a general assessment of the form of the granodioritic basement surface. As well as the normal gravity reductions (Section 6.8), an additional correction had to be applied for a strong regional anomaly generated by the crustal root underlying the Andes mountain range, the associated gravity gradient being derived from a small scale gravity map of the

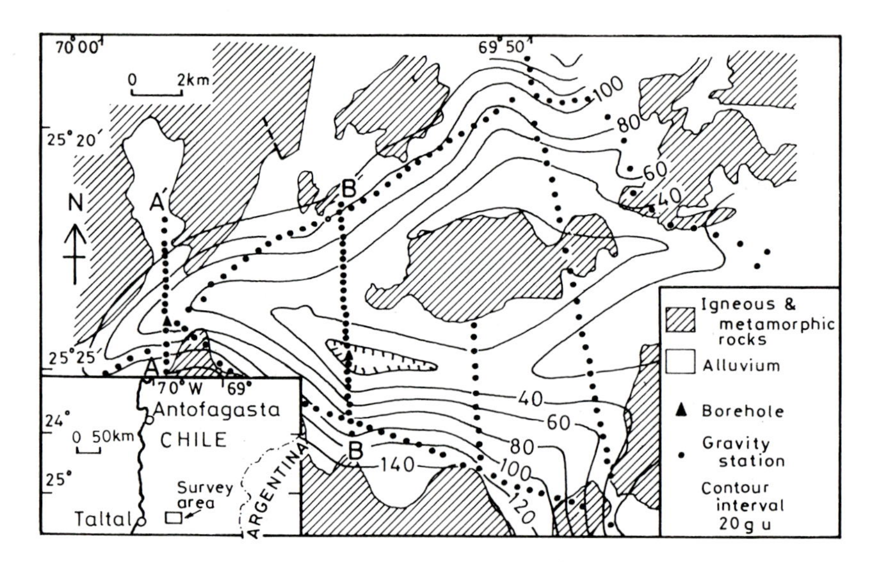

Fig. 10.17. Geological map of an area near Taltal, Chile, showing location of gravity and seismic refraction profiles and contoured Bouguer anomalies. (Redrawn from Van Overmeeren 1975.)

surrounding region. Minima on the gravity contour map (Fig. 10.17) revealed the presence of two buried valleys which meet in the west of the area.

More detailed information on the form of the basement surface was obtained by seismic refraction experiments. Initially the interpretation was performed using the plus-minus method (Section 5.4). This revealed the shape of the basement surface (Fig. 10.18(c)), and a velocity layering of some 0.60 km/s for the uppermost dry, weathered layer, 1.20 km/s for the underlying dry alluvial deposits, 2.20 km/s for water-saturated overburden and 4.00 km/s for the basement. The latter velocity is rather low for igneous rock and probably indicates that the granodiorite is deeply weathered. A more detailed interpretation of the seismic data, using a wavefront tracing method, revealed the presence of two faults in the basement (Fig. 10.18(b)) and these were included in the gravity interpretation.

The principal problem in the gravity interpretation was the highly variable density of the valley-fill deposits. A mean density contrast was, however, obtained by comparing two-dimensional models constructed using a range of density contrasts with the seismic interpretation. This indicated a mean density contrast of -500 kg m^{-3} between alluvial deposits and bedrock.

On the basis of the geophysical results, two boreholes (Fig. 10.17) were sunk in the deepest parts of the valley-fill and located groundwater ponded in the bedrock depressions.

This study indicates that the gravity method can provide a rapid and cheap means of determining the configuration of the bedrock surface over an area of recent sediment cover, while the refraction method reveals more precise, but areally restricted, details of bedrock surface relief and the additional information

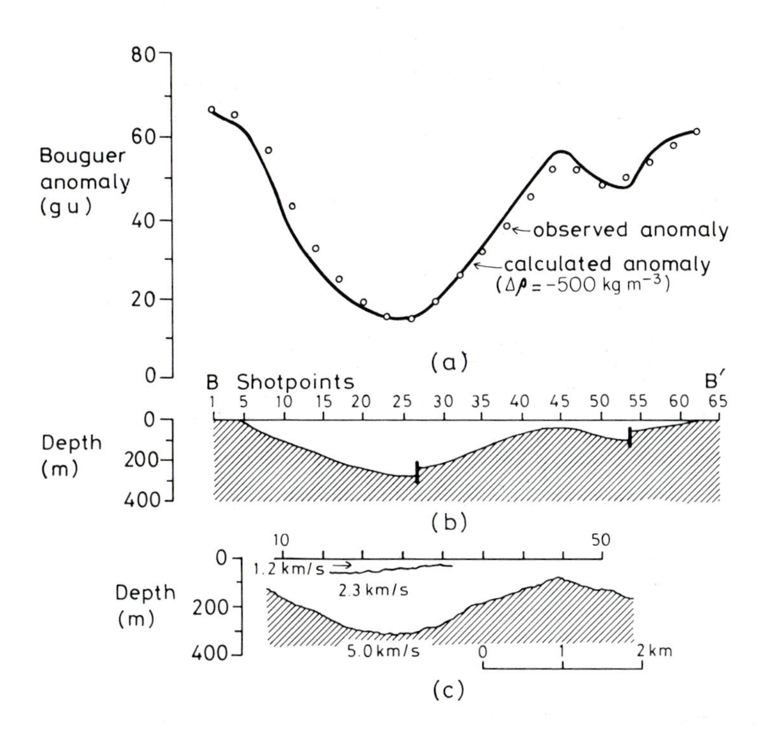

Fig. 10.18. Profile B-B′, Taltal area, Chile (see Fig. 10.17 for location). (a) Observed Bouguer anomaly and calculated anomaly for a model with a density contrast ($\Delta\rho$) of -500 kg m^{-3}. (b) Gravity interpretation. (c) Seismic refraction interpretation. (Redrawn from Van Overmeeren 1975.)

of depth to water table. Used in conjunction, the two techniques represent a powerful means of investigating the form and potential of certain types of aquifer.

Merkel (1972) has described the use of resistivity methods in the delineation of contaminated mine discharge in Pennsylvania. A large quantity of acid mine drainage originates from deep and shallow coal mining operations in this area. It renders the water unfit for recreational use and necessitates costly treatment before it can be used for industrial purposes. The sources of such contamination are usually difficult to identify since the acid water, after mixing with natural groundwater, can travel considerable distances before emergence and the location of many of the old workings is unknown. In the area studied (Fig. 10.19) bedrock consists of a sequence of shales, clays, sandstones and coal, the latter having been extensively mined.

Investigation of the local mine waters revealed a linear relationship between log resistivity and log ion concentration (Fig. 10.20). The groundwater becomes more conductive as the level of contamination increased, fresh water having a resistivity of 60–200 ohm m and acid water 6–12 ohm m. The polluted groundwater normally flows either in the highly jointed coal seams or, if their basal clays are breached, in the underlying sandstones. The coal consequently has a high resistivity when above the water table, a moderate resistivity when associated

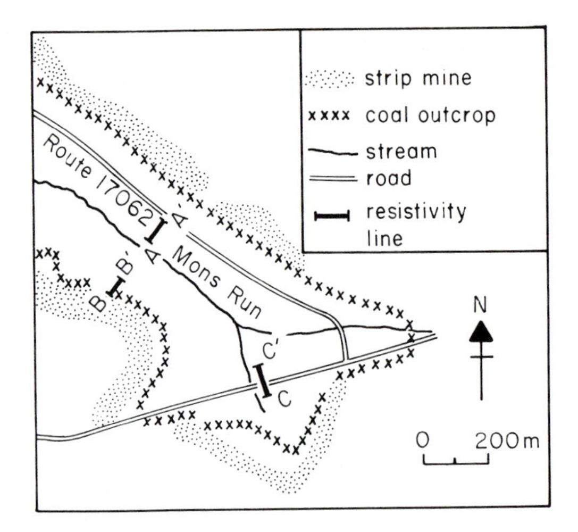

Fig. 10.19. Map of investigated area near Kylertown, Pennsylvania, USA. (Redrawn from Merkel 1972.)

with groundwater and a significantly lower resistivity if the groundwater is contaminated, according to Archie's formula (equation (8.2)).

Three geoelectric sections were constructed in the area from a series of vertical electrical soundings taken normal to the illustrated profiles at 10 m intervals. A modified Wenner array was used with a maximum electrode spacing of 25 m.

Section A-A′ (Fig. 10.21) was constructed near an area of active opencast mining. There is normally a time-lag between mining activity and the appearance of polluted groundwater so that no major contamination was expected. The geological interpretation of the geoelectric section was controlled by a number of

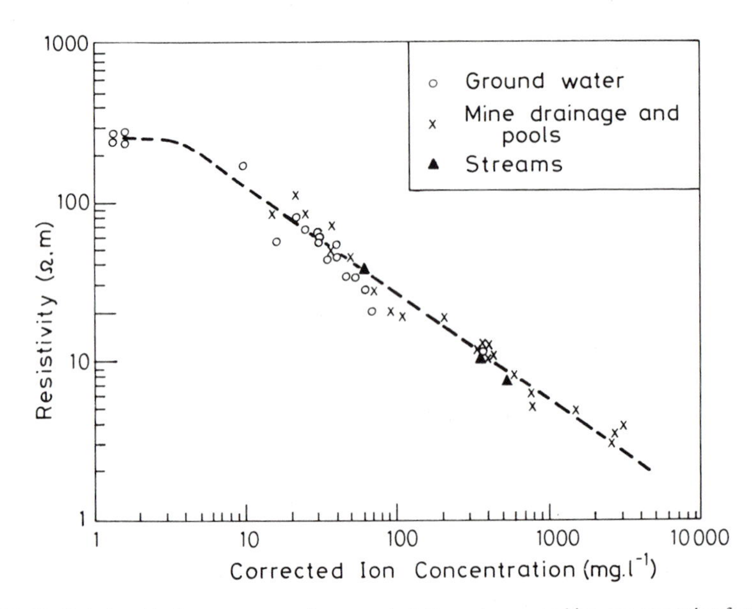

Fig. 10.20. Relationship between groundwater resistivity and corrected ion concentration for local mine waters near Kylertown, Pennsylvania. (Redrawn from Merkel 1972.)

boreholes, which revealed that the water table was at a depth of 18 m below A′ and 13 m below A. With the exception of a saturated coal seam, none of the layers had a resistivity of less than 100 ohm m indicating that no pollution was present.

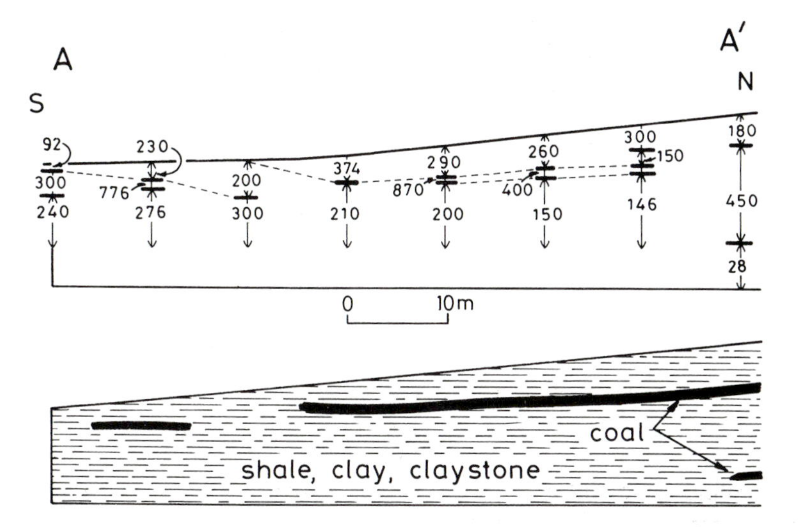

Fig. 10.21. Geoelectric section and geological interpretation of profile A-A′, Kylertown, Pennsylvania (see Fig. 10.19 for location). Numbers refer to resistivity in ohm m. (Redrawn from Merkel 1972.)

Section B-B′ (Fig. 10.22) was located near the point where acid groundwater generated in a disused strip mine emerged at the surface from a sandstone layer, and also crossed the coal outcrop. The section showed that the low resistivity layer corresponding to the contaminated sandstone aquifer could be traced back under the coal seam to the source of pollution in the strip mine. Section C-C′ (Fig. 10.23) crossed a road which had modified the groundwater flow pattern. However, again it was possible to trace the low resistivity zone of contaminated water,

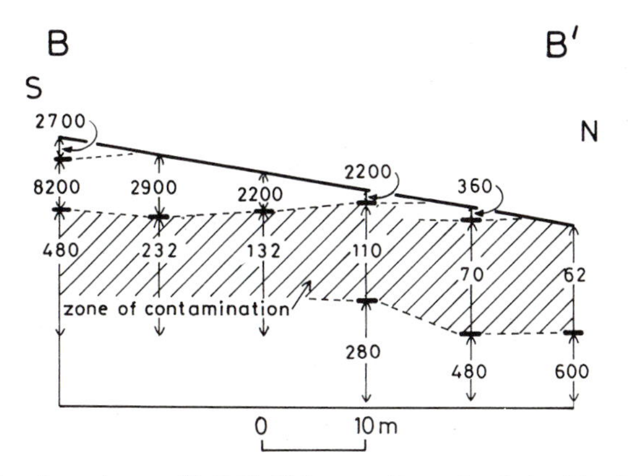

Fig. 10.22. Geoelectric section, profile B-B′, Kylertown, Pennsylvania (see Fig. 10.19 for location). Shaded area shows zone of contamination. Numbers refer to resistivity in ohm m. (Redrawn from Merkel 1972.)

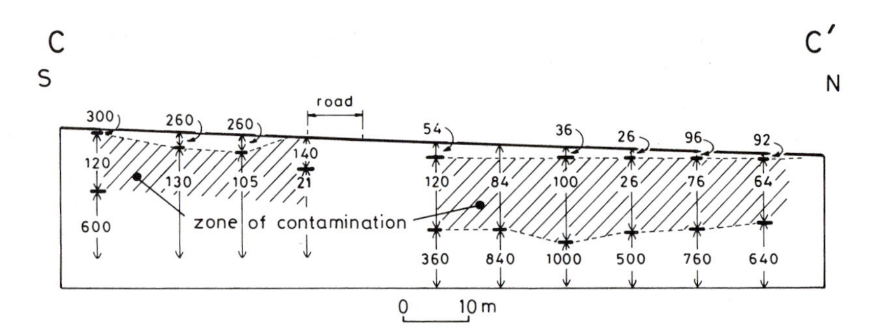

Fig. 10.23. Geoelectric section, profile C-C′, Kylertown, Pennsylvania (see Fig. 10.19 for location). Shaded area shows zone of contamination. Numbers refer to resistivity in ohm m. (Redrawn from Merkel 1972.)

which surfaced at C′, to its source as it flowed through the porous sandstone between a thin surface layer and impermeable shales.

The resistivity technique is thus effective in locating polluted groundwater and tracing its source. The contaminating ions cannot be identified by such methods, but if the geology is sufficiently well known the degree of contamination can be determined. If, then, electrodes were sited in a borehole penetrating the water table, periodic measurement of the resistivity could be used to reveal the onset and extent of acid mine drainage and to monitor the degree of contamination by use of the type of relationship illustrated in Fig. 10.20.

10.4 Geophysics in engineering geology

10.4.1 Land engineering applications

Geophysical methods are frequently used in an initial site investigation to determine subsurface ground conditions prior to excavation and construction work.

Both seismic refraction and vertical electrical soundings are routinely employed in the determination of overburden thickness for foundation purposes. In addition, the use of both P-wave and S-wave sources in a seismic survey and the consequent determination of both body wave velocities allows calculation of Poisson's ratio, together with the elastic moduli if the rock density is also known, for the *in situ* rock (see p. 47).

Magnetic surveys are occasionally used to delineate zones of faulting in bedrock, and may also be employed in the location of buried, metallic, man-made structures such as pipelines or old mine workings. Standard gravity surveying is rarely employed as its definition is normally too coarse. However, specialized microgravimetric methods may be used to detect subsurface cavities, buried valleys, faults within bedrock, underground workings and various archaeological features. The required accuracy in such surveys is much greater than normal, about ± 0.05 gu. This may be achieved by the use of special gravimeters and requires extreme accuracy in the determination of station locations and elevations. In addition, the application of accurate terrain corrections may be necessary to account for both the local topography and any buildings in the vicinity of the site.

In coalfield areas the possible presence of old mine workings is a frequent problem in planning new construction work. Old mine shafts, in particular, are often not capped effectively and are consequently potential sites of collapse. The position of old mineworkings is rarely known with any accuracy. For example, in Great Britain it became the statutory duty of owners to inform the Secretary of State of the location of abandoned workings only after 1873. Workings prior to that date often do not feature on any map and it is commonly the case that later workings have not been plotted to the accuracy necessary for their relocation. The diameter of old shafts is usually less than about two metres so that direct drilling investigations would be prohibitively expensive and disturbing to the site.

The geophysical methods used to locate mineshafts fall into two main categories depending upon the properties of the shaft that they exploit. Microgravimetric and resistivity methods are used to detect the presence of the subsurface void of the shaft which constitutes both a mass deficiency and a highly resistive zone. Magnetic and electromagnetic methods detect metallic objects associated with either the capping, infilling or lining of the shaft. The latter two methods are usually preferred since they are more rapidly executed. A recent development in this field is the use of a ground based radar transmitter which provides, essentially, a shallow penetration continuous profile of the subsurface similar to a seismic section. The technique has proved highly successful on certain sites (Leggo 1982).

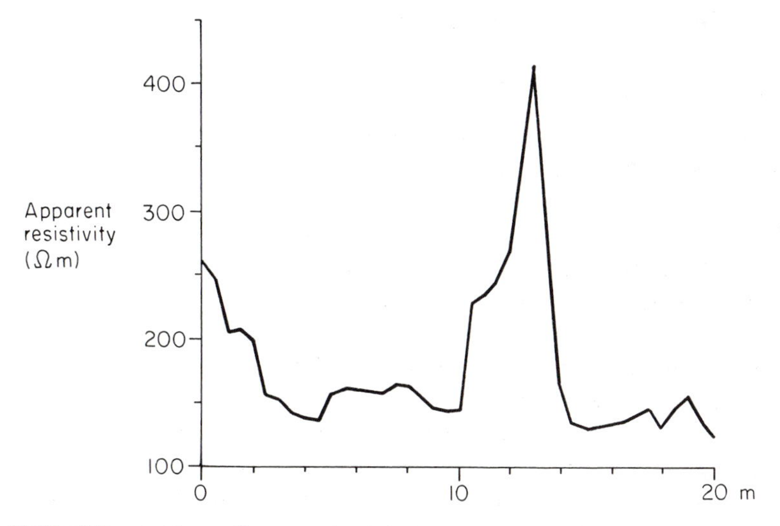

Fig. 10.24. CST resistivity profile across a buried mineshaft. (Redrawn from Aspinall & Walker 1975.)

Aspinall & Walker (1975) have presented the results of electric profiling over shallow subsurface features. Fig. 10.24 shows the anomaly associated with a suspected buried mine shaft, whose air-filled void gives rise to a zone of high apparent resistivity. Comparison of the anomaly with model curves indicated a depth of about 4 m. Fig. 10.25 shows a series of parallel traverses across a buried ditch which subsequent examination revealed to be cut some 3 m into chalk.

Barker & Worthington (1972) have used electric profiling techniques to produce an apparent resistivity contour map of an area of old mine workings in Warwickshire, England. A shaft was believed to be present at a depth of some 1.5 m. Consequently the profiling was performed using a Wenner configuration with an electrode spacing of 1.5 m. The contour map (Fig. 10.26) shows a marked anomaly *A* above the local noise level with a form characteristic of a buried vertical cylindrical void. Subsequent excavations revealed a brick-lined shaft of 1.8 m diameter, 0.15 m below the surface.

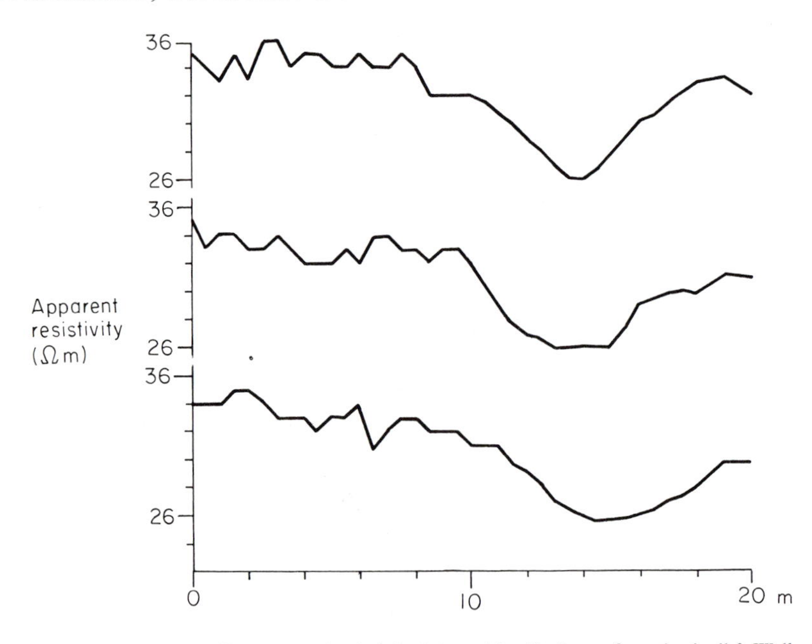

Fig. 10.25. Resistivity profiles across a buried ditch 4 m wide. (Redrawn from Aspinall & Walker 1975.)

Hooper & McDowell (1977) have described the location of buried mine shafts and wells using magnetic surveying methods. Fig. 10.27 shows a total magnetic field contour map of the site of a proposed apartment block in Bristol, England. The area had been mined for coal in the past and it was suspected that shafts and buried workings might be present beneath the surface. A proton magnetometer was used to take readings at 1.2 m intervals on traverses 1.8 m apart. Several anomalies are apparent. The magnitude of anomaly *D* suggested a shallow magnetic source and subsequent excavation revealed a 2 m diameter brick-lined shaft assumed to contain metallic debris. A second, stone-lined shaft, 1 m in diameter, was located at *A* beneath 1.3 m of overburden. The other isolated anomalies e.g. *B* and *C* were known, or suspected, to be associated with buried metallic objects.

Arzi (1975) has described the use of microgravity surveying in the investigation of the site of a cooling tower serving a nuclear power plant. The

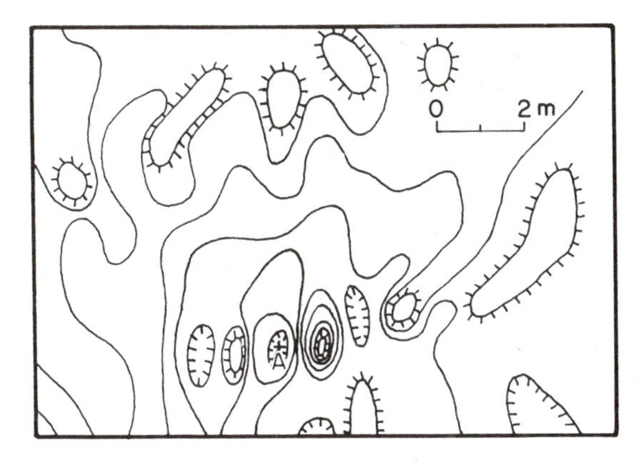

Fig. 10.26. Apparent resistivity contour map of an area of old mine workings in Warwickshire, England. Contour interval 10 ohm m. (Redrawn from Barker & Worthington 1972.)

geology of the site consisted of glacigenic silty clays overlying a horizontally stratified bedrock of argillaceous dolomite with interbeds of anhydrite or gypsum. Solution cavities in the bedrock had been found in the surrounding region.

Microgravity anomalies in the area might result from lateral density inhomogeneities in bedrock or soil, undulations of the bedrock surface, or local rock defects resulting from the presence of cavities filled with rubble or air. Interest

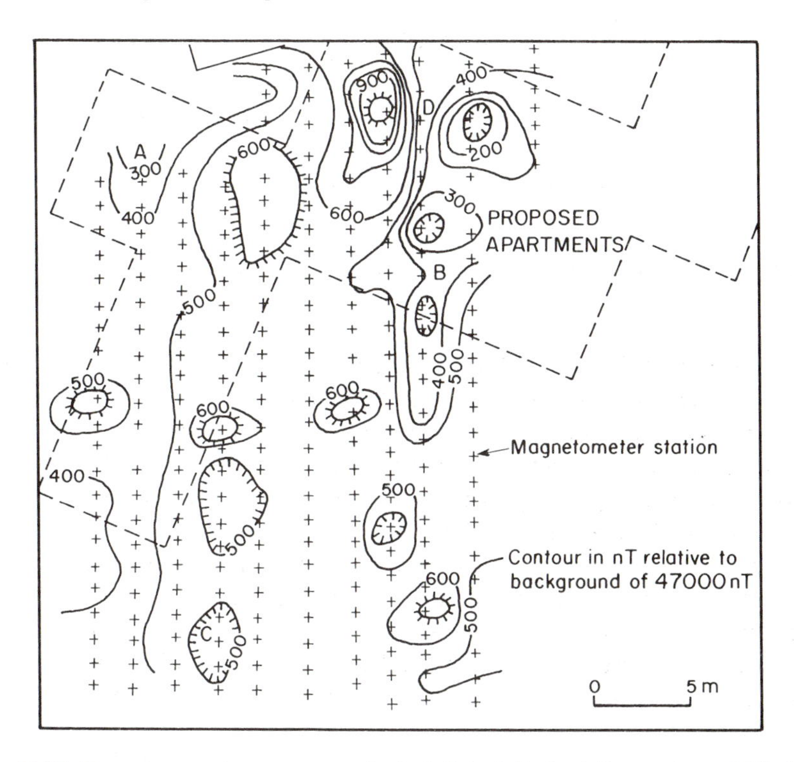

Fig. 10.27. Magnetic anomaly contour map of a site in Bristol, England. Contour interval 100 nT. (Redrawn from Hooper & McDowell 1977.)

was focused on the latter cause, with the other effects providing local 'geological noise'. Model calculations (Fig. 10.28) using buried horizontal cylinders indicated that buried cavities might produce anomalies as small as 0.4 gu. The microgravity survey was performed after preliminary excavation work had been completed so that topographic effects remained constant. Locations were determined on a 15 m square grid to within 0.3 m, and elevations to 3 mm. Base readings to monitor drift and tidal gravity variations were made at 40 minute intervals. The local soil thickness had previously been determined from several probes so that the gravitational effect of the soil cover could be computed and 'stripped' from the observations to remove the gravity variations caused by undulating bedrock topography.

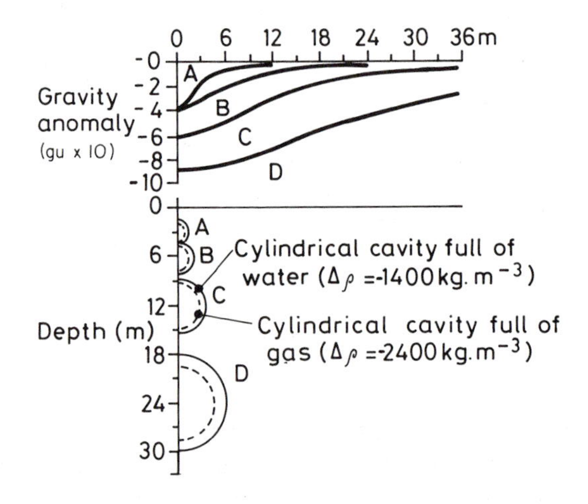

Fig. 10.28. Gravity anomalies over horizontal cylinders. (Redrawn from Arzi 1975.)

The Bouguer anomaly map of the site, uncorrected for terrain effects, is shown in Fig. 10.29. The gravity low in the SSW is almost certainly a terrain effect arising from a soil slope. However, in the northeastern part of the site, an area in which preliminary drilling had indicated rock defects, there are a number of sharp anomaly minima near the originally proposed perimeter of the cooling tower. Model calculations indicated that these anomalies were too large to result from a simple variation in soil density. It was thus concluded that the anomalies resulted from buried cavities and this was confirmed by subsequent drilling, which revealed extensive communicating cavities up to 1 m in diameter at a depth of 4.5 m below the bedrock surface.

Remedial work comprised the injection of grouting material into the cavities. The quantity of grout required was determined by an excess mass calculation using Gauss' theorem (Section 6.11.3). A repeat gravity survey after the grouting operation was then used to verify that the cavities had been filled according to

$$M = \frac{\rho_g}{(\rho_g - \rho_v)} \frac{1}{4\pi G} \Sigma \Delta g \Delta a$$

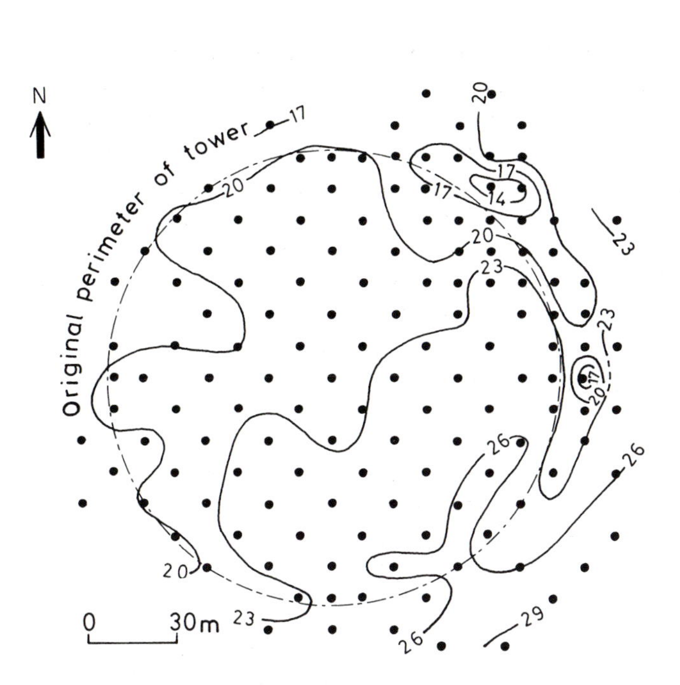

Fig. 10.29. Bouguer anomalies, uncorrected for topographic effects, over the cooling tower area. Contour interval 0.3 gu. (Redrawn from Arzi 1975.)

where M is the mass of injected grout of density ρ_g, ρ_v the density of material originally filling the cavities (air or water), G the gravitational constant and Δg the change in gravity due to grouting over an area element Δa.

Finally it may be noted that all the above survey techniques find application in archaeological investigations, where they may be used in the delineation of buried buildings, walls, tombs, and other artifacts.

10.4.2 Offshore engineering applications

Many major engineering activities are carried out at sea, both in the coastal zone and in deeper offshore waters. Such activities include the construction of harbours, tidal barrages and offshore platforms, the laying of submarine pipelines, and dredging. In all such offshore work, site survey using geophysical techniques has a major role to play.

Offshore construction usually requires detailed information on the nature of the sea bed and the thickness of any unconsolidated sediment layers. Dredging, which may be carried out either to establish and maintain a navigation channel in the approaches to a harbour or to extract sand or gravel from offshore banks, similarly requires information on the thickness and distribution of sediment layers. Such information can be used, for example, to locate a navigation channel in an area of soft sediment in order to avoid the need to blast a channel through solid bedrock.

Shallow-penetration, high resolution seismic reflection profiling (Section

4.12) is an ideal survey method for the above purposes. Single-channel reflection surveys using, for example, a pinger or precision boomer source (see Section 3.8.2) can be carried out from a small launch equipped with a suitable electrical supply to power the seismic profiling equipment. The necessary accurate position fixing can be achieved by the use of a portable high-precision radio-navigation system or ranging system (McQuillin & Ardus 1977).

10.5 Geophysics in the investigation of the Earth's crust

The crust is defined as that part of the Earth lying above the Mohorovičić discontinuity (Moho), below which the velocity of compressional seismic body waves increases abruptly to about 8.0 km/s. It is a layer of distinctive chemical composition at the top of the lithosphere, the mechanically rigid outer shell of the Earth that is composed of a series of lithospheric plates in relative motion. Geophysical surveying provides our main source of information on the internal constitution of the crust. It also provides insight into the plate tectonic processes by which the lithosphere is generated and destroyed.

Large-scale seismic refraction surveys (Chapter 5), using explosions as seismic sources, have been carried out to study crustal structure in most continental areas. An example is the LISPB experiment which was carried out in Britain in 1974 and produced the crustal section for northern Britain reproduced in Fig. 10.30. Such experiments show that the continental crust is typically 30–40 km thick and that it is often internally layered. It is characterized by major regional variations in thickness and constitution which are often directly related to changes of surface geology. Thus, different orogenic provinces are often characterized by quite different crustal sections. Upper crustal velocities are usually in the range 5.8–6.3 km/s which, by analogy with velocity measurements of rock samples in the laboratory (see Section 3.4), may be interpreted as representing mainly granitic or granodioritic material. Lower crustal velocities are normally in the range 6.5–7.0 km/s and may represent any of a variety of igneous and metamorphic rock types, including gabbro, gabbroic anorthosite and basic granulite. The latter rock type is regarded as the most probable major constituent of the lower crust on the basis of experimental studies of seismic velocities (Christensen & Fountain 1975).

Marine refraction surveys, usually single-ship experiments (see Section

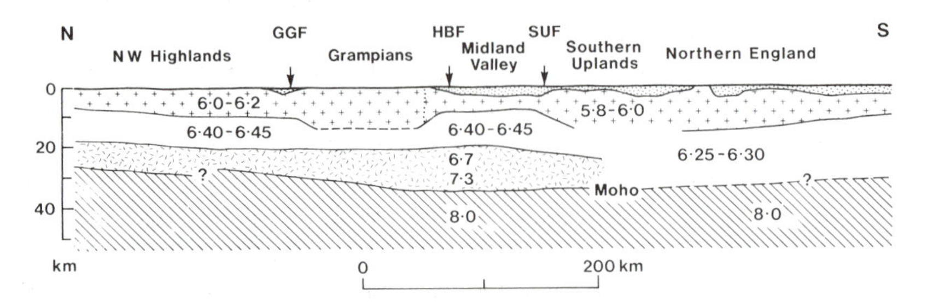

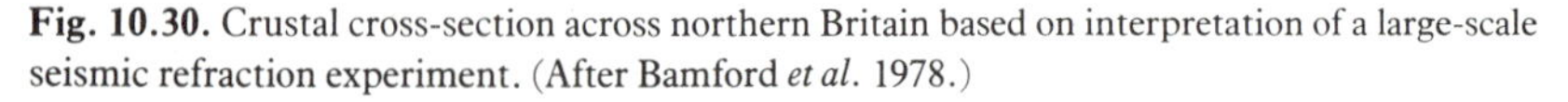

Fig. 10.30. Crustal cross-section across northern Britain based on interpretation of a large-scale seismic refraction experiment. (After Bamford *et al.* 1978.)

5.8.1), show the ocean basins to have a crust only 6–8 km thick, composed of three layers with differing seismic velocities. This thickness and layering is maintained over vast areas beneath all the major oceans. The results of deep-sea drilling, together with the recognition of ophiolitic complexes exposed on land as analogues of oceanic lithosphere, have enabled the nature of the individual seismic layers of the oceanic crust to be identified (see Table 10.1).

Table 10.1. Seismic layering of oceanic crust.

Layer	Typical thickness (km)	Velocity (km/s)	Rock type
1	0–1.0	1.6–2.5	sediments
2	1.0–2.0	4.0–6.0	pillow lavas (with intercalated sediments)
3	4.5–5.5	6.5–7.0	sheeted dolerite dykes and gabbro intrusions
	seismic Moho		

The seismic refraction method provides generalized models of continental and oceanic crustal structure with good velocity information but it is unable to provide a detailed picture of the internal structure of the crust. The latter is best achieved by seismic reflection profiling (Chapter 4), which is becoming increasingly important as a method of studying the geology of the crust and uppermost mantle. Following on from the extensive use of single-channel and multichannel reflection profiling to investigate the geology of oceanic areas, national programmes such as the US COCORP project (Consortium for Continental Reflection Profiling; Brewer & Oliver 1981) and the British BIRPS project (British Institutions Reflection Profiling Syndicate; Brewer 1983) are now producing seismic sections through the entire continental crust. These projects utilize modifications of the commercial survey methods that have been so successfully applied to the search for hydrocarbons. Spectacular results have been achieved, as exemplified by the Moine and Outer Isles Seismic Traverse (MOIST) of the BIRPS programme (Smythe *et al.* 1982). This was a marine reflection line established north of Scotland across the western margin of the Caledonian orogenic belt, and it revealed major Caledonian thrusts extending down through the entire crust (Fig. 10.31).

Gravity surveying has also made a major contribution to crustal geology. Classic early studies estimated regional variations of crustal thickness on the basis of variations in the level of the Bouguer anomaly field (Woollard 1969). An early result of gravity surveying was the discovery that the greater part of the Earth's continents exhibits isostatic anomalies close to zero, implying the widespread existence of isostatic equilibrium. This finding suggests that the Earth's lithosphere is not capable of sustaining significant loads and yields isostatically to any change of surface loading resulting, for example, from magmatism, tectonism, glaciation, deposition or erosion. Gravity surveying in oceanic areas has shown that oceans, also, exhibit widespread isostatic equilibrium. Near-zero free-air anomalies over ridges and rises indicate that the mass of these submarine mountain chains is compensated isostatically by a zone of mass deficiency in the underlying mantle (Fig. 10.32). This mantle zone, which can be mapped seismically by virtue of its low seismic velocity, is interpreted as a region of partial melting and, perhaps, hydration.

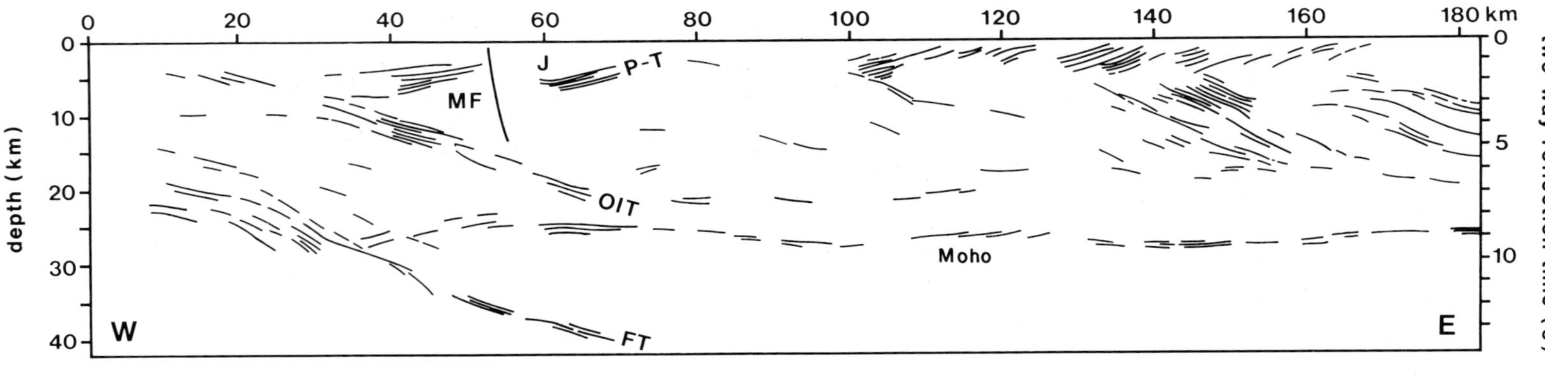

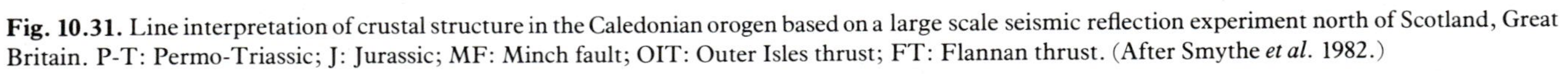

Fig. 10.31. Line interpretation of crustal structure in the Caledonian orogen based on a large scale seismic reflection experiment north of Scotland, Great Britain. P-T: Permo-Triassic; J: Jurassic; MF: Minch fault; OIT: Outer Isles thrust; FT: Flannan thrust. (After Smythe *et al*. 1982.)

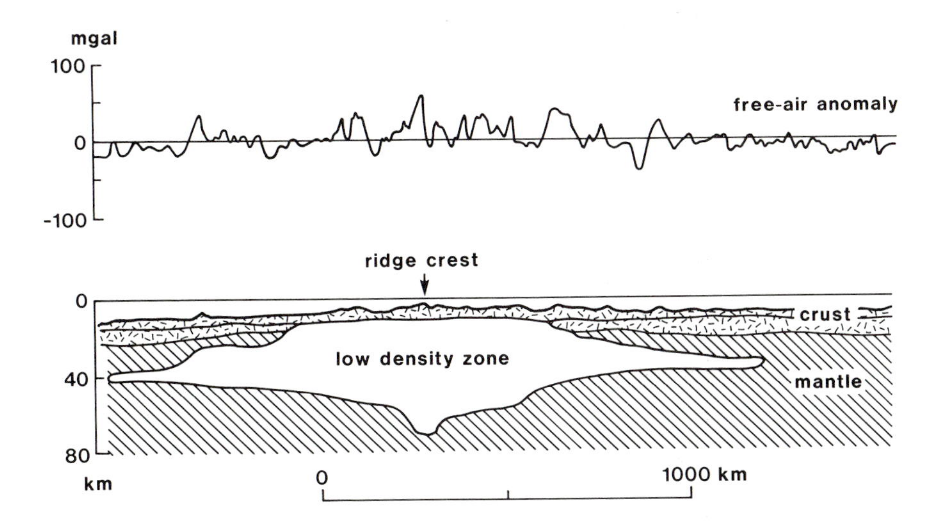

Fig. 10.32. Free-air anomaly profile across the mid-Atlantic ridge. (After Talwani *et al.* 1965.)

A further use of gravity surveying in crustal studies is to investigate the concealed form of large-scale geological features such as granite batholiths. For example, gravity surveys in southwest England (Bott *et al.* 1958) have revealed a belt of negative Bouguer anomaly of large amplitude overlying a zone of outcropping Variscan granites (Fig. 10.33). Modelling of the Bouguer anomalies

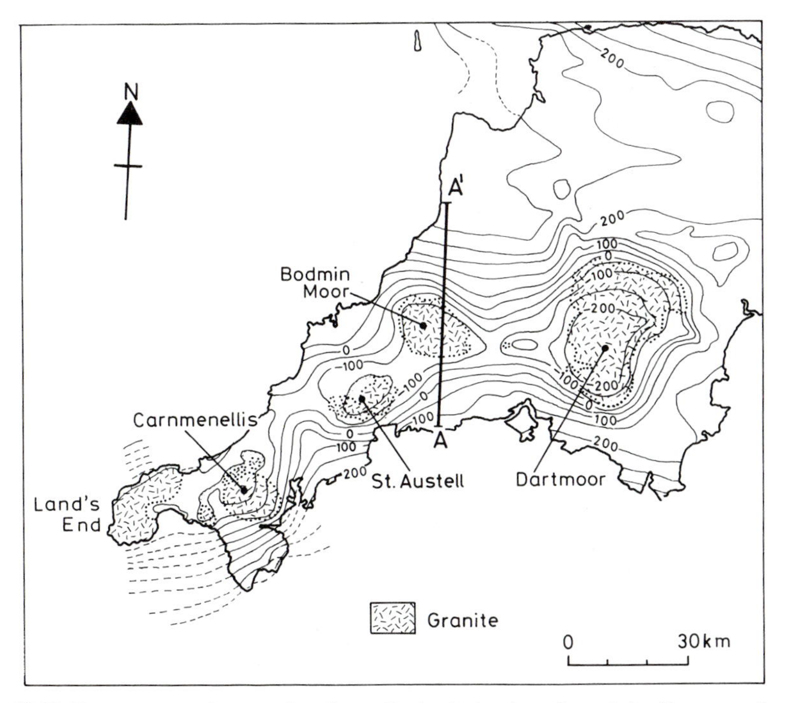

Fig. 10.33. Bouguer anomaly map of southwest England, showing a linear belt of large negative anomalies associated with the zone of Variscan granite outcrops. Contour interval 50 gu. (Redrawn from Bott & Scott 1964.)

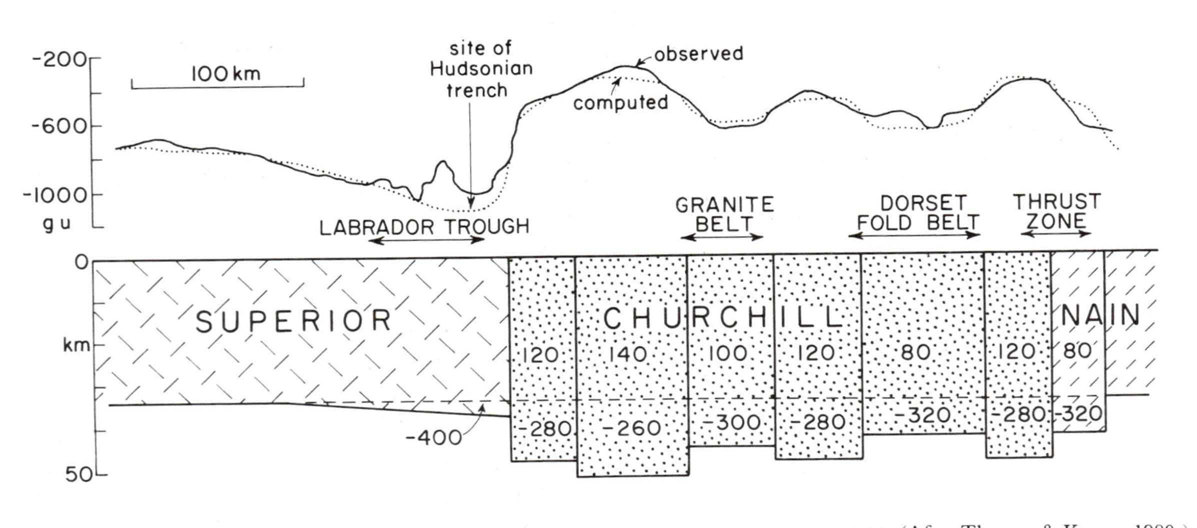

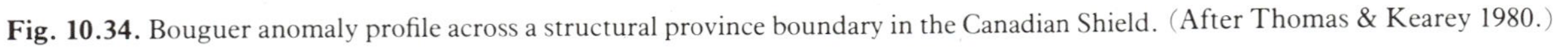
Fig. 10.34. Bouguer anomaly profile across a structural province boundary in the Canadian Shield. (After Thomas & Kearey 1980.)

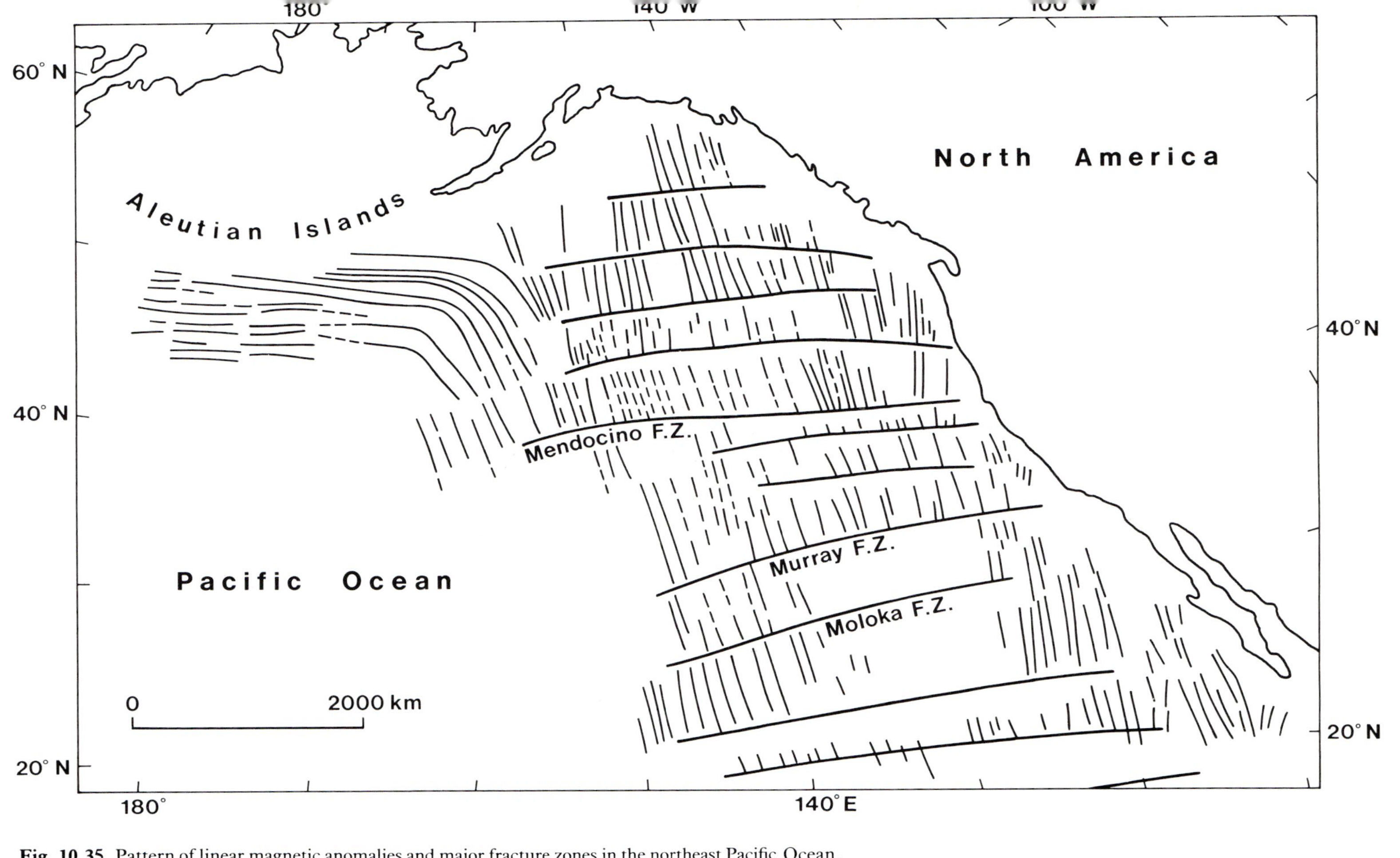

Fig. 10.35. Pattern of linear magnetic anomalies and major fracture zones in the northeast Pacific Ocean.

(Fig. 6.22) has led to the postulation of a continuous batholith chain perhaps 10–15 km thick underlying southwest England (see, e.g. Brooks *et al.* 1983). Gravity surveying can also be used to study ancient suture zones, interpreted to represent the sites of former plate boundaries within continental lithosphere. Such zones are often characterized by major linear gravity anomalies resulting from the different crustal sections juxtaposed across the sutures (Fig. 10.34).

Magnetic surveying (Chapter 7) can similarly be used for the regional investigation of large-scale crustal features, though the sources of major magnetic anomalies tend to be restricted to rocks of basic or ultrabasic composition. Moreover, magnetic surveying is of limited use in the study of the deeper geology of the continental crust because the Curie isotherm for common ferrimagnetic minerals lies at a depth of about 20 km and the sources of major anomalies are consequently restricted to the upper part of the continental crust.

Although the contribution of magnetic surveying to knowledge of continental geology has been modest, magnetic surveying in oceanic areas has had a profound influence on the development of plate tectonic theory and on views of the formation of oceanic lithosphere. Early magnetic surveying at sea showed that the oceanic crust is characterized by a pattern of linear magnetic anomalies (Fig. 10.35) attributable to strips of oceanic crust alternately magnetized in a normal and reverse direction (Mason & Raff 1961). The bilateral symmetry of these linear magnetic anomalies about oceanic ridges and rises (Vine & Matthews 1963) led directly to the theory of sea floor spreading and the establishment of a time scale for polarity transitions of the geomagnetic field (Heirtzler *et al.* 1968). Consequently, oceanic crust can be dated on the basis of the pattern of magnetic polarity transitions preserved within it.

Transform faults disrupt the pattern of linear magnetic anomalies (see Fig. 10.35) and their distribution can therefore be mapped magnetically. Since these faults lie along arcs of small circles to the prevailing pole of rotation at the time of transform fault movement, individual regimes of spreading during the evolution of an ocean basin can be identified by detailed magnetic surveying. Such studies have been carried out in all the major oceans and show the evolution of an ocean basin to be a complex process involving several discrete phases of spreading, each with a distinct pole of rotation.

References

Al-Chalabi, M. (1972) Interpretation of gravity anomalies by non-linear optimisation. *Geophys. Prosp.*, **20,** 1–16.

Al-Sadi, H.N. (1980) *Seismic Exploration*. Birkhauser Verlag, Basel.

Anstey, N.A. (1965) Wiggles. *J. Can. Soc. Exploration Geophysicists*, **1,** 13–43.

Anstey, N.A. (1966) Correlation techniques – a review. *J. Can. Soc. Exploration Geophysicists*, **2,** 55–82.

Anstey N.A. (1977) *Seismic Interpretation: The Physical Aspects*. IHRDC, Boston.

Anstey, N.A. (1981) *Seismic Prospecting Instruments, Vol. 1: Signal Characteristics and Instrument Specifications*. Gebrüder Borntraeger, Berlin.

Anstey, N.A. (1982) *Simple Seismics*. IHRDC, Boston.

Arzi, A.A. (1975) Microgravimetry for engineering applications. *Geophys. Prosp.*, **23,** 408–25.

Aspinall, A. & Walker, A.R. (1975) The earth resistivity instrument and its application to shallow earth surveys. *Underground Services*, **3,** 12–15.

Bamford, D., Nunn, K., Prodehl, C. & Jacob, B. (1978) LISPB–IV. Crustal structure of northern Britain. *Geophys. J.R. astr. Soc.*, **54,** 43–60.

Baranov, W. (1975) *Potential Fields and Their Transformations in Applied Geophysics*. Gebrüder Borntraeger, Berlin.

Baranov, V. & Naudy, H. (1964) Numerical calculation of the formula of reduction to the magnetic pole (airborne). *Geophysics*, **29,** 67–79.

Barker, R.D. (1981) The offset system of electrical resistivity sounding and its use with a multicore cable. *Geophys. Prosp.*, **29,** 128–43.

Barker, R.D. & Worthington, P.F. (1972) Location of disused mineshafts by geophysical methods. *Civil Engineering and Public Works Review*, **67,** No. 788, 275–6.

Barraclough, D.R. & Malin, S.R.C. (1971) *Synthesis of International Geomagnetic Reference Field Values*. Inst. Geol. Sci. Rep. No. 71/1.

Bayerly, M. & Brooks, M. (1980) A seismic study of deep structure in South Wales using quarry blasts. *Geophys. J.R. astr. Soc.*, **60,** 1–19.

Berry, M.J. & West, G.F. (1966) An interpretation of the first-arrival data of the Lake Superior experiment by the time term method. *Bull. seismol. Soc. Am.*, **56,** 141–71.

Bertin, J. (1976) *Experimental and Theoretical Aspects of Induced Polarisation*, Vols. 1 and 2. Gebrüder Borntraeger, Berlin.

Bhattacharya, P.K. & Patra, H.P. (1968) *Direct Current Electrical Sounding*. Elsevier, Amsterdam.

Birch, F. (1960) The velocity of compressional waves in rocks to ten kilobars, Part 1. *J. geophys. Res.*, **65,** 1083–102.

Birch, F. (1961) The velocity of compressional waves in rocks to ten kilobars, Part 2. *J. geophys. Res.*, **66,** 2199–224.

Boissonnas, E. & Leonardon, E.G. (1948) Geophysical exploration by telluric currents with special reference to a survey of the Haynesville Salt Dome, Wood County, Texas. *Geophysics*, **13,** 387–403.

Bolt, B.A. (1976) *Nuclear Explosions and Earthquakes: The Parted Veil*. Freeman, San Francisco.

Bolt, B.A. (1982) *Inside the Earth*. Freeman, San Francisco.

Bott, M.H.P. (1973) Inverse methods in the interpretation of magnetic and gravity anomalies. *In:* Alder, B., Fernbach, S. & Bolt, B.A. (eds.), *Methods in Computational Physics*, **13,** 133–62.

Bott, M.H.P. (1982) *The Interior of the Earth*. Edward Arnold, London.

Bott, M.H.P., Day, A.A. & Masson-Smith, D. (1958) The geological interpretation of gravity and magnetic surveys in Devon and Cornwall. *Phil. Trans. R. Soc.*, **251A,** 161–91.

Bott, M.H.P. & Scott, P. (1964) Recent geophysical studies in southwest England. *In:* Hosking, K.F.G. & Shrimpton, G.H. (eds), *Present Views of Some Aspects of the Geology of Devon and Cornwall.* Royal Geological Society of Cornwall.

Brewer, J.A. (1983) Profiling continental basement: the key to understanding structures in the sedimentary cover. *First Break (European Association of Exploration Geophysicists),* **1,** 25–31.

Brewer, J.A. & Oliver, J.E. (1980) Seismic reflection studies of deep crustal structure. *Ann. Rev. Earth planet. Sci.,* **8,** 205–30.

Brigham, E.O. (1974) *The Fast Fourier Transform.* Prentice-Hall, New Jersey.

Brooks, M. & Ferentinos, G. (1984) Tectonics and sedimentation in the Gulf of Corinth and the Zakynthos and Kefallinia Channels, western Greece. *Tectonophysics,* **101,** 25–54.

Brooks, M., Mechie, J. & Llewellyn, D.J. (1983) Geophysical investigations in the Variscides of southwest Britain. *In:* Hancock, P.L. (ed.), *The Variscan Fold Belt in the British Isles.* Hilger, Bristol, England, Ch. 10, 186–97.

Brown, L.F. & Fisher, W.L. (1980) *Seismic Stratigraphic Interpretation and Petroleum Exploration.* AAPG Continuing Education Course Note Series No. 16.

Bugg, S.F. & Lloyd, J.W. (1976) A study of freshwater lens configuration in the Cayman Islands using resistivity methods. *Q.J. Eng. Geol,* **9,** 291–302.

Cady, J.W. (1980) Calculation of gravity and magnetic anomalies of finite-length right polygonal prisms. *Geophysics,* **45,** 1507–12.

Cagniard, L. (1953) Basic theory of the magnetotelluric method of geophysical prospecting. *Geophysics,* **18,** 605–35.

Červený, V., Langer, J. & Pšenčik, I. (1974) Computation of geometric spreading of seismic body waves in laterally inhomogeneous media with curved interfaces. *Geophys. J.R. astr. Soc.,* **38,** 9–19.

Červený, V. & Ravindra, R. (1971) *Theory of Seismic Head Waves.* University of Toronto Press, Toronto.

Christensen, N.I. & Fountain, D.M. (1975) Constitution of the lower continental crust based on experimental studies of seismic velocities in granulites. *Bull. geol. Soc. Am.,* **86,** 227–36.

Claerbout, J.F. (1976) *Fundamentals of Geophysical Data Processing.* McGraw-Hill, New York.

Cook, K.L. & Van Nostrand, R.G. (1954) Interpretation of resistivity data over filled sinks. *Geophysics,* **19,** 761–90.

Cunningham, A.B. (1974) Refraction data from single-ended refraction profiles. *Geophysics,* **39,** 292–301.

Dehlinger, P. (1978) *Marine Gravity.* Elsevier, Amsterdam.

Day, G.A., Cooper, B.A., Anderson, C., Burgers, W.F.J., Rønnevik, H.C. & Schöneich, H. (1981) Regional structural maps of the North Sea. *In:* Illing, L.V. & Hobson, G.D. (eds.), *Petroleum Geology of the Continental Shelf of NW Europe.* Heyden & Son, London, Ch. 5, 76–84.

Dey, A. & Morrison, H.F. (1979) Resistivity modelling for arbitrarily shaped two-dimensional structures. *Geophys. Prosp.,* **27,** 106–36.

Dix, C.H. (1955) Seismic velocities from surface measurements. *Geophysics,* **20,** 68–86.

Dix, C.H. (1981) *Seismic Prospecting for Oil.* IHRDC, Boston.

Dobrin, M.B. (1976) *Introduction to Geophysical Prospecting.* McGraw-Hill, New York (3rd edn).

Duncan, P.M. & Garland, G.D. (1977) A gravity study of the Saguenay area, Quebec. *Can. Jour. Earth Sci.,* **14,** 145–52.

Elsasser, W.M. (1958) The Earth as a dynamo. *Sci. Am.,* **198,** 44–8.

Fitch, A.A. (1976) *Seismic Reflection Interpretation.* Gebrüder Borntraeger, Berlin.

Fitch, A.A. (1981) *Developments in Geophysical Exploration Methods, Vol. 2.* Applied Science Publishers, London.

Garland, G.D. (1951) Combined analysis of gravity and magnetic anomalies. *Geophysics,* **16,** 51–62.

Garland, G.D. (1965) *The Earth's Shape and Gravity.* Pergamon, Oxford.

Giese, P., Prodehl, C. & Stein, A. (eds.) (1976). *Explosion Seismology in Central Europe.* Springer-Verlag, Berlin.

Grant, F.S. & West, G.F. (1965) *Interpretation Theory in Applied Geophysics.* McGraw-Hill, New York.

Gregory, A.R. (1977) Aspects of rock physics from laboratory and log data that are important to seismic interpretation. *In:* Payton, C.E. (ed.), *Seismic Stratigraphy–Applications to Hydrocarbon Exploration.* Memoir 26, American Association of Petroleum Geologists. Tulsa, 15–46.

Griffiths, D.H. & King, R.F. (1981) *Applied Geophysics for Geologists and Engineers*. Pergamon, Oxford.

Gunn, P.J. (1975) Linear transformations of gravity and magnetic fields. *Geophys. Prosp.*, **23,** 300–12.

Habberjam, G.M. (1979) *Apparent resistivity and the use of square array techniques*. Gebrüder Borntraeger, Berlin.

Hagedoorn, J.G. (1959) The plus-minus method of interpreting seismic refraction sections. *Geophys. Prosp.*, **7,** 158–82.

Heirtzler, J.R., Dickson, G.O., Herron, E.M., Pitman, W.C. & Le Pichon, X. (1968) Marine magnetic anomalies, geomagnetic field reversals, and motions of the ocean floor and continents. *J. geophys. Res.*, **73,** 2119–36.

Hooper, W. & McDowell, P. (1977) Magnetic surveying for buried mineshafts and wells. *Ground Engineering*, **10,** 21–3.

Hubbert, M.K. (1934) Results of Earth-resistivity survey on various geological structures in Illinois. *Trans. Am. Inst. Mining Met. Engrs.*, **110,** 9–29.

Hubral, P. & Krey, T. (1980) *Interval Velocities from Seismic Reflection Time Measurements*. Society of Exploration Geophysicists, Tulsa.

Hutton, V.R.S., Ingham, M.R. & Mbipom, E.W. (1980) An electrical model of the crust and upper mantle in Scotland. *Nature, Lond.*, **287,** 30–3.

I.A.G. (International Association of Geodesy). (1971) *Geodetic Reference System 1967*. Pub. Spec. No. 3 du Bulletin Géodésique.

Illing, L.V. & Hobson, G.D. (eds.) (1981) *Petroleum Geology of the Continental Shelf of NW Europe*. Heyden & Son, London.

Jewell, T.R. & Ward, S.H. (1963) The influence of conductivity inhomogeneities upon audio-frequency magnetic fields. *Geophysics*, **28,** 201–21.

Johnson, S.H. (1976) Interpretation of split-spread refraction data in terms of plane dipping layers. *Geophysics*, **41,** 418–24.

Kanasewich, E.R. (1981) *Time Sequence Analysis in Geophysics*. Univ. of Alberta. (3rd edn).

Kanasewich, E.R. & Agarwal, R.G. (1970) Analysis of combined gravity and magnetic fields in wave-number domain. *J. geophys. Res.*, **75,** 5702–12.

Keller, G.V. & Frischnecht, F.C. (1966) *Electrical Methods in Geophysical Prospecting*. Pergamon, Oxford.

Kilty, K.T. (1984) On the origin and interpretation of self-potential anomalies. *Geophys. Prosp.*, **32,** 51–62.

Kleyn, A.H. (1983) *Seismic Reflection Interpretation*. Applied Science Publishers, London.

Knopoff, L. (1983) The thickness of the lithosphere from the dispersion of surface waves. *Geophys. J. R. astr. Soc.*, **74,** 55–81.

Koefoed, O. (1968) *The Application of the Kernel Function in Interpreting Resistivity Measurements*. Gebrüder Borntraeger, Berlin.

Koefoed, O. (1979) *Geosounding Principles, 1–Resistivity Sounding Measurements*. Elsevier, Amsterdam.

Kulhánek, O. (1976) *Introduction to Digital Filtering in Geophysics*. Elsevier, Amsterdam.

Kunetz, G. (1966) *Principles of Direct Current Resistivity Prospecting*. Gebrüder Borntraeger, Berlin.

LaCoste, L.J.B. (1967) Measurement of gravity at sea and in the air. *Rev. Geophys.*, **5,** 477–526.

LaCoste, L.J.B., Ford, J., Bowles, R. & Archer, K. (1982) Gravity measurements in an airplane using state-of-the-art navigation and altimetry. *Geophysics*, **47,** 832–7.

Le Tirant, P. (1979) *Seabed reconnaissance and offshore soil mechanics*. Editions Technip, Paris.

Leggo, P.J. (1982) Geological implications of ground impulse radar. *Trans. Instn. Min. Metall. (Sect. B: Appl. earth sci.)*, **91,** B1–6.

Logn, O. (1954) Mapping nearly vertical discontinuities by Earth resistivities. *Geophysics*, **19,** 739–60.

Marshall, D.J. & Madden, T.R. (1959) Induced polarisation: a study of its causes. *Geophysics*, **24,** 790–816.

Mason, R.G. & Raff, R.D. (1961) Magnetic survey off the west coast of North America, 32° N to 42° N. *Bull. geol. Soc. Am.*, **72,** 1259–66.

Mayne, W.H. (1962) Common reflection point horizontal stacking techniques. *Geophysics*, **27,** 927–8.

Mayne, W.H. (1967) Practial considerations of the use of common reflection point techniques. *Geophysics*, **32,** 225–9.

McQuillin, R. & Ardus, D.A. (1977) *Exploring the Geology of Shelf Seas*. Graham & Trotman, London.

McQuillin, R., Bacon, M. & Barclay, W. (1979) *An Introduction to Seismic Interpretation*. Graham & Trotman, London.

Merkel, R.H. (1972) The use of resistivity techniques to delineate acid mine drainage in groundwater. *Groundwater*, **10,** *No.* 5, 38–42.

Mitchum, R.M., Vail, P.R. & Thompson, S. (1977) Seismic stratigraphy and global changes of sea level, Part 2: The depositional sequence as a basic unit for stratigraphic analysis. *In:* Payton, C.E. (ed.), *Seismic Stratigraphy–Applications to Hydrocarbon Exploration*. Memoir 26, American Association of Petroleum Geologists, Tulsa, 53–62.

Mittermayer, E. (1969) Numerical formulas for the Geodetic Reference System 1967. *Bolletino di Geofisca Teorica ed Applicata*, **11,** 96–107.

Morelli, C., Gantor, C., Honkasalo, T., McConnell, R.K., Tanner, J.G., Szabo, B., Votila, V. & Whalen, C.T. (1971) *The International Gravity Standardisation Net*. Pub. Spec. No. 4 du Bulletin Géodésique.

Musgrave, A.W. (ed.) (1967) *Seismic Refraction Prospecting*. Society of Exploration Geophysicists, Tulsa.

Nafe, J.E. & Drake, C.L. (1963) Physical properties of marine sediments. *In:* Hill, M.N. (ed.), *The Sea. Vol. 3*. Interscience Publishers, New York, 794–815.

Nair, M.R., Biswas, S.K. & Mazumdar, K. (1968) Experimental studies on the electromagnetic response of tilted conducting half-planes to a horizontal-loop prospective system. *Geoexploration*, **6,** 207–44.

Neidell, N.S. & Poggiagliolmi, E. (1977) Stratigraphic modelling and interpretation–geophysical principles. *In:* Payton, C.E. (ed.), *Seismic Stratigraphy–Applications to Hydrocarbon Exploration*. Memoir 26, American Association of Petroleum Geologists, Tulsa, 389–416.

Nettleton, L.L. (1971) *Elementary Gravity and Magnetics for Geologists and Seismologists*. Society of Exploration Geophysicists, Tulsa, Monograph Series, No. 1.

Nettleton, L.L. (1976) *Gravity and Magnetics in Oil Exploration*. McGraw-Hill, New York.

O'Brien, P.N.S. (1974) Aspects of seismic research in the oil industry. *Geoexploration*, **12,** 75–96.

Orellana, E. & Mooney, H.M. (1966) *Master Tables and Curves for Vertical Electrical Sounding Over Layered Structures*. Interciencia, Madrid.

Orellana, E. & Mooney, H.M. (1972) *Two and Three Layer Master Curves and Auxilliary Point Diagrams for Vertical Electrical Sounding Using Wenner Arrangement*. Interciencia, Madrid.

Palacky, G.J. (1981) The airborne electromagnetic method as a tool of geological mapping. *Geophys. Prosp.*, **29,** 60–88.

Palmer, D. (1980) *The Generalised Reciprocal Method of Seismic Refraction Interpretation*. Society of Exploration Geophysicists, Tulsa.

Parasnis, D.S. (1966, 1973) *Mining Geophysics*. Elsevier, Amsterdam.

Parasnis, D.S. (1979) *Principles of Applied Geophysics*. Chapman & Hall, London.

Parkhomenko, E.I. (1967) *Electrical Properties of Rocks*. Plenum, New York.

Payton, C.E. (ed.) (1977) *Seismic Stratigraphy–Applications to Hydrocarbon Exploration*. Memoir 26, American Association of Petroleum Geologists, Tulsa.

Peddie, N.W. (1983) International geomagnetic reference field–its evolution and the difference in total field intensity between new and old models for 1965–1980. *Geophysics*, **48,** 1691–6.

Pemberton, R.H. (1962) Airborne EM in review. *Geophysics*, **27,** 691–713.

Peters, J.W. & Dugan, A.F. (1945) Gravity and magnetic investigations at the Grand Saline Salt Dome, Van Zandt Co., Texas. *Geophysics*, **10,** 376–93.

Ramsey, A.S. (1964) *An Introduction to the Theory of Newtonian Attraction*. Cambridge University Press, Cambridge.

Rayner, J.N. (1971) *An Introduction to Spectral Analysis*. Pion, England.

Reilly, W.I. (1972) Use of the International System of Units (SI) in geophysical publications. *N.Z. J. Geol. Geophys.*, **15,** 148–58.

Robinson, E.A. (1983) *Migration of Geophysical Data*. IHRDC, Boston.

Robinson, E.A. (1983) *Seismic Velocity Analysis and the Convolutional Model*. IHRDC, Boston.

Robinson, E.A. & Treitel, S. (1967) Principles of digital Wiener filtering. *Geophys. Prosp.*, **15,** 311–33.

Robinson, E.A. & Treitel, S. (1980) *Geophysical Signal Analysis*. Prentice-Hall, London.

Sato, M. & Mooney, H.H. (1960) The electrochemical mechanism of sulphide self potentials. *Geophysics*, **25**, 226–49.

Schramm, M.W., Dedman, E.V. & Lindsey, J.P. (1977) Practical stratigraphic modelling and interpretation. *In:* Payton, C.E. (ed.), *Seismic Stratigraphy–Applications to Hydrocarbon Exploration*. Memoir 26, American Association of Petroleum Geologists, Tulsa, 477–502.

Seigel, H.O. (1967) The induced polarisation method. *In:* Morley, L.W. (ed.), *Mining and Groundwater Geophysics*. Econ. Geol. Report No. 26, Geol. Survey of Canada, 123–37.

Sengbush, R.L. (1983) *Seismic Exploration Methods*. IHRDC, Boston.

Sharma, P. (1976) *Geophysical Methods in Geology*. Elsevier, Amsterdam.

Sheriff, R.E. (1973) *Encyclopedic Dictionary of Exploration Geophysics*. Society of Exploration Geophysicists, Tulsa.

Sheriff, R.E. (1978) *A First Course in Geophysical Exploration and Interpretation*. IHRDC, Boston.

Sheriff, R.E. (1980) *Seismic Stratigraphy*. IHRDC, Boston.

Sheriff, R.E. (1982) *Structural Interpretation of Seismic Data*. American Association of Petroleum Geologists Continuing Education Course Note Series No. 23.

Sheriff, R.E. & Geldart, L.P. (1982) *Exploration Seismology Vol 1: History, Theory and Data Acqusition*. Cambridge University Press, Cambridge.

Sheriff, R.E. & Geldart, L.P. (1983) *Exploration Seismology Vol 2: Data-processing and Interpretation*. Cambridge University Press, Cambridge.

Smith, R.A. (1959) Some depth formulae for local magnetic and gravity anomalies. *Geophys. Prosp.*, **7**, 55–63.

Smythe, D.K., Dobinson, A., McQuillin, R., Brewer, J.A., Matthews, D.H., Blundell, D.J. & Kelk, B. (1982) Deep structure of the Scottish Caledonides revealed by the MOIST profile. *Nature, Lond.*, **299**, 338–40.

Spector, A. & Grant, F.S. (1970) Statistical models for interpreting aeromagnetic data. *Geophysics*, **35**, 293–302.

Stacey, F.D. & Banerjee, S.K. (1974) *The Physical Principles of Rock Magnetism*, Elsevier, Amsterdam.

Stacey, R.A. (1971) Interpretation of the gravity anomaly at Darnley Bay, N.W.T. *Can. Jour. Earth Sci.*, **8**, 1037–42.

Stoffa, P.L. & Buhl, P. (1979) Two-ship multichannel seismic experiments for deep crustal studies: expanded spread and constant offset profiles. *J. geophys. Res.*, **84**, 7645–7660.

Sumner, J.S. (1976) *Principles of Induced Polarisation for Geophysical Exploration*. Elsevier, Amsterdam.

Talwani, M. (1965) Comparison with the help of a digital computer of magnetic anomalies caused by bodies of arbitrary shape. *Geophysics*, **30**, 797–817.

Talwani, M. & Ewing, M. (1960) Rapid computation of gravitational attraction of three-dimensional bodies of arbitrary shape. *Geophysics*, **25**, 203–25.

Talwani, M., Le Pichon, X. & Ewing, M. (1965) Crustal structure of the mid-ocean ridges 2. Computed model from gravity and seismic refraction data. *J. Geophys. Res.*, **70**, 341–52.

Talwani, M., Worzel, J.L. & Landisman, M. (1959) Rapid gravity computations for two-dimensional bodies with applications to the Mendocino submarine fracture zones. *J. geophys. Res.*, **64**, 49–59.

Taner, M.T. & Koehler, F. (1969) Velocity spectra–digital computer derivation and applications of velocity functions. *Geophysics*, **34**, 859–81.

Tarling, D.H. (1983) *Palaeomagnetism*. Chapman and Hall, London.

Telford, W.M., Geldart, L.P., Sheriff, R.E. & Keys, D.A. (1976) *Applied Geophysics* Cambridge University Press, Cambridge.

Thomas, M.D. & Kearey, P. (1980) Gravity anomalies, block-faulting and Andean-type tectonism in the eastern Churchill Province. *Nature, Lond.*, **283**, 61–3.

Thornburgh, H.R. (1930) Wave-front diagrams in seismic interpretation. *Bull. Am. Assoc. Petrol. Geol.*, **14**, 185–200.

Vacquier, V., Steenland, N.C., Henderson, R.G. & Zeitz, I. (1951) Interpretation of aeromagnetic maps. *Geol. Soc. Am. Mem.* **47.**

Vail, P.R., Mitchum, R.M. & Thompson, S. (1977) Seismic stratigraphy and global changes of sea level, Part 3: Relative changes of sea level from coastal onlap. *In:* Payton, C.E. (ed.), *Seismic Stratigraphy–Applications to Hydrocarbon Exploration*. Memoir 26, American Association of Petroleum Geologists, Tulsa, 63–81.

Vail, P.R., Mitchum, R.M. & Thompson, S. (1977) Seismic stratigraphy and global changes of sea level, Part 4: Global cycles of relative changes of sea level. *In:* Payton, C.E. (ed.), *Seismic Stratigraphy–Applications to Hydrocarbon Exploration*. Memoir 26, American Association of Petroleum Geologists, Tulsa, 83–97.

Van Overmeeren, R.A. (1975) A combination of gravity and seismic refraction measurements, applied to groundwater explorations near Taltal, Province of Antofagasta, Chile. *Geophys. Prosp.*, **23,** 248–58.

Vine, F.J. & Matthews, D.H. (1963) Magnetic anomalies over oceanic ridges. *Nature, Lond.*, **199,** 947–9.

Wait, J.R. (1982) *Geo-Electromagnetism*. Academic Press, New York.

Waters, K.H. (1978) *Reflection Seismology–a Tool for Energy Resource Exploration*. Wiley, New York.

Webb, J.E. (1966) The search for iron ore, Eyre Peninsula, South Australia. *In: Mining Geophysics*, Vol. 1, Society of Exploration Geophysicists, Tulsa, 379–90.

Westbrook, G.K. (1975) The structure of the crust and upper mantle in the region of Barbados and the Lesser Antilles. *Geophys. J.R. astr. Soc.*, **43,** 201–42.

White, P.S. (1966) Airborne electromagnetic survey and ground follow-up in northwestern Quebec. *In: Mining Geophysics*, Vol. 1, Society of Exploration Geophysicists, Tulsa, 252–61.

Willmore, P.L. & Bancroft, A.M. (1960) The time-term approach to refraction seismology. *Geophys. J. R. astr. Soc.*, **3,** 419–32.

Wood, R. & Barton, P. (1983) Crustal thinning and subsidence in the North Sea. *Nature, Lond.*, **302,** 134–6.

Woodland, A.W. (ed.) (1975) *Petroleum and the Continental Shelf of North-West Europe, 1, Geology*. Applied Science Publishers, London.

Woollard, G.P. (1969) Regional variations in gravity. *In:* Hart, P.J. (ed) *The Earth's Crust and Upper Mantle*. Amer. Geophys. Un. Monograph 13.

Yüngül, S. (1954) Spontaneous polarisation survey of a copper deposit at Sariyer, Turkey. *Geophysics*, **19,** 455–58.

Ziolkowski, A. (1983) *Deconvolution*. IHRDC, Boston.

Index

THE STRUCTURE OF AMERICAN MEDICAL PRACTICE 1875–1941

THE STRUCTURE OF AMERICAN MEDICAL PRACTICE 1875–1941

GEORGE ROSEN

CHARLES E. ROSENBERG
Editor

University of Pennsylvania Press
Philadelphia • 1983

Design by Adrianne Onderdonk Dudden

Photo credits:
All photographs which appear in this volume are from the Historical Collections at the College of Physicians of Philadelphia and are used with permission.

Library of Congress Cataloging in Publication Data

Rosen, George, 1910–
The structure of American medical practice, 1875–1941.

Bibliography: p.
Includes index.
1. Medicine—United States—History. I. Title.
[DNLM: 1. History of medicine, 19th century—United States. 2. History of medicine, 20th century—United States. WZ 70 AA1 R79s]
R152.R67 1983 610'.973 83-10461
ISBN 0-8122-7898-4
ISBN 0-8122-1153-7 (pbk.)

Printed in the United States of America

CONTENTS

FOREWORD

In 1976, the University of Pennsylvania Department of History inaugurated the Richard H. Shryock Lectures. Professor Shryock, a native of Philadelphia and longtime teacher at the university, had between the 1930s and 1960s been a dominant figure in making medical history a part of the standard canon of academic history. The lectures established in his name were intended to provide occasions for exploring the paths Shryock had so imaginatively pioneered. The Department of History felt particularly honored when George Rosen, professor of epidemiology, public health, and the history of Medicine at Yale University, agreed to give the initial series. He too enjoyed an international reputation, had known Shryock, and like him had always emphasized the social, economic, and cultural context of past medicine. The subject also seemed particularly appropriate: The Structure of American Medical Practice, 1875–1975.

After delivering the lectures, Professor Rosen set immediately to work expanding the text of his lectures into a book. But after completing all but the final section, he died suddenly in the summer of 1977. He had apparently left no draft of his concluding section, that covering the period between 1941 and 1975. The idea of asking another scholar to complete this section was suggested but ultimately rejected. It seemed best to allow a scholar of Rosen's stature to speak for himself, especially in view of the tangled underbrush of political and policy interests that pervades contemporary medical care debates. The body of this volume is thus almost exactly as George Rosen left it. I have made only minor textual revisions in his manuscript, bearing in mind that he had not been able to make a final literary revision. In addition, I have provided a brief introductory chapter, describing the shape of medical practice in the early nineteenth century and thus setting the stage for Rosen's analysis of the seismic changes which were to take place in the years between 1875 and 1941. I have also added a brief bibliographical note, emphasizing the scholarship of the past decade.

Medical history has too frequently reflected a concern with technique and intellectual innovation, with ideas and formal institutions, at the expense of the behavior of ordinary physicians and patients.[1] It was one of George Rosen's achievements never to have lost sight of the fact that medicine was a social function, not simply a chain of ideas and innovators. The former was a point of view natural enough in the 1930s, when Rosen grew to intellectual maturity, and it is certainly widespread today (if not entirely welcomed by every member of the medical community); Rosen's contextual outlook was little honored in the years between 1941 and the mid-1960s. Yet the social context of medicine was always in the forefront of his work, and it is only appropriate that the subject of his last major synthetic statement should be the structure of American medical practice and especially its fundamentally economic aspect.[2] Rosen's attempt to demonstrate the key role of the medical marketplace and the determined efforts by medical leaders to maintain that marketplace for the exclusive access of the fee-for-service solo practitioner will be almost as controversial today as it would have been in the past. I hope that this book not only will have served to inaugurate a lecture series but will help introduce a new generation of readers to a century-old but still lively debate.

CHARLES E. ROSENBERG

PROLOGUE: THE SHAPE OF TRADITIONAL PRACTICE, 1800–1875

Medical care in the Western world had in many ways changed little in the two millennia before 1800. The assumptions and practices of George Washington's physicians would in most respects have been immediately understandable to the medical men who treated Julius Caesar. It was not until the last half of the nineteenth century that the shape of traditional medicine began to alter, although then it did so with ever-increasing rapidity. By the First World War, the outlines of late twentieth-century medicine had already taken shape. Practice was shifting from a primarily family-oriented form, normally undertaken in a private home and rationalized in terms understandable to any educated person, to a therapeutics guided by concepts increasingly inaccessible to the ordinary patient, aided by tools that had not existed

a half-century before, and often conducted in an impersonal institutional setting. The medical world of 1800 had been a very different place.

Medicine in 1800

Medical practice in the new nation was, as it had always been, primarily an affair of shared concepts and personal relationships. Many of the fundamental institutions we associate with medicine were marginal or non-existent—hospitals, specialism, third-party payment, an elaborate and rigid course of medical training. Most strikingly, perhaps, much medical care was provided by individuals who did not think of themselves and were not thought of by others as physicians. In every family and in every community, individuals—many, though by no means all, women—were accustomed to treating the sick, often with the aid of folk knowledge passed on from generation to generation. This was often supplemented by the guidance of books and pamphlets written precisely to aid the "domestic practitioner."[1] A large portion of such popular medicine still rested on a knowledge of native plants, their habitats and physiological effects, and their preparation and administration; the ancient traditions of botanic medicine were still very much alive in 1800.[2]

Although many American lay persons still practiced medicine, and although Americans of every class shared an endemic skepticism toward medical men and medical ideas, they nevertheless turned in times of need to trained and self-identified physicians.[3] The question was when, not whether, to call a doctor. Not surprisingly, the answer varied a great deal, depending on whether one lived in a town, a city, a rural village, or an isolated farm. It depended as well on one's class, to an extent on one's cultural background, and, especially in the case of "female complaints," on the nature of the ailment itself and the social attitudes which surrounded it. Diseases in infants rarely led to a physician's call, for most people felt there was little that could be done in such cases: either they were too trivial or, paradoxically, too abrupt, elusive, and fatal to warrant the physician's intervention. Some ailments never entailed a visit from the doctor. Rheumatism, mild diarrhea, influenza, malaria in areas where it was endemic, for example, were conditions almost invariably treated at home—in the case of influenza because the patient almost always recovered and in the case

of rheumatism because he so rarely did. At the other end of the spectrum of severity were severe trauma or a debilitating "dropsy"; such ills seemed almost by definition to exceed the lay practitioner's capacity. Thus for each family and each ailment a different configuration of factors dictated not only the likelihood of a physician's being called, but also the point in the course of the illness at which that consultation would take place. The poor would be slower than the rich to call in a physician, the dweller on an isolated farm slower than his town-dwelling cousin, an eastern Pennsylvania Episcopalian sooner to send for a physician than his German Pietist neighbor.

A similarly varied group of practitioners paralleled this structure of possibilities. One cannot think of "physicians" in this period as constituting a monolithic group, homogeneous in training, status, and mode of practice. Nor did medical men appeal equally to an undifferentiated universe of patients. Economic realities as well as social and class attitudes defined socially appropriate matings of doctor and patient. At the top of the profession were a handful of successful urban practitioner-consultants. These physicians had enjoyed the best available medical training: apprenticeship with a prominent doctor, formal medical education, often hospital experience, possibly a sojourn in Edinburgh or London. The rank-and-file physician, on the other hand, would have had a brief and frequently casual apprenticeship; he might have no other training, have never seen the inside of a lecture hall or dissecting room, and almost certainly have had no hospital experience.[4]

The small circle of elite clinicians enjoyed the patronage of America's social and economic elite. They also monopolized teaching positions in the few medical schools and the handful of institutional posts as attending physicians in the young nation's hospitals and almshouses. As a group these prominent physicians were intellectually active. Many sought to publish and to keep in touch with the newest and most prestigious of British medical thought. Only the time it took a sailing ship to cross the Atlantic separated the medical community of Philadelphia or New York from that of London or Edinburgh. But even the elite among urban practitioners were not immune from competition; they vied energetically for the patronage of their community's wealthiest families. They competed as well to attract those would-be doctors whose families were able to pay stiff apprenticeship fees—an important supplement to the medical man's income.

Below this small but influential elite was a much larger number of

regularly trained physicians who were able to make a secure but hardly opulent living in medicine. Competing with these "respectable"—if only apprentice-trained—physicians were a varied assortment of other practitioners: midwives, apothecaries, bone-setters, and itinerant "specialists" in eye ailments or "secret diseases." These shaded imperceptibly into the numerous laymen who provided primary care for their family and neighbors but who did not think of themselves as physicians. The rigid line which we have become accustomed to draw between the trained, licensed, and socially authoritative physician and any other pretender to medical skill makes little sense in understanding medical care during the presidencies of John Adams and Thomas Jefferson.

There was, indeed, a bustling medical marketplace, with almost all care outside the family being on a fee-for-service basis. But it was a confused and confusing marketplace, more like a Middle Eastern bazaar than the orderly world of neat and interchangeable profit-maximizing transactions assumed in economic theory. The rules were complex and most practitioners' chances for success were small. Economic security for a physician turned on his ability to attract and hold the loyalty of a sufficient number of fee-paying families. The world of specialists and consultations was far in the future. In every class, as I have suggested, and in almost every location, the number of would-be practitioners outnumbered the supply of fee-paying families who might potentially employ their services.

The most important aspect of medical practice centered on the family; in theory and usually in reality a physician treated an entire family. His relationship to its members was personal, permanent, and continuous, not episodic and impersonal. Consistently enough, early codes of ethics emphasized the sacredness of the personal ties between doctor and patient; medical theory, similarly, emphasized the significance of individual and family idiosyncrasies in the making of clinical judgments.[5] Both ethics and medical theory enshrined and legitimated the primacy of the relationship between the physician and the family he treated and thus created formal constraints preventing the worst excesses of competition.

"Family" must, moreover, be construed broadly. It included husband, wife, and children, of course, but in addition servants, apprentices, and in the South slaves as well. (On larger plantations separate contracts might be made for the medical care of the slave force.)[6] Care meant every aspect of therapeutics and, aside from an

occasional consultation in a matter particularly grave or obscure (and in a family more than ordinarily solvent), the solo physician served a family as general practitioner, surgeon, pediatrician, gynecologist, and obstetrician. Most surgery in this era before anaesthesia and asepsis was minor: the opening of an abscess, the dressing of an ulcer, the setting of an uncomplicated fracture. Trauma or an occasional difficult childbirth might present grave and unavoidable challenges, but the physician was presumed able (with God's help) to manage any clinical situation. Most rural and some urban doctors dispensed their own drugs as well. Few towns, in fact, boasted specialized druggist-apothecaries; the local shopkeeper dispensed drugs as well as salt, cloth, and needles—and often provided the local medical man's only pharmaceutical competition.

The physician's account books neatly mirrored this reality. Accounts tended to be kept under the name of the head of a family, and entries included wife, children, and servants. Though each visit occasioned a separate charge, bills tended to be settled at annual or semiannual intervals. It was, in fact, considered the mark of a businesslike physician to submit statements at such regular—if infrequent—periods; many apparently allowed years to pass before submitting accumulated bills. Indeed, such accounts were often ignored until an estate was settled, and then ordinarily at some fraction of the nominal total. It was not accidental that physicians throughout the nineteenth century were always careful to make a distinction between amounts billed and sums actually collected and that practices were routinely evaluated in such terms; a cautious young medical man tried to avoid communities notorious for the penury with which its citizens contemplated the doctor's bill. This chronic irregularity of payment mirrored both the personal ties between doctor and patient and the scarcity of cash in many areas. Account books kept by physicians in rural areas and small towns indicate that until the very end of the nineteenth century medical men often received payment or part-payment in kind. Bushels of wheat, firewood, oats for the physician's horse, even days of work in fields or stable might serve to compensate the medical man for his services.[7]

Medicine was no easy trade in which to earn a living. Especially in rural areas, physicians were often shopkeepers, small farmers, or land speculators as well as doctors, and many drifted in and out of practice as local conditions dictated. Educational requirements were casual, licensing unenforced where it existed, the amount of capital necessary

to enter the trade relatively small (even in a capital-starved society). Given the relative accessibility of medicine as a vocational option and the paucity of paying customers, it was inevitable that the healing art should have been an intensely, even ruthlessly, competitive field. The unscrupulous might seek to malign their colleagues and undermine the loyalties of a fellow physician's patients. Physicians were often hesitant to suggest a consultation—even in cases where prudence dictated that responsibility be shared—for fear that the consultation might prove the occasion for an aggressive practitioner to attract even a faithful patient away from his regular medical man. In a period before specialization and elaborate referral networks, every physician was potentially a competitor for the same family practice. Competition might also assume the form of cost-cutting, but local medical societies soon began to draw up fee-bills, agreed-upon schedules of minimum charges for the more common services.[8]

Formal institutions played a relatively small role in medical practice. Historians have studied the role of medical schools, of the vicissitudes of state licensure, and the formation of pioneer medical societies. In retrospect they are significant, because such organizational developments seem to foreshadow the shape of twentieth-century medicine. And such developments do constitute an important thread in the profession's evolution; yet they nevertheless played a minor role in shaping the *practice* of medicine in 1800. Society membership and licensing were elements in the public presentation of elite physicians, but they were hardly preconditions for practice. Like writing for medical journals or serving as a hospital physician, such institutional activities were instrumental in helping a physician define his identity as a gentleman of learning and intellect. The physician's individual poise, self-assurance, and social connections determined success, not formal credentials or an abundant ratio of patients to providers. One could survive economically with little education and few social pretensions; but to enjoy a prestigious urban or town practice, the style of a gentleman and the tact of a diplomat were required.

The place of classroom teaching was still small in medical education, the place of hospital wards and clinics even smaller, except for that handful of young men well-connected and prosperous enough to seek and occupy hospital staff positions. Internship and clinical training requirements were not to become universal until the twentieth century. Many doctors did not even bother to apply for licenses and, as has been pointed out, much practice was undertaken by individuals

who did not think of themselves as physicians. Professional nursing, like hospital care, was hardly contemplated by respectable Americans in 1800. The site of practice remained in most cases the patient's home; no available technology dictated the hospital's superiority, and social values defined it as a recourse of the poor and dependent. Nursing was done by family members or local women who regarded it as one of the skills implied by their sex. The network of high-technology-oriented hospital facilities, of specialists and subspecialists, was entirely alien to this period in American medicine—as alien as systematic research, government subventions, and third-party payment.[9]

A Unified Vision

Ideas, not institutions, were fundamental in shaping early nineteenth-century medical practice. Far more important than societies, statutes, or formal educational requirements was a shared heritage of intellectual assumptions and a mode of practice tightly integrated with them. It was a pattern of shared assumptions that turned on a particular view of the body in health and disease, a view hallowed by antiquity and also based on the physician's limited access to the body's internal realities and his consequent need to create a pathology and therapeutics based on those aspects of the body he could see and feel.[10]

In 1800 the physician had little more in the way of diagnostic aids than his forerunner in classical antiquity. He could evaluate the patient's own account of his or her symptoms, and he could inspect his or her tongue or feel the pulse for abnormalities, observe the patient's facial appearance, and feel his or her face for indications of an elevated temperature. He could observe the patient's urine and feces or, if the medical man drew blood, could observe it for clues to the body's internal state.[11] This dependence on the patient's external appearance and metabolic "outputs" underlined the ways in which the continued integrity of the body must somehow assure health and its disturbance constitute disease; all provided clues to present and future states of health. Prognosis was a key to demonstrating the physician's skill; it was at least as significant as—and in many ways indistinguishable from—his therapeutic abilities.

Both prognosis and therapeutics were still based on a widely held view of the body and its physiological nature. It was a view, let me emphasize, shared by medical men and educated laymen. The body

was seen metaphorically as a kind of equilibrium system in constant interaction with its environment. Food, air, exercise, and rest were variable environmental forces impinging on the body; excretion, perspiration, and defecation were outputs necessary to the maintenance of a body's particular equilibrium. Balance constituted health, imbalance illness, and the symptoms of disease could be thought of as efforts by the body to restore itself to a proper balance. Diarrhea, for example, might indicate that the body was seeking to relieve itself of fluids; thus, instead of seeking to assuage this particular symptom, the physician might well prescribe a cathartic to help the body along in its healing efforts. Similarly, the vomiting in whooping cough or the perspiration associated with an acute fever might also indicate an innate healing mechanism. All such symptoms could be regarded as evidence for the existence of what many physicians had since classical antiquity termed "the healing power of nature."[12]

There was comparatively little place for specific disease entities and specific causative agents in this holistic and individualistic way of understanding disease. Sickness was fundamentally a particular state of a particular organism—not the acting out in that organism of a patterned ailment with a specific cause and course.[13] Two subsidiary assumptions followed naturally from these holistic assumptions. First, every part of the body was related inevitably and inextricably to every other part. A distracted mind could curdle the stomach; a dyspeptic stomach could agitate the mind. Local lesions might reflect imbalances of nutrients in the blood; systemic ills might be caused by fulminating local lesions. Physicians were consistently suspicious of drugs which were alleged to have a specific efficacy in specific ailments; they were equally suspicious of any practitioner who made such claims—indeed, they were presumptive evidence of quackery. Drugs exerted their effects through their ability to modify the body's ongoing efforts to maintain or restore a health-defining equilibrium. Almost all of the drugs, accordingly, in the physician's medicine bag demonstrated their plausibility through a visible physiological activity. Some caused copious urination (diuretics), others defecation (cathartics) or perspiration (sudorifics), thus demonstrating a tangible efficacy which could be understood and rationalized in terms of the age-old physiological model that has been described. (Textbooks routinely listed drugs by category of physiological activity; the medical man would presumably be in search not of a specific drug, but of some drug capable of producing a desired effect.) Bleeding (or the related practices of

cupping and leeching) could be seen in the same terms, as efforts to help the body restore itself to health by relieving it of some component present in excess. Such modes of therapy might make patients uncomfortable, but the discomfort was understandable to most patients and thus tolerable.

The effectiveness of this mixture of speculative pathology and therapeutic practice was attested to by its centuries-long grip on medical theory, and it was intensified, common sense suggests, by the social setting of doctor-patient encounters. Normally conducted, as has been seen, in the patient's home and under the aegis of a medical man well known to the family, the administration of physiologically active drugs and their visible consequences not only legitimated the physician's authority but also mobilized individual and family psychological resources in support of the patient's efforts to heal himself. In this sense, the traditional medicine of the early nineteenth century can be seen as a kind of healing ritual, one acted and witnessed by all its participants: patient, family members, and physician.

In the second half of the twentieth century, the relationship between doctor and patient has been much altered; its context has, in the great majority of cases, shifted from the home to some institution. The healer is in many cases unknown, or known only casually, to the patient. Even the place of drug therapeutics has changed, not only in the sense that the efficacy of most drugs is beginning to be understood, but also in the social ambience which surrounds their use. The patient still maintains a faith in the physician's prescription (often, indeed, demands such a prescription), but it is a rather different kind of faith than that which shaped the interaction of physician, patient, and therapeutics at the beginning of the nineteenth century.

Clearly the physician and the great majority of his patients no longer share a similar view of the body and the mechanisms which determine health and disease. Differing views of the body and the physician's ability to intervene in its mysterious opacity divide groups and individuals, rather than unifying them, as the widely shared metaphorical view of body functions had still done in 1800. Physician and patient are no longer bound together by the physiological activity of the drugs administered. In a sense, almost *all* drugs now act as placebos, for with the exception of certain classes of drugs, such as diuretics, the patient experiences no perceptible physiological effect. He does ordinarily have faith in the efficacy of a particular therapy, but it is a faith based not on a shared nexus of belief and participation in the kind of

experience I have described, but rather on confidence in the physician and his imputed status (even when that physician is a stranger), and, indirectly, in that of science itself.[14]

Seeds of Change: New Values and Institutions

Although the basic realities of medical care in a still largely rural society had little place for institutional care, hospitals and dispensaries played an important role for two groups, small in numbers but increasingly significant as the century progressed. These were the dependent urban poor and the elite among urban physicians. Workingmen and women—even when employed—often could not afford the fees of private physicians. When ill for more than a few days, most urban working people could not be treated at home. In many households both husband and wife had to work so as to provide an adequate income; hiring nurses was hardly an option, especially when the family's precarious income was interrupted. Often the household lacked a suitable sickroom as well as nursing help. Thus, in the event of sickness there was no one available to care for the patient. Even short stretches of incapacitating illness could mean economic crisis. Single men were at even greater risk than married men, who could turn to other family members.[15]

If the worker had only a minor ailment, he might consult a local druggist or go to a dispensary, a free-standing institution founded to provide outpatient care for the needy. Though such care was normally provided at a clinic, some dispensaries dispatched physicians to the patient's home in the event of serious illness.[16] These "community health centers" provided a valued opportunity for young physicians to hone their clinical skills. But that was a secondary goal of the philanthropic Americans who supported these institutions; they founded them primarily as a first line of medical defense against the costly and stigmatizing recourse of hospitalization.

Should the individual become gravely or chronically ill, the hospital or the almshouse provided his or her only refuge. In the first quarter of the nineteenth century, only a few American cities boasted voluntary hospitals—the oldest being Philadelphia's Pennsylvania Hospital, chartered in 1751—but every town of any size possessed an almshouse, a multipurpose facility in which all of the dependent, the crippled, the blind, the foundling, and the aged as well as the sick were held.[17] The

sick wards in these buildings were to evolve into municipal hospitals in the larger cities. In both almshouse and voluntary hospital, however, prominent local doctors and their private pupils had already begun to use the "clinical material" for teaching purposes.[18] The hospital also had other functions for the ambitious urban physician. In a period before board certification, research grants, and other badges of status, the holding of a hospital physicianship defined one as belonging to the profession's elite. And for those intellectually as well as socially ambitious, the hospital provided a steady source of case materials for publication.

The prominence of the hospital in teaching, in publication, and in the definition of status was already well established by 1800. With increasing immigration and urbanization, the hospital was to become steadily more important in the delivery of medical care, though not until the twentieth century did it become a place in which middle-class Americans might expect to find themselves when sick. In the years before 1875, the hospital was still almost exclusively a refuge for the dependent poor.

Events in the intellectual history of medicine were, however, to underline the hospital's centrality to the medical profession far earlier. The hospital was to become increasingly an agent of intellectual change and the transmission of innovation, as well as a place where a poor man could find a bed in which to recover or die. For the first half of the nineteenth century was a period during which traditional views of the nature of disease were beginning to break down, and the hospital was at the very center of this trend.

Instead of the traditional holistic view of the body, in which disease was seen ultimately as an individual state of physiological imbalance, a new kind of clinical viewpoint was being elaborated, one emphasizing the specificity of disease and the relationship of that specificity to characteristic lesions observable at autopsy. Paris was the leader in shaping this new way of looking at disease, and the hospital was the Parisian medical man's clinical laboratory; only in its extensive wards and easily available postmortem rooms could physicians systematically correlate patterns of symptoms during life with lesions after death.[19] America did not possess the sprawling hospitals of Paris or Vienna, nor were its poor as docile in submitting to the physician's examinations, but, as I have pointed out, it did have an eager and ambitious group of young physicians and a scattering of public hospitals in which they could seek to re-create the clinical and pathological achievements

of Paris. Mid-nineteenth century America boasted as well an assortment of medical schools in which the new doctrines could be taught.

Medicine's age-old conceptual bases had already begun to break down, while the knowledge which impelled these changes was becoming increasingly specialized. The gap between lay and medical understanding of disease and therapeutics grew inexorably wider. Therapeutic practice too began to change, if rather more slowly than the doctrines which had so long rationalized it. By the middle of the nineteenth century, the institutional foreshadowing of twentieth-century patterns could already be discerned as well. The beginnings of specialism and of the growing significance of institutional care and institutional connections, and the burgeoning of medical publication, were already present. The practice of medicine remained, however, a stern mistress; competition continued to be intense, perhaps even exacerbated by the flourishing of marginal and inexpensive medical schools in the century's second quarter. The intellectual and institutional world of medicine was beginning to change, but the ordinary doctor's path to economic security remained as treacherous and problematic as it had been in 1800.

1

COMPETITION IN THE MEDICAL MARKET: THE NEED FOR RATIONALIZATION AND REGULATION 1875–1910

Over the past fifty years a marked characteristic of the organized medical profession in the United States has been a generally negative attitude toward innovation in the organization, financing, and provision of medical care. Such changes have not been welcomed by the national, state, or local bodies speaking in the name of the profession, and have often been rejected outright or, at best, accepted with grudging toleration. An example is voluntary health insurance for medical service, initiated and developed under the threat of a compulsory national system. Physicians have exhibited similar reactions to hospitalization insurance, prepayment group practice, union health centers, and above all government action in the health field. To interpret this

negativism simply as a form of human perversity would be to misconstrue its meaning completely. It should be understood rather as a consequence of a process extending over a period of roughly a century, during which the medical profession has fought to maintain a medical market which would safeguard the basic structure of private practice.

From the middle of the nineteenth century to the present, this aim has been central in the opposition of clinical practitioners to specialization, dispensaries, contract practice, public health clinics, and various other forms of medical activity regarded as interfering with the existing medical market and thus with the access of physicians to patients. Although such opposition has not prevented the occurrence in medical practice of highly significant changes during the past hundred years, the fact remains that fee-for-service solo practice as the major system for the provision of medical care still exerts a pervasive determining influence in the medical market. Clearly, the considerable degree of success achieved by the medical profession in retaining basic aspects of the medical market relating to private practice is an important factor in the current medical care situation. For this reason the earlier circumstances within which physicians practiced, the problems that beset them, their endeavors to cope with these conditions and the consequent developments, and the attitudes which evolved as a result, as well as the contribution of these elements to the present situation, invite the historian's attention. The focus of such attention must be the structure of medical practice—that is, whom the physician treats, how he obtains patients, where the physician works and whether by himself or in an institution with others, what scientific and technical resources he can deploy, and how he is paid—for these factors taken together determine what the physician does and how he does it. This is the framework for the analysis of continuity and change in American medical practice from the last quarter of the nineteenth century to World War II which will be presented in this book.

Structure of the Profession

The census of 1860 found 55,055 medical practitioners in the United States and its territories. This represented a ratio of 175 practitioners per 100,000 population for the country as a whole, or one for every 572 persons. A decade later there were 64,414 practitioners, and by

1900 the number had risen to 132,002. The increase in the number of practitioners tended generally to keep pace with the growth of population, as indicated by a ratio of one physician to every 578 persons in the United States as a whole in 1900. During the following decade, however, as proprietary medical schools began to close down, the physician-population ratio declined, reaching one to 610 persons by 1910, and this trend continued for about another twenty years.[1]

Medical practice is obviously affected by the number of available practitioners, but more important than the mere number in proportion to population may be their education and training, the type of practice engaged in, the economic framework of practice, the geographical location, and the transportation technology, as well as public attitudes to medicine and its practitioners. To a greater or lesser degree these factors affected the practice of medicine during most of the nineteenth century and have continued to exert an influence in the present century.

Regarded from this point of view, what was the state of medicine one hundred years ago, and who were its practitioners? From the perspective of the present, one can see that medicine in 1875 was about to begin its rise from the low estate to which it had declined by the middle years of the century. To concerned physicians and laymen at the time, however, the condition of medicine appeared to be one of utter confusion and anarchy. As one observer noted in 1864, the practice of medicine was seen by many as an absurd hodgepodge comprising "allopaths of every class in allopathy; homeopaths of high and low dilutions; hydropaths mild and heroic; chronothermalists, Thomsonians, Mesmerists, herbalists, Indian doctors, clairvoyants, spiritualists with healing gifts, and I know not what besides."[2] Six years earlier, in 1858, Thomas C. Brinsmade, a physician of Troy, New York, had expressed a similar view of the situation in medicine, although in more sober terms. "I believe," he said, "there is not a town or city in this State [New York], in which there is not more money paid to irregular practitioners of various names, and for nostrums and patent medicines, than is received by regularly educated physicians of these towns and cities, and I am inclined to think the *number* of diseases partially treated and maltreated in this random way is fully as great as those submitted to more rational management."[3]

Both descriptions tend to exaggerate in one respect. Although many credulous, desperate, or nervous individuals visited a variety of healing sects and cults, irregular practitioners were a minority, even if one of

considerable size. Estimates indicate that in the 1850s about 10 percent of all practitioners in the nation were irregulars. Based on this ratio, there were, among the 40,755 practitioners enumerated in the 1850 census, some 37,000 regular physicians and less than 4,000 others. Fifty years later, the data do not reveal any marked change in the ratio of professionally trained physicians to irregular practitioners; in 1900, there were about 110,000 physicians, 10,000 homeopaths, 5,000 "eclectics," and over 5,000 other healers of various kinds. By that time the homeopaths were being assimilated into the regular profession and may therefore be grouped with other physicians. On this basis the irregulars comprised about 9 percent of all practitioners at the turn of the century.[4]

In light of these figures, it is realistic to assume that during the later nineteenth century the majority of sick people were treated by regular physicians. The significance of the irregular practitioners lay not so much in their numbers as in their activities—competition with and denigration of physicians, claims of sure and pleasant cures, etc., which tended to exacerbate the situation of a profession beset by problems and divided within itself.

Nostalgic illusions frequently lead to the misconception of a past golden age when current problems did not exist and all was for the best in the best of all possible worlds. Such a yearning for the good old days probably motivated F. W. Mann, a Maine physician, to wonder in 1925 "if the medical man is held in as high esteem, or is as important a factor in the community, as was his confrere a half century ago?" As might be expected when such rhetorical questions are raised, his answer was that the situation of the physician in 1825 was much better than that of his successor fifty years later.[5] Yet, as Shryock pointed out in 1946, the available evidence leads to a different conclusion. All was not well with the profession in the 1840s, nor was the situation much better in the 1870s. In fact, there was no clear improvement in the situation of the profession until the turn of the century, when some of the problems plaguing the practice of medicine began to yield to changing circumstances.

The American physician around the middle of the nineteenth century was an individual entrepreneur who was sufficient unto himself, except as he adhered to rules of professional behavior established by his colleagues and employed methods of diagnosis and treatment more or less accepted by them. Differences of status, wealth, and

influence existed within the profession and were evident to perceptive contemporaries. As John Shaw Billings observed in 1876, the profession had . . .

> a very few men who love science for its own sake, whose chief pleasure is in original investigations, and to whom the practice of their profession is mainly, or only, of interest as furnishing material for observation and comparison. Such men are to be found for the most part only in large cities where libraries, hospitals and laboratories are available for their needs, although some of them have preferred the smaller towns and villages as fields of labor . . .
>
> We have in our cities, great and small, a much larger class of physicians whose principal object is to obtain money, or rather the social position, pleasures, and power, which money only can bestow. They are clear-headed, shrewd practical men, well educated, because "it pays", and for the same reason they take good care to be supplied with the best instruments, and the latest literature. Many of them take up specialties because the work is easier, and the hours of labor are more under their control than in general practice. They strive to become connected with hospitals and medical schools, not for the love of mental exertion, or of science for its own sake, but as a respectable means of advertising, and of obtaining consultations. They recite and lecture to keep their names before the public, and they must do both well, or fall behind in the race. They have the greater part of the valuable practice, and their writings, which constitute the greater part of our medical literature, are respectable in quality, and eminently useful. [Furthermore, they are] the active working members of municipal medical societies, the men who are usually accepted as the representatives of the profession . . .
>
> There is another large class, whose defects in general culture and in knowledge of the latest improvements in medicine, have been much dwelt upon by those disposed to take gloomy views of the condition of medical education in this country. The preliminary education of these physicians was defective, in some cases from lack of desire for it, but in the great majority from lack of opportunity, and their work in medical school was confined to so much memorizing of textbooks as was necessary to secure a diploma. In the course of practice they gradually obtain from personal experience, sometimes of a disagreeable kind, a knowledge of therapeutics, which enables them to treat the majority of their cases as successfully, perhaps, as their brethren more learned in theory. Occasionally they contribute a paper to a Journal, or report to a medical society; but they would rather talk than write, and find it very difficult to explain how or why they have succeeded, being like many excellent cooks in this respect.

> They are honest, conscientious, hardworking men, who are inclined to place great weight on their experience, and to be rather contemptuous of what they call "booklearning and theories."[6]

Despite the differences of status and power described by Billings, the medical profession by the last quarter of the nineteenth century was still largely undifferentiated in its organization. Although specialism was well established in New York and to a lesser extent in other urban centers, most physicians were general practitioners. Some had developed special competence in a particular branch of medicine, but exclusive specialism was still infrequent. In 1875 an editorial writer in the *Virginia Clinical Record* remarked that "there are very few 'specialists' simon-pure, this side of New York."[7] Thomas Addis Emmet recalled that in the 1860s he decided to limit his practice to gynecology when he found that he could not successfully combine it with general practice as most of his colleagues did.

> I was at that time in a good general practice among some of the best people in the city, when I became convinced my work did not give as satisfactory results, as formerly, and from some cause my surgical work was not always what it should have been. I promptly decided it was my duty to give up general practice and to attend no obstetrical cases, and this I promptly did, notwithstanding I thus lost the greater portion of my means of support. I also gained the mistrust of the profession generally, in thus as it was claimed, putting myself on the level, as it were, with a quack, who from necessity held his position as to the treatment of *one* disease from want of the necessary training and knowledge to treat everything. No one seemed to recognize that the unusual hospital advantages which I had experienced, and this advantage alone, had fitted me for being a specialist.[8]

With specialism still in embryo form there were few facilities for training in the clinical specialties. Of necessity, therefore, specialists tended to be physicians who, while engaged in general practice, became interested in a particular field of medicine, and by acquiring more than average competence in this work made it their specialty, a pattern of differentiation which persisted well into the twentieth century. As Emmet put it, they were especialists [sic], who from taste, or from the force of circumstances, had gradually given more attention to the treatment of some one special class of diseases than to another."[9] On the whole, this line of development did not greatly alter the relative homogeneity of the profession, which remained undifferentiated and largely

decentralized throughout the nineteenth century. The concern aroused by specialization after 1850 had little to do with such problems as the fragmentation of medical practice and the need for coordinating patient care, matters of considerable moment at present. Hostility to medical specialism centered basically, no matter what the arguments advanced, on the competitive danger of specialists to general practitioners, thus pointing up the crucial significance of the basic economic institution around which medical practice was organized: the market in which physicians competed for patients.[10]

The Medical Marketplace

The physician in the middle of the nineteenth century generally shared the social and economic values, opinions, and attitudes widely held in American society. As far as possible, it was felt, the individual should be free of governmental interference and given full scope for initiative, self-assertion, and the development of self-interest. Each man was held to be entitled to carve out the largest possible stake and to dispose of it as he wished. After all, the United States was a free society, in which competition was the lifeblood not only of trade but of the professions as well.

Competition among practitioners was an accepted fact of professional life. Throughout most of the nineteenth century, the regular medical profession faced the competition of irregular practitioners and outright quacks. Within the profession, young physicians attempting to establish themselves had to compete with practitioners who had already developed a secure practice. Increasingly, as the nineteenth century neared its end, general practitioners also competed with specialists on terms that were to become more and more unfavorable to the former in the twentieth century.

Acceptance of competition as natural was based on the assumptions that egalitarianism reigned within the profession and that it would enable the most competent physicians to achieve the greatest success. As Billings had written in 1876, professional competence and success were directly related, and physicians who did not maintain a superior level of achievement would "fall behind in the race." A decade later, in an address before the British Medical Association, he was even more explicit, asserting that "the successful men are the survivors of a struggle in which there has been keen and incessant competition."[11] Several

interacting factors produced this intense competition, which was most evident in the cities.

One was that the medical profession had an excessive number of practitioners in its ranks.[12] Although the number of medical schools increased after 1850, the dislocations caused by the Civil War impeded the founding of new ones. By the 1870s, however, medical schools began to be established at an unprecedented rate, and they continued to proliferate rapidly during the last quarter of the century. In 1860 there were 53 regular schools; by 1900 the number had risen to 126. Similar increases, though on a smaller scale, took place among homeopathic and eclectic schools. Overall, the number of schools in the United States increased by 16 percent from 1860 to 1870, by 23 percent from 1870 to 1880, and by 34 percent from 1880 to 1890.[13]

Furthermore, though the schools varied considerably in quality, most of them had relatively low entrance requirements. Since degrees were readily obtained, medicine was an attractive and popular profession, and the numbers of medical graduates increased even more rapidly than the schools. This situation also provided opportunities for ambitious young people from the lower middle and working classes to achieve higher social status.

Inadequate control of licensure was another factor that helped increase the number of medical men.[14] Regulation of entry into the profession by establishing examining boards began slowly and hesitantly in the 1870s. By 1893, eighteen states required those applying for licensure to take an examination; in seventeen other states a diploma from a medical school "in good standing" conferred the right to practice. But not until the early twentieth century did these efforts to upgrade standards have any serious impact, and then only because they were supported and reinforced by the development of scientific medicine and the related movement to reform and to standardize medical education.

Last, but certainly not least, competition among practitioners was a consequence of the limited effectiveness of their practice.[15] Ignorance of the etiology and pathogenesis of most diseases meant that there was little they could do on the basis of valid knowledge. Diagnosis and treatment were based in considerable measure on clinical experience and judgment, supported by etiological theories emphasizing environmental, constitutional, and familial factors. For this reason, the physician's ability was likely to be judged largely by his prognostic skill and accuracy. Austin Flint observed in 1866 that "people in general are apt

to estimate [the practitioner's] knowledge and ability by the correctness of his judgment in this regard."[16] Medical care in these terms required the physician to become acquainted with the familial, domiciliary, and social circumstances of the patient, since therapy and management of the case involved not just medication but frequently required advice about nutrition, work habits, sex relations, travel, and other factors that might influence an individual's health or illness.

Throughout most of the nineteenth century, the opinion prevailed that this kind of care by an experienced practitioner was most advantageous for the patient and was provided most effectively through family practice—that is, by attending all the members of a household at home. One advantage of this pattern of practice was that the physician would become aware of familial tendencies and characteristics of family members, since it was not unusual for several generations to be present in a household or a locality. In an address to the graduating class of Bellevue Hospital Medical College delivered in 1871, Oliver Wendell Holmes said: "The young man knows his patient, but the old man also knows his patient's family, dead and alive, up and down for generations. He can tell beforehand what diseases their unborn children will be subject to, what they will die of if they live long enough, and whether they had better live at all, or remain unrealized possibilities . . ."[17]

Within this context, an experienced physician could, by the manner in which he prescribed and advised, deal with anxieties and psychological tensions and satisfy the expectations of professional behavior held by the patient and the family. This fact of medical practice is well portrayed by Henry James in his novel *Washington Square*. The vignette of Dr. Austin Sloper, the fashionable practitioner about the middle of the century, was just as appropriate for the period when the book was written, in the winter of 1879–1880. As portrayed by James, an important element in Dr. Sloper's reputation was . . .

> that his learning and his skill were very evenly balanced; he was what you might call a scholarly doctor, and yet there was nothing abstract in his remedies—he always addressed you to take something. Though he was felt to be extremely thorough, he was not uncomfortably theoretic; and if he sometimes explained matters rather more minutely than might seem of use to the patient, he never went so far (like some practitioners one had heard of) as to trust to the explanation alone, but always left behind him an inscrutable prescription. There were some doctors that left the prescription without offering any explanation at all; and he did not belong to that class

> either, which was after all the most vulgar. [Thus, by meeting the expectations of the circle in which he practiced,] he daily justified his claim to the talents attributed to him by the popular voice.[18]

The conditions of family practice made it possible for the physician to compensate for the limitations of his knowledge by impressing patient and family with his behavior and personality. Family practice and the rationale which justified it continued to prevail throughout the last quarter of the nineteenth century not only in the United States but also in Europe. As late as 1896, Jacob Wolff, a German physician, insisted that "office practice is less well suited than treatment at home to demonstrate [the physician's] ability to deal with patients. The physician [in his office] lacks knowledge of domestic conditions and cannot control the implementation of his instructions."[19] By his bedside manner the experienced practitioner could also obtain a favorable verdict from "the juries of matrons," who, as Weir Mitchell commented acidly in the late 1880s, "do so much to make or mar our early fates."[20] Though conditions had changed in various respects by 1912, apparently the situation noted by Mitchell had not, for young doctors were still being informed that "a physician's practice depends to a great extent upon the ladies . . ."[21]

Moreover, as long as the methods of examination and the armamentarium of the practitioner remained relatively simple, attendance on the patient at home continued to be practical. Mitchell commented in 1887: "Men yet live who can remember when all of our knowledge of disease was acquired by the unaided use of the eye, the ear, and the touch. The physician felt the pulse, and judged of fever by the sense of warmth. He looked at the skin and tongue and the secretions, and formed conclusions, more or less just in proportion to the educated acuteness of his senses and the use he made of these accumulations of experience."[22] Even the introduction of more precise instrumental aids did not immediately alter the situation. Instruments were inconvenient to use in some cases, and in others the physicians were unable to accustom themselves to new ways of practicing and the use of new instruments (a situation not unknown at present). When John Shaw Billings entered the Union army as a contract surgeon in 1861, he wrote that he "had three things . . . that none of the other surgeons had: A set of clinical thermometers . . . a straight one and one with a curve; a hypodermic syringe, and a Symes staff for urethral stricturotomy. The hypodermic syringe was in constant requisition. The clinical

thermometer was troublesome and was not used very much."[23] In the 1870s and 1880s Robert B. Pusey, a country doctor in Kentucky, employed the following instruments in physical diagnosis: a stethoscope, a clinical thermometer, an exploratory syringe, a cannula and trocar, and several specula. For obstetrics and surgery, he had the instruments and other technical equipment used at that time. But Pusey did not have a microscope, nor was he completely at ease with the stethoscope, preferring direct auscultation of the chest wall.[24]

The pattern of family practice could persist as long as the physician was able to bring the equipment required for his work to the patient's home, and as long as he remained sufficient unto himself without any need for recourse to an institution such as the hospital. The surgeon even operated in private homes, a practice which persisted well into the twentieth century.[25] Within this framework, competition among practitioners rested on the assumption of a medical market within which physicians could have relatively equal access to patients and the latter to them. Disturbing influences were envisaged as arising only from irregular healers or from regular practitioners who engaged in unfair competition. Comparatively few hospitals existed in the United States by the third quarter of the nineteenth century, and the organized medical market took little note of them. A survey published in 1873 revealed 178 hospitals in the country as a whole. As late as 1900 there were only 2,070 in existence, but by 1909 the number had risen to 4,359 (with a bed capacity of 421,065), and by 1923 the number had increased to 6,830 (with 755,722 beds).[26] The greatest increase in the number of hospitals occurred toward the end of the nineteenth century and especially after 1900, so that most physicians practicing between 1875 and 1900 had few contacts with hospitals.

Practice within a competitive market went hand in hand with recognition of the need to maintain a situation which would enable physicians to participate under conditions of relative equality, to deal with the excessive zeal displayed by some competitors, and to avoid or patch up possible abuses. To a certain extent this was done informally. All physicians were expected to engage in clinical practice, and those who stepped out of line could expect the disapproval of their peers.[27] The lay public, as well as the profession, regarded the physician as primarily a practitioner and tended to penalize those who did not conform to this image. When Weir Mitchell began to practice in 1851, at the same time undertaking a program of research, he was not encouraged by his older colleagues. In fact, Samuel Jackson warned him that every exper-

iment in the laboratory meant the loss of a patient.[28] The persistence of this attitude in the nineteenth century is attested to by Mitchell's comment in 1887 that "the wise physician, who is fond of etching or botany, the brush or the chisel or the pen, or who is given to science, does well to keep these things a little in the background until he is securely seated in the saddle of professional success."[29]

Moreover, practitioners were expected to avoid actions that could lead to invidious comparisons among members of the profession and denigration of their colleagues. Nevertheless, in a situation in which medical opinion could not be supported by valid knowledge and demonstrable evidence, physicians frequently disagreed. This state of affairs was depicted by G. M. B. Maughs in 1880: "So long as physicians depended upon rational signs for their diagnosis, all was uncertainty and doubt . . . Then, when all, illuminated by an uncertain light, groped their way to conclusions which could not be demonstrated, a difference of opinion was in many cases inevitable, hence that 'doctors disagree' has passed into an adage."[30] Under such circumstances, competition for patients led some practitioners to attack the professional ability and reputation of rivals; others resorted to overt or less obvious forms of advertising.[31]

But while the medical profession accepted competition in principle, these and other practices were condemned because they led to instability of the market. To protect and promote the economic interests of the profession, various medical organizations attempted in a number of ways to set limits on competition so as to maintain a stable market. Fee bills were adopted to standardize and regulate service charges and to discourage underbidding by rival physicians.[32] Furthermore, rules governing competition were formulated in codes of professional ethics.[33] Most important was the Code of Ethics adopted by the organizing convention of the American Medical Association in 1847. Its stabilizing intent is clear from the prediction by Nathan Smith Davis in 1883 that it "will probably continue to be the guide of the great mass of intelligent medical men through the centuries to come."[34]

What may be considered desirable in theory is not always possible in practice. Although the needs to regulate competition and to impose uniform standards in practice were recognized, the achievement of these aims was impeded by the prevailing decentralized entrepreneurial system. Moreover, despite the intentions expressed in fee schedules and codes of ethics, the economic, social, scientific, and technological changes in the late nineteenth century began to affect the structure of

medical practice, increasing the instability of the medical market. The period encompassing the last thirty years of the nineteenth century and the first decade of the twentieth was one of transition, marked by the emergence of new forms of medical organization, education, and practice. Conflicting and converging factors and interests were involved in this transition. As far as possible, practitioners endeavored to retain a competitive market in which all recognized practitioners could have equal access to patients, since this was the basis of their livelihood. In terms of this objective, any factor or element that tended or threatened to draw off consumers of medical service—that is, patients—was hotly resented. From this viewpoint, all arrangements for the provision of medical care differing from fee-for-service private practice were seen as disruptive of the market structure. At the same time, the strengthening of professional standards regulating the practice of medicine was envisaged as a means of reversing the trend toward easy entry into the profession and uncontrolled competition among practitioners. This tendency toward restriction of numbers by upgrading standards for education and licensure became increasingly strong toward the turn of the century, as a firmer scientific basis for medical practice provided objective criteria.

The Shift to Office Practice

But while physicians engaged in general or family practice were attempting to safeguard their interests, the ground on which they stood was shifting under their feet, in several directions. One was the trend away from family practice to office practice. The treatment of ambulatory patients in the physician's office at hours set aside for the purpose was an innovation that appeared in European countries as well as in the United States in about the middle of the nineteenth century.[35] Although initially considered marginal and inferior to family practice, office practice had by the beginning of the twentieth century become as important as attendance at the patient's home. This trend was no sudden development but occurred due to the interaction of factors in society with others operating within the profession.

As long as urban communities were small and the number of patients was not too large, the physician made his house calls on foot. By midcentury, however, and increasingly thereafter, urban growth and shifting residential patterns within the city made it more difficult to

maintain this mode of visiting patients. Busy practitioners began to employ horse-drawn vehicles, and from the 1850s to the early twentieth century the horse and buggy was commonly used by urban physicians and their colleagues in small towns and rural areas.[36] In either situation, the physician's daily round allowed little time for office practice. Keeping regular office hours was still not customary among small town and rural physicians in the 1870s and 1880s. Robert A. Pusey, for example, was of the opinion "that regular office hours would be out of the question for him; that if he stayed around in the morning waiting to see people, he would never get through his work; and that if he did not have work to do he would be available anyway," and so he had no office hours.[37] By the last decade of the century this attitude had vanished, and doctors in small towns and rural areas were maintaining regular office hours. Office practice had long since established itself in cities such as New York, Philadelphia, Chicago, and San Francisco, where it was fostered by the growth of specialism.[38]

The American experience in this respect paralleled a similar trend in Europe. A retrospective analysis in 1906 led a Nuremberg clinician, Gottlieb Merkel, to conclude that "increasing specialization of the medical art" was a major cause of the growing preference for office visits over house calls. Furthermore, a consequence of the expansion of medical knowledge and the development of increasingly refined and precise methods for research was the introduction into medical practice of new instruments and techniques for diagnosis and therapy, which were very difficult if not impossible to bring to the patient. As Merkel put it, "Since the mountain cannot come to Mohammed, he must come to the mountain, that is, the patient must see the physician in his office."[39]

Perspicacious American physicians were aware that these changes were taking place and that the pattern of medical practice would necessarily be altered. In 1865, Benjamin E. Cotting, a leading Boston clinician, predicted that "the time is coming, perhaps it is nearer than we are aware of, when the public shall no longer consider the proper care of the sick (their true *cure*) to consist in a mysterious and indispensable administration of drugs, but in rationally and understandingly attending to all their necessities."[40] The achievement of this goal, he asserted, would require a greater knowledge of disease and a more painstaking and thorough investigation of the patient than were then possible. But it was precisely these elements that became available in the late nineteenth century. To be sure, neither discoveries in medical

science nor instruments for more precise and accurate examination of patients were immediately accepted by many physicians. Cotting quoted with approval a comment by another physician deploring the ignorance of recent advances in medical knowledge which was widespread among members of the profession.[41] In 1887 Weir Mitchell decried the traditional faith in the virtues of the family practitioner and compared him unfavorably with the newer modern physician. The family doctor, he observed, is supposed to have "some mysterious knowledge" of the patient's constitution, but frequently he fails in this respect because he does not obtain a good history and examines the patient superficially, thus leading to inadequate and inaccurate judgments based on carelessness. Mitchell advised his readers that a union of precision and thoroughness was the hallmark of the best type of practitioner, and that "the methodical manner in which a physician of modern training goes over the case" should be the criterion for judging him.[42] This standard of practice had become possible through the introduction of instruments and methods of precision which made it possible to examine a patient with greater accuracy, obviating to a considerable degree the need to rely on subjective impressions for diagnosis.

Medicine, like other fields of human activity based on scientific knowledge, does not affect all practitioners in the same way at the same time. Specialists led the way in using new diagnostic techniques, but general practitioners gradually learned to employ these impressive tools. When Mitchell commented in 1891 that "the use of instruments of precision . . . has tended to lift the general level of acuteness of observation," he was not referring alone to specialists.[43] In many ways specialists were an avant-garde of medicine, but numerous instruments and techniques introduced by them were eventually assimilated by general physicians. For example, when the Chicago physician James B. Herrick opened his office in 1890, in the basement of his home, his equipment included not only the usual instruments for diagnosis and minor surgery, but also an ophthalmoscope, a laryngoscope, and an ear speculum. In addition, Herrick relates, "Installed a storage battery and cauterized hypertrophied turbinates and snared off nasal polypi. With a guillotine-like tonsillotome I cut off the tops of tonsils—the big ones that nearly met in the center of the throat. The thorough exsection of diseased tonsils was then scarcely known."[44] Herrick carried on a general practice for ten years, seeing patients in the office and at home for a variety of complaints, among them conditions of the eye,

ear, nose, and throat. Though not a specialist, he employed the instruments and procedures of the ophthalmologist and the otolaryngologist to remove a foreign body from the eye, to puncture an eardrum for middle ear suppuration, or to lance a peritonsillar abscess.

Increasingly, emphasis was placed on accuracy of diagnosis as a prerequisite for appropriate therapy and on the need to acquire the skills to employ newer knowledge and methods for this purpose. An anonymous Connecticut physician informed his colleagues in 1902 that "one of the most essential conditions for the treatment of disease is an exact diagnosis," and that "diseases that every family doctor thought he could tell with but little difficulty are shown to require unusual skill and special training."[45] It was no longer sufficient to rely on the indications of the tongue, the pulse, or the temperature; one had to use the microscope, staining reagents, and test tubes. A well-trained physician in a well-equipped office should be able, for instance, to prepare and examine sputum for tubercle bacilli, to count red and white blood cells, and to use the hemoglobinometer and the hematocrit. At the same time, the Connecticut doctor deplored the fact that "not one physician in a thousand can estimate the percentage of hemoglobin, or count the erythrocytes and the leucocytes, or estimate the relative volumes of corpuscles and plasma, or make a spectroscopic or bacteriological examination of the blood."[46]

Evidence of change in this situation was apparent by the second decade of the twentieth century. The newer medical knowledge of infectious diseases as well as developments in surgery and their significance for practice were being transmitted through medical societies, journals, and books, and were being demonstrated in hospitals, dispensaries, and physicians' offices. This change was reflected in the way physicians entering practice were furnishing their offices. Young physicians, in 1912, were advised, if financially feasible, to purchase a microscope with an immersion lens, various kinds of electrical apparatus, an X-ray machine, and instruments and equipment for the examination of urine, sputum, and blood.[47]

The shift of the locus of medical practice from the patient's home to the physician's office was furthered not only by advances in knowledge and by the ability to apply it effectively at the latter site, but by other advantages as well. Despite the alleged benefits of visiting patients at home, this form of practice entailed several disadvantages. Not infrequently, family members and neighbors tended to interfere.[48] Furthermore, a family practitioner with a large practice led a strenuous

life. On a single Sunday in 1864, Weir Mitchell saw twenty city patients and four in the country—and this before the day of the automobile.[49] This situation was not peculiar to the United States. Urban and rural practitioners in Europe worked under similar conditions at the same period, and they exhibited a similar trend toward office practice, indeed perhaps somewhat earlier than in this country. The Swiss physician Laurenz Sonderegger, for example, after carrying on a town and country practice for fourteen years decided in 1863 that he could no longer meet its onerous and fatiguing demands, and so he limited his work to town patients. By 1894, Sonderegger had only an office practice.[50] In short, office practice was a way of economizing time and resources. This lesson was repeatedly brought home to physicians. A doctor could see and deal with half a dozen patients during the time it would take to visit one patient at a distance of four or five miles, and he would use much less effort in doing so. Thomas G. Atkinson, in 1916, urged a related reason for cultivating office practice; he wrote

> . . . that it constitutes a provision against the inevitable approach of that time when the practitioner can no longer live the strenuous life; when he is obliged to leave to young men the long, hard rides and the answer to the night 'phone. If, when this time arrives, he finds himself in possession of a well-built office practice, he can still remain in active and profitable performance of the work to which he has devoted his life.[51]

Not only was office practice physically less burdensome, but it also had financial advantages for physician and patient. The fee for an ordinary office visit was generally less than the customary fee for a house call. To judge from remarks by several European physicians, this economic factor played a significant part in bringing affluent as well as poor patients to the practitioner's office during the latter nineteenth century.[52] August Kreutmair, a Nuremberg physician of this period, claimed in 1887 that lower prices were a major factor in bringing the patient to the doctor.[53] A comment in 1867 by the Swiss physician Sonderegger that his friend Bänziger's "waiting room is full of glacé gloves" reveals that well-to-do patients were beginning to visit the physician's office rather than have him make a house call.[54] Although the shift to office practice may have begun somewhat earlier in Europe, the trend on both sides of the Atlantic continued in the same direction despite differences in the organization of the profession in various

countries. The shift to office practice was slower in some branches of medicine than in others—for example, in pediatrics. In Nuremberg, until the last decade of the nineteenth century it was not customary to bring children to the physician's office.[55] In certain areas of the United States as late as 1943, home visits by pediatricians were greater in proportion to office visits than was the case among general practitioners.[56] Today the house call has virtually disappeared as an element in the pattern of medical practice in the United States, a development to be considered later.

The trend to office practice was fostered as well by an increasing ability on the part of the physician to provide diagnostic and therapeutic services in his office which he could not perform in the patient's home—for example, minor surgery, gynecological treatments, laboratory and electrical tests, and visual examination by refraction. In part, this was also a generational process, as young physicians more receptive to newer modes of practice replaced older colleagues reluctant to change their accustomed ways.

Anticipation of economic advantage was a further incentive to develop office practice. The profession was overcrowded by the turn of the century; complaints were particularly common in urban centers.[57] The consequence was intensified competition. A California physician observed in 1913 that "medical competition, which includes price-cutting, and other cheapenings of our professional position, has made general practice almost impossible"; and the same year, a New Hampshire physician reported that "cut-throat work is constantly being done, and it seems to be a case of anything to get patients."[58] To achieve professional and financial success under such conditions, medical practitioners endeavored to acquire competitive advantages. The appearance, around the turn of the century, of a movement to help physicians put their practice on an efficient business basis was not a chance occurrence but a response to current problems.

In 1877 a Boston practitioner emphasized that, although medicine is not a trade, "it is an occupation by which men gain a livelihood, and therefore it should be conducted on strict business principles."[59] By 1900 this theme was being repeated with increasing frequency. In 1895 a Chicago physician urged his colleagues to recognize that, while medical practice was primarily concerned with healing the sick, it was also an activity by which practitioners earned a living and should therefore be carried on as efficiently as any business.[60] An anonymous Connect-

icut physician in 1902 stated the need for efficiency and business principles in medical practice very clearly:

> A physician cannot live on air, or hope, and, as the nature of his profession requires activity and labor, both of brain and body, he needs both brain and body sustenance. He and his family must be clothed and fed, and he is entitled to a just remuneration for his services and the medicine used. . . . The physician who is a poor collector of his just dues does not appreciate his own services at their full value, and under such circumstances his patients are very apt not to appreciate them either. In this intensely utilitarian and commercial age physicians must adopt sound business methods.[61]

In 1907 the American Medical Association urged business training for medical students, a proposal in line with the experience and needs of practicing physicians.[62] A survey of graduates of the Harvard Medical School in the classes of 1901 to 1910, carried out in 1913, revealed that many of them advocated a business course which would prepare physicians to manage a practice in an orderly, systematic way.[63] To satisfy this need, numerous books and articles offering practical business methods appeared, under such titles as *The Physician's Business and Financial Adviser, Dollars to Doctors, Building a Profitable Practice, Successful Office Practice—The Key to More Business,* and *Business Efficiency in Medicine.*[64]

When Abraham Jacobi discussed medicine as a career in 1909, he emphasized how much the general physician could do for the patient in the office. "With modern methods," he said, "the opportunities of the doctor improve, and the young doctor matures in knowledge at an earlier age than formerly."[65] At the same time, the practitioner was advised that the application of newly available knowledge and methods in office practice not only made it possible to compete more successfully but also could be more remunerative. Fees for house calls were customarily charged on a per-visit basis, with no leeway for extra charges. On the other hand, the established fee for a routine office visit was less than the price of a house call, but office practice offered more flexibility in charging for special services. "Any special work," advised Thomas F. Reilly of Fordham University in 1912, "such as opening abscesses, injecting morphine, passing sounds, electrical treatment, etc., is worth a double office fee, and most people will gladly pay it . . . Don't be afraid to ask the regular price of the visit or call. Which

shall it be? People would not employ you if they thought that your advice was of no value. So don't flinch."[66]

Office practice thus seemed to afford an excellent way to apply the newer knowledge and techniques of medicine for the benefit of both patient and physician. Furthermore, in cultivating office work, the physician might also develop some form of special practice in his office to supplement his general practice, and eventually he would perhaps become a full-time specialist. The extent to which physicians were actually engaging in this pattern of practice may be judged from the 1913 survey of Harvard Medical School graduates mentioned previously. Of 312 physicians, 36 were in general practice only, 134 carried on a general practice combined with a specialty, and 142 were full-time specialists.[67] That these figures are not unrepresentative is shown by the fact that for the period 1900–1924, less than half (forty-seven percent) of medical graduates were in general practice—and a number of these were partial specialists.[68]

Partial specialism emerged as a significant element in the structure of medical practice during the later nineteenth and early twentieth centuries. On the one hand, it was a way station for physicians who eventually became full-time specialists. Owing to the paucity of facilities for special training, the specialist, as I have noted, tended to be a man who, while carrying on his general practice, had become particularly interested or especially proficient in some limited branch of medicine. This extended and to some extent arduous pattern of training became a part of contemporary medical mores.[69] James B. Herrick, for example, was a general practitioner from 1890 to 1900, before limiting his activity to internal medicine. During that period, however, he gradually gave up treating certain kinds of cases—surgical, gynecological, and obstetrical, as well as conditions of the eye, ear, nose, and throat.[70] On the other hand, partial specialism made it possible for a considerable number of general practitioners to meet the competition of specialists on more nearly equal terms.

Institutional Competition

Whether partial specialism developed for reasons of self-interest or because of scientific advancement, this type of practice was facilitated by work in free-standing dispensaries or in hospital outpatient departments. Physicians were needed to man these establishments, and

doctors were eager to take advantage of the opportunity for professional development. As Abraham Jacobi commented in 1871, patients "serve as our material for observation and study, they furnish us with experience and knowledge . . ."[71] The studies in electrotherapy that George M. Beard and A.D. Rockwell were then carrying out on patients from the DeMilt dispensary, with which they were connected, are an instance in point.[72] To be sure, very few of these physicians actually developed into full-fledged specialists. Nevertheless, the majority acquired a certain competence in a particular branch of practice without relinquishing their general work. Furthermore, appointment to the staff of a clinic gave the general practitioner some standing as a specialist among potential patients, enabling him to compete more effectively with the specialist. Such appointments were desirable for yet another reason. A position as physician to outpatients could be the first step leading eventually to an appointment with full attending privileges in a hospital, and achievement of increasing significance as hospitals grew in importance for medical practice after the turn of the century.[73] Finally, partial specialism was also effective in blunting the edge of the initial diehard opposition to specialization by providing a partial solution to the problem of defining the relations between the general practitioner and the specialist, a problem which exercised the medical profession from 1870s to the 1930s.

Not all physicians, however, saw dispensaries and other clinics in the same light. While some viewed these facilities as rich sources of suitable, interesting cases for learning and teaching, others who practiced outside these institutions resented them as unfair competitors that took away the patients on whom they depended for a livelihood. Dispensaries and outpatient departments were medical charities providing free care to those who could not afford to pay. During the nineteenth century, particularly during its later decades, urban workers and their families depended upon medical charity—dispensaries for the sick poor and the outpatient departments of hospitals.[74] (If their slender means permitted, they might turn to physicians whose charges were low, often young doctors just building a practice.) The growth of dispensaries and other ambulatory care facilities was initially concentrated in some eastern states and cities, and it was there that the first negative reactions of the medical profession appeared, when a number of practitioners raised the issue of the "abuse of medical charity," as represented by the "dispensary evil." As early as 1868, the *Boston Medical and Surgical Journal* commented editorially on the needless

increase of free dispensary and hospital treatment, pointing out that it was a vital injury to young physicians who had to live on the small fees obtainable from "just those middling classes of the community whom the dispensary system invites to a gratuitous treatment."[75] Many poor people found private medical care a luxury they could not afford and sought medical attention where it was free or heavily subsidized. Physicians to whom such patients might otherwise have turned for medical care regarded with suspicion and hostility the provision of free care to those who in their opinion should be paying for it. As one physician defined the issue in 1877, free treatment of all who applied diverted "the legitimate profits of the ordinary practitioners."[76]

The economic attitudes of physicians were buttressed by contemporary social thought, which held that the provision of gratuitous treatment to patients who could afford to pay would demoralize and, as contemporary jargon had it, pauperize them.[77] From the 1880s onward, medical charity and its abuse were discussed with increasing frequency in medical circles, and the question became even more prominent in the first decades of the present century. Underlying all the discussion of the "dispensary evil" and the "abuse of medical charity" were conflicting issues and interests. On the one hand, there had emerged the problem of medical indigence, particularly acute in urban centers, due to the circumstance that a considerable portion of the population lived from hand to mouth and would be unable to cope with the financial and social exigencies arising from illness and possible unemployment. The provision of ambulatory medical care for these people of small means was an early stage in the process which by the early twentieth century led to a major social concern with the financing and organization of health care. From the practitioner's viewpoint, on the other hand, the facilities that provided free care were engaging in unfair competition and were depriving him of paying patients. Though some physicians recognized that there were people who could pay for most necessities yet not for medical care, most others would probably have agreed in principle if not in substance with a Philadelphia practitioner who wrote in 1903: "When people in this land, except recent immigrants, are so poor that they cannot pay a moderate fee to some struggling physician, their poverty is a crime deserving punishment rather than encouragement."[78] As a consequence, efforts were made to reduce competition from the dispensaries, clinics, and outpatient departments by insisting that economic criteria be established to determine who could or could not pay for medical care and that such

standards be applied in these facilities. Indeed, in New York state the agitation was so strong that a law was enacted in 1899 licensing dispensaries and placing them under the general supervision of the State Board of Charities. Under this law, applicants for service who falsely represented themselves as indigent could be punished, although fifteen years after its enactment no violators had been prosecuted.[79]

Attacks on the "dispensary evil" continued unabated after the turn of the century, but for the most part they remained ineffective and served mainly to give vent to the hostility felt by many practitioners. Nevertheless, the debate over the "abuse of medical charity" reflected a real problem and had significant consequences. The development of the ambulatory care facility, insofar as it drew patients away from individual practitioners, was seen as disrupting the market structure which the profession was endeavoring to stabilize. Aggravating the situation was the fact that the professional efforts of a large proportion of physicians were yielding little economic reward. Estimates of the annual incomes of some 46,000 physicians made around the turn of the century showed that 20,000 earned $2,000 or less. Another 8,000 averaged $3,000 a year. The next higher group comprised 5,000 whose incomes ranged from $3,500 to $7,000, with an average of about $5,000. This group included surgeons and other specialists in larger cities. Some 2,000 physicians with incomes over $7,000 a year were also located in large cities, and of these about 1,500 averaged $10,000 annually. Finally, the remaining 500, a group consisting of successful surgeons, gynecologists, ophthalmologists, and other specialists able to charge large fees, had incomes ranging from $20,000 upward, with the highest group of 50 averaging $45,000 annually.[80]

A judgment on the significance of these and other contemporary data on income must take account of various factors, such as urban or rural location, length of time in practice, character of the population served, and social and professional relations of the practitioner. Since living expenses in rural areas were one-half or one-third what they were in urban areas, a $1,000 practice in the country could be considered equivalent to a $3,000 practice in New York City. Furthermore, a physician entering practice generally had to count on a deficit at the end of the first year and possibly for two years. In 1894 Theodore W. Schaefer, a Kansas City physician, asserted that "the young physician, in the first five years of his practice, hardly earns his board, and his income often does not amount to fifteen dollars a month in cash!"[81] Young physicians were aware that they could not even expect to cover

their living expenses during the first years of practice. The survey of Harvard graduates covering the first decade of the present century further illuminates this situation. On the average, members of this group were not able to cover their expenses and earn a reasonable living until they had been in practice for about three years. As one physician remarked:

> The lack of small salaried jobs to enable one to live while getting a start is really the great difficulty. It is almost necessary to have capital or a wealthy wife to get a start anywhere within twenty-five miles of Boston. If there could be a system of paid dispensary or contract work similar to the Jewish Lodge system . . . , it would be a tremendous help to the future graduates who have to earn their livings from the start."[82]

Indeed, it was this situation that did lead numerous young practitioners in urban centers into "lodge doctoring."[83]

Thus, during the last decades of the nineteenth century and the early years of the twentieth there were numerous practitioners whose economic existence was not entirely satisfactory and even precarious. These physicians considered themselves injured by unfair competition from various sources, and it was within this context that a very large segment of the medical profession developed a hostile attitude toward the institutions and facilities that they regarded as undermining their ability to compete in the medical market. Arising from the desires to retain professional autonomy and to maintain a stable medical market, this hostile attitude would be carried over in the reactions to various alternative forms for organizing and paying for medical care: contract practice, lodge practice, health department clinics, pay clinics, group practice, and health insurance. To many physicians it was clear that the profession could not hope to improve its situation and to achieve its objectives unless it upgraded the standards for admission to its ranks and limited the number of those who entered its ranks. The achievement of these goals early in this century made it possible to stabilize the medical market for a time and to come to terms with the circumstances and institutions that then affected the character of medical practice.

Jacob Dunham

TABLE OF FEES,

AND

RATES OF CHARGING,

For sundry articles and services in Medicine and Surgery, as agreed upon and established, June 12, 1810, by the Medical Society of New-Jersey, for the government of its Members.

VISITS.

Visits per mile or shorter distance	$ 00 25
do. In the night, exclusive of medicine.	1

CONSULTATIONS.

Every first visit and opinion by the consulted Physician or Surgeon, exclusive of travelling fees,	4
Do. of the attending Physician,	4
Every succeeding advice by consulted Physician,	1 25

CHIRURGICAL OPERATIONS, &c.

(Exclusive of visits and travelling charges.)

Phlebotomy,	25

57430

Pre-bellum medical societies sought to control the dangers of competition by agreeing on a minimum fee; such schedules were often ignored (Medical Society of New Jersey, Fee Bill, 1810).

LARGE FEES

AND HOW TO GET THEM

A BOOK FOR THE PRIVATE USE OF PHYSICIANS

BY

ALBERT V. HARMON, M.D.

WITH INTRODUCTORY CHAPTER BY

G. FRANK LYDSTON, M.D.

W. J. JACKMAN, PUBLISHER

121-127 PLYMOUTH COURT

CHICAGO

Turn-of-the-century physicians were still participants in a highly competitive medical market. They provided the clientele for a number of such how-to-do-it manuals.

All photos courtesy of the Historical Collections, College of Physicians of Philadelphia

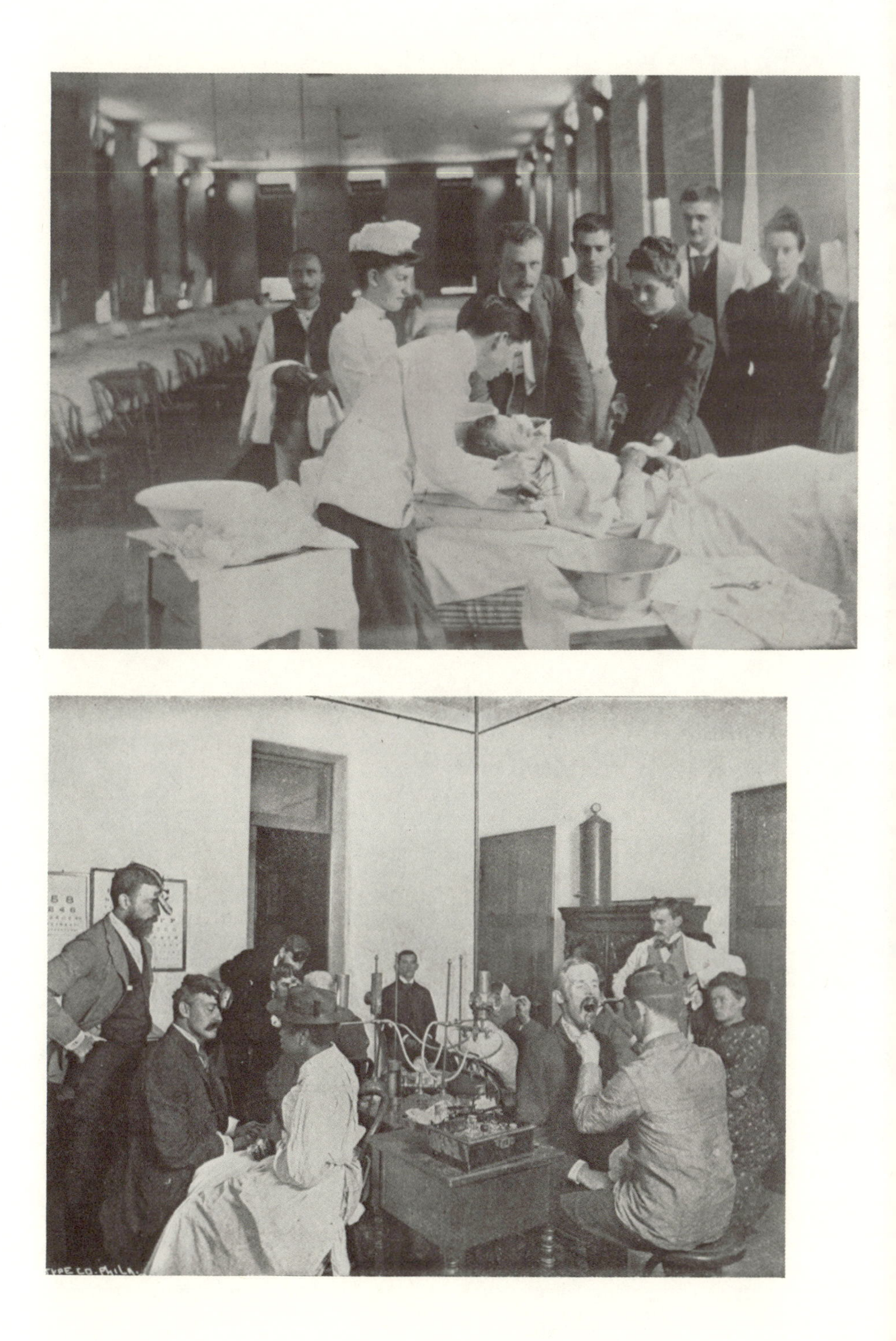

LEFT TOP: *Clinical teaching began to find a place in the late nineteenth-century hospital (Philadelphia General Hospital, 1889).*

LEFT BOTTOM: *The demands of a growing specialism imposed a new shape on America's hospitals and dispensaries (Philadelphia Polytechnic, Annual Report, 1895).*

X-ray and chemical pathology laboratories added to hospital costs, yet solidified the physician's scientific image (Philadelphia Polytechnic, Annual Report, 1896).

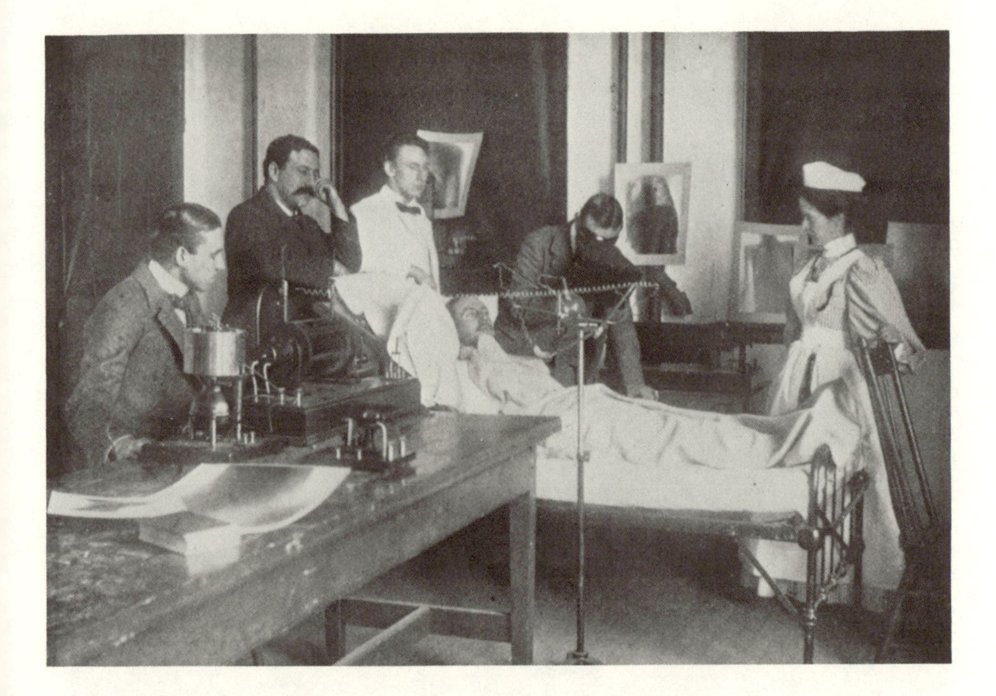

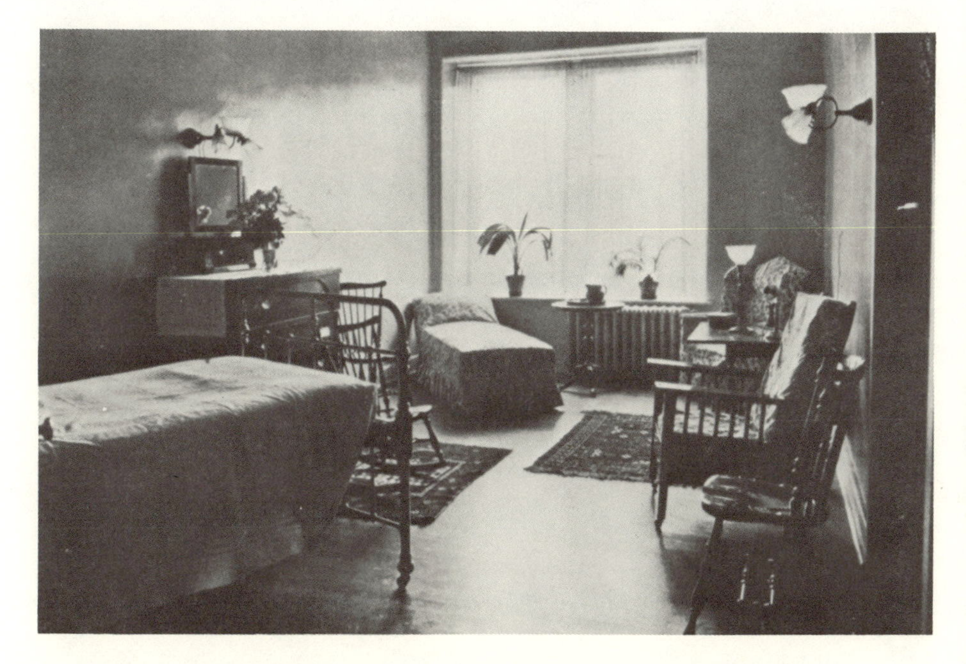

In the era of the Flexner Report, voluntary hospitals sought to attract paying patients by providing and publicizing their "home-like" facilities (Hospital of the University of Pennsylvania, Annual Report, 1910).

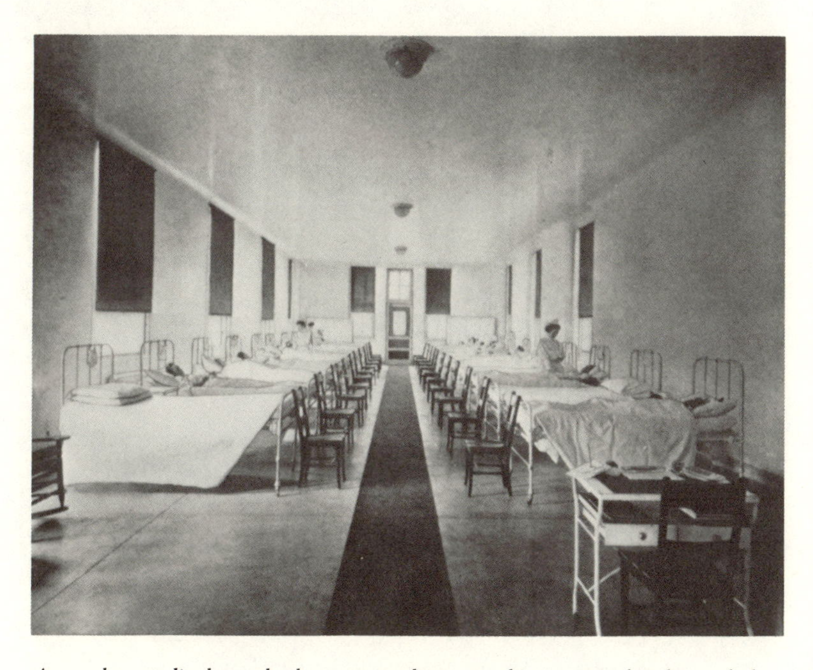

A regular medical ward; the contrast between this image of order and the private patient's room is apparent (Jefferson Hospital, Annual Report, 1915).

2

MEDICAL SCIENCE, PROFESSIONAL CONTROL, AND THE STABILIZATION OF THE MEDICAL MARKET 1910–1940

What a physician does in his practice, the kinds of patients he treats, what can be done for them and in what circumstances, depend to a large extent on the practitioner, but there are other conditions over which he has little or no control that nevertheless affect the practice of medicine. An examination of medical practice in the city of Providence, Rhode Island, as described in 1915 by Stephen A. Welch, president of the Rhode Island Medical Society, illuminates the influence of such factors.[1] A primary factor is population and the changes which it undergoes in numbers and character. In 1865, more than half (56 percent) of the population of Providence was of American parentage and 44 percent of foreign parentage; by 1914, only 28 percent were American and 72 percent were foreign.

These trends reflect three intertwining developments in American society during the last decades of the nineteenth century and the early decades of the twentieth, developments which produced the economic and social world in which medicine was practiced just before and after the World War I. Industrial expansion, urban growth, and a flood of immigration combined to produce the social conditions associated with poverty and ill health. Between 1860 and 1910, the urban portion of the population of the United States rose from 19 to 45 percent of the total, due chiefly to the immigrants who poured into the cities and industrial towns, where workers were in demand. Thus, around the turn of the century and in the subsequent years, the growing cities of the United States had an increasing number of people whose economic and social circumstances radically affected the practice of medicine, for when they were sick they sought to obtain medical care as cheaply as possible from institutions or nonmedical personnel.

A second factor affecting practice is the number of physicians actively providing service. Providence directories for the fifty years from 1865 to 1914 show that practitioners increased in number from 116 to 391, though there was little change in the physician-population ratio: there was one practitioner to 496 people in the decade ending in 1874, and one to every 591 in the decade ending in 1914. During this period, however, there was a shift in the composition of the practitioners. Before 1882, Providence directories divided practitioners into three groups: members of the Rhode Island Medical Society (regular physicians), homeopathic physicians, and a third mixed group of irregulars lumped together as "other physicians."[2] For many years, the third group outnumbered the regular and homeopathic physicians together. By 1914, however, the irregulars had declined to one-sixth of the total number of practitioners, despite the addition of sixteen osteopaths and eleven Christian Scientists.

At the same time, there was a steady trend toward specialization. In 1865, one physician in forty was a specialist; in 1885, one in thirty; in 1905, one in nine; and in 1913, one in six. To judge from an observation by a Harvard graduate practicing in Providence in the early years of the century, a considerable number of the specialists did not limit their practice completely to a specific branch of medicine. He noted that "most of the surgeons do medical work. The so-called gynecologists do medical work, and the medical men do obstetrics and minor surgery. Everybody is afraid he will miss something. Hence it is hard for a younger man to limit himself to the work he has trained

himself for, without starving."[3] Another trend, paralleling that of specialization, was the growing number of medical practitioners associated with hospitals. Beginning with one hospital in 1847, the number of public and voluntary hospitals increased to seven by 1914. In addition, a number of private hospitals were established. Five physicians in Providence had hospital connections in 1865; by 1914 the number had risen to 185, a total including not only practitioners but also consultants and administrators. Twenty-five percent of the practitioners in Providence did some hospital work in 1868; by 1914, the proportion had risen to 50 percent, and there were indications that this trend would continue upward.

Just as more physicians associated themselves with hospitals and devoted a steadily increasing proportion of their time to hospital work, so also a growing proportion of patients turned to hospitals for medical care. For the decade ending in 1874, one resident of Providence out of every forty-seven sought medical attention in a hospital. During the decade ending in 1894, the proportion was one in nineteen, and during the ten years ending in 1914, it was one in ten. Another trend indicative of the increasing use of hospitals was the nature of obstetrical care. In 1893, one delivery in forty-five took place in the Providence Lying-In Hospital. By 1903, the ratio was one in twenty-eight, and in 1913, one in eight.

Finally, an increase in the number of physicians employed by the Providence Health Department is an early indication of an emerging trend toward diversification of professional activities. For a number of years before 1904, three physicians had been sufficient to handle the work of the Health Department. In 1904, the number was increased to six, and ten years later it had risen to eighteen. Though the motivation for such employment may have been partly to supplement income derived from practice, it is also an early indication of an emerging heterogeneity in the profession due to more opportunities outside of clinical practice. This point is supported by the survey of Harvard medical graduates made in 1913. Somewhat less than eight percent were not clinical practitioners. Of these, the largest numbers, twenty-three and fourteen respectively, were in clinical laboratory work or in public health. A scattering of others were in hospital administration, military medicine, research, or insurance.[4]

Public Health and Medical Practice

The state of medical practice in Providence in 1915, and the changes that had taken place during the preceding fifty years, reflected nationwide trends. The growth of hospitals and their increasing use by physicians and patients, as well as the expansion of public health activities, were expressions of a changing medical and social environment. Advances in medical science and technology during the late nineteenth century made it possible to prevent, diagnose, and treat many diseases, thus greatly enhancing the effectiveness of medicine. Light was cast on the etiology of communicable diseases and traumatic infections by demonstrating the specific causative organisms of many of them, including cholera, diphtheria, dysentery, erysipelas, tetanus, tuberculosis, typhoid, and wound sepsis. As a result, by the early decades of the twentieth century, there was a solid basis of knowledge and technique for the diagnosis, prevention, and control of a number of infectious conditions, derived mainly from bacteriology and immunology but increasingly from biochemistry. Although Americans contributed only a limited degree to the growth of microbiological knowledge during this period, they were alert to its practical applications, an awareness which gave rise to a new public health unit, the diagnostic laboratory for the application of bacteriology and immunology. Recognizing that the application of microbiology held rich promise for the control of communicable disease, almost every state and practically all large cities in the United States had established a public health laboratory within a few years after the turn of the century. Through these laboratories, health departments to a considerable extent took over the task of diagnosing communicable diseases, and in order to control these conditions provided free biological products to physicians in clinical practice and to public health officers.[5] The growing influence of microbiology and the salient importance of the public health laboratory are clear from the establishment in 1899 by the American Public Health Association of a Section on Bacteriology and Chemistry (later the Laboratory Section). Upon its creation, almost 100 persons joined the Section.[6] This event may be seen as symbolizing the emergence of a trend that developed with increased momentum in the first three decades of the present century, reflecting the widening scope and growing activism of the new public health based on an expanding knowledge of disease prevention and the promotion of health.

Initially health department programs had been relatively modest,

concerned with the effects of the physical environment on the health of the community. Application of the newer knowledge of disease meant a major shift in the orientation of community health action, a shift of attention from environmental aspects of the community to the people within it. Contemporary observers noted the implications of this reorientation. According to Lewellys F. Barker, professor of medicine at Johns Hopkins University during the early years of the century, "private practitioners of medicine" were "most useful in the diagnosis and treatment of disease and in the advising of individuals and families regarding disease prevention and health promotion." However, he added, though "it is only relatively recently that the public-health services have entered the fields of diagnostic and curative medicine (left formerly entirely to private practice) . . . experience with public medical and nursing services for maternal and child welfare; for the health of school children; for the early recognition and proper treatment of certain infectious diseases . . . would indicate that public-health workers will from now on become even more active in diagnostic and curative domains."[7]

Although a *modus vivendi* was achieved between private practitioners and public health physicians, confrontations occurred at various times, ranging in severity from simple disagreement to violent controversy. A number of such conflicts reveal starkly their economic basis. On January 19, 1897, the New York City Board of Health adopted an ordinance requiring the reporting of cases of pulmonary tuberculosis by physicians. It did not take long for the storm to break. In February 1897, an editorial in the *Medical Record,* which took a leading part in opposing the action of the health authorities, struck directly to the heart of the controversy. After pointing out that "there is no objection to the reports of pulmonary cases for statistical purposes," the editor continued:

> It is, however, the extra missionary work assumed by the board which is the ominous and threatening quantity in the equation—the desire to assume official control of the cases after they have been reported, thus not only by means of alarming bacteriological edicts, directly interfering with the physician in the diagnosis and treatment of the patient, but in the end, by the creation of a public suspicion of his ignorance, possibly depriving him of one of the means of a legitimate livelihood.[8]

Other instances of opposition to the activities of public health agencies reveal with equal clarity a concern with the economics of practice.

Baby health stations were set up in New York City in 1910, at first with private funds and later with public support, to sell bottled pasteurized milk and to teach mothers how to care for their infants. Josephine Baker, first director of the Bureau of Child Hygiene in the New York City Health Department, relates that after such stations had been established and were "doing well in the Brownsville section of Brooklyn, a petition was forwarded to my desk from the mayor's office, signed by 30-odd Brooklyn doctors, protesting bitterly against the Bureau of Child Hygiene because it was ruining medical practice by its results in keeping babies well, and demanding that it be abolished in the interests of the medical profession."[9] Opposition to maternal and child health programs was not limited to New York. In the 1920s, activities to promote the welfare of mothers and children in Massachusetts were acceptable to most physicians only with the provisions that government did not become involved in any aspect of medical care and there was no interference with the relationship between physician and patient.[10]

In objecting to public health activities, physicians in private practice were opposing what they considered actual or potential encroachment on the area of medical practice and its economic aspect. Linked with and buttressing the economic interest of practitioners were other values and attitudes, such as an insistence on the confidential nature of the patient-doctor relationship, a highly negative valuation of government, and a fear of centralized public authority. This stance is not surprising in view of the strong emphasis in our culture on individualism and the associated view of health as a private matter. When coupled with a situation in which physicians developed and maintained a practice in a highly competitive social environment, any act by government which appeared in some way to interfere with the access of physicians to patients in the medical market, or to divert patients to other sources of medical care, was a hostile act to be condemned and opposed as unfair competition.

This did not mean, of course, that private practitioners would not accept measures intended to protect the public health. Such acceptance, however, tended to be hedged about by conditions to safeguard the area of private practice, an attitude that continued to characterize most private practitioners in the decades preceding the outbreak of World War II. Based on their studies of a midwestern town (Muncie, Indiana) during the decade 1925–1935, the Lynds commented in 1937 that "the attitude of the majority of Middletown's doctors

regarding private practice and public health facilities remains substantially that described in 1925. All proposals to develop clinical facilities of whatever type in Middletown still operate within the straitjacket of insistence by the majority of the medical profession that nothing shall be done to make Middletown healthier that jeopardizes the position of the doctors."[11] As long as preventive and curative medicine were kept relatively separate, an accommodation could be reached between medical and public health practitioners. This situation was unstable, but it could be maintained as long as no major disturbing factors intruded. To the extent to which public health limited its province of action by maintaining relatively distinct boundaries, coexistence endured. Accommodation was possible because the problems with which public health practitioners were concerned (vital statistics, environmental sanitation, communicable disease control, laboratory services, public health nursing, and health education) did not impinge crucially on private practice, or affected specific groups of the population, most of them poor and unable to pay for private care, or because the official health agency provided services useful to the private physician.

It is true that there was a certain amount of overlap in disease prevention and health promotion, so that physicians in private practice and public health practitioners provided some of the same services. Immunizations could be administered in private practice, by general physicians or pediatricians, or in clinics run by health agencies. Similarly, periodic checkups of pregnant women, infants, and children could be provided privately as well as by community agencies. Clinical practitioners could supervise patients with tuberculosis or venereal diseases, but the cases had to be reported, and checks were made by the official health agency to try to avoid transmission. This was also the case with such acute communicable diseases as diphtheria, scarlet fever, or typhoid fever. Although the participants in these activities did not always see eye to eye, and conflicts sometimes occurred, on the whole the accommodation worked.[12] Basically, this pattern of activity remained unchanged until after the Second World War.

The Impact of the Hospital

At the same time that private physicians were endeavoring to protect their interests and arrive at a *modus vivendi* with public health agencies, they were also coming to terms with other institutional develop-

ments. The effect of advances in medical science on the public health agency was paralleled by their impact on the hospital. During the nineteenth century, hospitals were considered charitable institutions providing medical care to the needy, and public views of hospital care reflected this attitude.[13] The well-to-do were treated at home, and the poor feared admission to a hospital as a way station to a pauper's funeral. As John Brooks Wheeler relates, when he was appointed to the attending surgical staff of the Mary Fletcher Hospital in Burlington, Vermont, in 1883, it had only about thirty beds, and it was very rare for all to be filled. "There was a feeling," Wheeler observed, "that nobody but paupers were treated at hospitals. Paupers and people who, though self-supporting were too poor to pay much of anything for medical attendance, thought that they would be 'experimented on'. They were filled with the idea that hospitals existed for the sole purpose of 'experimenting' on people; but I never could find out just what they thought 'experimenting' was."[14]

By the second decade of this century, however, the situation had changed completely, due largely to the introduction into surgical practice of antiseptic techniques and then their replacement by aseptic procedures. The consequence of this development was an unprecedented expansion of the scope of surgical activity, a decline in operative morbidity and mortality, and an increasing acceptance of institutional care as fear of the hospital began to break down. Surgery and obstetrics in particular created a growing demand for hospital care for all classes of the population, with the result that hospitals no longer provided principally for the destitute, and paying patients were cared for in greatly increased numbers.

This trend was further strengthened and intensified by the development of clinical pathology and the introduction of diagnostic laboratory procedures. In the United States, clinical pathology and its application to medical diagnosis began in the late 1870s in New York City, where it was practiced by William H. Welch, T. Mitchell Prudden, and their students. According to George Dock, at the time that he entered the medical department of the University of Pennsylvania in 1881, hospitals had no laboratories.[15] This situation is exemplified by the circumstance that when William Osler was clinical professor at the University of Pennsylvania Hospital in the 1880s, he had the only microscope in the hospital as well as the first blood-counting apparatus in Philadelphia.[16] David Edsall was a medical student at the University of Pennsylvania and an intern at the Mercy Hospital in Pittsburgh

from 1890 to 1894. At that time, he recalled later, "the diagnostic use of bacteriological methods was already possible in a few ways, but was little comprehended or employed. For instance, with that great disease, tuberculosis, ever about and in many cases suspected to be the cause of obscure symptoms, my able chiefs never once asked for an examination for tubercle bacilli during the whole time that I was a hospital intern, though that was already feasible."[17]

Nonetheless, change was taking place. Following his return to Philadelphia from Europe in 1887, Dock fitted out a laboratory at the University Hospital, with funds provided by William Osler and John Musser. In 1891, when he went to the University Hospital at Ann Arbor as head of medicine, he was able to develop a proper unit for clinical pathology, despite cramped quarters. During his seventeen years at Ann Arbor, Dock established the practice of making routine laboratory examinations. On admission all patients were given complete urine and blood tests, and in almost all cases the stomach contents and stools were examined. In addition, all sputa, vomitus, exudates, and fluids obtained by puncture were examined. Dock was also allowed to take patients with pulmonary tuberculosis for diagnosis, and "these patients," he said, "served as my first subjects for fractional stomach tests." Also, "for all patients with appendicitis, or suspected peritonitis, immediate leucocyte counts were made and surgical consultations held."[18]

During the same period, between 1897 and 1910, when he left Philadelphia, David Edsall was opening up the field of laboratory investigation of clinical disease, using relatively new chemical techniques. Much of this work took place at the William Pepper Laboratory of Clinical Medicine, opened in 1895, where Edsall studied problems of metabolism.

Similar developments were taking place in both academic and nonacademic hospitals. Lankenau Hospital in Philadelphia is reputed to have created a bacteriological and chemical laboratory in 1889 and the city's first X-ray laboratory in 1896.[19] When Harper Hospital in Detroit reorganized its staff in 1884, establishing separate departments, it had two pathologists and a microscopist to deal with pathological examinations; when a new table of organization took effect in 1910, the department of pathology and research comprised three divisions; bacteriology, pathology, and radiography, with adequate facilities.[20] At the Boston City Hospital clinical pathology began with the establishment of a primitive laboratory for the diagnosis of diphtheria.

In 1891, with the appointment of W. T. Councilman and Frank B. Mallory as pathologist and assistant pathologist respectively, a department with adequate facilities was created. In May 1896, Francis H. Williams, a staff physician at Boston City, set up an X-ray laboratory for treatment and diagnosis; nine years later, in 1905, a formal department was established.[21] The impact of the medical advances around the turn of the century is also reflected in the action of the board of administrators of the New Orleans Charity Hospital in 1894 urging the expansion of the pathology department and the establishment of a bacteriological section. In 1906, the board reported the procurement of "an electrical apparatus for the use of the Roentgen Ray in the diagnosis of obscure surgical conditions."[22] In 1896, a new clinical laboratory was built at the Johns Hopkins Hospital, to be used in teaching students as well. At the time, Henry M. Hurd, superintendent of the hospital, boasted that "no more convenient or serviceable clinical laboratory has been erected in connection with any hospital or medical school."[23]

The consequences of these trends soon became evident in several ways. After 1900 hospital construction boomed and the number of hospitals increased rapidly. In 1875 there were 661 hospitals in the United States, and as late as 1900 there were only 2,070. From 1900 to 1929 the rate of establishment of hospitals was almost 200 per year. In 1909 the American Medical Association (AMA) listed 4,359 hospitals with an estimated bed capacity of 42,065, an average of 96 per hospital. By 1923 there were 6,830 hospitals with 755,722 beds, and in 1928 the 6,852 hospitals registered by the AMA had a bed capacity of 892,924, or an average of 130 per hospital. In subsequent years, the number of hospitals declined but those in existence grew in size, so that by 1932 they had a total of more than a million beds.[24]

Not only did hospitals increase in number and size, but they also became more complex in their organization. After 1870 large hospitals increasingly recognized the existence of specialties by creating special departments, a process already noted. By the first decade of the present century, the results of this trend were clearly evident. In 1901, for example, plans for the new building of the Mt. Sinai Hospital in New York included provision for departments of general medicine, general surgery, gynecological surgery, ophthalmology and otology, neurology, dermatology and pediatrics. There were also to be separate departments of anesthesiology, pathology, radiology, and physical therapy (hydrotherapeutics), as well as a service for tuberculosis, a ward for

accident cases, and an outpatient unit.[25] The table of organization which took effect at the Harper Hospital in Detroit in 1910 comprised five main departments: medicine; surgery; gynecology and obstetrics; eye, ear, nose, and throat; and research. However, each of these departments was to be subdivided, thus opening the road to a more advanced stage of specialization. Medicine, for example, had five divisions: internal medicine, infectious diseases, gastroenterology, neurology, and dermatology. Surgery comprised general and special surgery, the latter including gynecology, orthopedics, neurosurgery and thyroid surgery, genitourinary surgery, and proctology. In effect, the Harper Hospital medical staff was to consist of specialists, with no place for the general practitioner, a policy that led inevitably to future problems.[26]

The creation of special departments furthered the growth of specialism by providing an institutional structure for the development and application of knowledge as well as for the training of personnel. The trend toward the institutionalization of medicine and the associated continuing advance of specialism were given further impetus by the expansion of medical knowledge and its application. Research on endocrines, vitamins, chemotherapeutic agents, metabolic processes, and immune states had important consequences for prevention, diagnosis, and therapy. X-ray procedures for diagnosis and therapy became a part of medical practice in the hospital and the office. By 1912, physicians entering practice were being advised that an X-ray machine was "a valuable adjunct."[27] When Sinclair Lewis has Arrowsmith open his office in a small town around 1910 and purchase "a small Röntgen-ray outfit" after one year in practice, the detail is historically accurate.[28] Soon after the work of Carl Landsteiner in 1901 on blood agglutinins and blood groups placed blood transfusion on a scientific and reasonably safe basis, it became a common hospital procedure, which was further advanced when Oswald H. Robertson demonstrated in 1918 that citrated blood could be preserved under refrigeration as long as a month.[29] Other diagnostic and therapeutic modalities—electrocardiography, the measurement of the basal metabolic rate, intravenous therapy, fever therapy, endocrine products such as insulin, thyroid and sex hormones, vitamins, and chemotherapeutic agents such as the heavy metals and the sulfonamides—expanded the horizons of medicine and enhanced the power of the physician to act effectively.

The increasing complexity of medical care, particularly within the hospital setting, coupled with the growth of the hospitals themselves,

also led to an expansion of auxiliary services and personnel. Among hospitals registered with the AMA between 1929 and 1933, slightly over 4,000 had clinical laboratories, a somewhat larger number had X-ray departments, and about 2,000 had dental and physiotherapy departments. About 1,000 had social service units.[30] As hospitals accommodated more functions, requiring additional personnel, facilities, and equipment, their organization, records, and business procedures also grew increasingly complex and their operation more costly.[31]

A consequence of these developments was that the hospital moved inexorably from the periphery of medical care to its center and became more and more indispensable for the practice of good medicine. Physicians spent more time in hospital service, and a hospital connection became essential to successful medical practice. By the mid-1920s, practitioners in large centers such as New York and Chicago were spending up to 30 percent of their time in hospitals and clinics.[32] A survey of private practitioners in Philadelphia in 1929 revealed that a full-time specialist spent on the average about 36 percent of his time in hospital practice. Out of a work week of about fifty hours, sixteen were devoted to hospital practice and nine to clinics.[33] Thus, during the decades preceding World War II, the hospital was becoming a workshop for the highest type of professional practice, and the physician was being brought into a more intricate relationship with this institutionalized collective enterprise. The changing relationship between the practitioner and the hospital both within and outside the institution had a profound impact on the medical market.

The character of this impact and the response it evoked were to a considerable extent determined by the structure of the voluntary nongovernmental hospital inherited from its eighteenth-century origins in Great Britain and the United States. Initially the hospital was an autonomous institution, managed by its own governors or trustees and staffed by physicians and surgeons who gave their services gratis. In return, the medical staff was enabled to use suitable and interesting cases for teaching and research. Later, as the public began to accept the advantages of hospitalization, practitioners began to use the hospital for their private patients. Hospitals initially accepted these paying patients as a courtesy to the medical practitioners who were providing free care for indigents. When private patients were first admitted to hospitals, they entered as guests who paid for their accommodations. In 1868, St. Vincent's Hospital in New York City had "several spacious and comfortable private apartments, where ladies and gentlemen, espe-

cially strangers in the city, who may chance to be stricken by sickness during their sojourn, can find all the conveniences of home united to the advice and services of the first physicians and surgeons of the city; and what is still more important, the kind and devoted attention of the Sisters, whose great study is the comfort and welfare of those under their care. The revenues arising from the admission of pay-patients is the principal means of support."[34] Fifty years later, the policy of selling services rendered by hospital personnel and facilities to paying patients had been generally accepted, so that by 1932, as Rufus Rorem noted, "the paying patients of private physicians have become the most important source of the revenue for the non-governmental hospitals of the United States."[35]

Though the physician was needed for the provision of medical care in the hospital, he had no financial or administrative responsibility beyond the maintenance of professional standards of competence and performance. For some time, the physician retained his position as an independent entrepreneur who used hospital facilities for his own purposes and conducted his business within the institution. This situation was possible as long as there was a sharp separation between paying and nonpaying patients. However, as the advantages of medical care in the hospital received wider recognition, new patterns of utilization emerged, leading to several kinds of competition between the hospital and the private practitioner. With regard to paying patients, separate charges for hospital care and for the services of the attending physician could at times result in sharp competition for a share of the patients' resources, depending on the length of stay or type of accommodations in the hospital.

More significant was the threat which general practitioners, especially those just beginning to practice and those who had no hospital connections, saw in the care given by hospitals in their outpatient departments or wards to persons who might otherwise have been private patients. Complaints of hospital abuse were essentially a prolongation of the older controversy of dispensary abuse. Typical was the view expressed in 1910 by William E. Darnall, a New Jersey physician, at a symposium on the economic influence of hospitals:

> Since the rapid development of hospitals, not alone in the large centers but also in most cities of smaller size, there has been added to the usual difficulties of keen competition, high cost of living, youth, inexperience, lack of adequate acquaintance and the other drawbacks the young practi-

> tioner has to face—the serious competition of hospitals and dispensaries. It used to be that when the young physician began his professional career his practice was made up of the poorer classes, the slum patients and those charity patients whom some well-established practitioner turned over to his care, because of not having time to give to them. While it might have been true that they did not remunerate him richly in money they gave him a start. . . . As the situation now is, this class of patients goes straight to the local hospital or dispensary, receiving free treatment, and daily encouragement in the pauperizing not only of themselves but the profession too. The young physician waits and waits for work while the hospital eats up his bread and butter.[36]

As other physicians were aware, however, for the patient there were definite advantages in hospital medicine, in terms of the cost and the quality of care. At the same symposium in 1910, William F. Waugh, dean of the medical school at Loyola University, explained that patients resorted

> to the hospital for two reasons—to secure cheaper attendance or better attendance. Those to whom the former appeal is most potent, feel that when they pay the charges of the institution they are entitled to board, nursing, the use of such appliances and apparatus as the hospital affords, and the service of the intern. To a certain proportion of the paying patients, varying with the institution, the additional expense of a paid medical adviser seems under ordinary circumstances superfluous. It is only when something extraordinary in the way of treatment is required that they look with equanimity on a charge above the regular weekly payment. . . . Even when the attending physician receives fees for his visits, these are needed less frequently on account of the service of the unpaid interns. The case as such is less remunerative to the physician, or the patient would not care to go for treatment to the hospital.[37]

Neither the hospital nor its outpatient clinics created the problem discussed by Darnall and Waugh. It arose basically from the existence of a large group in the population, predominantly wage earners, whose economic situation was precarious and for whom illness could be catastrophic.[38] Even under favorable circumstances many of these people found it difficult to pay moderate amounts for medical service. Periods of unemployment due to economic depression or to seasonal work made their situation worse.[39] Industrial development in the United States produced unemployment at periodic intervals, resulting in the panics of 1893, 1907, 1914, 1920, and of course the Great

Depression that began in 1929. Moreover, in the 1920s, improved methods of production, the introduction of labor-saving machinery, and new forms of industrial organization produced a large number of unemployed workers.[40] Consequently, throughout the period from the turn of the century to the outbreak of World War II, a considerable proportion of the American population, especially in the cities, sought free or subsidized medical care from institutions such as hospitals and clinics as an alternative to private care. An analysis in the mid-1920s of clinic visits in fifteen large cities revealed that the group receiving ambulatory medical care probably represented 20 to 25 percent of the population receiving medical attention in those communities.[41] The increasing use of these facilities is reflected in the absolute numbers of individuals who visited outpatient departments as well as in the proportion of persons who did so. In 1921 an estimated 3 million persons sought patient care in such facilities; by 1929 the estimate was 6.6 million. Proportionately this increase is even more striking. In 1921 it was estimated that one American in thirty-five visited a hospital outpatient department. By 1929 the proportion had risen to one in eighteen and in 1933 to one in thirteen.[42] No doubt the latter figures reflect Depression conditions.

Paying the Doctor's Bill

The trend represented by these figures was a consequence of the increasing divergence between the economic needs and expectations of medical practitioners and those of large numbers of actual or potential patients. It is indicative of the emerging problem of financing and organizing personal health services which was to have a major impact on the structure of medical practice. Contemporary observers were aware of the problem and its implications. In 1910 Frank F. Dow, a physician from Rochester, New York, pointed out that

> there is a large class of self-respecting industrious people whose income in health affords only a fair yet respectable living. When sickness of a character requiring trained nurses, or hospital care, comes, they are unable to provide payment in advance for either because the expense of nurse or hospital equals or exceeds the average wage. The suggestion of charity is properly repugnant and often unnecessary. . . . According to existing rules for pay patients this large class of industrious and frugal people are practi-

> cally denied hospital or trained assistance. To provide for such as these is the great problem before society at the present time.[43]

Several years later, in 1914, Michael M. Davis, director of the Boston Dispensary, noted that there was a gap in services for persons of moderate means. Since they were self-supporting, they were not eligible for the free medical care given to the indigent, yet they were frequently unable to pay for the services of specialists or to bear the burden of prolonged illness.

> Dispensaries and out-patient departments are ordinarily supposed to treat only "the poor", as are the ward beds in the hospital. "The poor" and the well-to-do thus have easy access to the best medical service today. The self-supporting middle classes, including the great majority of the population of cities and towns, have no access to such service; nor have rural communities. We may expect in the near future to see the need for better medical service to this middle class met on an organized basis. Groups of physicians representing different and complementary specialties will probably organize in groups, so as to make high-grade, cooperative diagnosis and treatment accessible to those who call themselves neither "worthy" nor "poor", and who can pay moderate fees. This will not mean cheap practice, but organized practice.[44]

What Davis envisaged was an organized unit based on group practice. By 1918 he had gone beyond the idea of the dispensary or the outpatient department as an institution for medical charity, or as a place to obtain medical experience, and had developed a conception of the ambulatory care unit as "an organization for the efficient rendering of medical service, either curative (clinical) or preventive (public health) service." Just as the hospital had become a large-scale organization requiring a more explicitly structured division of labor and more efficient management, so a similar change was necessary for the provision of ambulatory health service. He went on to observe:

> The cooperative work of specialists in the modern hospital has impressed many thoughtful physicians with the advantages of the joint use of expensive laboratories and other equipment, and of the opportunity for consultation and team play which can be best afforded when the various specialists are at work in the same building at the same time. The Dispensary represents the same principles of organization applied to ambulatory instead of to bed cases. Can we not, in the Dispensary, test out the possi-

> bilities of organized medical service, as distinguished from the traditional individualistic service of private practice?[45]

Shortly thereafter, beginning in 1921, this approach was developed in New York at the Cornell Clinic, which provided a model for others. By the early 1920s Davis had already created a model of comprehensive medical care provided through organized group practice and supported by pooling of fees from patients, and he had demonstrated its practicality.[46] This variation from the traditional pattern of medical practice based on the individual practitioner foreshadowed a line of development which emerged more prominently after the Second World War in the form of prepayment group practice and more recently as health maintenance organizations.

Location of Medical Practice

Though Davis had conceived of group practice as a way of bringing the benefits of modern scientific medicine to persons of moderate means, the concept was not limited in its application, and it was adopted more generally after 1918 due in part to experience with cooperative military practice during the First World War. There had been a few private group clinics before the war, most notably the Mayo Clinic. By 1930 there were about 150 such groups in the United States, in which approximately 1,500 to 2,000 physicians were engaged in medical practice, with a marked concentration in the Middle West and the West, and almost none in the eastern states.[47] The majority of physicians in such groups were specialists, which is not surprising since the increasing specialization of medicine appeared to lead logically to a need for coordination of medical services.

Despite its logic, specialist group practice in its most developed form was not acceptable to a large number of practitioners, who preferred a looser, more autonomous type of relationship with their colleagues. In fact, a salient economic trend in private practice during the period between the two World Wars was the shift of the physician's office from his home to a professional or business building, generally in the downtown business area. This development was not local but national. In the late 1920s almost one-half of the 500 practitioners in Dallas, Texas, had offices in one building. In Memphis, Tennessee, about 100 physicians rented space in a building erected by a local hospital and

situated close to four large general hospitals.[48] Sanford R. Gifford, a Chicago physician, has recalled that from 1917 to 1929 the office occupied by his father, a practitioner in Omaha, Nebraska, together with his associates, was located on the sixth floor of the Brandeis Theatre Building, an office building with elevators, "located one block from the busiest corner in town."[49] An inquiry in 1932 into the distribution of physicians' offices in Philadelphia revealed that doctors were located largely in the downtown office and business center and along the main transportation arteries.[50] James H. S. Bossard, the sociologist who directed the study, noted that this was a natural development related to the growth of the city and the development of its transportation system. A similar pattern could be seen for the churches and other services operating on a private basis. What this meant was that as cities grew and their populations moved within them, medical practitioners followed their paying patients.

Some physicians used the facilities and equipment of hospitals in the care of their patients, but this was relatively uncommon. A more usual arrangement was for groups of physicians to reduce their overhead costs by jointly utilizing office space, waiting rooms, and equipment, as well as technical and clerical personnel.[51] Such informal arrangements were facilitated by locating private practices in buildings where facilities and expenses could be shared, while retaining individual financial relationships with patients.

This trend was a response not only to changes in medicine but also to shifts in land use following the First World War. Large buildings were the most distinctive feature of nonresidential construction of the postwar period, and the office building was the most conspicuous development of this type.[52] Data on office buildings constructed in Chicago between 1901 and 1930 are indicative of this tendency. During the period 1901–1910, Chicago had twenty-nine office buildings, with a total floor space of 4,001,822 square feet. Between 1911 and 1920, the number of buildings increased to thirty-eight, with a floor area of 5,530,572 square feet. However, during the following decade, 1921–1930, the number of office buildings more than doubled to a total of 80, while the total floor space increased almost two and a half times to 13,283,339 square feet.[53]

Employing a principle initially embodied in the department store in the 1890s, hospitals and physicians began to concentrate and to conduct their operations in fewer but larger buildings. By housing related services under a common roof and by substituting vertical for

horizontal transportation, office or professional buildings made possible a saving of time for practitioners and their patients. Proximity in such situations also enabled greater ease of contact for discussions and consultations among physicians.[54]

To be sure, not all physicians located their practices in office buildings. A considerable number continued to house their office in their residence. Others rented space on or near the street level in large apartment houses along certain main streets. This was quite common in New York City and other urban centers. According to one medical commentator in 1924, on Park Avenue it was not uncommon "to see offices the rental of which is anywhere from $5000 to $7000 a year. Such rooms are near the street level and are undesirable for living quarters anyway . . . The result is that no one man can afford such an outlay, and therefore it is quite usual to see a cluster of signs at the doorways. Some office schedules are so arranged that no physician dares run over his allotted time without infringing on the time of the colleague who follows."[55] Similar developments occurred in other sections of New York; in Brooklyn, for example, Eastern Parkway and adjoining streets had concentrations of physicians' offices. Not infrequently these locations were within easy reach of hospitals where the doctors had staff appointments.

The tendency to concentrate in certain urban areas can be seen as an extension within the city of the progressive exodus of medical practitioners from rural areas and their aggregation in the larger urban and industrial centers. The movement of physicians to urban centers during this period to a very considerable degree corresponded to the shift of population from rural to urban areas. However, medical practitioners concentrated in cities more rapidly than did the rest of the population. In their classic study of the costs of medical care which appeared in 1933, Falk, Rorem, and Ring showed that the movement of doctors from towns of 5,000 and less had been going on continuously for about a quarter of a century. They estimated that in 1906 these communities contained 60 percent of the population of the United States and 48 percent of the physicians. By 1929, however, although they still had about 48 percent of the population, only 31 percent of the physicians were located in them. An opposite trend is to be found in cities of 100,000 population and over. In 1906 these communities had about 22 percent of the population and 30 percent of the physicians. At the end of the third decade of the century, they contained 30 percent of the population and 44 percent of the

doctors.[56] It was already evident in the 1920s that this trend would continue and probably intensify. A study of medical school graduates of 1920 and 1925 revealed that six years after graduation slightly more than 50 percent were practicing in cities of over 100,000 population, although only 30 percent of the population lived in these communities. Less than 20 percent were located in communities of 5,000 or fewer people, despite the fact that almost half the population still lived in these places.[57] As viewed by Mayers and Harrison, who in 1924 reported on the distribution of physicians in the United States, "the city always has been and still is overcrowded with physicians."[58]

Underlying this growing concentration were important medical, economic, and social factors. Owing to the declining occurrence of communicable diseases during the early decades of the century, there were significant changes in the pattern of disease and thus in the clinical problems with which practitioners had to deal. Characteristic are the comments of B. B. Bagby, a physician in the small Virginia town of West Point, who kept a careful record of all his patients since the beginning of his practice.

> During the five summer months of 1909, I saw 158 town patients. During the same period of 1922 I saw 202 town patients. Of the 158 patients seen in 1909, 96 had well defined cases of malaria, with chills, fever, sweats, etc., 15 had cholera infantum, ileocolitis, or dysentery, with two deaths, and 7 had typhoid fever, making a total of 108 cases out of 158 that should have been prevented.
>
> During the five summer months of 1922 I did not have in town a single typical case of malaria, typhoid fever, or cholera infantum . . . I had only one case of ileocolitis that lasted over five days, and this was the only case of dysentery or infectious diarrhea in town this summer. There has not been a case of typhoid fever in West Point since February, 1919. Dr. A. S. Hudson, the other physician in West Point, says he has not had a case of malaria, cholera infantum, or typhoid fever this summer. So malaria, typhoid, and infantile diarrhea have about disappeared in West Point. . . . I have not found a case of hookworm in five years. When I began to practice medicine in 1904, some sections in King and Queen County showed a hookworm infection of nearly 100 percent among the school children, and many adults were sallow, anemic, sick, and thin. Thanks to the State Board of health these same people are now healthy, prosperous, and happy. I know of several families of prosperous farmers that are now enjoying touring cars of their own, who a few years ago, on account of hookworm, were more or less dependent on charity.[59]

These changes were not limited to Virginia. A practitioner in Nebraska reported in 1924 that during the early years of his practice, a major portion of it had been the treatment of typhoid fever, particularly during the first two or three months of each fall. Now, he said, "owing to the lessening of the prevalence of contagious disease; the competition of irregular cults, many of whom have scant preparation; and the better education of the lay public in questions of hygiene, diet, etc., the volume of business in any community is necessarily limited."[60]

Based on such observations, the opinion was prevalent among a large proportion of the medical profession during the 1920s that disease prevention was eliminating a considerable portion of the physician's practice and consequently decreasing his earnings. This view was succinctly expressed in 1923 by John M. Dodson, executive secretary of the Bureau of Health and Public Instruction of the AMA. "Preventive medicine," he said, "resulting in a very substantial decrease in the incidence of disease [is] an appreciable and growing factor in reducing the average income of the physician."[61] This attitude was in part responsible for the effort by the organized medical profession from 1922 into the early 1930s to promote periodic health examinations as a measure for health protection. The general practitioner was urged to stimulate "interest in periodic medical examinations on the part of his own patients so that he may practice preventive medicine, by the use of his own diagnostic skill, in his own private office."[62] On its face, this proposal seemed an appropriate means of enhancing the position of the practitioner in the community and of providing him with another source of income. However, for a number of reasons related basically to the existing structure of medical practice, the periodic health examination had little or no effect on the economic situation of physicians.[63]

The shift from rural to urban areas was partly a matter of medical generations. In 1906, 33 percent of all graduates from 1901 to 1905 inclusive were located in communities of more than 50,000 population. By 1923, 58 percent of the graduates from 1916 to 1920 had settled in places of this size.[64] The tendency of young physicians coming out of medical school to locate in urban centers meant that physicians in rural areas were not being replaced; their number was declining as a result of attrition. An indication of this process is that the average age of physicians practicing in smaller towns and rural areas was higher than that of practitioners in larger cities. A study by W. A. Pusey published in 1925 showed that the average age of physi-

cians practicing in rural counties was well over fifty years. Only 9 percent of the 4,410 physicians in the 283 counties included in the study had graduated during the preceding ten years; in 35 percent of the counties, there were no recently graduated physicians at all.[65] The small town or rural practitioner of this period on whom Pusey reported is well portrayed in the person of Dr. George Bull of New Winton, Connecticut, the protagonist of James Gould Cozzens' novel *The Last Adam,* published in 1933.

Basically, however, the changes in the distribution of physicians during the early decades of this century were a matter of economics. The statistician Raymond Pearl demonstrated in 1925 that, in general, the wealthier the section of the country, the greater was the concentration of physicians.[66] The same point was made ten years later by the Bureau of Medical Economics of the AMA, when it noted that there was a distinct relationship between the number of physicians in an area and the proportion of individuals who earned enough to pay a federal income tax.[67] This is not surprising if one keeps in mind that the physician in private practice earns his living by the practice of his profession, thereby inevitably participating in the economic organization of society and responding to its influences. In short, medical practitioners cannot be set wholly apart economically and socially from the rest of their society, and their behavior has to be viewed within this context. As Pearl observed, "physicians behave, in the conduct of life, about as any group of sensible people would be expected to. They do business where business is good, and avoid places where it is bad. Like any other normal person, the physician wants to make a decent living. Experience has shown him that it is harder to do this in the country than it is in the city. Therefore, he either sticks to or moves to the city."[68] Pearl clearly showed that the exodus of physicians from rural areas during the years before and shortly after the First World War was associated with "a definite and marked decline in the per capita *real* value of farm property. In short, it is seen once again that the behavior of the physician in the conduct of his affairs betokens a considerable degree of good economic sense."[69]

The changing rural-urban distribution of medical practitioners was, in an economic sense, a response to a decline in the quality of rural living that was already evident during the first decade of the present century. A concern with this problem had led President Theodore Roosevelt in 1908 to appoint a Country Life Commission to investigate economic, social, and sanitary conditions in rural areas. Condi-

tions in rural areas were also a consequence of changes in the American economy and in agriculture. During the sixty years from 1870 to 1930, the number of persons gainfully employed in agriculture dropped from 47 to 18 percent of the population.

Though economic circumstances were most important in leading medical practitioners to locate in cities, other factors tended to reinforce the trend. The introduction of the telephone, the automobile, and improved roads brought country and city closer together in many areas. In 1925 the Massachusetts Department of Health reported on health personnel and facilities in 118 towns in the state with fewer than 2,500 people. Only eighteen of these towns were found to be more than six miles from a physician, and only five were more than ten miles. Only eight towns with about 3,000 people were more than twenty miles from a hospital.[70] Moreover, the automobile and better roads enlarged the physician's radius of practice from the previous three to five miles to twenty to thirty miles, thus making medical care geographically more accessible, particularly in areas contiguous to larger communities. As a result, practitioners in the larger communities provided medical services for a large number of people living outside the immediate cities, especially for ambulatory care and in various specialties.[71]

Physicians concentrated in urban centers for the same general reason that other Americans were doing so in this period: the city seemed to offer better opportunities. Country living was less attractive because the city held out the promise of greater financial rewards and better facilities for practice. As N. P. Colwell, secretary of the Council on Medical Education of the AMA, diagnosed the situation in 1923, "the objectionable features of country practice are loss of patients, loss of income, increased expenses, long drives, bad roads, hard work, poorer facilities for practice, no hospitals, no libraries, no laboratories, few churches, poorer schools and loss of time or opportunity for professional or personal development. . . . Was it surprising," he asked, "that so many country doctors who entered the government service during the World War did not return to the country after they had obtained their discharge?"[72]

The war experience seems indeed to have made a profound impression on a number of physicians. As one doctor said, "They caught us and sent us to medical officers' training camps, and there the best scientific minds in the profession put us to our lessons again. Everything that was newest was learned. Every latest and best invention,

contrivance and discovery for the diagnosis, relief and prevention of disease was at our command. . . . For a year or more we lived in this atmosphere of high pressure, intensive training. The average middle-age country doctor got more training out of three months in camp than he had out of his whole college course."[73] This point is highly relevant to an understanding of the changing structure of medical practice between 1910 and 1940. A number of physicians practicing during the 1920s and the early 1930s had been graduated several decades earlier from medical schools of which a considerable proportion were inferior. Unable to obtain the benefits of graduate or continuing education, they were not as well equipped as more recent graduates for the scientific practice of medicine. Changes in the character of medical practice required an institutional environment which cities were more likely to provide. Many small communities lacked adequate hospitals and other facilities deemed essential by the 1920s for the best medical practice. Consequently, more recent medical graduates, trained to depend on hospitals, laboratories, and specialized equipment, did not want to practice without the facilities which they had come to consider as necessities. Similarly, many physicians who had served in the armed forces during the World War I preferred, on their return to civilian life, to settle in a city, where they could apply the knowledge and skills they had acquired and where facilities were available for further professional development, often as specialists.

As the scope of the rural medical market shrank due to economic, demographic, and technological changes, and as the practice of medicine was altered by the enormous increase in medical knowledge and technology which could be applied for prevention, diagnosis, and treatment, the urban market expanded and became increasingly important to the medical practitioner. The progressive urbanization of the physician and of the market within which he practiced was furthered by the advancing institutionalization of medicine and by profound changes in medical education. Although these developments derived basically from a need to cope with an enormous expansion of medical knowledge, they functioned also as a means for sustaining both a stable market system for the private practitioner and his professional autonomy.

Reform of the Medical Market

Around 1900 many physicians were convinced that the profession could not hope to improve its position in society as long as it was overcrowded and harbored within its ranks many poorly educated and incompetent practitioners. Both the public and the profession would benefit, it was held, if medical education were upgraded and the number of those becoming physicians reduced. A Connecticut physician asserted in 1902 that "the public can be protected from incompetent practitioners only by raising the standards of education and by legal barriers to the practice of such." He went on to note with approval that "the present four years' course now being adopted by the best schools, with less lecturing and more laboratory and clinical work, is a great step in advance. Harvard and Johns Hopkins in the East and Michigan and Minnesota in the West have magnificent equipment in the way of laboratories and expert teachers, and some of the other schools have begun to fall into line as fast as possible . . ." Furthermore, he insisted:

> No medical college has any valid reason for its existence unless it can show two things; first that it is thoroughly equipped to give a complete medical education of the most advanced type; second, that its graduates present evidence, in their careers as physicians, that they possess the liberal general education of a scientific man and the thorough training of the modern physician. It is now fully in the hands of State Boards to raise the standards of requirements to a degree that would efficiently exclude all but the best-fitted morally and mentally to practice medicine. With the raised standard the smaller medical colleges would either come up to the requirements or fall victims to their own impotence.[74]

Between 1900 and 1930 this agenda was carried out by a coalition headed by the AMA, its state and county societies, and the Association of American Medical Colleges, the National Confederation of State Medical Examining and Licensing Boards, and the American Academy of Medicine. Efforts to upgrade medical education and to develop higher standards for entrance into the profession had been made since the 1870s.[75] The basic problem, however, had been a lack of agreement on what the standards should be. Furthermore, effective action required the development of public support for the position that demanding standards and legal regulation were necessary to serve the public interest in the practice of medicine. Stephen Smith, a leading

New York surgeon and public health reformer, had seen this clearly as early as November 1874, when, addressing the American Public Health Association, he emphasized the need for the reform of medical education and regulation of medical practice. "Before the law," Smith said, "medicine occupies the position of the most ordinary handicraft, and is subjected to the same legal restrictions and obligations. The qualifications of the physician have been and still are regarded as of no greater importance than the qualifications of the artisan or laborer." For this situation to be altered, "there must be established in the public mind a fixed belief that their welfare requires the incorporation of certain new powers and functions into the civil administration, and these new powers and functions must, in their practical application, accomplish the reform desired." However, reform will not advance nor be permanent until "there will be established in the public mind a fixed and unalterable faith in the power of scientific medicine to protect them from pestilence, whether foreign or domestic; to discover and remove the causes of disease within and around their homes, and to promote the general health of communities."[76]

By the first decade of the present century these conditions had in considerable measure been achieved. The realization that many previously obscure diseases could be prevented, or diagnosed and treated successfully, was spreading within the medical profession and among the general public. As a result, confidence in the ability of physicians to deal with health problems was rising. This trend coincided with the reform spirit and efforts of the Progressive era, characterized by a commitment to social change on the basis of knowledge and standards developed by experts, an approach compatible with the desires and needs of a large part of the medical profession. Within this context the AMA was able to act effectively by virtue of a reorganization after 1900, inspired in part by the success of the British Medical Association in a similar endeavor. Structural reforms introduced shortly after the turn of the century joined the Association more closely to state and local societies, thus making certain that control of policy on matters of concern to practitioners would reside with the organized profession, which could thus present a united front in dealing with other interest groups.[77]

As part of its agenda, the AMA turned to the improvement of medical education as a matter of high priority, and in 1904 it created the Council on Medical Education. Along with other aspects of its program, the Council inspected the nation's medical schools, complet-

ing a report by 1907. Of the 161 schools then in existence, 82 were found to be acceptable, 47 were described as doubtful, and 32 were unsatisfactory.[78] During the period from 1905 to 1910, the Council developed standards for medical education, including a high school education and a year's training in basic sciences (chemistry, biology, and physics) as prerequisites for admission to a medical college, and a four-year graded course, with adequate access to clinical material in hospitals, to be followed by a year's internship. By 1910, after two tours of inspection, the Council was able to report substantial gains, though it stated that progress could be greatly accelerated.[79] A 25-point program entitled "Essentials of an Acceptable Medical College," issued to guide medical schools interested in improving the quality of their performance, set forth the goals to be achieved. As a result of the Council's efforts, which were supported by several state licensing boards, the number of schools began to decline, and it was evident that more would close in the near future.[80]

The process was greatly accelerated by the catalytic effect produced by the muckraking report prepared for the Carnegie Foundation for the Advancement of Teaching by Abraham Flexner and published in 1910.[81] Organized in 1906, the Foundation was headed by Henry S. Pritchett, an astronomer who had reorganized the Coast and Geodetic Survey (1897–1900) and had then served as president of the Massachusetts Institute of Technology (1900–1906). Pritchett was interested in problems of higher education in the United States, including preparation for the professions. In line with this interest and probably at the suggestion of the Council on Medical Education made sometime in 1908, the Carnegie Foundation at the end of that year undertook a survey of American medical schools. The study was made by Abraham Flexner (1866–1959), an educator who had come to Pritchett's attention through his critical assessment, *The American College,* published in 1908.[82] Flexner's survey, initiated in December 1908, capitalized on the earlier studies of the Council on Medical Education, and the fact that the secretary of the Council, N. P. Colwell, accompanied him on his visits to the schools indicates a continuing close association. Flexner also sought the counsel of William H. Welch, W. S. Halsted, and other leading faculty members of the Johns Hopkins Medical School, the standard against which he judged other institutions. Furthermore, his role as a foundation representative with a potential influence on the distribution of funds to institutions strengthened Flexner's position vis-à-vis the medical colleges, particularly among educators and adminis-

trators who recognized that money was needed if the schools were to be improved.[83]

The Flexner report spotlighted the lows and the highs of American medical education, exposing conditions that were appalling, condemning institutions that could under no circumstances be countenanced, pointing to those worthy of emulation, and indicating where salvage was possible and how it could be accomplished. Fundamentally, Flexner emphasized three principles: (1) Progress in medical education required a reduction in the number of medical schools, chiefly through the elimination of proprietary schools and those poorly financed and equipped. (2) Medical education should be articulated with general education, so that students would begin their professional training with adequate basic preparation, particularly in science. (3) Effective teaching of clinical medicine required the affiliation of the medical school with one or more hospitals where educational control could be exercised through a university of which the medical institution was an integral unit. Within this framework, Flexner envisaged a full-time faculty engaged in research as well as teaching, and students brought into direct contact with medical science through laboratory experience and with clinical medicine through a hospital clerkship.

The goal of reform was to produce fewer, more highly trained physicians, with emphasis on scientific excellence. Furthermore, by reducing the number of physicians, medical practice could be rationalized and made more efficient, thus removing conditions leading to inordinate competition and all the evils of an unstable market. As Pritchett observed in his introduction to Flexner's report:

> It is evident that in a society constituted as are our modern states, the interests of the social order will be served best when the number of men entering a given profession reaches and does not exceed a certain ratio. . . . In a town of two thousand people one will find in most of our states from five to eight physicians where two well-trained men could do the work efficiently and make a competent livelihood. When, however, six or eight ill trained physicians undertake to gain a living in a town which can support only two, the whole plane of professional conduct is lowered in the struggle which ensues, each man becomes intent upon his own practice, public health and sanitation are neglected, and the ideals and standards of the profession tend to demoralization.[84]

Ideologically, this position was in tune with the time. Frederick W. Taylor began to develop his concept of scientific or efficient manage-

ment in 1895, but it was not until 1910–1911 that the notion of efficiency caught the American imagination and spread like wildfire. Efficiency was in the public eye, and as so often happens when a trend becomes popular, it acquired an aura of approval which was applied to a variety of activities, among them improvement of medical education.[85] Furthermore, the socioeconomic interests of the medical profession clearly interlocked with its educational goals.[86] For the physician in practice, fewer graduates meant a less crowded profession and therefore less competition.

The consequences of the Flexner report were drastic. Within a decade there was a precipitous decline in the number of medical schools. In 1910 there were 131 schools with 4,440 graduates; by 1922 there were 81 schools with 2,529 graduates. This decrease was due largely to mergers among the better schools and closing of the poorer ones. While the number of schools continued to drop, reaching 76 in 1929, the number of graduates slowly rose again to 4,936 in 1932, even though the number of medical colleges remained stationary. Thus, after two decades, the number of graduates was less than what it had been in 1906, when 5,364 students received medical degrees.[87] These trends were influenced by actions taken by the AMA and the organizations collaborating with it. In 1916 the Association of American Medical Colleges agreed to refuse membership to any school unless it required two full years of college for admission. Of the 90 schools in existence, 48 already met this requirement, and many others soon followed. Pressure was also exerted by state examining boards; sixteen boards refused licensure to candidates lacking this prerequisite. Concurrently, the AMA used its standards to approve or disapprove the medical schools, publishing this information annually in its *Journal*.[88]

Equally if not more important in maintaining and supporting the momentum for change in medical education was foundation subsidy. The contributions of foundations to medical education over the preceding twenty-odd years were estimated in 1938 to total about $154 million. From 1914 to 1928 the General Education Board of the Rockefeller Foundation alone distributed $61,000,000 among twenty-five medical schools.[89] Milton Winternitz, former dean of the Yale Medical School, writing in 1955, acknowledged the General Education Board's millions, strategically deployed, as one of the three major factors in the transformation of American medical education, the other two being the model provided by Johns Hopkins and Flexner's

report.[90] Behind this course of action was the guiding hand of Flexner himself, who joined the Board's staff in 1913 and played a leading role in its medical school program until 1928.[91] As Flexner recalled later, "in less than ten years—between 1919 and 1928—operating with something less than fifty million dollars, the General Education Board had, directly and indirectly, added half a billion dollars or more to the resources and endowments of American medical education."[92]

Money provided by the General Education Board, the Carnegie Corporation, and other foundations was spent on physical plant and equipment, improvement of relations between medical schools and hospitals, and establishment of full-time teaching staffs, particularly for the basic medical sciences. Full-time clinical professorships were also established by some schools, but this trend met with resistance and advanced more slowly.[93] On the whole, however, the schools benefited greatly from the installation of good laboratories, the linkage with expanded or newly constructed hospitals, and the reorganization and improvement of medical faculties.[94]

By 1920 the basic goals of the movement to reform medical education and to improve the status of the medical profession had been achieved. A standard educational system intended to produce a smaller number of well-trained practitioners had been created. Control of this system was to be exercised through the interlocking efforts of the AMA, the Association of American Medical Colleges, and the state medical examining and licensing bodies. In this system the AMA occupied a strategic position, since the institutional accreditation agency, the Council on Medical Education, was an integral element within its structure. Moreover, their joint annual meeting clearly pointed to the close collaboration of the state licensing bodies with the Council.

Social Composition of the Profession

The effects of these changes were felt in several ways. Although the population of the United States increased throughout the period from 1910 to 1960, the number of physicians did not increase proportionately. From 1910 to 1930 the physician-population ratio declined steadily, reaching a low of 125 per 100,000 in 1930. Then it rose slightly, but from 1940 to 1960, except for a slight rise during the Second World War, it stabilized at 132–133 per 100,000.[95] Several consequences of the reform in medical education are reflected in these

trends. The reduction in the number of schools and the elevated educational requirements for admission to those that remained made it possible to control the number of potential practitioners. One factor reinforcing this development was the increased cost of a medical education. More stringent premedical education requirements, the lengthened medical curriculum, and the increased emphasis on the value of an internship meant that a medical education was being priced out of reach of low-income students. This consequence of the Flexner program had been envisaged as a possibility when the report was issued, and by the 1920s the prediction was being confirmed. In May 1924 an editorial in the journal *American Medicine* noted that "medical education is exceedingly expensive these days and requires years of effort. More and more is being demanded of the candidate for the degree of doctor of medicine . . ."[96] Another editorial in the July 1924 issue of the same periodical reported the opinion of Walter L. Niles, dean of the Cornell University Medical School, that students with limited financial resources should not go into medicine, since the demanding character of the medical course in terms of time and energy made it difficult if not impossible for a student from a poor family to work his way through medical school and do well academically.[97]

However, economic circumstance was not the only reason advanced for discouraging certain medical school applicants. The July 1924 editorial also declared:

> In these days children of the foreign born acquire sufficient education thru [*sic*] the public school system to obtain a medical student's certificate and, therefore, they enter medical school with satisfactory standing but they uniformly lack culture. There is no background of good breeding and their aims are often commercial. They consider medicine a rather easy and genteel way of making money. Such matriculates ought to be discouraged. But there is a type of poor who should be welcomed, such as the sons and daughters of educators, clergymen, physicians and others, where the upbringing has been quite correct and the mental and moral education furnishes a satisfactory background for such an exalted profession as medicine.[98]

These remarks reflect the prejudices and animosities that marked the "tribal twenties."[99] Racial and religious prejudice had existed in American society before the World War I and was shared by large sections of the medical profession. In 1912, for example, physicians were

advised that "it is better to treat colored people and Italians in their own houses, unless they are very cleanly, as it makes a bad impression on the patients to have your office smell like a dispensary or a hovel."[100] After a temporary lull during the war, these hostile attitudes were revived and intensified by economic depression and a fresh wave of immigration, reaching a crescendo in the mid-1920s. The xenophobic ferment found an outlet in discrimination against those considered "inferior," chiefly the immigrants from the Mediterranean area and eastern and northern Europe and their children who were endeavoring to improve their social and economic position. Prime targets were Jews, Italians, Poles, Negroes, and Catholics, even though many of the latter had arrived in the United States several generations earlier.

Discrimination in the educational sphere took the form of outright exclusion or of quotas, the infamous *numerus clausus*. The former affected blacks particularly. At the beginning of this century there were eight medical schools for black students. Between 1910 and 1923 six of these schools were closed, leaving only the Howard University Medical School in Washington, D.C., and Meharry Medical College in Nashville, Tennessee, to train black physicians. In view of the discriminatory practices of the traditionally white medical schools, the result of this reduction was to limit severely the further production of black physicians. What this meant can be seen from the fact that as late as 1938–1939 there were only 45 black students in all of the eighty traditionally white medical schools, compared to 305 in Howard and Meharry.[101]

Although discrimination was not applied exclusively to any single European immigrant group, probably the most intense and sustained hostility leading to social and economic discrimination was directed at the Jews. Among the immigrant groups, the Jews were characterized by a dynamic endeavor to rise from the slum and to improve their economic and social condition by establishing themselves either in business or in some profession. When parents were unable to achieve these goals themselves, they tried to provide an education for their children which would open paths to desirable careers, one of which was in medicine. Consequently, they sent a relatively large proportion of their children to colleges and universities. But as Jews strove to improve their position in society, they were accused of being unmannerly and of being given to vulgar social climbing and greed, and obstacles were thrown in their way. After World War I, a number of eastern colleges and universities, among them Harvard, Columbia, and

New York University, imposed numerical restrictions on the admission of Jewish students.[102] This practice soon spread to other educational institutions, including medical schools. As soon as the medical colleges found that the number of applications from students eligible for admission exceeded the number that could be accepted, selection procedures were instituted. In addition to scholarship, emphasis was placed on personality and character as criteria for selection. Such subjective considerations could easily be used to exclude applicants regarded as undesirable. In addition, some institutions established specific quotas. At the Bellevue Hospital Medical College, for example, it was decided to accept only a certain percentage of applicants from the College of the City of New York.[103] Since the number of students applying from the latter institution was greater than from any other college, and since a very large proportion was Jewish, the result in practice was an admissions quota for this ethnic group. A consequence of such restrictive practices was that by the end of the 1920s a substantial number of American students began to seek admission to foreign medical schools, at first in Great Britain and later on the Continent, particularly in Austria, Germany, Switzerland, and Italy.[104] A very large number of these students were Jewish, but Americans who studied in Italian medical schools were with few exceptions the children of Italian immigrants. The number of Americans studying medicine abroad continued to increase during the 1930s, but this trend was terminated by the outbreak of World War II in 1939.

Women desiring to study medicine were also subjected to quotas. It is hardly a matter of chance that the proportion of women graduating from American medical schools did not change appreciably over a period of several decades. Women accounted for 3 percent of the 1915 graduates and for 3.7 percent of the students who graduated in 1920. In 1932, they comprised 4.2 percent of medical graduates, and five years later they still accounted for only 4.4 percent of the graduates.[105]

Limiting the production of physicians by reducing the number of medical schools inevitably affected the composition of the medical profession. Although academic excellence was ostensibly the primary criterion for admission to medical education, the use of other criteria tended to hold down the numbers of students considered undesirable as future members of the medical profession. Women, blacks, Jews, Italians, and members of other ethnic groups did enter the medical profession, but for the most part in small numbers, and they generally remained peripheral to the centers of professional power. At the same

time, the increased cost and time required for a medical education tended to eliminate possible candidates from low-income groups, except for the few whose ability and strong motivation enabled them to surmount the obstacles. The result was a medical profession composed overwhelmingly of white middle-class males, few of whom came from the families of recent immigrants.

Admission to Medical School

The restrictive practices initiated in the wake of the Flexner report were intensified after 1929 as the medical profession began to feel the impact of the Great Depression. Today it is perhaps difficult to imagine large numbers of practitioners relatively idle because patients were unable to afford private medical care, yet a large number of physicians did experience economic hardships during this period. In 1929 the average net income of nonsalaried private practitioners was $5,467. However, the median was $3,700, with more than 60,000 practitioners having incomes below this level. Indeed, one-quarter had incomes below $2,300. To put the situation in another way, one-half of the private practitioners received less than 17 percent of the total net income of all active practitioners, while 36 percent went to the most prosperous tenth of the profession. The full force of the depression did not strike the medical profession until well into 1930, but as it increased in severity, the economic situation of many physicians, especially those on the lower economic levels, became quite precarious. Practitioners in rural areas and small towns were in the most unenviable economic position. By 1933 the average net income of physicians had declined by about 40 percent, so that those at the bottom of the economic ladder had their already meager incomes reduced even further.[106]

A major aim of the organized medical profession in urging reform of medical education had been to reduce the number of physicians so as to limit competition and thus improve the economic position of physicians. During the relative prosperity of the 1920s, this aim appeared in large measure to have been achieved, and the issue of overcrowding in the profession became quiescent although it did not completely disappear. However, the economic stringency of the early 1930s, when many practitioners saw themselves struggling for a living, led to a vigorous revival of this theme.

Efforts by medical colleges to limit the number of admissions were now intensified by the AMA and particularly by its Council on Medical Education and Hospitals. In 1925 the Association of American Medical Colleges created a Commission on Medical Education to study the "relationships of medical education to general and university education and to the shifting problems of medical practice, community health needs, and medical licensure . . ."[107] The Commission's *Final Report,* when it appeared in 1932, concluded that the production of physicians must be drastically reduced. The report stated:

> It is clear that in the immediate past, there has been a larger production than necessary and that at the present time we have an oversupply of physicians, although they are not well distributed in relation to the population and the medical needs of certain areas . . . An adequate medical service for the country could probably be provided by about 120,000 active physicians, if they were distributed in relation to the medical needs and their efforts were properly correlated. On such an assumption, there is an oversupply of at least 25,000 physicians in this country. . . . if the present rate of supply is continued the numbers of physicians in excess of indicated needs will increase. . . . An oversupply is likely to introduce excessive economic competition, the performance of unnecessary services, an elevated total cost of medical care, and conditions in the profession which will not encourage students of superior ability and character to enter the profession.[108]

Although not spelled out, the *Final Report* implicitly recommended a restrictionist policy to limit not only entrants to recognized medical schools but also physicians trained in foreign countries.

An examination of the reasoning by which the Commission arrived at its conclusions reveals its shortcomings. In the first place, the claim that medical care could be provided by 25,000 fewer physicians than the total number available at that time did not take into account those engaged in teaching, research, public health, hospital administration, insurance work, and other activities not directly related to private practice. Had this been done, the alleged oversupply would probably have been considerably reduced. Furthermore, though the Commission noted the maldistribution of physicians in relation to medical need, it did not suggest an examination of the implied possibility of a more rational distribution of medical personnel and of the means by which it might be achieved. Instead, the Commission simply reported the continuing trend toward urban concentration of physicians and

increasing specialization within the profession. That this trend was a consequence of a system in which geographic location, size of community, specialization, and experience largely determined the income and professional position a practitioner might expect to obtain was clear from the data presented by the Commission (and as Raymond Pearl had already shown in 1925), but any suggestion that the situation might be basically altered was rejected out of hand.[109] Despite a passing reference to regional planning, as well as statements indicating an awareness of "defects in the methods of practice in this country," the *Report* strongly emphasized the "fundamental advantages in the American scheme which ought to be retained and extended."[110]

Though not necessarily for the same reasons, the *Final Report* of the Commission on Medical Education was welcomed by both the medical schools and the organized medical profession as justifying a policy of restricting the production of physicians. Flexner had urged a combination of teaching and research in medical schools so that medical education could properly take its place as a university function. As with other aspects of the Flexner program, the possibility that implementation of the requirement for scientific excellence might have unanticipated and unwanted consequences was apparently not envisaged. In fact, increasing emphasis on clinical and laboratory instruction, in conjunction with advances in medical science, stimulated clinical investigation and fostered a growing research orientation in the schools. Furthermore, the availability of foundation funds not only accelerated the trend toward research but also advanced the process of medical specialization. These tendencies shaped the medical schools and other faculties throughout the 1920s and 1930s into the period of the Second World War, leading ultimately to the appearance of medical specialists wholly or largely occupied with teaching and research.

This line of development influenced the character of the leadership of organized medicine after the 1920s and led to antagonisms between nonacademic practitioners and medical academicians. From the end of the nineteenth century through the early decades of this century, the scientific and educational leaders of American medicine were prominent in the leadership of the AMA.[111] But as the leaders of medical education became increasingly professionalized and the schools tended to produce graduates oriented more to the scientific problems of medicine and to specialized clinical disciplines rather than to the pragmatic concerns of practitioners, a divergence of interests developed within the medical profession. Although there was no absolute separation of

interests, the academies, the colleges, and the societies organized by scientific and clinical leaders tended to concentrate on professional matters, while organized medicine as represented by the AMA became more and more preoccupied with political and economic issues and questions of public policy. The leaders of the AMA and of its constituent state societies were mostly physicians practicing in cities and larger towns, among whom a large proportion were specialists.[112]

Nonetheless, despite the divergency of interests which became evident by the early 1930s, there was agreement on the "need for professional birth control" and where it should be practiced.[113] In September 1933, Walter L. Bierring, an internist who was then president-elect of the AMA, presented his views on the "social dangers of an oversupply of physicians" and on ways of dealing with this threat.[114] Quoting with approval the comment of the dean of the University of Colorado Law School that "an oversupply in any branch of learning usually results in the development of price cutting, irregular trade practices and a fringe of casualties, losses and waste at an unsuccessful margin," leading to a "temptation to low ethical standards produced by desperation," Bierring proposed a course of action to prevent these evils or at least reduce their severity. His solution was essentially simple: limit the production of physicians by making it harder to enter a medical school.

> The responsibility to a large extent rests with the medical schools, for they hold in their hands the power to control the supply of physicians for the future. In the trades and handicrafts it has been the invariable practice sanctioned by years of tradition to limit the entry. In the medical profession similar limitations will have to be recognized, but medical standards must be maintained to assure adequate medical service to the public. University medical schools are already committed to a policy of restricting the number of students, and rather than fix the quota of admission, it will be more logical to bring about a reformation of the methods of entry as well as the curriculum of study in the preparatory or premedical colleges.[115]

The rationale for this proposal had a twofold basis, demographic and economic. Between 1850 and 1910 the medical profession grew at about the same rate as the population in general. From 1910 to 1920, however, the number of physicians decreased by 4 percent while the population increased 15 percent. During the following decade, 1920 to 1930, the medical profession gained 6 percent but the population grew by 16 percent. However, other demographic developments of this

period seemed to belie the apparent significance of the shifting physician-population ratio. As a consequence of World War I and the subsequent imposition of quotas, immigration had almost stopped. At the same time the birth rate had been falling, and the opinion prevailed among many demographers and others studying population trends that the period of rapid population growth in the United States was coming to a close or had already ended.[116] Henceforth the population would be stable, with little growth or even with a decline, so that an increase in the number of medical practitioners would be unnecessary. Instead, the trends called for a reduction in the size of the medical profession. "With the pendulum swinging toward a balanced population," said Bierring, "it will be clearly evident that the time is coming, and not far distant, when the principal function of medical service can be effectively furnished by half the number of physicians actually practicing in the United States."[117] Furthermore, Bierring argued, the economic welfare of present and future practitioners required a reduction in the number of physicians. "Even at the present time [i.e., 1933]," he said, "it is reported that in many urban communities, there are two doctors for every call and many can barely earn a decent living."[118] How much worse, then, the situation would become if the tendency to overcrowd the profession continued unchecked.

To Bierring and many other physicians who shared his views, action was clearly necessary, and in 1933 the Council on Medical Education and Hospitals of the AMA initiated a campaign to decrease the number of medical graduates. To achieve this objective the Flexner approach was again employed. Between 1934 and 1936, the Council, the Federation of State Medical Boards, and the Association of American Medical Colleges jointly sponsored an evaluative survey of the seventy-seven medical schools in the United States. It was directed by Herman G. Weiskotten, dean of the Syracuse University Medical School, and ostensibly its objective was to evaluate the quality of medical education by reducing the size of the student body in each school. It is obvious, however, that this mechanism was also intended to limit the supply of physicians. An explicit avowal of such an intent was made in 1936 by the Federation of State Medical Boards:

> While the main purpose [of the survey] is the improvement of the existing methods of training for the future practice of medicine, other beneficial results are becoming evident. Careful estimates made during the past year indicate that 500 fewer students registered in acceptable medical schools in

> 1935, compared with the previous year. If this condition should prevail for twenty years, it would mean approximately 10,000 fewer graduates entering the practice of medicine. The implication of this on the welfare of the future practitioner is evident, and considered as one of the results of the survey would make the expense and efforts devoted to it worthwhile.[119]

The substantial effect of this endeavor is evident from the considerable reduction in the size of the entering classes between the academic years 1933–1934 and 1939–1940. Of 12,128 applicants in the earlier period, 7,578 were accepted, or 62.5 percent; during the later period, among 11,800 applicants there were 6,211 acceptances, 52.6 percent. As a result of these cuts in enrollment, there were fewer medical students in 1940 than there had been a decade earlier.[120]

Not all members of the profession agreed that limitation of enrollment in medical schools was desirable. The warning against such action issued by Hugh Cabot in 1935 reflects this dissent.[121] Pointing out that information on a desirable supply of physicians was inadequate, he argued that the endeavor to restrict the size of the profession was actually intended to protect the economic welfare of the private practitioner by preventing "overcrowding" and "unfair competition." Cabot did not deny the economic problems of physicians, but he emphasized that these were a consequence of external factors and could not be handled without reference to their larger social context. The crucial fact was that the number of persons able to pay for medical care had declined greatly during the depression years, so that the profession was actually trying to adjust its size to a shrunken market and thus to stabilize it. Furthermore, Cabot called attention to the inconsistency of the demands being made by the profession. It was urging limitation of numbers at the same time that it was insisting on the maintainance of unfettered competition. "I do not myself believe in the doctrine of free competition," he said, "whether in this or in any other field, and I gravely doubt whether any such situation has ever existed. On the other hand, it seems to me that to clamor for limitations ill behooves those who support the doctrine of free competition. They cannot have it both ways."[122]

Physicians from Abroad

Logical consistency was not, however, of major concern to those who made these demands. Their central objective was to stabilize the medi-

cal market and to give it an organizational structure which would ensure the economic viability of private solo practice. The effort to reduce the supply of physicians at its source was only one of several means employed to achieve this objective. As in the pre-Flexner period, the state medical boards, collaborating with the Council on Medical Education and Hospitals, endeavored to limit the entry into practice of those considered undesirable, but now attention was turned chiefly to persons who had received their medical education abroad. After Hitler's accession to power in 1933, a number of Jewish physicians as well as others opposed to the Nazi regime migrated to the United States from Germany, and later from Austria and Czechoslavakia. Regulations introduced by licensing boards in various states had the effect of making their entry into practice more difficult. Some states required citizenship as well as licensure for the practice of medicine. Others also required applicants to demonstrate proficiency in English before being admitted to the licensing examination.

On the whole, however, more attention was devoted to American students trained in foreign schools, since their admission to practice would nullify the purpose of reducing enrollment in medical schools. Between 1930 and 1933, there were 4,371 Americans studying medicine in Europe, and it was anticipated that they would return to the United States to practice their profession.[123] As a result, steps were taken to prevent any increase in the size of this group as well as its eventual entry into the American medical profession.

The *Final Report* of the Commission on Medical Education had already noted that most of the Americans studying medicine in Europe had been unable to obtain admission to a medical school in the United States, and it had also pointed out that graduates of many European schools had done poorly in the examinations of the state medical boards. The implication was that the latter experience referred to the American students, though no supporting evidence was presented.[124] Efforts to stamp the American students abroad as inferior continued throughout the 1930s, but the tenuous nature of the evidence is clear from a discussion of the foreign medical graduate by Harold Rypins, secretary to the Board of Medical Examiners of New York State.[125] The experience of foreign graduates in licensing examinations was used by Rypins in a somewhat subtler manner. Instead of directly labeling the American students abroad as inferior, he did so with faint praise.

> It would, of course, be grossly unfair to stigmatize the individuals of this group as inferior, and doubtless it contains a not inconsiderable number of very excellent students. Taken by and large, however, it cannot be denied that this group is made up of two thousand or more students who were rejected on the basis of competitive selection for admission to our medical schools, and to this extent, at least, they must be considered as constituting a relatively inferior group. In addition, not a few of these students have previously been admitted to American medical schools and having failed to keep up with their work they have been dropped or been asked to withdraw.[126]

This characterization was reinforced by associating it with the poor performance of foreign graduates in state licensing examinations. In light of this experience, Rypins concluded, "we must frankly face the question as to whether or not at least some of these students may not be exposed to a professional training which is in many ways inferior to that now required in the United States, and which does not adequately train American students for the proper practice of their profession in this country."[127]

Based on arguments and evidence of this nature, steps to scotch the problem at its source were recommended in February 1933 at a meeting attended by representatives of the Council on Medical Education and Hospitals, the Federation of State Medical Boards, the National Board of Medical Examiners, the Association of American Medical Colleges, and the Board of Regents of the University of New York. First of all, American students matriculating in foreign medical schools after 1932–1933 would be required to present satisfactory evidence of premedical education equivalent to the requirements of the Council or the Association and of graduation from a European school after completing a medical course comprising at least four academic years. More crucial was the second recommendation, which would require graduates of a foreign medical school to obtain a license to practice medicine in the country where the school was located. This was a "Catch-22" recommendation, since the committee urging its adoption knew that, with the exception of Great Britain, countries with American medical students (Germany, Austria, Switzerland, France, and Belgium) did not license foreigners to practice. In fact, to make doubly sure that Americans would not benefit from attendance at foreign medical schools, the committee had conferred with governmental and medical representatives of several of these countries so as to paint a

bleak picture of the alleged quality of such would-be medical students.

For some advocates of a diminution in the size of the medical profession, even these measures were not enough. In 1936 Arthur Dean Bevan of Chicago, a leader of the AMA, urged special action against a specific group, namely, Jewish applicants from New York City, on the ground that their number had increased steadily over a period of ten years, rising from about 10 percent of all applicants to about 20 percent in 1936. According to Bevan, "In the present over-crowded condition no group should be permitted to enter medicine in such numbers as to crowd out of medicine the members of other groups who desire to enter. No group should be discriminated against. The members of all groups must have the same rights and opportunities."[128] To make certain, however, that these brave principles led to the desired result, Bevan proposed that the medical schools and the state boards unite to prevent any further increase in the size of the profession by allowing no more than about 4,000 new practitioners to be added annually, in essence just enough to replace those who had retired or died. The intent of Bevan's proposal is clear: to achieve zero growth in the profession by setting a ceiling on admissions to medical schools and creating quotas for certain categories of applicants, particularly Jews. Although Bevan's suggestions were not explicitly adopted, there is no doubt that efforts to reduce the supply of physicians were pursued vigorously and in the same spirit. Admission standards were raised and class size decreased so that the number of medical graduates declined. Stimulated by the economic problems of the depression, the leaders of the medical profession used the movement initiated by the Flexner reforms as well as the organizations associated with them to further the creation of a seller's market.

Hospital Training

The Flexner program had focused on the production of a scientifically oriented competent general physician by a school centered in a university. It was soon apparent, however, that academic education alone was inadequate; it was necessary to have a period of practical experience under supervision in a hospital before engaging in independent practice.

Hospital training as a part of medical education began in England and on the continent in the seventeenth century and became more

common in the eighteenth century, when house surgeons and physicians were actually apprentices.[129] Since many of the leading physicians of the colonial and early republican period in the United States were graduates of European schools, particularly the University of Edinburgh, it was natural that they would favor the pattern of training they had observed abroad. Throughout the nineteenth century, since the number of hospitals was much smaller than the number of medical schools, the great majority of graduates failed to obtain any hospital training; they learned the practical aspects of medicine from experience acquired with their patients. There were some interns, but they were appointed on the basis of competitive examinations and thus formed a small elite group.[130] Under these circumstances, the internship was important in providing the first rung to be mounted on the ladder of professional advancement. An internship almost always led to further staff appointments in the same hospital, beginning close to the bottom of the ladder with a position in the outpatient department, from which continued ascent could lead to full membership on the attending staff and then to the ultimate goal of ranking attending physician or surgeon. Those who attained these positions were recognized as specialists in their chosen fields and enjoyed high professional status.

By the end of the nineteenth century, the internship had established itself as a period of postgraduate education comprising at least one year of hospital experience, but it still involved only a small number of medical graduates. This situation changed radically after 1900, due to the interplay of developments already discussed. The growing complexity of medical care, particularly within the hospital setting, as a result of the diversification of functions and multiplication of activities arising from advances in medical science, increased the need for the services of a larger house staff. Because of the expansion of the hospital through the introduction of the clinical laboratory, aseptic surgery, roentgenology, and various medical specialties, a small number of interns could no longer care for a large number of patients. Interns performed laboratory procedures for diagnostic workups, assisted at operations, served as anesthetists, assisted in the outpatient department and specialty clinics, and rode the ambulance, thus providing the hospital with a sorely needed work force.[131] As a result, the number of internships mushroomed, although little attention was given to supervision and educational quality.[132]

This trend coincided with the increasing recognition of the need to

supplement undergraduate medical education with a period of hospital training. At its first annual conference in 1905, the Council on Medical Education had adopted an "ideal standard" for American medical schools which included one year as an intern. Although the Council had examined the internship situation as early as 1904, it was not until 1912 that it initiated a thorough survey of hospital facilities for advanced training, and in 1914 the first list of "Approved Internships" appeared in the *Journal of the American Medical Association*. In 1904, probably 50 percent of the medical graduates received further training in a hospital; ten years later, about 70 or 80 percent became hospital interns. This development continued unabated in the following decades, due not only to the increasing demands of medical practice but also to pressures from state medical licensing boards and medical schools. In 1914 Pennsylvania made an internship a requirement for licensure, and by 1934 sixteen other states as well as the District of Columbia and the territory of Alaska had followed its example. During this period, a number of medical schools instituted a similar requirement as a prerequisite for the M.D. degree. In 1915 the University of Minnesota became the first school to require an internship, and by 1934 fourteen other schools had adopted such a requirement.[133] Though the majority of the medical schools did not require an internship for graduation, their students often went on to at least one year of hospital experience after completing the medical curriculum.

By the 1920s the internship was a firmly established element in the pattern of American medical education. Accreditation of hospitals for internship was assumed by the Council on Medical Education and Hospitals, and in the following years an increasing number of hospitals was approved for internship. In 1923, with 3,120 graduates and 3,119 internships available, demand and supply were in balance. Thereafter, however, until the outbreak of the Second World War, as medical school enrollments declined and then remained at the same level for about two decades, a growing excess of approved hospital positions appeared. Until 1940, these excess positions were filled by graduates of foreign medical schools.[134]

The introduction and acceptance of the internship as a part of the physician's education had significant consequences for the medical profession. Despite the inadequacies of the internship system as it developed during the period 1915–1940, it did contribute positively to the preparation of a larger group of competent practitioners than had previously engaged in the practice of medicine.[135] Furthermore, the

internship extended the period prior to licensure, thus delaying entry into the medical market by one to two years and for a time making possible a more stable market structure.

This latent function of the internship was carried still further by the residency, which emerged in the later 1920s as the route to medical specialism. Throughout the greater part of the nineteenth century, a long training period, carried on simultaneously with the general practice of medicine, was the rule before one became a specialist. Abraham Jacobi, for example, though primarily a physician who concentrated on children's diseases, also carried out surgical procedures. He performed tracheotomy in cases of diphtheria, resected ribs for empyema, and operated for cancer of the esophagus.[136] If the aspiring specialist wanted to go abroad to the great European medical centers to secure training not available in America, a considerable financial outlay was involved, and relatively few could afford it.[137] By the beginning of the twentieth century, these extended and to some degree arduous patterns of specialist training had become an accepted part of medical mores. Moreover, as previously noted, recognition as a specialist was integrated with the structure of the hospital staff, and it was expected that all newcomers would cover the full course rather than briskly proceeding to a lucrative special practice. Nevertheless, several developments began during this period that undermined this system and led to its replacement by a different pattern of specialist training.

Specialism and Certification

The first of these was the resident system of training surgeons that had been inaugurated by W. S. Halsted in 1889 at the Johns Hopkins Hospital. Halsted's aim was to create a school of surgery which would train not merely competent surgeons but "surgeons of the highest type, men who would stimulate the first youth of our country to study surgery and to devote their energies and their lives to raising the standards of surgical science."[138] For this purpose he adopted the German system of *Assistenzärzte,* in which a professor selected as members of his resident departmental staff those students whom he considered gifted and promising and who could be trained as teachers and researchers. Halsted's staff consisted of three assistant residents and one resident, all carefully selected. The average length of time spent as residents was between two and two and a half years, although some

men stayed for periods as long as seven years. Assistant residents remained for periods ranging from one to three years. An analogous system was developed later in Minnesota, when in 1915 the Mayo Clinic, a private group practice, was affiliated with the University of Minnesota to become the graduate school for its medical college. Assistants in the medical or surgical sections of the Clinic participated in a three-year training plan which prepared them for specialized work in clinical medicine or in medical research.[139]

Outside these and other major medical centers, however, residencies for specialist training were uncommon until after World War I. Some physicians, eager to scale the economically inviting heights of specialism, did not hesitate to skip what they considered to be unnecessary steps in the training process. The Commission on Medical Education concluded in 1932 that "many specialists are self-named; many are not fully trained in their limited field and still less well equipped in the broad fundamentals of medicine; some are frankly commercial. At present about 35 percent of recent graduates in this country limit their work to a specialty and most of them have not had a sufficiently broad clinical experience." To guarantee that those claiming to be specialists were actually experts in their fields, the Commission urged the creation of "a particular identification for those who profess to be specialists . . . to be granted only on evidence of the successful completion of a training which in the opinion of competent physicians is adequate to prepare the individual for practice in the limited field."[140]

The problem delineated by the Commission and the course of action which it recommended were not new. Concern over the educational preparation and competence of those claiming specialist skills had already emerged early in the century as specialties became increasingly prominent. The rapid expansion of surgery, its increasing division into subspecialties (urology, neurosurgery, etc.), and its unrestricted practice and attendant evils (incompetent surgery, fee splitting) indicated the need for some form of professional regulation based on standard criteria that would establish boundaries among the several branches of surgery and between specialists and general physicians and that would reduce the harmful impact of poorly trained practitioners.[141] Largely through the efforts of Franklin H. Martin, a leading Chicago surgeon, the American College of Surgeons came into being early in 1913. The College assumed the function of drawing a line between the qualified surgeon and the general practitioner by certifying

the basic competence of the former. Prerequisites for the certifying examination given by the College included a one-year internship, three years as an assistant in surgery, fifty abstracts of cases treated by the applicant, and a number of visits to surgical clinics. By outlining a standard course of training to be followed by those desiring to be certified as qualified surgeons, the College established a pattern for other specialties to follow. Furthermore, the requirements set by the College further strengthened the linkage between the hospital and the surgeon. Hospital appointments were necessary to become proficient in the required operative procedures and techniques. This situation, in turn, required that attention be given to the quality of hospital care and that standards be established for institutions that would train specialists, and these tasks, too, were assumed by the College.

The example set by the American College of Surgeons was not lost on physicians in other special fields. Considering internal medicine as much a specialty as surgery, Heinrich Stern, a New York internist, together with a number of like-minded colleagues, established the American College of Physicians in 1915. Although the College provided an organizational focus for internal medicine, it had no certification mechanism until the American Board of Internal Medicine was established in 1936.[142]

More successful were the ophthalmologists who, as early as 1864, provided a professional center for their specialty by organizing the American Ophthalmological Society.[143] By the turn of the century, the ophthalmologists were well established on the medical scene, but like the surgeons they were faced with several significant issues. On the one hand, they faced the competition of optometrists who claimed to be more competent in correcting refractive errors than medical practitioners, and who could advertise their services, since they were not restrained by the ethical code of the medical profession.[144] On the other hand, numerous general practitioners, after attending a short course of several weeks, maintained that they were trained refractionists and capable of treating eye diseases. Some were general practitioners, others were partial specialists, but there were no established criteria by which to identify a qualified ophthalmologist and to clarify the relations between the specialist and other practitioners. Moreover, ophthalmology was more than just the correction of refractive errors. Following the introduction of the ophthalmoscope in the nineteenth century, clinical ophthalmology had expanded greatly through the study of fundal pathology in relation to various systemic diseases and

other disorders, leading to closer contact with other branches of medicine. Ophthalmologists keenly felt the need for a clear demarcation of their specialty and a definition of their identity, and after the turn of the century various proposals to achieve these aims were vigorously discussed.[145] Eventually, a professional examining board sponsored by the major ophthalmological organizations was accepted as a mechanism for certifying competence on the basis of uniform standards, expressed in requirements and examinations. The American Board for Ophthalmic Examinations was established in 1916 (its name was changed to the American Board of Ophthalmology in 1933), and its first examinations were held at the end of the year.[146]

At about the same time the Council on Medical Education also began actively to consider the problem of specialist education and training by setting up a committee to investigate graduate medical teaching.[147] Though the possibilities of state licensure or a university degree for specialists were discussed over the following years, by the mid-1920s professional opinion was shifting to the need for some form of certifying examination based on a definite period and kind of training and administered by a national voluntary organization representing the specialty.[148] In 1924 the Council investigated the hospitals offering residencies, and its first list of approved residencies, published in 1927, comprised 1,699 positions in 270 hospitals.[149] It was apparent that facilities for training specialists were inadequate qualitatively as well as quantitatively and that positive action would have to be taken to find ways of improving the situation.

During this period several specialty groups followed the ophthalmologists in creating boards to examine and certify those who wished to be recognized as qualified specialists. A National Board of Examiners in Otolaryngology was established in 1924, representing the joint interests of the five national organizations in this field.[150] Obstetricians and gynecologists were beginning to consider a similar course of action, and in 1928 a committee was appointed for this purpose. Two years later the American Board of Obstetrics and Gynecology was established.[151] This action was followed in 1932 by the formation of the American Board of Dermatology and Syphilology, and a number of other groups of specialists considered similar steps.

In this situation, in which the further uncoordinated creation of specialty boards could be anticipated, the absence of a central controlling body was painfully apparent. Action to remedy this lack was taken in 1933 by the AMA in authorizing the Council on Medical Education

to formulate standards for the certification of specialists based upon those of the specialty boards then in existence. Paralleling this development was the creation of an Advisory Board for Medical Specialties by a coalition representing chiefly the existing boards, the medical schools, the state licensing boards, and the hospitals, with links to the AMA. Cooperation between the Council and the Advisory Board led to the adoption in June 1934 of a statement outlining "Essentials for an Approved Special Examining Board."[152] Future boards would be organized under the joint auspices of the AMA and the various national specialty societies, and they would have to be approved by the Advisory Board. Among the requirements for candidates applying for certification were no less than three years of training after an internship of at least one year, a license to practice, and membership in the AMA or in an accepted equivalent organization.[153] With the adoption of standards and an appropriate mechanism, specialty boards were organized in rapid succession. Between 1933 and 1940, twelve boards were approved: pediatrics (1933), radiology (1934), psychiatry and neurology (1934), orthopedic surgery (1934), colon and rectal surgery (1934), urology (1935), pathology (1936), internal medicine (1936), anesthesiology (1937), plastic surgery (1937), surgery (1937), and neurological surgery (1940). Together with the four previously organized boards, then, there were a total of sixteen organized certifying bodies at the outbreak of the Second World War.

To meet requests for certification by younger applicants (established practitioners were certified in most cases under a grandfather clause), training facilities required expansion, a need met chiefly through an increase in the number of hospital residencies. As in the case of internships, recognition of the advantages accruing to institutions accepted for the training of specialists led many hospitals to seek approval, and as the number of specialty boards increased a growing number of hospitals were approved for residencies. In 1927, there were 270 hospitals with a total of 1,699 approved residencies; in 1930, there were 338 and 2,028 respectively; and in 1935, the number of hospitals reached 410, with 2,840 positions. By 1939, the number of hospitals had risen to 518 with 3,951 residencies, and in 1941 there were 5,256 residencies in 610 approved hospitals.[154]

The establishment of the boards and the growth of residency programs for the various medical specialties carried significant implications for the future structure of medical practice and the social relations of the medical profession. Specialization which had been

developing for over fifty years was now formally recognized as an essential characteristic of modern medicine. Evidence of this may be observed by comparing the distribution of physicians among various specialties in 1903 and 1932. A study of physicians graduated from the Harvard Medical School from 1892 to 1901 revealed a group of 172 specialists. Two-thirds of the group practiced in five specialties: surgery; obstetrics and gynecology; ophthalmology; ear, nose, and throat; and internal medicine. In 1932, two-thirds of all specialists (complete and partial) were distributed among the same specialties plus pediatrics, which had risen to the rank of a specialty during the intervening three decades.[155] This comparison likewise reveals that the trend toward specialization was strongest in surgery and fields characterized by a marked surgical component, a trend which emerged even more clearly with the formal recognition of various surgical specialties.

The creation of the specialty boards was justified not only in terms of raising the educational status of specialists and evaluating their competence but also as a means of restricting entry into and practice in the specialties. It was hoped that the effect of the boards would be to reduce the numbers of specialists to small cadres of high professional competence, thus strengthening the position of general practice which felt itself to be threatened. A. C. Christie, professor of clinical radiology at the Georgetown University Medical School, who had been a member of the Committee on the Costs of Medical Care, claimed in 1935 that the boards would "gradually bring about a reform in specialism which will greatly improve the quality of work performed by specialists, lessen the number of specialists and directly effect a reduction in the total costs of medical care." At the same time, he urged the general practitioner to become better prepared

> to do many of the things which are now in the field of the specialist . . . It is only by so preparing himself that the general practitioner can reclaim the field that is naturally his. It is essential that he should do so in order that he may again become the central and dominating figure of medicine. The order of development which must be forced within the profession if it cannot be attained otherwise is that specialists become fewer and be restricted to those highly trained and of extraordinary ability while the general practitioner assumes many of the duties now erroneously thought to belong to the specialist.[156]

Walter Bierring, a leader in the American College of Physicians who led the movement for a board in internal medicine, expressed a similar

opinion in 1936. Emphasizing that "the medical profession has solved another of its important problems," he envisaged a future situation in which a small, highly qualified group of specialists would practice within clearly defined areas, while the rest of medical practice would remain the realm of the general physician.[157]

The Triumph of Specialism

However sincere and well-intentioned the proponents of these views may have been, their understanding of the dynamic factors in the situation they hoped to stabilize was woefully inadequate. Expectations that the creation of standards for professional accreditation of specialists would tend to limit their number took little account of the powerful forces—economic, social, educational, and institutional—operating to foster and to accelerate the trend toward specialization. The great importance of economics in the acceptance of specialization by the medical profession is quite evident. It did not take physicians long to realize that specialists could command much larger fees than general practitioners. As William Brodie, president of the AMA, stated frankly in 1886, the members of the profession "began to realize that by devoting themselves to one branch instead of working up a general practice, they would often do more good, earn more money, and have less arduous work to perform."[158] A nationwide survey of physicians in private practice conducted in 1931 by the Committee on the Costs of Medical Care to determine the effect of specialization on earnings revealed striking disparities in the level of financial returns for partial specialists as contrasted to general practitioners and for full-time specialists in contrast to both the preceding groups. The average gross and net incomes were lowest for general practitioners, higher for partial specialists, and highest for full specialists. The last-mentioned had an average annual net income of $10,000, as compared with $6,100 for partial specialists and $3,900 for general practitioners. The investigators concluded that "with due allowance for pride in professional ability and absorbing interest in some particular field of medicine, the economic factor must be a strong motivating force to induce physicians to become specialists."[159]

These observations were corroborated by information obtained in a survey conducted by the journal *Medical Economics* in 1939. The survey revealed that the average annual gross income of general prac-

titioners was $6,605, of partial specialists $7,411, and of full specialists $10,057. The average net incomes of the three groups were $3,969, $4,507, and $6,184 respectively. Not only did specialists receive larger gross and net incomes; they also collected a higher percentage of their accounts. On an average, the general practitioner collected 77 percent of his accounts, while the specialist averaged 83 percent.[160] Small wonder then that physicians tended to become specialists, especially when the obvious pecuniary advantages were reinforced by public attitudes and opinion and the more general orientation of the culture.

Linsly R. Williams, executive director of the New York Academy of Medicine, observed in 1928:

> During the past twenty-five years there has been a most astonishing change in the manner of living in this country, luxuries have become necessities and as each new luxury becomes a necessity the cost of this necessity becomes a tax on all the people by an increase in the cost of living . . . This increase in cost and in standard has influenced our habits of eating, dressing and general living, causing a larger expenditure than necessary. Naturally the physician . . . has been affected and his desire to advance his economic level is only to be expected. He has learned of the earnings made by lawyers, engineers and chemists, to say nothing of the earnings of men busy in the financial and industrial world.
>
> Should not the physician keep up too? Efficiency costs him a pretty penny and although costs of maintenance may be charged off on the income tax yet the money must be spent to keep a proper office and play the role of the modern practitioner.[161]

Moreover, in our society where "a considerable element of prescriptive expensiveness" is observable in most consumption, it would be too much to expect that medical care would be sought on a more rational basis.[162] For many people there is a high degree of social prestige in being treated by a high-priced specialist, and few will forgo if they can afford it.

The effect of these tendencies on medical practice, already clearly evident in the 1920s, was noted by various observers in the profession. As Linsly Williams saw the situation, one effect of specialization was to fragment medical care and to weaken the patient-physician relationship, at least as it had existed for those who had been served by a general physician. Confronted by a multiplicity of specialists, the patient was becoming an Alice in a Wonderland of special practition-

ers, whose interest was limited for the most part to the specific field of their competence. "One sometimes gains the impression in some of our larger cities," Williams commented, "that in the wealthier sections of society the general practitioner no longer exists and the sentiment has been expressed on more than one occasion that there are fifty-seven different varieties of specialists to diagnose and treat fifty-seven different varieties of diseases and lesions but no physician to take care of the patient."[163]

William A. Pusey, president of the AMA in 1924, viewed the growing tendency of young physicians to specialize in a broader context. In a series of articles published in the *Journal* of the AMA in 1925, he linked this trend as well as other changes in medical practice to the reform of medical education.

> The difficulties in medical service in the cities are seen in the way our young men are seeking the special careers. The great expression of this fact is the way our present graduates show a preponderant tendency to go into the specialties. They are not going into general practice. The situation in the cities is not acute, because the supply of physicians of the older generation leaves for the present enough of that generation to meet the demands of general practice. But it is evident that, unless we can do something to change the trend, the time is not far distant when the problem of the general practitioner as we have always known him—the family doctor for the man of ordinary means—will be a serious one even in the cities.[164]

In Pusey's opinion, this trend resulted from a number of factors inherent in the reformed medical schools. One was the increased cost of medical education. "The chief reason for our new difficulties," he said, "is that medical education has made the license to practice medicine so expensive that those who are not paying the high price for this license exact a correspondingly high price for their services under it."[165] In short, increasing the cost of production led to a rise in the price paid by the consumer for the services of a physician. The schools were turning out not only a costly type of practitioner but also one who increasingly elected to obtain advanced training in a clinical specialty. The decision to specialize was no doubt influenced significantly not only by the factors cited above but also by the members of the medical faculty who provided role models for students. This situation was itself a consequence of the emphasis in the post-Flexner period on clinical instruction in a teaching hospital under control of the faculty. As hospitals became affiliated with medical schools for clinical teaching, their

organization of services by specialty soon led to a similar structure in the medical faculties, especially as the latter assumed greater control of the hospital staffs. The result was that for about three decades, from the 1920s through World War II, the clinical specialist increasingly dominated the medical school, shaping its students to a greater or lesser degree in his image.

Under these circumstances it could hardly be expected that the creation of specialty boards would stem the trend toward specialization, reduce the number of specialists, and strengthen the position of the general practitioner in numbers, prestige, and quality of service. In fact, precisely the opposite happened. By establishing a clearly defined and accepted institutional pattern for the training and accreditation of specialists, the boards facilitated the growth and spread of special practice. Throughout this century, the proportion of general practitioners in relation to specialists has been declining without interruption, and there is no evidence that the boards had any retarding effect on this development.[166] On the contrary, an acceleration of specialization during World War II and in the postwar period was possible because of the existence of a system of board certification based on specified requirements.[167]

At the same time, the creation of specialty boards did nothing to improve the situation of the general practitioners. Indeed, the malaise felt by physicians in general practice since the turn of the century was intensified even further. As has already been described, during this period many physicians, particularly general practitioners, saw themselves as threatened by developments arising from the expansion and application of medical science, a marked increase of institutional practice, and a rising awareness of the social problems associated with ill health as well as a recognition of the need to provide adequate medical service at a reasonable cost. These threats were conceived of by practitioners in terms of loss of income, loss of prestige, and limitation of autonomy. Whether actual or potential, the dangers which appeared to menace the generalist were, however, but the overt aspects of a more deep-seated problem: What was the proper role and function of the general practitioner in the changing organization of American medical practice?

An awareness of this problem was already apparent in the later nineteenth century. In 1886, John S. Apperson, a Virginia physician, reporting to his colleagues on advances in medical practice, noted:

> A few years ago . . . the practice of medicine included, not only everything pertaining to etiology, therapeutics and prophylaxis of general diseases . . . , but it held dominion over distinct organs and applied its principles to the whole organism . . . [But now] the practice of medicine as a separate and distinct department is being rapidly reduced to narrower limits than was perhaps anticipated by any of our profession who lived in the preceding generation . . . Under this new régime . . . the old-time practitioner of medicine, who in days agone toiled through the prescribed course . . . and came forth wearing the title doctor of medicine . . . must to-day acknowledge some bewilderment when he sees the rapidly multiplying divisions springing up around him, and admit that but little of the important whole has been left to his care.[168]

Forty years later this situation had changed only in becoming more acute. Some observers of the state of medicine in the period between 1910 and 1940 felt that general physicians could handle the great majority of cases without calling in specialists. The Council on Medical Education and Hospitals estimated in 1923 that 80 to 90 percent of all patients could be cared for satisfactorily by general practitioners, and 90 percent could receive proper care at home or in the doctor's office. Despite these claims, however, general practice was being increasingly limited in terms of numbers of practitioners and scope of practice. Pusey charged in 1924 that medical graduates were being trained to be specialists and to practice chiefly in hospitals and that the schools were not "producing men to do the ordinary service of medicine for ordinary people in the cities or the country."[169] At the same time, general physicians were, in varying degree, limiting their service to patients. As an example, Peebles noted in his survey of Shelby County, Indiana, in 1929, that the rural physicians had "fewer self-imposed restrictions upon the scope of their practice than the physicians of Shelbyville," the county seat.[170] Some of those who restricted their practice could be considered partial specialists.

Although general practitioners complained that patients did not consult them for minor and major conditions that the physicians felt they were competent to diagnose and treat, it was evident to various observers that the nature of general practice was changing inexorably and that the process was irreversible. Pusey observed in 1924 that nostalgia for the old-time family physician and extolling the virtues of general practice were inadequate responses to the growing need for practitioners who could provide high-quality primary care. Nor would

the problem be solved by tinkering with the medical curriculum. He commented:

> We are striving to meet the difficulties of service in the cities by urging the importance and dignity of the family practice of medicine, by trying to devise changes in our curriculum that will make general practitioners and not specialists, by proposing that graduates be required to engage for a given number of years in general practice and by other proposals to try to overcome artificially the natural tendencies that are produced by the training that we are giving our graduates. These expedients are, I believe, as futile as all other efforts to make water flow up hill . . .[171]

Moreover, Pusey recognized that medicine was becoming increasingly complex not only through specialization among medical practitioners but simultaneously through the proliferation of new types of personnel who were taking over tasks previously performed by physicians. In this connection he was also touching on a trend with profound implications for the future structure of medical practice. "In the way we are undertaking to find substitutes for the physician," Pusey said, "by filling medical and quasimedical positions with laymen, by the employment of nonmedical clinical assistants as substitutes for interns, by having nurses take over many of the functions of the physician, we are indirectly confessing that much of the work that was formerly done by physicians must under our present policy be done by someone else."[171] Clearly a situation was developing in which even a well-qualified practitioner could no longer work alone.

This trend was also apparent in practice outside the hospital. Not only did the practitioner in the later 1920s need a larger armamentarium than had been employed by his counterpart at the turn of the century, but, as one observer summed up the situation, "medicine cannot be practiced today without the assistance of a nurse at times, without well equipped offices and without making use of special technical methods of diagnosis, primarily in the clinical and x-ray laboratories."[172] Concurrently it was also obvious that numerous urban and rural practitioners were not able to provide adequate medical care because they had no hospital appointments and lacked laboratory facilities.[173] To a considerable degree, these were practitioners who upon graduation lacked the financial resources to equip an office satisfactorily, to employ assistants, and to pursue postgraduate study in a laboratory or outpatient department—the kind of career pattern that eventually led to a hospital position. Forced to earn a living from the

outset, it was not uncommon for such physicians to find themselves in a rut after several years, becoming superficial, careless, and disinclined to use thorough, painstaking methods of diagnosis. By working hard, most obtained a limited, if sufficient, income, and though they might consider specializing, very few achieved this aim within the first ten years of their practice.[174]

The state of mind of a physician in this situation was well captured by the novelist Gerald Green in his portrayal of a Brooklyn general practitioner who had graduated from Bellevue Medical College in 1912. Looking back in 1925, Dr. Samuel Abelman, his protagonist,

> was forced to admit that he had been standing still. His income had remained almost stable. He saw the usual quota of new faces in his office, yet seemed to lose the same amount to new men, to specialists, even to a fat chiropractor . . . The way his finances were working out, he had more than enough to cover expenses, to pay for a vacation . . . He had been in the practice now for thirteen years, and while he still loved the work, the challenge, the mystery, the rude camaraderie of the hospitals and the dressing room, he often wondered whether he was right in remaining in general practice . . . It was no secret that the big money, the easy regular hours, the dignity of being a "professor" lay in the magic of specialization. X-ray work fascinated him, and he often thought of doing postgraduate work at Bellevue . . . But it would be no easy chore. It would mean giving up several nights a week, and with new MD's popping up all over Brownsville, he'd be sure to lose patients. Besides, he was two years shy of forty, and he had a family to worry about, bills to pay.[175]

And so, like many of his real-life colleagues, Dr. Abelman remained a general practitioner.

The uncoordinated drift toward specialism and medical complexity did not bode well for the future of the general practitioner. The disappearance of the general physician of the nineteenth century was already accepted by the second decade of the twentieth. Addressing the members of the New York Academy in 1912 on the doctor's future, Charles L. Dana, a prominent New York physician, offered a bittersweet eulogy of the departing practitioner:

> There has been much said about the disappearance of the old fashioned family physician and general practitioner. He was a splendid figure and useful person in his day; but he was badly trained, he was often ignorant, he made many mistakes, for one cannot by force of character and geniality

> of person make a diagnosis of appendicitis or recognize a streptococcus infection. I think the ill educated old family doctor, idealized perhaps less by our memories than in our story books, is going, and it is a good riddance.[176]

Variations in the Medical Market

But who would replace the vanishing general practitioner of the earlier period, and how could that be done? The need to redefine the status and the function of the general physician emerged as a basic problem in the period between the two World Wars, and a number of alternatives were proposed to deal with it. Dana predicted that a new type of family doctor would replace the old-fashioned general practitioner. This physician of the future would be a well-trained internist capable of handling most medical problems but who also could recognize when a specialist was needed. Dana knew, however, that this prediction was by no means certain of fulfillment. "Here is where my prophecy limps," he admitted, "my hope and belief begin."[177]

Eight years later, in 1920, at a time when the concept of the health center was popular, Frank Billings, dean of Rush Medical College, in Chicago, and a leading figure in organized medicine, examined the future of private medical practice and considered possible ways of improving the status of the general practitioner.[178] After reviewing group medical practice, compulsory health insurance, and federal and state action in relation to certain diseases, which he generally rejected, Billings proposed the establishment of local health centers to promote the efficiency and competence of general physicians. What he had in mind he said, was

> the erection of a modern hospital or the adaptation of an existing hospital with provision for all sick and injured who require hospital care . . . , adequately equipped diagnostic and clinical laboratories of chemistry, pathology, bacteriology, serology, and radiology; a personnel of a qualified superintendent, assistants, clinical and medical social nurses and laboratory technicians; a medical and health reference and circulating library; suitable rooms for medical and social welfare meetings, and adequate provision for an out patient and diagnostic clinic.[179]

Financial support for construction, equipment, and operation of the center would be obtained through tax funds and/or bond issues as

well as state subsidies. Responsibility for administration of the center would be lodged in an elected or appointed board of trustees, while a board composed of medical practitioners appointed by the district or county medical society would be responsible for the professional work of the facility. The function of the center would be to offer medical service to a population ranging in size from 50,000 to 200,000 within definite political jurisdictions, such as city wards, townships, or counties, by providing the medical profession of the area with proper facilities for diagnosis and hospital treatment of patients who required it. Physicians could be assisted in the care of their patients by medical social workers. Billings also envisaged the health center as helping practitioners to keep abreast of new developments in medical knowledge and their application, thus promoting postgraduate study and professional improvement.

The contemporary health center movement undoubtedly stimulated Billings' proposal, since he mentioned attempts to establish health or community centers intended to provide adequate medical care for rural residents and other medically disadvantaged groups. In fact, he called attention specifically to the Sage-Machold bill, introduced in the New York state legislature in March 1920, which embodied the health center program devised by Hermann M. Biggs, the state health commissioner.[180] Although the health center plans proposed by Biggs and Billings had a number of similar objectives, they differed basically on the issue of administrative responsibility, an issue which has been crucial in a variety of efforts to alter the structure of American medical practice. Billings was critical of the New York bill because in his opinion it gave insufficient consideration to "the welfare of the medical profession," by which he meant an emphasis on centralized administration through the state Health Department. In contrast, he wanted the state medical society, as the representative body of the organized profession, to regulate the relations between county and district medical societies and the proposed health centers. "These relations," he pointed out, "include the appointment or election by the county or district medical society of the medical board of management of the center, the promotion of postgraduate medical work in the center, the compilation of a uniform rational fee bill, and other objects. The branch, county or district medical society should be left free to administer its own medical affairs."[181]

The course of action advocated by Billings was not new. Opposition to changes in the organization and provision of medical services that

might limit the autonomy of the private physician and encroach upon his practice was a long-standing principle of organized medicine. Essentially, this principle was employed to maintain a system of practice functioning within a medical market where, in theory, individual practitioners competed on an equal basis according to rules embodied in a code of ethics and provided medical care to patients under a fee-for-service arrangement. This concept of the structure of medical practice in the United States never entirely coincided with the circumstances in which physicians actually worked, even in the mid-nineteenth century. As the consequences of scientific discovery and application, of industrialization and urbanization, impinged on the medical profession and its institutions, especially after the turn of the century, the gap between concept and actuality widened to such an extent that the former could have meaning only as a myth to be used as an ideological weapon to prevent or retard change. But since change could not be held back indefinitely, the negative principle of opposition had a positive corollary. If changes in the organization, financing, and delivery of medical services had to be made, it would be in the best interests of both patients and physicians to have them carried out by members of the organized medical profession.

Although expressed in different ways, this basic standpoint has not changed appreciably since the turn of the century. In 1912, Dana voiced the view that he and his colleagues did "not want to become as a profession lodge doctors, government doctors, institutional doctors. There must be opportunity for talent to rise, and genius to reach its proper level. We cannot conceive of this if we are a body of civil servants at so much per annum. But this is the trend and *we must help guide it* wisely."[182] Slightly more than two decades later, in 1934, during the Great Depression, Nathan B. Van Etten, head of the Bronx County Medical Society in New York City, still fulminated against medical charity and proposed that the Society take over a number of public health activities. According to Van Etten, its members could examine food handlers and school children, provide service in tuberculosis clinics and child health stations, and administer immunizations in their offices.[183] These activities were already being carried out by physicians within the framework of health department programs both in clinics and in private practice. The aim of Van Etten's proposal was to assert the primacy and the autonomy of the medical profession in dealing with the organization, financing, and delivery of health care, and his position was shared by a large proportion of his colleagues. As

the head of another county medical group put it, "organized medicine is the logical trustee of society in the care of public health."[184] Behind this insistence on the primacy of physicians in deciding matters of health care and on the right of the medical profession to control its own affairs was actually a desire to protect the economic position of the practicing physician. Achievement of this aim required the maintenance of a medical market limited to individual private practitioners remunerated on the basis of a fee for service rendered. Any modification of this system was regarded as potentially or actually disruptive and so was to be rejected or if necessary brought under the control of the organized profession.

As has been pointed out, resistance to change in the basic pattern of practice was evident in the later nineteenth century, but it became more intense in the 1920s and hardened in the 1930s and 1940s in opposition to proposals for insuring medical expenses. About the turn of the century, before the issue of health insurance threw its shadow over medical practice, physicians concerned with the economics of medicine were far more worried about free clinic services and various arrangements for contract practice. Opposition to dispensaries, health department clinics, pay clinics, and health demonstrations has been discussed above, but at this point it is worth noting efforts by state or county medical societies in various parts of the United States to dictate, often successfully, what health services should be provided by groups and agencies other than the medical profession. In 1928, for example, the Kansas Medical Society forced the University of Kansas Medical School to close its orthopedic clinics in the state. Furthermore, clinics sponsored by local health departments, school boards, or hospitals were attacked if they had not been approved and were not supervised by the local county medical society.[185] The situation in Middletown (Muncie, Indiana) about the same time was very similar. Another example may be cited from California. In 1919, J. L. Pomeroy, the health officer of Los Angeles County, had undertaken to develop an ambitious program of health centers. Originally these centers included clinics staffed by physicians, nurses, and social workers to provide preventive and curative services on an ambulatory basis. The clinics were available to the poor whose eligibility was established by a means test. Due largely to the complaints of physicians that medical care was being given to patients who could afford to go to private practitioners, by 1935 almost all the centers had been closed and their activities turned over to the county general hospital and the Welfare Depart-

ment.[186] The health center program foundered on the argument that undeserving individuals were abusing a service intended only for the indigent, but behind the rhetoric was a more fundamental reality.

Contract Practice

The great majority of physicians felt that programs such as those that have just been described competed unfairly with the private practitioner. Ideologically, the medical profession favored competition; in practice, however, this meant competition only on its own terms. As the depression deepened, this attitude intensified and became an uncompromising resistance to any significant change in the prevailing structure of medical practice. Physicians did not object in principle to prepayment for medical service. At issue was the possible loss of control over conditions of practice, particularly the scope of service and the arrangements for setting and collecting fees.

This position had developed initially in reaction to contract practice, a term designating an arrangement by which a physician agreed to provide medical service to groups of patients, such as members of benevolent organizations, fraternal lodges, or employees of industrial companies, for a fixed fee per annum. Contract practice apparently grew out of an earlier retainer arrangement, by which a physician accepted an annual fixed sum of money for services rendered to an individual or to a family, without regard to the amount of service. Describing the situation of the profession in the District of Columbia in the years immediately preceding the Civil War, Samuel G. Busey, a leading Washington practitioner, noted that such a system had long been in vogue among physicians practicing in Georgetown.[187] Nor was the system limited to a particular section of the United States. In 1869, on a motion by J. S. Moore, a Mississippi physician, the AMA condemned the contract system as "contrary to medical ethics," and it resolved "that all contract physicians, as well as those guilty of bidding for practice at less rates than those established by a majority of regular graduates of the same locality, be classed as irregular practitioners."[188] That such condemnation had little or no effect is indicated by the denunciation of contract practice in 1877 by the California state medical society and the Los Angeles County society. Fourteen years later, in 1891, the contract system was again attacked by the California state society, and its members were urged to sign an agreement not to

engage in this type of practice. The failure of this approach is evident from a proposal made in 1897 that the county society expel members practicing on a contract basis, but the attempt lacked support and no action was taken.[189]

After the turn of the century, the problem of contract practice aroused increasing concern. The situation was particularly acute in urban areas, where contract practice was most prominent. As Robert A. Allen, a physician, observed in 1914, "There is scarcely a city in the country in which medical societies have not issued edicts against members who accept contracts for lodge practice."[190] S. S. Goldwater, health commissioner of New York City, noted the following year, "In many localities medical care by lodge doctors is the chosen or established method of dealing with sickness among the relatively poor." To indicate the prominent role of this form of practice, he reported that in North Adams, Massachusetts, a city "with a population of 22,000, 8,000 persons are in the care of lodge physicians to whom the members pay an annual stipend for medical care."[191]

A detailed picture of contract practice in Rhode Island and in its largest city, Providence, was drawn in 1909 by George S. Mathews, a physician of that city, in a report presented to the American Academy of Medicine as part of a panel discussion of the problem.[192] In considering the distribution of contract practice in Rhode Island, he emphasized that the "lodge doctor" was almost unknown in the rural areas and small towns, and even in cities where such practice was common it was found only in some sections. However, in these areas Mathews reported "it is almost as rampant as it is in the East Side of New York City." Contract practice was common in communities and sections of cities inhabited by workers and their families, many of them immigrants or the children of immigrants. Overwhelmingly, they belonged to the working poor who were or might become medically indigent and whose situation has already been discussed above, in connection with complaints of dispensary abuse.[193]

According to Mathews, contract practice involved three types of organizations: lodges and fraternal groups, factory and shop organizations, and private clubs generally organized by physicians. As in other parts of the United States, there were in Rhode Island many branches of lodges and fraternal organizations which offered some kind of medical care among the benefits that were available to members. Mathews reported:

> The English, Irish, Scotch, Germans, French-Canadians, Jews, have clubs employing the contract doctor. The Manchester Unity, Foresters, Sons of St. George, Eagles, Owls and others are in this number. The rates for the physician vary from $1 to $2.50 per member per annum. In Providence one of these lodges numbers about 1200 members. This lodge pays its doctor $2.00, but this price includes medical attendance on the entire family. In this instance the physician's clientele must be between 4,000 and 5,000. Surgery and obstetrics are not included.
>
> Among the Jewish people of Providence it is estimated that one-third have contract doctors. in the Olneyville and Mount Pleasant districts it is estimated that 50 percent of the wage-earning men are members of lodges employing contract doctors. In the populous Pawtucket Valley mill towns at least six medical men . . . are engaged in lodge practice.[194]

A number of factories and shops also had organizations of this kind, limited to their own employees. Mathews describes two such "clubs" in one factory, the larger having a membership of about 700, the smaller 400. The larger club paid its physician $2.25 a year per member, inclusive of medicines. For similar services the other club paid $2.00 per person annually. Apparently it was more exclusive, since its membership was limited to workers who earned at least $12 and as high as $30 per week. (The average weekly wage in the larger club was from $10 to $15 a week.) In the factory where the members were employed, their requests for office or house calls were posted on a slate, and the practitioners hired by the clubs called daily to note who wanted to see them. In the larger club there were between fifteen and thirty-five office visits and two to three house calls a day.

The third type of contract practice was in the form of groups organized by medical practitioners. They were small but numerous among the foreign-born in some sections. Ten or a dozen families might be brought together, with each family paying the physician three to five dollars. Under this arrangement medical attendance was provided for all members of the family, but surgery, obstetrics, and medicaments were excluded. The largest of these clubs solicited members in factories and stores. For one dollar a year, medical and minor surgical conditions were treated at the office. Home visits were not included, and a small fee was charged for medicines.

Mathews' comment that contract practice was exceedingly widespread in New York's lower East Side is supported by other sources. Newly arrived immigrants crowded together in separate neighbor-

hoods in the large cities, where they soon organized institutions within which they could establish a sense of community and which would provide various forms of assistance when needed. Jewish immigrants in New York City, for example, formed benevolent associations, progressive societies, or similar groups. Such a lodge or, to use the generic Yiddish term, *landsmanschaft* was initially made up of persons from the same town or region in the country from which they had emigrated. The great majority of these lodges were formed during the first decade of the century at the height of the great migration. Though estimates of their number vary from some 3,000 to almost double that figure, the lower estimate is probably closer to reality.[195] In 1910 Morris J. Clurman, a physician practicing on the East Side, reported that "there are in existence downtown somewhere between 1,500 and 2,000 lodges, societies and benevolent associations founded mainly by the poorer class of workingmen for a double purpose; namely social intercourse and mutual aid or benevolence."[196]

Among the benefits provided by the lodges to their members was attendance by a physician when illness occurred. As Clurman put it, "An iron-bound practice or custom has arisen for each society to elect some physician to take care of the health of the society members—for a consideration." The society paid the physician on a capitation basis. In New York in 1910 the average rate of remuneration was one dollar per annum for an unmarried member and three dollars for a married member and his family, but not infrequently the retainer was smaller. What this meant in practice has been graphically described by Samuel Silverberg, who had been a lodge doctor in this period. He recalled in 1972 that

> [the] society would pay me a certain amount for coverage for a certain number of patients—fifty cents for a single member every three months, seventy-five cents or a dollar for a family. Every member had a right to come to my office and ask me to call at his house. I took the job because in that way I was sure of being able to pay the rent for my office. On my own I took in very little . . . I delivered babies in the house and would get a practical nurse to follow up for a week or so. The society member paid extra for the delivery of babies, something like ten or fifteen dollars, as I remember.
>
> The society member would recommend the doctor to his friends, and in that way you could build up a practice. But it was hard, lots of running

> up and down tenement stairs. When I moved my office to the Grand Concourse, I gave up the society.[197]

There were advantages as well as disadvantages in such schemes, for both practitioner and patient. Physicians undertook lodge practice as a means of obtaining a reputation and a clientele. The lodge doctor found that at the bedside he came in contact with a large group of patients who otherwise would not have called upon his services. After establishing a reputation as a busy practitioner, in part due to the recommendations of numerous families whom he had attended as a lodge doctor, and in part as a result of the normal growth of practice, the physician tended to disengage himself from lodge practice.

For needy young practitioners, as well as for older physicians with meager earnings, a position as a lodge doctor meant a relatively assured minimum income. As a result, competition for such posts was keen. Since the members of the lodge or society elected the physician who would look after their health needs, candidates frequently electioneered for votes. At an election where there were several candidates, it was not unusual for each one to come prepared with printed ballots bearing his name for distribution among lodge members. Some used more devious means to obtain the desired position. Given these circumstances one is not surprised to find a considerable degree of ambivalence in attitude and behavior on both sides of the physician-patient relationship. The situation is well summed up by Silverberg: "Some doctors were devoted," he observed, "many not. Some patients took advantage of the system and it wasn't always very pleasant. Most society members treated their doctor with respect, but some said, 'A society doctor? What can he know?' For more serious illnesses, they'd go to another doctor."[198] In short, how competent could a physician be who offered his skills the way hucksters cried out their wares?

This attitude had its counterpart in the feeling of many physicians that lodge members, for the most part workers, were hardly the best judges of the professional merits of a medical practitioner. Patient behavior in numerous instances tended to reinforce this view. The physician was obligated to see professionally as many patients as requested his services. Sometimes patients called for apparently trivial reasons, arousing the resentment of the physician, who felt his time was being wasted. Practitioners who served more than one lodge, in some instances as many as seven or eight, probably saw thirty or more lodge patients in the office or at home in the course of an average day.

At the same time, these physicians endeavored to establish or to maintain a separate practice free of the demands and uncertainties of lodge doctoring. Under these conditions it was practically inevitable that the quality of care would be uneven, and in numerous instances poor. Superficial and cursory examination was not unusual, and more often than not therapy was directed toward relief of symptoms.

But there were also competent and conscientious physicians whose practice was not slipshod, who did not make snap diagnoses, and who were painstaking in the care of their patients and treated them to the best of their ability. Yet even the best lodge doctors had definite limitations. They were general practitioners who worked individually, equipped with whatever knowledge and skills they had acquired through education and experience.[199] As long as medical practice was relatively simple, a diligent physician could care for lodge patients in a reasonably satisfactory manner. For lodge members, the provision of sickness benefits inclusive of medical care offered some protection against the burden of financial loss through illness. Furthermore, without a doctor available, a considerable number of lodge members would probably have consulted a pharmacist and received some patent medicine, or would have used traditional home remedies, rather than incur the expense of a visit to a physician and the filling of his prescription. Very likely, some lodge members used all these resources, as well as dispensaries.

Contract practice had developed out of the need to provide medical care for groups whose social and economic circumstances made it difficult if not impossible to do so individually. For this reason, a number of physicians urged the acceptance of contract practice by the medical profession, subject to the establishment of conditions which would be equitable for both patients and practitioners. In 1899, a program for prepaid medical care on a contract basis was proposed to the Illinois State Medical Society.[200] Families would be urged to contract for annual medical care, with fees to be paid quarterly or monthly. The authors of the plan argued that physicians had to take account of the impact on medical practice of the scientific and social changes that were taking place and would continue. In view of the increasing significance of preventive medicine, its application in medical practice would be facilitated by the proposed arrangement.

In 1890, two Chicago physicians, J. K. Crawford and Oscar De Wolf, the latter a former health commissioner, started a prepaid program called the Mutual Medical Aid Association of Chicago. Its

objective was to "secure to those of limited means prompt and efficient medical and surgical treatment in cases of sickness or accident, by a corps of competent physicians and surgeons, at nominal cost." This last phrase was somewhat exaggerated, since the dues paid by members of the association ranged from twelve dollars per annum for a single person to twenty dollars per annum for a family of five or more. Membership dues, paid quarterly, covered all professional services except obstetrics, which cost an additional ten dollars. Salaried physicians employed by the association provided the medical care. The organizers of the plan insisted that it was intended to provide health protection for workers in factories and plants and that it was founded on a well-established insurance principle, but the Chicago Medical Society nevertheless condemned their activities.[201]

Yet the real problem which Crawford and De Wolf tried to solve did not disappear. In Chicago, as elsewhere in the United States, there was a growing feeling that some form of properly organized contract practice would be preferable to the existing situation. A committee of the Chicago Medical Society, appointed to study contract practice, reported in 1907 that "many of the men working under these various contracts are desirous of improving the conditions of things, that they are not wanton violators of the ethical codes and that they are willing to cooperate in any amicable solution of the question."[202] Furthermore, recognition that the economic necessities of many patients and physicians made some form of prepaid medical care inevitable became increasingly prevalent by the second decade of the present century. Indeed, by 1913 the Judicial Council of the AMA concluded that "lodge practice under certain circumstances is one of health insurance that must be accepted and controlled, not condemned and shunned." In 1914 Robert A. Allen, surgeon for the A. C. White Lumber Company of Idaho, said it was not surprising that "workingmen who have learned to organize themselves in trade unions should also unite for mutual protection against accident and illness. It is not surprising that American workingmen should be following in the footsteps of their European confrères with their Friendly Societies and their Krankenkassen." Pointing to the social legislation of various European countries, particularly Great Britain and Germany, Allen noted that there were "many indications that we are rapidly approaching the time when similar government insurance will be adopted in the United States." It was only a question of time before Americans would also have "state insurance against sickness, non-employment and old

age."[203] The following year S. S. Goldwater also saw the United States moving toward sickness insurance, and he warned that team work was essential for good medical care. This goal could be achieved through hospital clinics and dispensaries for ambulatory patients, where "the medical work will no longer be charitable work, but part of a scheme of social insurance," in which "adequate compensation will have to be provided for the doctors."[204]

In that same year Stephen A. Welch, in his presidential address to the Rhode Island Medical Society, confronted this prospect and raised a number of specific questions that were to be of considerable future significance. If a prepayment scheme (or "cooperative practice," as he termed it) were instituted, who would be included, and what would these individuals pay? How many persons should be assigned to a given contract physician? What is the proper remuneration for a practitioner caring for a given number of people? Should the medical society negotiate an agreement on what the fees shall be, whether with the state, with an insurance company, or with whomever would pay the practitioners providing service? Should the medical society provide a list of suitable or approved physicians for the persons covered by the scheme, and how would they be selected? Welch emphasized the need for the organized profession to study the problem of sickness insurance in terms of such questions, otherwise "it may pass into practice without such reasonable regulations as physicians are best qualified to suggest."[205]

Others beside Welch had also recognized the complexity of the problem and identified major issues. Commenting on the discussion of contract practice at the annual meeting of the American Academy of Medicine on June 7, 1909, Charles McIntire, editor of the Academy's *Bulletin,* listed among the many factors at issue, "the principle by which the physician should charge for his services;" and he asked, "Is the usual way of the family physician, the only way, or in fact the best way?" Equally salient was the question of determining adequate compensation for professional services, particularly since a major objection to contract practice was the inadequate remuneration of the physician, especially by benevolent societies, fraternal orders, lodges, clubs, and the like. To deal with these issues as well as with the problem of the medically indigent, which McIntire recognized, he asserted that "a plan by which the wage earner may receive proper medical attention and the physician be properly paid for his services" could be achieved only on a cooperative basis. He insisted that, to obtain a

secure foundation for such an enterprise, "facts must be known, and as the life insurance actuary prepares his tables of the expectancy of life, so must tables be prepared to determine the expectancy of disease. Upon these tables the fee to be paid to the lodge or society must be determined. In the second place the proper fee to recompense physicians for such services should be determined." As a standard, McIntire suggested that a sum equal to the average salary of physicians who held salaried positions in medical schools, or who were health officers, or researchers in hospital laboratories, be paid to the physician for an equal period of time in a lodge contract. Another question raised by McIntire concerned the average number of families to be cared for by a physician so that proper attention could be given to each person who might be ill without making such excessive demands on the practitioner that he would have no time for study and recreation. Finally, McIntire urged that the contracting organization, the lodge or society, "should charge its members a fee sufficiently large to pay for the entire time of as many physicians as may be necessary to give each one a proper clientele, and pay a fitting salary. Under such conditions, no odium could attach to Contract Practice."[206]

Five years later, in 1914, Albert T. Lytle, a physician of Buffalo, New York, addressed himself to the same questions, particularly to the problem of the physician's remuneration. Lytle attributed this problem to the chaotic state of the medical market, resulting from changes in the organization of medical practice. "The rapid development of surgery," he said, "the startling growth of the specialties, the immense and expensive scientific equipment required, the great cost of a medical education, the pharmacist and the nurse, all have had a disastrous influence upon the former well-balanced remuneration of the physician."[207] To correct the situation, Lytle made several proposals to be acted upon by state and county medical societies. One was the creation of a committee on medical economics to study and to make recommendations on all economic questions of concern to practitioners. Accepting the inevitability of contract or prepaid practice, Lytle urged that such a committee, as one of its major priorities, ought to establish the "service-value" or cost of specific services rendered by physicians. These determinations could then be used in negotiations with insurance companies, government agencies, or lodges, and they would make it possible to set "minimum medical remuneration, or selling prices." To insure compliance with these standards by all physicians, Lytle proposed unionization of the profession by bringing "all eligible prac-

titioners into the medical societies" which would represent them vis-à-vis other groups. Finally, to put these proposals into effect, to work out the problems of medical service costs and remuneration, and eventually to obtain valuable statistical data, he suggested the organization of a Mutual Health and Accident Insurance Society, which would be incorporated, financed, and managed by physicians. The policies written by the society would not provide for cash indemnity in case of illness, but would "furnish medical and surgical services and treatment, medicines, dressing, hospital accommodations and nursing attention during the period of a disability occurring during the life of the policy. The physicians and surgeons rendering such service should be stockholders or members of the association and they should receive from the association fees according to a proper schedule."[208] To minimize fraud, a few high-salaried full-time physicians of established competence should be appointed as inspectors or supervisors. Based on a population of 500,000 served by 500 physicians, Lytle estimated that an annual premium of ten dollars would produce enough income to cover costs and to provide for a sinking fund and dividends. Since the society would be a nonprofit organization, any dividends would be used as far as possible to improve the quality of care and to reduce costs while providing the physicians with a secure and respectable living.

As revealed by the discussions of contract practice, the decades immediately before and after the turn of the century were a period of ferment during which alternative methods of financing and providing medical service were considered by the medical profession. Embedded in these discussions are premises that have remained central to positions taken since then by the majority of the medical profession on questions of health policy and medical economics.

According to these premises, the needs and the welfare of patient and physician coincide. Since the economic security of the profession is essential to any health program, medical activities that do not assure the economic position of the private practitioner are fundamentally unsound. Consequently, society through its various agencies must not compete with the practicing physician in any way that would impair the structure of individual medical care or the economic status of the practitioner. As representative agencies of the profession concerned with public benefit, medical societies should determine whether a contract or prepayment arrangement for medical care is in the best interests of the patient and the physician. Logically, therefore, any plan

not approved by a medical society would be contrary to the public welfare.

The heart of this ideology was thus the control of health activities by the medical society, reflecting the desire of physicians, particularly general practitioners, not only to maintain a medical market based on individual fee-for-service practice but also to prevent various institutions and agencies from encroaching on the domain of medical practice and competing with the individual practitioner. This position is understandable if seen from the point of view of a practitioner who felt exploited and increasingly threatened economically by health departments, workmen's compensation and industrial medicine, hospitals, free and pay clinics, and last but certainly not least, compulsory health insurance.

The Struggle Over Sickness Insurance

Just as physicians were dissatisfied with the benefits they received from contract medical practice, so others concerned with the welfare and the health problems of American wage earners found that most of the prepayment schemes provided inadequate protection for their presumptive beneficiaries. At the first annual meeting of the American Association for Labor Legislation, held in Madison, Wisconsin, in December 1907, delegates discussed a program of social legislation dealing particularly with the problems of wage earners. They were well aware that some means of insurance was needed to assist the wage-earner during periods of non-work related sickness."[209] In June 1911, at the National Conference of Charities and Corrections in Boston, the future Supreme Court justice, Louis D. Brandeis, addressed the attending social workers on the need for social insurance against sickness, old age, disability, and unemployment. A year later the conference adopted a report favoring legislation to achieve this aim, and it sought to have the basic points of the report embodied in the plank on social legislation of the Progressive party for the election of 1912.

That same year, in December, the American Association for Labor Legislation appointed a Committee on Social Insurance. Its membership consisted of four economists, an actuary, two insurance company statisticians (one of whom, I. M. Rubinow, was also a physician), a leading social worker, and an official of the U.S. Bureau of Labor. Shortly afterward the Committee added to its membership two prom-

inent physicians: Alexander Lambert, professor of clinical medicine at the Cornell University Medical School, and S. S. Goldwater, who had been health commissioner of New York City and was then superintendent of Mt. Sinai Hospital. At the same time, Lillian D. Wald, head of the Henry Street Settlement and a leader in public health nursing, also joined the committee.

In December 1915, after almost three years of work, the committee presented a "standard bill" for compulsory health insurance that would cover every employed person earning $1,200 or less. Special provisions were included to cover casual workers and those who worked at home. Cash and medical benefits were specified, the latter comprising "all necessary medical, surgical and nursing attendance treatment," which would be provided from the first day of illness for a period of up to twenty-six weeks in any one year. Maternity benefits were included for insured women and the wives of insured men. Surgical supplies, medicines, eyeglasses, and related materials would be provided up to a maximum of fifty dollars for any one patient in a single year. These services were to be provided through insurance carriers, created in a state and supervised by a state social insurance commission, consisting of three members, one of whom would be a physician. Services could be supplied by one of three different methods or by a combination of the three. The first was to set up a panel which all legally recognized physicians would have a right to join and from which the patient could choose freely, though the physician had a right to refuse a patient on specified grounds. To insure high-quality care, no practitioner was to have on his list more than 500 insured families or more than 1,000 insured individuals. A second method was the employment by the insurance carriers of salaried physicians, from among whom insured patients could also choose the doctor they wanted. Thirdly, arrangements could be made for district medical officers to care for insured persons in specifically designated areas. To satisfy the insistence of the medical profession on autonomy, the bill included a provision for a medical advisory board, chosen by the state medical society, that would be consulted on all medical matters and would deal with medical disputes that might arise.[210]

Beginning in 1916, a well-organized campaign got under way to have state legislatures enact bills establishing compulsory sickness insurance. Between 1916 and 1920, the standard bill was introduced in fifteen state legislatures, nine of which appointed commissions to study the proposed program. In five of these, the commissions reported

in favor of compulsory sickness insurance, but no legislation was enacted.[211] By 1920 the campaign for compulsory sickness insurance was over, defeated by the combined opposition of employers, insurance companies, organized labor, and the medical profession and a few allied professional groups such as dentists and pharmacists.

Throughout the period from 1900 to 1920, the organized medical profession, represented by the AMA and its constituent state and county societies, had taken several different positions on prepayment schemes intended to deal with the problem of meeting the costs of medical care. Although opposed to contract practice of the kinds discussed above, the representatives of the medical profession were willing to consider alternatives, including government-sponsored sickness insurance. In 1915 an editorial in the *Journal of the Kansas Medical Society* accepted the likelihood "that in a few years at least, the medical profession will have an opportunity to try out some plan of sickness insurance under federal supervision."[212] This readiness to discuss proposals and plans is also evident in the pages of the *Journal* of the AMA. On July 18, 1914, for example, the *Journal* carried an article on the socialization of medicine by James P. Warbasse, a physician of Brooklyn, New York, who advocated the Rochdale principles of cooperation as a basis for the organization and delivery of medical care. "The value of cooperation in science is proved," he wrote. "Medical practice withholds itself from the fields of science as long as it continues a competitive business."[213] Another instance was the publication early the following year, on January 30, of an article on social insurance and the medical profession by I. M. Rubinow, a member of the Committee on Social Insurance and a leading advocate of compulsory sickness insurance.[214]

This relatively open-minded attitude at the national level was also due in considerable measure to the leadership of the AMA. Alexander Lambert, who had also served as a member of the Committee on Social Insurance, was chairman of the AMA's Judicial Council. In January 1916, Henry B. Favill, a leading Chicago internist, who was chairman of the Council on Health and Public Instruction, proposed to Lambert the creation of a committee to study social insurance in its relation to the medical profession, since "in all possibility this subject would be, in the next few years, one of the most important that would occupy the attention of the profession both from an economic and a purely medical and surgical point of view." The committee was to consist of Favill, Lambert, and Frederic J. Cotton, who had been chair-

man of the Committee on Compensation of the Massachusetts State Medical Society. On February 9, 1916, the AMA's Board of Trustees onfirmed the appointment of this committee, charging it with the compilation of information concerning "social or health insurance and the relation of physicians thereto; and to do everything in their power to secure such constructions of the proposed laws as will work the most harmonious adjustment of the new sociologic relations between physicians and laymen which will necessarily result therefrom . . ."[215] Rubinow was engaged as executive secretary of the committee, and an office was opened in New York City.

At a previous meeting of the AMA, Lambert had reported on social insurance in a number of European countries, with particular emphasis on Great Britain. Further reviews of various aspects of social insurance (workmen's compensation, invalidity, old age, unemployment, and sickness insurance) were carried on throughout 1916 and 1917, and articles and letters discussing compulsory sickness insurance continued to appear in the AMA *Journal.* Many of the letters, in particular, were favorable to the principle of insurance and to the system proposed for the United States. Around 1917, however, signs of rising opposition began to appear.

Although the Committee on Social Insurance did not formally recommend that the AMA endorse compulsory sickness insurance, it did propose to the Association's House of Delegates a set of principles to be included in a system acceptable to the medical profession. Legislation setting up a sickness insurance system, it said, should "provide for freedom of choice of physician and by the insured; payment of the physician in proportion to the amount of work done; the separation of the functions of medical official supervision from the functions of daily care of the sick, and adequate representation of the medical profession on the appropriate administrative bodies."[216] At the same time, in a statement which in large measure forecast future developments, the committee underscored the need for the profession to participate in working out problems of medical service under social insurance legislation.

> These problems must be faced by it if the profession is to obtain the justice that is due it, if it is to protect its economic position in the community. Laymen, however, just do not understand the point of view of the profession and do not understand the peculiar problems that the professional life of a physician and surgeon brings forth. There is no question but that the intense reaction in many states against sickness insurance or any further

> expansion of the social insurance laws in this country is due to the injustice and the cold-blooded unfairness by which physicians and their patients have been treated under the workmen's compensation laws. To work out these problems is a most difficult task. The time to work them out, however, is when the laws are molding, as now, and the time is present when the profession should study earnestly to solve the questions of medical care that will arise under the various forms of social insurance. Blind opposition, indignant repudiation, bitter denunciation of these laws is worse than useless; it leads nowhere and it leaves the profession in a position of helplessness if the rising tide of social development sweeps over them. The profession can, through its influence in the community, prevent for a time these laws being passed and it can, by a refusal to cooperate, still further retard them, but in the end the social forces that demand these laws and demand an improvement in the social existence of the great mass of the people of the nation will indignantly force a recalcitrant profession to accept that which is unjust to it and that which is to its detriment.[217]

Although the House of Delegates approved the report of the committee and endorsed the proposed principles, the prophetic warning which accompanied them was not heeded. The entry of the United States into World War I in April 1917 soon turned the attention of the medical profession and of the public to wartime problems and away from such issues as social insurance. When the subject of compulsory sickness insurance was again raised within the American Medical Association in 1919, it encountered a much chillier reception. The Committee on Social Insurance was reorganized, and though Alexander Lambert continued as chairman, I. M. Rubinow left his position as its secretary. Furthermore, several new appointees to the committee were unsympathetic or opposed to compulsory sickness insurance. Thus the tenuous link between the medical profession and the advocates of compulsory sickness insurance was disrupted.

Meanwhile large groups in state and county medical societies had become increasingly vehement in their opposition to compulsory insurance as well as to activities by governmental and voluntary health agencies. On January 16, 1920, at a meeting of the Cleveland Academy of Medicine, the practitioners of that city were warned that unless they united to protect their interests they would find themselves

> *in the near future* working *only* for the State and receiving such fees as the state may see fit to pay you; that the profession of medicine as now practiced is in danger of being engulfed in State medicine *at the present session* of the State Legislature; that unless the profession . . . is properly fore-

> warned and prepared the Doctor is about to become a mere tool of Capital, Labor and the Politicians, to be used and cast aside as fits their fancy; and that this pauperization of the profession can be prevented by concerted and united action of its individual members.[218]

Only through the compelling influence of an aggressive, efficient organization would the profession be able to make its desires known to the public and to the legislative branches of government and be in a position to exert influence on matters concerning medicine. With even more flamboyant rhetoric, the Illinois State Medical Society in 1920 instructed its delegates to the annual meeting of the AMA in New Orleans to "oppose state medicine, compulsory health insurance, county and state health agencies, and 'allied dangerous Bolsheviki schemes.' "[219]

Deeply imbued with these sentiments, which were particularly strong not only in Illinois and Ohio but also in California, Michigan, and New York, a large and powerful group of delegates went to New Orleans "to get Lambert," who was then president of the Association, and to establish a basic policy of opposition to compulsory sickness insurance on the part of organized medicine. This goal was achieved on April 27, 1920, when the House of Delegates adopted a resolution stating that "the American Medical Association declares its opposition to the institution of any plan embodying the system of compulsory contributory insurance against illness, or any other plan of compulsory insurance which provides for medical service to be rendered contributors or their dependents, provided, controlled, or regulated by any state or Federal government."[220]

The die had been cast, and over the next two decades the hostility of organized medicine to any form of governmental action in the interest of health, except for certain limited areas, grew ever more implacable and was expressed with increasing stridency. Members of the AMA were pressured to conform to the official policy. Frank Billings, a leading Chicago internist and a trustee of the AMA who also served twice as its president, had initially been open-minded on the issue of compulsory sickness insurance. By 1921, he found himself under attack on that score, and he soon embraced the official line.[221] Nor was he an isolated case in accepting the policy of attacking all forms of "state medicine." Adopted as the code word for various kinds of health action not sanctioned by the organized medical profession, state medicine was defined in 1922 by the AMA as:

> any form of medical treatment, provided, conducted, controlled or subsidized by the federal or any state government, or municipality, excepting such service as is provided by the Army, Navy and Public Health Service, and that which is necessary for control of communicable diseases, the treatment of the indigent sick, and such other services as may be approved by and administered under the direction of or by a local county medical society, and are not disapproved by the state medical society of which it is a component part.[222]

In the same year the Sheppard-Towner Act, which provided grants-in-aid to states for maternal and child health programs, was passed by the Congress, and even though it did not involve compulsion it was officially disapproved by the House of Delegates because it was viewed as opening the door to government encroachment on private practice. Opposition by the AMA to the World War Veterans Act of 1924 was based on the same ground.[223] Only when activities by health agencies appeared to benefit the private practitioner and not to diminish his autonomy was there no opposition from the medical profession. Fundamental to the position taken by the organized medical profession in the 1920s and the 1930s were two questions: What should be the structure of medical practice?, and, Who should control it? These basic issues and the responses of the majority of the profession to various efforts to deal with them are major themes in the evolution of medical practice in the United States during the past fifty years.

The advent of the depression in 1929 and its devastating impact in the succeeding years rendered the issues more acute and intensified the opposition of organized medicine. In 1934, with the adoption of ten principles governing health insurance, the position of the AMA assumed the form it was to maintain in subsequent decades. The crucial points remained unaltered: All aspects of medical service had to remain under the control of the medical profession, and no third party could be permitted to intervene between patient and physician. At the same time, the Association mobilized all the resources at its disposal and developed sanctions to be used against those opposed to its policies on the economic and political aspects of medical practice. The code of ethics became a weapon for punitive action against dissenters. An example is the Mundt Resolution, passed by the House of Delegates in 1934, which urged that the staffs of hospitals accredited for internships be limited to physicians who belonged to their local medical society. In essence, this meant that practitioners who defied the Association in matters of medical economics and were

consequently denied membership in or expelled from county societies on "ethical" grounds would not be able to obtain appointments in these institutions and would thus be denied facilities for the care of their patients. Nor was this an idle threat.[224] The crucial position occupied by the AMA, acting through the Council on Medical Education and through its connections with the specialty boards, in the approval of schools for undergraduate medical education and of hospitals for internships and residencies, had enabled it to acquire a concentration of power undreamed of at the beginning of the century. By 1940 the power of organized medicine could be and was exercised over a wide range of medical affairs. Nowhere, however, was this power more evident or applied more actively than in the efforts of organized medicine to retain the existing structure of medical practice or, if change could not be avoided, to introduce only modifications that were under professional control. The major objectives were to determine the conditions of practice and remuneration and to prevent or restrain expansion of governmental involvement in health affairs.

Conclusions

Just before World War II, as a consequence of developments during the preceding forty years, the medical profession had achieved most of the economic and political advantages envisaged at the turn of the century. Advances in medical science had made possible improvement in the quality of medical education and had lengthened the period of training by requiring internships for all graduates and in addition residencies for aspiring specialists. During the same period licensing was tightened. By these means restrictions were placed on entry into the profession and practice within it. The number of physicians being educated was limited. At the same time, scientific and clinical knowledge and expertise were used to establish criteria with which to circumscribe areas of specialization, to validate the competence of those who wished to practice in them, and to exclude self-styled specialists. These developments went hand in hand with, and were intimately related to, the expanding and changing functions of the hospital. The institutionalization of medicine, specifically in the hospital, fostered an increasingly hierarchical structure of medical practice, involving not just physicians but also growing numbers of auxiliary paramedical personnel. By establishing itself in a strategic position relative to these devel-

opments, the AMA, representing the majority of the medical profession, was able to avert major changes in the structure of medical practice and to achieve a sufficient degree of control over the medical market to produce a stable situation for medical practice, thus creating conditions which would be economically advantageous for private practitioners.

However, these changes in the situation of the medical profession did not occur in isolation from the rest of society but were closely intertwined with political, economic, and social developments. Both the movement to reform medical education and the campaign for compulsory sickness insurance were products of the Progressive era and exhibit many of its characteristics. As with so many of the reform efforts of this period, there was a desire to bring order into areas of social life that were in disarray, to develop mechanisms by which these aims could be attained, and to develop criteria and standards for the evaluation of these achievements. The varying fates of these endeavors can be attributed to the circumstances and the times in which they occurred. Reform of medical education was undertaken when Progressivism was at its peak, and with the support of most of those involved—organized medicine, medical schools, licensing authorities, and foundations. On the other hand, the movement for compulsory sickness insurance came at a time when Progressive liberalism received a mortal blow from the effects of World War I. As peace returned, projects pushed by Progressives crumbled before their eyes, among them compulsory insurance.

Opposition to sickness insurance was also fueled by other developments outside the medical profession. By 1920, frightened at the turn taken by the Russian Revolution, large segments of the American public reverted to conservatism, looking askance at anything that smacked of radical social innovation. The rhetoric of the Illinois State Medical Society cited above reflected this atmosphere and illustrates how it was used as a weapon against change in medicine. This attitude, shared by a majority of the medical profession, especially in the heartland of the country, also reflected the rearguard effort by rural and small-town America to oppose the social and economic trends that were inexorably altering their world.[225] Like other groups in American society, a large body of medical practitioners experienced difficulty in adjusting to the rapid pace of social and professional change, preferring to hark back to a past and to hold on to a present which they knew, rather than to plunge into an unknown and uncertain future.

These tendencies were further strengthened by an increasing separation between academic physicians, oriented toward teaching and research, and those physicians who were oriented toward private practice.

Nevertheless, despite the high degree of control of the medical market and the relative stability of the structure of practice achieved by organized medicine by 1940, there remained unsolved problems and uncontrolled factors capable of upsetting the unstable equilibrium upon which they rested. Of major importance in this connection was the continuing scientific revolution in medicine and its impact on medical education and practice. Closely related were the changes in the demographic structure of the American people and the emergence of new problems of medical care. Intertwined with these developments were the problems of financing, organizing, and delivering health services. At the outbreak of World War II the effects of these trends had not yet been fully felt, but over the subsequent thirty years they would upset the fragile balance that had been achieved and begin to alter the structure of medical practice in the United States.

ABBREVIATIONS

Am. J. Clin. Path.	American Journal of Clinical Pathology
Am. J. Med. Sci.	American Journal of Medical Science
Am. J. Pub. Health	American Journal of Public Health
Bull. Am. Acad. Med.	Bulletin of the American Academy of Medicine
Bull. Hist. Med.	Bulletin of the History of Medicine
J. A. M. A.	Journal of the American Medical Association
J. Assn. Am. Med. Coll.	Journal of the Association of American Medical Colleges
J. Hist. Med.	Journal of the History of Medicine
J. Social Hist.	Journal of Social History
N. Y. State J. M.	New York State Journal of Medicine
Persp. in Biol. and Med.	Perspectives in Biology and Medicine
Proc: Nat. Conf. Char. and Corr.	Proceedings of the National Conference on Charities and Corrections
Trans. A. M. A.	Transactions of the American Medical Association

NOTES

FOREWORD

1. See, for example: E. H. Ackerknecht, "A Plea for a 'Behaviorist' Approach in Writing the History of Medicine," *J. Hist. Med.* 22 (1967): 211–14; George Rosen, "People, Disease, and Emotion: Some Newer Problems for Research in Medical History," *Bull. Hist. Med.* 41 (1967): 5–23; and Charles E. Rosenberg, "The Medical Profession, Medical Practice, and the History of Medicine," Edwin Clarke, ed., *Modern Methods in the History of Medicine* (London: Athlone, 1971), pp. 22–35.

2. For evaluations of Rosen's work, see Charles E. Rosenberg, "George Rosen and the Social History of Medicine," and Saul Benison, "George Rosen: An Appreciation," both in Charles E. Rosenberg, ed., *Healing and History: Essays for George Rosen* (New York: Science History Publications, 1979), pp. 1–5, 242–251.

PROLOGUE

1. See, for example, Guenter B. Risse, Ronald L. Numbers, and J. W. Leavitt, eds., *Medicine without Doctors: Home Health Care in American History* (New York: Science History Publications, 1977); and John Woodward and David Richards, eds., *Health Care and Popular Medicine in Nineteenth Century England* (New York: Holmes & Meier, 1977). The most widely used and reprinted of such domestic guides was William Buchan's *Domestic Medicine;* published originally in Edinburgh in 1769; it was reprinted in scores of American editions by the mid-nineteenth century. See Charles E. Rosenberg, "Medical Text and Social Context: Understanding William Buchan's *Domestic Medicine,*" *Bull. Hist. Med.* 57 (1983): 22–42.

2. Manuscript "receipt" books kept by ordinary housewives routinely included directions for the use of botanic remedies. In 1800, the most popular guide to such herbal remedies was still Nicholas Culpeper's *English Physician*—an "astrologo-botanical" text originally published in the middle of the seventeenth century. The first full-length American herbal did not appear until 1801: Samuel Stearns, *The American Herbal, or Materia Medica. Wherein the Virtues of the Mineral, Vegetable, and Animal Productions are Laid Open . . .* (Walpole, N.H.: Printed by David

Carlisle, 1801). See also Christa M. Wells, "A Small Herbal of Little Cost, 1762–1778: A Case Study of a Colonial Herbal as a Social and Cultural Document" (Ph.D. diss., University of Pennsylvania, 1980).

3. Barnes Riznick, "The Professional Lives of Early Nineteenth-Century New England Doctors," *J. Hist. Med.* 19 (1964): 1–16.

4. The most detailed account of medical education in antebellum America is William F. Norwood, *Medical Education in the United States before the Civil War* (Philadelphia: University of Pennsylvania Press, 1944).

5. Charles E. Rosenberg, "The Practice of Medicine in New York a Century Ago," *Bull. Hist. Med.* 41 (1967): 230–231. See also Donald E. Konold, *A History of American Medical Ethics, 1847–1912* (Madison: State Historical Society of Wisconsin, 1962).

6. William Dosite Postell, *The Health of Slaves on Southern Plantations* (Baton Rouge: Louisiana State University Press, 1951); Todd L. Savitt, *Medicine and Slavery: The Diseases and Health Care of Blacks in Antebellum Virginia* (Urbana: University of Illinois Press, 1978).

7. These summary remarks are based on a reading of medical account books held by the State Historical Society of Wisconsin, Minnesota Historical Society, Ohio Historical Society, Duke University, Southern Historical Collection (University of North Carolina), and the South Caroliniana Collection (University of South Carolina). This research was undertaken as part of a more general study of medical care in America.

8. George Rosen, *Fees and Fee Bills: Some Economic Aspects of Medical Practice in Nineteenth Century America*, Supplements to the Bulletin of the History of Medicine, No. 6 (Baltimore: Johns Hopkins University Press, 1946).

9. In 1798, the United States Congress did establish a prepaid health scheme for merchant seamen—a plan which does constitute something of an exception to this generalization. Twenty cents a month was deducted from the seaman's wages and the proceeds were paid to local collectors of customs. This action remained atypical throughout the nineteenth century, both in the federal provision of hospital care and in the prepaid aspect.

10. The following pages are drawn largely from Charles E. Rosenberg, "The Therapeutic Revolution: Medicine, Meaning, and Social Change in Nineteenth-Century America," *Persp. in Biol. and Med.* 20 (1977): 485–506.

11. Diabetes could be diagnosed by the urine's characteristic sweetish taste. Since the medieval period, physicians had paid careful—and at times whimsically elaborate—attention to the observation of a patient's urine (uroscopy).

12. Still the best historical discussion of this phenomenon is that by Max Neuburger, *Die Lehre von der Heilkraft der Natur im Wandel der Zeiten* (Stuttgart: Ferdinand Enke, 1926).

13. For an important discussion of this general distinction, see Owsei Temkin, "The Scientific Approach to Disease: Specific Entity and Individual Sickness," in *The Double Face of Janus and Other Essays in the History of Medicine* (Baltimore: Johns Hopkins University Press, c.1977), 441–455.

14. The two preceding paragraphs are drawn from Rosenberg, "Therapeutic Revolution," pp. 503–504.

15. Hospital records indicate that, throughout the nineteenth century, single men were admitted in numbers far out of proportion to their numbers in the general population.

16. For a general discussion of the dispensary, see Charles E. Rosenberg, "Social Class and Medical Care in Nineteenth-Century America: The Rise and Fall of the Dispensary," *J. Hist. Med.* 29 (1974): 32–54.

17. William A. Williams, *America's First Hospital: The Pennsylvania Hospital, 1751–1841* (Wayne, Pa.: Haverford House, 1976); Charles Lawrence, *History of Philadelphia Almshouses and Hospitals* . . .([Philadelphia]: The Author, 1905); Charles E. Rosenberg, "From Almshouse to Hospital: The Shaping of Philadelphia General Hospital," *Health and Society* 60 (1982): 108–154. For a study emphasizing the twentieth century, see Harry Dowling, *City Hospitals: The Undercare of the Underprivileged* (Cambridge: Harvard University Press, 1982).

18. Genevieve Miller, "Medical Education and the Rise of Hospitals. I. The Eighteenth Century," *J. A. M. A.* 186 (1963): 938–942; Dale C. Smith, "The Emergence of Organized Clinical Instruction in the Nineteenth Century American Cities of Boston, New York and Philadelphia" (Ph.D. diss., University of Minnesota, 1979).

19. For general accounts of these developments, see Knud Faber, *Nosography in Modern Internal Medicine* (New York: Paul B. Hoeber, 1923); Erwin H. Ackerknecht, *Medicine at the Paris Hospital, 1794–1848* (Baltimore: Johns Hopkins University Press, 1967); Michel Foucault, *The Birth of the Clinic: An Archaeology of Medical Perception*, trans. A.M. Sheridan Smith (New York: Pantheon, 1973).

CHAPTER 1: COMPETITION IN THE MEDICAL MARKET

1. Eighth Census of the United States, Washington, D.C., pp. 670, 677, cited by Bernhard J. Stern, *American Medical Practice in the Perspectives of a Century* (New York: Commonwealth Fund, 1945), p. 63; Michael M. Davis, "The Supply of Doctors," *Medical Care* 2 (1942): 314.

2. T. L. Nichols, *Forty Years of American Life* (New York: Stackpole, 1937), p. 226.

3. Thomas C. Brinsmade, "Vice President's Address," in *Transactions of the Medical Society of the State of New York, 1858* (Albany, 1858), p. 257.

4. Joseph F. Kett, *The Formation of the American Medical Profession: The Role of Institutions, 1780–1860* (New Haven: Yale University Press, 1968), pp. 185–186; William G. Rothstein, *American Physicians in the Nineteenth Century: From Sects to Science* (Baltimore: Johns Hopkins University Press, 1972), pp. 344–345.

5. F. W. Mann, "The Public Status of Our Profession," *Journal of the Maine Medical Association* 16 (1925): 137, cited in Richard H. Shryock, "The American Physician in 1846 and 1946," *J. A. M. A.* 134 (1947): 417–424.

6. John Shaw Billings, "A Century of American Medicine, 1776–1876; Literature and Institutions," *Am. J. Med. Sci.* 72 (1876): 438–480. Also in *A Century of American Medicine, 1776–1876* (Philadelphia: H. C. Lea, 1876), pp. 289–366, and Frank Bradway Rogers, ed., *Selected Papers of John Shaw Billings, Compiled with a Life of Billings* (Chicago: Medical Library Association, 1965), pp. 24–75 (see pp. 72–74).

7. *Virginia Clinical Record* 3 (1875): 189.

8. Thomas A. Emmet, *Incidents of My Life* (New York: G. P. Putnam's, 1911), p. 203.

9. Ibid. See also John Janvier Black, *Forty Years in the Medical Profession 1858–1898* (Philadelphia: J. B. Lippincott, 1900), p. 53.

10. For an analysis of the process of specialization and the resistance aroused, see George Rosen, *The Specialization of Medicine* (New York: Froben, 1944), pp. 14–72 (reprinted by Arno Press, New York, 1972). Ideas developed in this and the following sections were first sketched in George Rosen, "The Impact of the Hospital on the Physician, the Patient and the Community," *Hospital Administration* 9 (1964): 15–33.

11. John Shaw Billings, "Medicine in the United States, and Its Relation to Co-operative Investigation," *British Medical Journal* 2 (1886): 299–307, in Rogers, *Selected Papers*, p. 191.

12. R. H. Shryock, *Medicine and Society in America, 1660–1860* (New York: New York University Press, 1960), p. 147; *Boston Medical and Surgical Journal,* 79 (1868): 112; ibid. 96 (1877): 180–181.

13. "Medical Education in the United States," *J. A. M. A.* 79 (1922): 629–633.

14. R. H. Shryock, *Medical Licensing in America, 1650–1965* (Baltimore: Johns Hopkins University Press, 1967), p. 48.

15. For a more detailed analysis of the relation of medical knowledge and theory to the pattern of practice in New York City in 1866, see Charles Rosenberg, "The Practice of Medicine in New York City a Century Ago," *Bull. Hist. Med.* 41 (1967): 223–253.

16. Austin Flint, *A Treatise on the Principles and Practice of Medicine* (Philadelphia: H. C. Lea, 1866), p. 102.

17. Oliver Wendell Holmes, *Medical Essays* (Boston: Houghton Mifflin, 1911), p. 377. For a similar opinion on rural practice in the same period, see William Allen Pusey, *A Doctor of the 1870's and 80's* (Springfield, Ill.: Charles C. Thomas, 1932), pp. 85–86.

18. Henry James, *Washington Square* (1881; reprint ed., New York: New American Library, 1964), pp. 5–6.

19. Jacob Wolff, "Über den Umgang mit Patienten," *Die Heilkunde* 1 (1896): 105.

20. S. Weir Mitchell, *Doctor and Patient,* 3rd ed. (Philadelphia: J. B. Lippincott, 1900), p. 53.

21. Thomas F. Reilly, *Building a Profitable Practice, Being a Text-Book on Medical Economics* (Philadelphia: J. B. Lippincott, 1912), p. 41.

22. Mitchell, *Doctor and Patient,* p. 32.

23. Fielding H. Garrison, *John Shaw Billings: A Memoir* (New York: G. P. Putnam's Sons, 1915), p. 20.

24. Pusey, *Doctor,* pp. 89–90, 105–109; see also H. M. F. Behneman, "Leaves from a Doctor's Notebook of Seventy Years Ago," *Military Surgeon* 86 (1940): 547–554.

25. My brother and I, ages 5 and 7 respectively, had our tonsils and adenoids removed at home in New York City in 1917. The otolaryngologist came with a nurse, who served as anesthetist, and his equipment; he performed the operation, returned later to check for bleeding, and then saw us in his office, not very far from our home. See also George Rosen, "Christian Fenger, Medical Immigrant," *Bull. Hist. Med.* 49 (1974): 143–144.

26. Joseph M. Toner, "Statistics of Hospitals in the United States, 1872–1873," *Trans. A. M. A.* 24 (1873): 314–333; C. Rufus Rorem, *Capital Investment in Hospitals* (Washington, D.C.: Committee on the Costs of Medical Care, 1930), p. 9.

27. Donald E. Konold, *A History of American Medical Ethics, 1847–1912*

(Madison: State Historical Society of Wisconsin, 1962), pp. 32–34.

28. Ernest Earnest, *S. Weir Mitchell: Novelist and Physician* (Philadelphia: University of Pennsylvania Press, 1950), p. 35.

29. Mitchell, *Doctor and Patient,* pp. 55–56. See also Oliver Wendell Holmes, *An Introductory Lecture Delivered before the Medical Class of Harvard University, November 6, 1867* (Boston: David Clapp, 1867), p. 30.

30. G. M. B. Maughs, "Medical Ultraisms," *Trans Missouri State Medical Association,* 23rd session (1880), p. 22.

31. John Brooks Wheeler, *Memoirs of a Small-Town Surgeon* (New York: Frederick A. Stokes, 1935), pp. 96, 176–283; Charles B. Johnson, *Sixty Years in Medical Harness, or the Story of a Long Medical Life, 1865–1925* (New York: Medical Life Press, 1926), pp. 154–155; Franklin H. Martin, *Fifty Years of Medicine and Surgery, An Autobiographical Sketch* (Chicago: Surgical Publishing, 1934), p. 172; *Medical Record* 18 (1880): 521–522; Philip Van Ingen, *The New York Academy of Medicine, Its First Hundred Years* (New York: Columbia University Press, 1949), pp. 60–61, 116, 144–145, 125, 222, 286, 297; J. Marion Sims, *The Story of My Life* (New York: D. Appleton, 1888), pp. 268–295.

32. George Rosen, *Fees and Fee Bills: Some Economic Aspects of Medical Practice in Nineteenth Century America* (Baltimore: Johns Hopkins University Press, 1946), pp. 36–90.

33. Konold, *Medical Ethics,* pp. 9–13.

34. N. S. Davis, Editorial, "Do Moral Principles Change?," *J. A. M. A.* 1 (1883): 57.

35. Ingrid Vieler, "Die deutsche Arztpraxis im 19. Jahrhundert" (doctoral diss., Johannes Gutenberg Universität, Mainz, 1958), pp. 2–3.

36. George Rosen, "Medical Care for Urban Workers and the Poor: Two 19th Century Programs," *Am. J. Pub. Health* 65 (1975): 300; George Rosen, "Five Years in the Practice of a Nineteenth Century New York Physician—Dr. William H. Van Buren," *N. Y. State J. M.* (April 15, 1949), pp. 932–935 (see p. 933); Pusey, *Doctor,* pp. 41–43, 48–50; James B. Herrick, *Memories of Eighty Years* (Chicago: University of Chicago Press, 1949), p. 94. For shifting residential patterns, see James, *Washington Square,* pp. 15–17, and also for its pertinent references to society, commercial development, and practical details of medicine.

37. Pusey, *Doctor,* p. 37.

38. Earnest, *S. Weir Mitchell,* p. 76; Herrick, *Memories of Eighty Years,* pp. 92, 99, 136.

39. Gottlieb Merkel, "Die arztlichen Sprechstunden," *Münchener Medizinscher Wochenschrift* 53^2 (1906): 2355.

40. Benjamin E. Cotting, *Medical Addresses* (Boston: David Clapp, 1875), pp. 51–52.

41. Ibid., pp. 51, 109.

42. Mitchell, *Doctor and Patient,* pp. 28, 37.

43. S. Weir Mitchell, *The Early History of Instrumental Precision in Medicine* (New Haven: Tuttle, Morehouse and Taylor, 1892).

44. Herrick, *Memories of Eighty Years,* pp. 91–92.

45. *How to Be Successful as a Physician* (Meriden, Conn.: Church Publishing, 1902), pp. 91, 96–97. See also Nathan E. Wood, *Dollars to Doctors, or Diplomacy and Prosperity in Medical Practice* (Chicago: Lion Publishing, 1903), p. 57.

46. *How to Be Successful,* p. 95.

47. Reilly, *Profitable Practice,* pp. 38–39, 115.

48. Wood, *Dollars to Doctors,* pp. 67, 187; Vieler, "Die deutsche Arztpraxis," pp. 27–28; Pusey, *Doctor,* p. 94.

49. Earnest, *S. Weir Mitchell,* p. 58.

50. Elias Haffter, ed., *Dr. L. Sonderegger in seiner Selbstbiographie und seinen Briefen* (Breuenfeld: J. Huber, 1898), pp. 45, 268. See also Arthur B. Emmons, *The Profession of Medicine* (Cambridge: Harvard University Press, [c. 1914]), p. 49.

51. Thomas G. Atkinson, *Successful Office Practice* (Chicago: American Journal of Clinical Medicine, 1916), p. 7.

52. This point was noted in 1906 by Merkel, "Die arztlichen Sprechstunden," p. 2355.

53. Hans Kirste, "Der Tageslauf eines Nurnberger praktischen Arztes um die Wende des 18. und 19. Jahrhunderts (Ein Beitrag zur Kulturgeschichte des praktischen Arztes)," *Münchener medizinischer Wochenschrift* 84^2 (1937): 1910–1912 (p. 1911).

54. Haffter, *Dr. L. Sonderegger,* p. 262.

55. Kirste, "Der Tageslauf," p. 1911.

56. Antonio Ciocco and Isidore Altman, "The Patient Load of Physicians in Private Practice: A Comparative Statistical Study of Three Areas," *Public Health Reports* 58 (1943): 1336; Antonio Ciocco and Isidore Altman, "Statistics on the Patient Load of Physicians in Private Practice," *J. A. M. A.* 121 (1943): 506–513.

57. *Boston Medical and Surgical Journal* 79 (1868): 112; ibid., 96 (1877): 180–181; *How to Be Successful,* pp. 54–70; Emmons, *Profession of Medicine,* pp. 57–59, 62, 102–104.

58. Emmons, *Profession of Medicine,* pp. 65, 69.

59. "Letter: The Abuse of Medical Charities," *Boston Medical and Surgical Journal* 97 (1877): 598–599.

60. "Americus Veritatus to the editor," *J. A. M. A.* 25 (1895): 336–338.

61. *How to Be Successful,* pp. 113–114.

62. *J. A. M. A.* 48 (1907): 48.

63. Emmons, *Profession of Medicine,* pp. 76–79.

64. C. R. Mabee, ed., *The Physician's Business and Financial Adviser,* 4th ed. (Cleveland: Continental Publishing, 1900); *Successful Office Practice—The Key to More Business* (Chicago: American Journal of Clinical Medicine, 1916); Thomas M. Dorsey, "Business Efficiency in Medicine," *American Medicine* 19 (1924): 96–102. These are only a few of the many publications on the subject.

65. A. Jacobi, *Collectanea Jacobi,* vol. 8, *Miscellaneous Addresses and Writings* (New York: Critic and Guide, 1909), p. 346.

66. Reilly, *Profitable Practice,* p. 115.

67. Emmons, *Profession of Medicine,* p. 38.

68. Anna E. Sevringhaus, "Distribution of Graduates of Medical Schools in the United States and Canada According to Specialties, 1900 to 1964," *Journal of Medical Education* 49 (1965): 733.

69. On the process of medical specialization, see George Rosen, *Specialization of Medicine,* pp. 30–72.

70. Herrick, *Memories of Eighty Years,* p. 92; see also Wheeler, *Small-Town Surgeon,* pp. 176–177.

71. Jacobi, *Miscellaneous Addresses,* p. 62.

72. A. D. Rockwell, *Rambling Recollections: An Autobiography* (New York: Paul B. Hoeber, 1920), p. 194; Earnest, *S. Weir Mitchell,* p. 34.

73. Stephen Smith, *Doctor in Medicine: and Other Papers on Professional Subjects* (New York: William Wood, 1872), pp. 247–250; Rockwell, *Rambling Recollections,* p. 215; Wheeler, *Small-Town Surgeon,* pp. 176–177, 179; Thomas N. Bonner, *The Kansas Doctor: A Century of Pioneering* (Lawrence: University of Kansas Press, 1959), pp. 65, 90–91; Joseph Hirsh and Beka Doherty, *The First*

Hundred Years of the Mount Sinai Hospital of New York, 1852–1952 (New York: Random House, 1952), p. 114.

74. For the context in which dispensaries developed, the nature of their operations, and their evolution from the end of the eighteenth century to the early twentieth century, see George Rosen, "Impact of the Hospital," pp. 15–33; George Rosen, "The Efficiency Criterion in Medical Care, 1900–1920: An Early Approach to an Evaluation of Health Service," *Bull. Hist. Med.* 50 (1976): 28–44; Charles E. Rosenberg, "Social Class and Medical Care in Nineteenth-Century America: The Rise and Fall of the Dispensary," *J. Hist. Med.* 29 (1974): 32–54.

75. "Duties of Hospital Physicians and Surgeons," *Boston Medical and Surgical Journal*, 78 (1868): 399.

76. "Abuse of Medical Institutions," *Boston Medical and Surgical Journal* 97 (1877): 393.

77. S. Humphreys Gurteen, *A Handbook of Charity Organization* (Buffalo: The Author, 1882), p. 99; George M. Gould, "Charity Organization and Medicine," *Bull. Am. Acad. Med.* 1 (1891–1895): 547–554.

78. Matthew Woods, *Medical News* 83 (1903): 1042. See also Charles P. Emerson, "Free Medical Care for the Poor," *Proc. Nat. Conf. Char. and Corr.* (1904), pp. 168–175; and Henry D. Chapin, "Home Treatment of the Sick Children of the Poor," *Charities* 7 (1901): 470–472.

79. Stephen Smith, "Uses and Abuses of Medical Charities," *Proc. Nat. Conf. Char. and Corr., 1898* (Boston, 1899), pp. 320–327; Michael M. Davis and Andrew R. Warner, *Dispensaries, Their Management and Development* (New York: Macmillan, 1918), p. 45.

80. *How to Be Successful*, pp. 24–25.

81. Theodore W. Schaefer, "The Commercialization of Medicine; or, the Physician as Tradesman," *Boston Medical and Surgical Journal* 131 (1894): 501–502.

82. Emmons, *Profession of Medicine*, pp. 66–67.

83. *How to Be Successful*, p. 115.

CHAPTER 2: MEDICAL SCIENCE, PROFESSIONAL CONTROL, AND THE STABILIZATION OF THE MEDICAL MARKET

1. Stephen A. Welch, "Some of the Conditions Affecting the Practice of Medicine in the City of Providence," *Providence Medical Journal* 16 (1915): 192–202.

2. These "other physicians" included eclectics, magnetic physicians, and bone setters, as well as unspecified healers.

3. Arthur B. Emmons, ed., *The Profession of Medicine: A Collection of Letters from Graduates of the Harvard Medical School* (Cambridge: Harvard University Press, 1914), p. 41.

4. Ibid., pp. 18, 38, 43.

5. For the development of microbiology and immunology and their application,

see George Rosen, *A History of Public Health* (New York: MD Publications, 1958), pp. 304–343, and George Rosen, *Preventive Medicine in the United States, 1900–1975* (New York: Science History Publications, 1975), pp. 20–54. For vaccines and sera, see H. J. Parish, *A History of Immunization* (Edinburgh: Livingstone, 1965). See also Barbara G. Rosenkrantz, *Public Health and the State: Changing Views in Massachusetts, 1842–1936* (Cambridge: Harvard University Press, 1972), pp. 113–126.

6. H. W. Hill, Chairman's Address, *Public Health Papers and Reports (APHA)* 32, Part II (1908): 108, and M. P. Ravenel, "The American Public Health Association. Past, Present, Future," in *A Half Century of Public Health* (New York: American Public Health Association, 1921), pp. 13–55.

7. Lewellys F. Barker, "Public Health and the Future Commonwealth," *Transactions of the Conference on the Future of Public Health in the United States and the Education of Sanitarians* (Washington, D.C.: Public Health Bulletin No. 126, 1922).

8. "The Health Board and Compulsory Reports," *Medical Record* 51 (1897): 305–306; C. E. A. Winslow, *The Life of Hermann M. Biggs* (Philadelphia: Lea and Febiger, 1929), pp. 144–146.

9. S. Josephine Baker, *Fighting for Life* (New York: Macmillan, 1939), p. 139; for a similar reaction in Philadelphia, see Joseph C. Aub and Ruth K. Hapgood, *Pioneer in Modern Medicine: David Linn Edsall of Harvard* (Harvard Medical Alumni Association, 1970), pp. 51–52. In Philadelphia some practitioners objected to purification of the city's water supply because it would eliminate typhoid fever and reduce their practice.

10. Rosenkrantz, *Public Health and the State*, pp. 149–154.

11. Robert S. Lynd and Helen M. Lynd, *Middletown in Transition: A Study in Cultural Conflicts* (New York: Harcourt, Brace, 1937), p. 395; see also Robert S. Lynd and Helen M. Lynd, *Middletown: A Study in Contemporary American Culture* (New York: Harcourt, Brace, 1929), pp. 443–444, 451.

12. George Rosen, "Some Substantive Limiting Conditions in Communication between Health Officers and Medical Practitioners," *Am. J. Pub. Health* 51 (1961): 1805–1816.

13. Henry J. Cammann and Hugh N. Camp, *The Charities of New York, Brooklyn, and Staten Island* (New York: Hurd and Houghton, 1868), pp. 1–138; James D. McCabe, Jr., *Lights and Shadows of New York Life* (Philadelphia: National, 1872), pp. 648–654; and *King's Handbook of Boston* (Cambridge, Mass.: Moses King, 1878), pp. 205–215.

14. John Brooks Wheeler, *Memoirs of a Small-Town Surgeon* (New York: Frederick A. Stokes, 1935), p. 179. See also Thomas N. Bonner, *The Kansas Doctor: A Century of Pioneering* (Lawrence: University of Kansas Press, 1959), pp. 90–91.

15. George Dock, "Clinical Pathology in the Eighties and Nineties," *Am. J. Clin. Path.* 16 (1946): 671–680.

16. Aub and Hapgood, *Pioneer in Modern Medicine*, pp. 14–15.

17. David L. Edsall, "The Transformation in Medicine," *Southern Medical Journal* 24 (1931): 1103–1113.

18. Dock, "Clinical Pathology," p. 673.

19. H. R. Somers and A. R. Somers, *Doctors, Patients and Health Insurance* (Washington, D.C.: Brookings Institution, 1961), p. 63.

20. Frank B. Woodford and Philip P. Mason, *Harper of Detroit: The Origin and Growth of a Great Metropolitan Hospital* (Detroit: Wayne State University Press, 1964), pp. 184, 186, 239.

21. John J. Birne, ed., *A History of the Boston City Hospital 1905–1964* (Boston City Hospital, 1964), pp. 303–305, 327–328.

22. John Duffy, ed., *History of Medicine in Louisiana,* 2 vols. (Baton Rouge: Louisiana State University Press, 1962), 2: 509–510.

23. Alan M. Chesney, *The Johns Hopkins Hospital and The Johns Hopkins University School of Medicine: A Chronicle,* 3 vols. (Baltimore: Johns Hopkins University Press, 1958), 2: 115–118.

24. C. Rufus Rorem, *Capital Investment in Hospitals* (Washington, D.C.: Committee on the Costs of Medical Care, 1930), p. 9; Rosemary Stevens, *American Medicine and the Public Interest* (New Haven: Yale University Press, 1971), p. 145, n. 36.

25. Joseph Hirsch and Beka Doherty, *The First Hundred Years of Mount Sinai Hospital of New York, 1852–1952* (New York: Random House, 1952), p. 114.

26. Woodford and Mason, *Harper of Detroit,* p. 239.

27. Thomas F. Reilly, *Building a Profitable Practice, Being a Text-Book on Medical Economics* (Philadelphia: J. B. Lippincott, 1912), p. 39.

28. Sinclair Lewis, *Martin Arrowsmith* (1925; reprint ed., London: Jonathan Cape, 1930), p. 181.

29. Richard M. Pearce, "The Hospital Laboratory—Its Purposes and Methods," *Modern Hospital* 6 (1916): 158–163.

30. Harry H. Moore, "Health and Medical Practice," in *Recent Social Trends in the United States: Report of the President's Research Committee on Social Trends* (New York: McGraw-Hill, 1933), pp. 1063–1064; Allan Peebles, *A Survey of Statistical Data on Medical Facilities in the United States: A Compilation of Existing Material* (Washington, D.C.: Committee on the Costs of Medical Care, 1929); "Hospital Service in the United States," *J. A. M. A.* 94 (1930): 929; ibid., 98 (1932): 2072; ibid., 99 (1933): 1017.

31. Niles Carpenter, *Hospital Service for Patients of Moderate Means: A Study of Certain American Hospitals* (Washington, D.C.: Committee on the Costs of Medical Care, 1930), pp. 9–20, 73.

32. Michael M. Davis, *Clinics, Hospitals and Health Centers* (New York: Harper, 1927), p. 5.

33. Nathan Sinai and Alden B. Mills, *A Survey of the Medical Facilities of the City of Philadelphia,* Publication No. 9, Committee on the Costs of Medical Care (Chicago: University of Chicago Press, 1931), p. 37.

34. Cammann and Camp, *Charities of New York,* pp. 46–47.

35. Michael M. Davis and C. Rufus Rorem, *The Crisis in Hospital Finance and Other Studies in Hospital Economics* (Chicago: University of Chicago Press, 1932), pp. 104–105. See also George Bugbee, "The Physician in the Hospital Organization," *New England Journal of Medicine 261* (1959): 896–901.

36. William E. Darnall, "The Hospital and the Young Physician," *Bull. Am. Acad. Med.* 12 (1911): 73.

37. William Francis Waugh, "The Economic Influence of the Hospital as Shown upon the Profession," *Bull. Am. Acad. Med.* 12 (1911): 79.

38. Frank H. Streightoff, *The Standard of Living among the Industrial People of America* (Boston: Houghton Mifflin, 1911), pp. 9–28, 121–135.

39. Moore, "Health and Medical Practice," pp. 107–121; Sue A. Clark and Edith Wyatt, *Making Both Ends Meet: The Income and Outlay of New York Working Girls* (New York: Macmillan, 1911), pp. 1–43, 89–147.

40. Edward C. Kirkland, *A History of American Economic Life* (New York: F. S. Crofts, 1940), pp. 523–524, 671, 701–707.

41. Commission on Medical Education, *Final Report of the Commission on*

Medical Education (New York: Office of the Director of Study, 1932), p. 77, Appendix Table 41. See also Davis, *Clinics, Hospitals and Health Centers,* pp. 5–9.

42. "Hospital Service in the United States," *J. A. M. A.* 96 (1931): 1021; *ibid.,* 102 (1934): 1010.

43. *Bull. Am. Acad. Med.* 12 (1911): 119.

44. Michael M. Davis, "The Functions of a Dispensary or Out-Patient Department," *Boston Medical and Surgical Journal* 171 (1914): 337n.

45. Michael M. Davis and Andrew R. Warner, *Dispensaries, Their Management and Development* (New York: Macmillan Company, 1918), pp. 20–21.

46. Committee on Dispensary Development, *The Cornell Clinic, 1921–1924: Medical Service on a Self-Supporting Basis for Persons of Moderate Means* (New York, 1925); W. L. Niles, "The First Five Months of the Cornell Clinic," *Modern Hospital* 18 (1922): 560–562.

47. C. Rufus Rorem, *Private Group Clinics: The Administrative and Economic Aspects of Group Medical Practice as Represented in the Policies and Procedures of 55 Private Associations of Medical Practitioners* (Washington, D. C.: Committee on the Costs of Medical Care, 1931), pp. 11–18.

48. I. S. Falk, C. Rufus Rorem, and Martha D. Ring, *The Costs of Medical Care* (Chicago: University of Chicago Press, 1933), pp. 388–389.

49. Sanford R. Gifford, *Garlic and Old Horse Blankets* (Chicago: Chicago Literary Club, 1943), p. 4.

50. James H. S. Bossard, *A Sociologist Looks at the Doctors, The Medical Profession and the Public: Currents and Counter-Currents* (Philadelphia: American Academy of Political and Social Science, 1934), pp. 7–9.

51. Falk, Rorem, and Ring, *Costs of Medical Care,* pp. 390–396.

52. *Recent Economic Changes in the United States: Report of the Committee on Recent Economic Changes of the President's Conference on Unemployment* (New York, 1929), 1: 240.

53. *Recent Social Trends,* p. 477.

54. The similarities to business management may be seen in R. M. Haig, "Towards an Understanding of the Metropolis," *Quarterly Journal of Economics* 40 (1926): 426–428.

55. "Solving the Office Rent Problem," *American Medicine,* n.s. 19 (1924): 424–426.

56. Falk, Rorem, and Ring, *Costs of Medical Care,* pp. 196–198, 604. See also Lewis Mayers and Leonard V. Harrison, *The Distribution of Physicians in the United States* (New York: General Education Board, 1924), pp. 164–166.

57. H. G. Weiskotten, "Tendencies in Medical Practice," *J. Assn. Am. Med. Coll.* 7 (1932): 65–85; Commission on Medical Education, *Final Report,* pp. 104–105.

58. Mayers and Harrison, *Distribution of Physicians,* p. xii.

59. B. B. Bagby, "Changes in a Small Town Brought About by the Health Department," *Public Health Reports* 38 (1923): 456–458.

60. *Nebraska State Medical Journal* 9 (1924): 411.

61. John M. Dodson, "The Growing Importance of Preventive Medicine to the General Practitioner," *J. A. M. A.* 81 (1923): 1423–1428.

62. Haven Emerson, "The Protection of Health by Periodic Medical Examinations," *Journal of the Michigan State Medical Society* 21 (1922): 399; also in his *Selected Papers* (Battle Creek, Mich.: W. K. Kellogg Foundation, [1949], pp. 158–171.

63. Rosen, *Preventive Medicine,* pp. 58–61.

64. Mayers and Harrison, *Distribution of Physicians,* p. 171.

65. W. A. Pusey, "Medical Education and Medical Service," *J.A.M.A.* 84 (1925): 281–285, 365–369, 437–441, 513–515, 593–595 (see especially p. 283). See also Peebles, *Data on Medical Facilities,* pp. 22–23, 66–67.

66. Raymond Pearl, "Distribution of Physicians in the United States. A Commentary on the Report of the General Education Board," *J. A. M. A.* 84 (1925): 1024–1028.

67. Bureau of Medical Economics, *Distribution of Physicians in the United States* (Chicago: American Medical Association, 1935), pp. 9–10.

68. Pearl, "Distribution of Physicians," p. 1027.

69. *Ibid.*, p. 1026.

70. Massachusetts State Department of Health, *Special Report of the Department of Public Health* (House No. 1075, January 1925), pp. 69–71. See also Commission on Medical Education, *Final Report,* pp. 106–107, 112.

71. Joseph S. Lawrence, "A Study of the Distribution of Physicians in the Rural Districts of New York State," *N.Y. State J. M.* 29 (1929): 996–1000.

72. *Bulletin, American Medical Association* 18 (1923): 463–468.

73. Samuel Hopkins Adams, "Why the Doctor Left," *Ladies' Home Journal* (November 1923), p. 26; see also "The Decline of the Country Doctor," *American Medicine,* n.s. 19 (1924): 271–272.

74. *How to Be Successful as a Physician* (Meriden, Conn.: Church Publishing, 1902), pp. 68–69.

75. See, for example, the efforts of the New York Medico-Legal Society in 1872–1874 and of other groups in New York state in subsequent years, in Seymour J. Mandlebaum, *Boss Tweed's New York* (New York: John Wiley, 1965), pp. 151–154.

76. Stephen Smith, "On the Reciprocal Relations of an Efficient Public Health Service and the Highest Educational Qualifications of the Medical Profession," American Public Health Association. *Reports and Papers Presented at the Meetings of the American Public Health Association in the Years 1874–1875* (New York: Hurd and Houghton, 1876), 2: 187–200.

77. James G. Burrow, *AMA: Voice of American Medicine* (Baltimore: Johns Hopkins University Press, 1963), pp. 27–32. For a detailed account of the reform of medical education within a broad social context, see Stevens, *American Medicine,* pp. 58–70.

78. *J. A. M. A.* (August 21, 1915): 717–718.

79. *J. A. M. A.* 45 (July 22, 1905): 269–270; ibid., 54 (June 11, 1910): 1970, 1974.

80. Morris Fishbein, *A History of the American Medical Association, 1847–1947* (Philadelphia: W. B. Saunders, 1947), pp. 896–897.

81. Abraham Flexner, *Medical Education in the United States and Canada* (New York: Bulletin No. 4, Carnegie Foundation for the Advancement of Teaching, 1910).

82. Abraham Flexner, *I Remember: The Autobiography of Abraham Flexner* (New York: Simon and Schuster, 1940), pp. 109–111, 113–130; Herman G. Weiskotten and Victor Johnson, *A History of the Council on Medical Education and Hospitals of the American Medical Association, 1904–1959* (Chicago: American Medical Association, 1959), pp. 4–10.

83. *Bulletin, A. M. A.* 3 (1907): 263.

84. Henry S. Pritchett, "Introduction," in Flexner, *Medical Education,* p. xiv.

85. Samuel Haber, *Efficiency and Uplift: Scientific Management in the Progressive Era 1890–1920* (Chicago: University of Chicago Press, 1964); George Rosen, "The Efficiency Criterion in Medical Care, 1900–1920: An Approach to an Evaluation of Health Service," *Bull. Hist. Med.* 50 (1976): 28–44.

86. William Pepper, *Higher Medical Education: The True Interest of the Public and of the Profession* (Philadelphia: J. B. Lippincott, 1894), pp. 29–30; Francis N. Thorpe, *William Pepper, M.D., LL.D. (1843–1898)* (Philadelphia: J. B. Lippincott, 1904): George H. Simmons, "Medical Education and Preliminary Requirements," *J. A. M. A.* 45 (1904): 1205; "An Overcrowded Profession—The Cause and the Remedy," *J. A. M. A.* 37 (Sept. 21, 1901): 775–76.

87. Willard C. Rappleye, "Medical Education," *Biennial Survey of Education, 1928–1930* (Bulletin 20, 1931), p. 547; Commission on Medical Education, *Final Report,* Table 104; "Medical Education in the United States and Canada," *J. A. M. A.* 91[2] (1928): 473–476; Irving S. Cutter, *The School of Medicine* (Boston: Ginn, 1930), pp. 287–289.

88. For brief accounts, see Stevens, *American Medicine,* pp. 55–74 and James G. Burrow, *Organized Medicine in the Progressive Era* (Baltimore: Johns Hopkins University Press, 1977), pp. 31–51.

89. Ernest V. Hollis, *Philanthropic Foundations and Higher Education* (New York: Columbia University Press, 1938), pp. 211–217; Raymond B. Fosdick, *Adventure in Giving: The Story of the General Education Board* (New York: Harper & Row, 1962), pp. 172, 328.

90. Cited in Fosdick, *Adventure in Giving,* p. 173.

91. Flexner, *I Remember,* pp. 257–322.

92. *Ibid.,* p. 321.

93. George Blumer, *The Modern Medical School: Its Relation to the Hospital and to the Profession* (Albany, 1916); G. Canby Robinson, *Adventures in Medical Education* (Cambridge: Harvard University Press, 1957), pp. 85–86, 130–137, 306–309.

94. Henry E. Sigerist, *American Medicine* (New York: W. W. Norton, 1934), pp. 139 ff.

95. General Education Board, *Annual Report, 1921-1922,* p. 17; General Education Board, *Annual Report, 1924–1925,* p. 8; U.S. Department of Health, Education, and Welfare, *Health Education and Welfare Trends* (Washington, D.C., 1960), p. 44.

96. "Decline of the Country Doctor," pp. 271–272.

97. "Shall the 'Poor Boy' Study Medicine?," *American Medicine,* n.s. 19 (1924): 426–427.

98. *Ibid.,* p. 426.

99. John Higham used the term "tribal twenties" to characterize the period from the end of the First World War to about 1925, during which social nativism and xenophobia reached a high point. See John Higham, *Strangers in the Land: Patterns of American Nativism 1860–1925* (New York: Atheneum, 1963), p. 264.

100. Thomas F. Reilly, *Profitable Practice,* p. 132.

101. Herbert M. Morais, *The History of the Negro in Medicine,* 2nd ed. (New York: Publishers Company, 1969), pp. 60–67, 89–90; Dietrich C. Reitzes, *Negroes and Medicine* (Cambridge: Harvard University Press, 1958), pp. 6–7.

102. Norman Hapgood, "Jews and College Life" and "Schools, Colleges and Jews," *Harper's Weekly* 62 (1916): 53–55, 77–79; Charles S. Bernheimer, "Prejudice against Jews in the United States," *Independent* 65 (1908): 1106–1107; Harry Starr, "The Affair at Harvard," *Menorah Journal* 8 (1922): 263–276; Heywood Broun and George Britt, *Christians Only: A Study of Prejudice* (New York: Vanguard Press, 1931), pp. 231–232; Stephen Thernstrom, *The Other Bostonians: Poverty and Progress in the American Metropolis, 1880–1970* (Cambridge: Harvard University Press, 1973), pp. 171–175; Stephen Steinberg, "How Jewish Quotas Began," *Commentary* 52 (1971): 67–76.

103. John Wyckoff, "Relation of Collegiate to Medical Student Scholarship," *Bulletin of the Association of American Medical Colleges* 1–2 (1926–1927); Burton D. Myers, "Report on Applications for Matriculation in Schools of Medicine for 1927–1928," *Bulletin of the Association of American Medical Colleges* 3 (1928): 193–199.

104. Commission on Medical Education, *Final Report,* p. 95 and Tables 101 and 103. For a personal account of the situation, see the autobiography of Theodore S. Drachman, *The Grande Lapu-Lapu* (New York: Abelard-Schuman, 1972), pp. 14–25, 30.

105. H. G. Weiskotten, "A Study of the Present Tendencies in Medical Practice," *Bulletin of the Association of American Medical Colleges* 3 (1928): 132; Commission on Medical Education, *Final Report,* p. 96; *Internships and Residencies in New York City, 1934–1937, Their Place in Medical Education* (New York: Commonwealth Fund, 1938), p. 101.

106. Maurice Leven, *The Incomes of Physicians: An Economic and Statistical Analysis,* Publications of the Committee on the Costs of Medical Care, No. 24 (Chicago: University of Chicago Press, 1933), pp. 199–211; Elton Rayack, *Professional Power and American Medicine: The Economics of the American Medical Association* (Cleveland: World, 1967), p. 73.

107. Commission on Medical Education, *Final Report,* pp. 2–3. The Commission was chaired by A. Lawrence Lowell, president of Harvard University, and its work was directed by Willard C. Rappleye, a physician who later headed the medical faculty of Columbia University. A *Preliminary Report* was issued in 1927, followed by several others and ending with a *Final Report* in 1932.

108. Commission on Medical Education, *Final Report,* pp. 93, 100, 89.

109. *Ibid.,* pp. 19–21.

110. *Ibid.,* p. 19.

111. J. G. Freymann, "Leadership in American Medicine: A Matter of Personal Responsibility," *New England Journal of Medicine* 270 (1964): 710–719.

112. Oliver Garceau, *The Political Life of the American Medical Association* (Cambridge: Harvard University Press, 1941), pp. 50–59.

113. The phrase "need for professional birth control" appeared in an editorial published in *J. A. M. A.,* 99 (August 27, 1932): 765.

114. Walter L. Bierring, "Social Dangers of an Oversupply of Physicians," *American Medical Association Bulletin* 29 (1934): 17–18.

115. *Ibid.,* p. 18.

116. Esther L. Brown, *Physicians and Medical Care* (New York: Russell Sage Foundation, 1937), p. 113; W. S. Thompson and P. K. Whelpton, *Population Trends in the United States* (New York: McGraw-Hill, 1933), pp. 262–291.

117. Walter L. Bierring, "The Family Doctor and the Changing Order," *J. A. M. A.* 102 (1934): 1997.

118. Bierring, "Oversupply of Physicians," p. 17.

119. "Resurvey of Medical Schools: Its Purpose and Objectives," *Federation Bulletin* (April 1936): 98.

120. Edward F. Potthoff, "The Future Supply of Medical Students in the United States," *Journal of Medical Education* 35 (1960): 224–225; Stevens, *American Medicine,* p. 177.

121. Hugh Cabot, *The Doctor's Bill* (New York: Columbia University Press, 1935), pp. 259–266. For other views opposed to restriction, see *American Medicine: Expert Testimony Out of Court,* 2 vols. (New York: American Foundation, 1937), 1: 249–250.

122. Cabot, *Doctor's Bill,* p. 261.

123. "Report of the Committee on Foreign Medical Students," *J. Assn. Am. Med. Coll.* 8 (1933): 361.

124. Commission on Medical Education, *Final Report,* p. 95.

125. Harold Rypins, "The Foreign Medical Graduate," *J. Assn. Am. Med. Coll.* 8 (1933): 92–96.

126. *Ibid.,* p. 94.

127. *Ibid.* Rypins did not state how many of the foreign graduates were Americans and how many were immigrants. In this connection it may be worth noting that there has been no follow-up study of American students of medicine in foreign schools between 1930 and 1939 to determine how many were licensed to practice after their return or what their later career patterns were. A retrospective analysis would be interesting to compare with the arguments presented in the 1930s.

128. Arthur Dean Bevan, "The Overcrowding of the Medical Profession," *J. Assn. Am. Med. Coll.* 11 (1936): 384.

129. For the history of the internship and the residency, see New York Committee on the Study of Hospital Internships and Residencies, *Internships and Residencies,* pp. 26–45; J. A. Curran, "Internships and Residencies, Historical Backgrounds and Current Trends," *Journal of Medical Education* 34 (1959): 873–884; James A. Campbell, "The Internship: Origins, Evolution, and Confusion," *J. A. M. A.* 189 (1964): 273–278.

130. J. M. Toner, "Statistics of Regular Medical Associations and Hospitals of the United States: Section II, Statistics of the Hospitals in the United States, 1872–1873, Derived from the Inquiries by the U. S. Bureau of Education," *Trans. A. M. A.* 24 (1873): 314–333; Pepper, *Higher Medical Education,* p. 74.

131. *The Story of the First Fifty Years of Mt. Sinai Hospital, 1852–1902* (New York: The Hospital, 1944), pp. 58–59, 67–70, 72–81, 83–87; *The Roosevelt Hospital, 1871–1957* (New York: The Hospital, 1957), pp. 85–103, 107–115, 117–151, 153–181.

132. "The Education of the Intern," *J. A. M. A.* 43 (1904): 469–470.

133. Brown, *Physicians and Medical Care,* pp. 70–72.

134. Curran, "Internships and Residencies," pp. 70–72.

135. New York Committee on the Study of Hospital Internships and Residencies, *Internships and Residencies,* pp. 228–230; Commission on Medical Education, *Final Report,* pp. 143–150.

136. *First Fifty Years of Mt. Sinai Hospital,* p. 58.

137. Thomas N. Bonner estimates that between 1870 and 1914 some 15,000 American medical students and physicians had some educational experience, varying in length from six weeks to several years, in German, Austrian, or Swiss medical centers. See T. N. Bonner, *American Doctors and German Universities: A Chapter in International Intellectual Relations, 1870–1914* (Lincoln: University of Nebraska Press, 1963), p. 23.

138. B. Noland Carter, "The Fruition of Halsted's Concept of Surgical Training," *Surgery* 32 (1952): 518. See also Samuel J. Crowe, *Halsted of Johns Hopkins: The Man and His Men* (Springfield, Ill.: Charles C. Thomas, 1957), pp. 47, 54–56; W. G. MacCallum, *William Stewart Halsted, Surgeon* (Baltimore: Johns Hopkins University Press, 1930), pp. 94–95, 121–138; M. M. Davis, "The History of the Resident System," *Transactions and Studies of the College of Physicians of Philadelphia* 27 (1959–1960): 76–81.

139. Helen B. Clapesattle, *The Doctors Mayo* (Minneapolis: University of Minnesota Press, 1941), pp. 505–559.

140. Commission on Medical Education, *Final Report,* pp. 125–126.

141. William J. Mayo, "The Medical Profession and the Issues Which Confront It," *J. A. M. A.* 16 (1906): 737–740; Franklin H. Martin, *Fifty Years of Medicine*

and Surgery: An Autobiographical Sketch (Chicago: Surgical Publishing, 1934), pp. 292–332; Franklin H. Martin, *The Joy of Living: An Autobiography,* 2 vols. (New York: Doubleday, Doran, 1933), 1: 392–442; Loyal Davis, *Fellowship of Surgeons: A History of the American College of Surgeons* (Springfield, Ill.: Charles C. Thomas, 1960), pp. 26–30, 42–64, 72–73, 121–128, 152–155, 481–482; J. M. T. Finney, "The Standardization of the Surgeon," *Transactions, Section on Surgery of the A. M. A.* (1914): 211–223.

142. William Gerry Morgan, *The American College of Physicians: Its First Quarter Century* (Philadelphia: American College of Physicians, 1940), pp. 1–8, 87–102.

143. George Rosen, "New York City in the History of American Ophthalmology," *N. Y. State J. M.* 43 (1943): 754–758.

144. Edward Jackson, "The Optometry Question and the Larger Issues Behind It," *J. A. M. A.* 57 (1911): 265–270; Charles F. Prentice, *Legalized Optometry and the Memoirs of Its Founder* (Seattle: Casprin Fletcher, 1927); M. J. Hirsch and R. E. Wick, *The Optometric Profession* (Philadelphia: Chilton, 1968), pp. 124–147.

145. Derrick T. Vail, "The Limitations of Ophthalmic Practice," *Transactions, American Academy of Ophthalmology and Otolaryngology* (1908): 1–6; A. A. Hubbell, "The Ophthalmic Qualifications Which Should Be Demanded of the General Practitioner and of the Specialist, Respectively," *Transactions, Section on Ophthalmology, A. M. A.* (1909): 9–20; Judd Beach, "American Ophthalmology Grows Up: Turbulent Years 1908–1915," *American Journal of Ophthalmology* 22 (1939): 367–374; Edward Jackson, "Report of the Committee on Education in Ophthalmology," *Transactions, Section on Ophthalmology, A. M. A.* (1914): 395–406.

146. "Report of the Board for Ophthalmic Examinations," *Trans. Amer. Acad. Ophthal. Otolaryn.* (1917): 23; F. C. Cordes and C. W. Rucker, "History of the American Board of Ophthalmology," *Amer. J. Ophthal.* 53 (1962): 243–264.

147. "Preliminary Report of the Committee To Investigate Graduate Medical Instruction," *A. M. A. Bulletin* 9 (1914): 313–318.

148. "Graduate Education in the Specialities," *A. M. A. Bulletin* 15 (1921): 17–82.

149. "Hospitals Approved for Residencies in Specialities," *J. A. M. A.* 88 (1927): 829.

150. W. P. Wherry, "The American Board of Otolaryngology: History, Plan of Operation, Purposes," *Annals of Otology* 37 (1928): 1067–1072.

151. Walter T. Dannreuther, "The American Board of Obstetrics and Gynecology: Its Organization, Function and Objectives," *J. A. M. A.* 96 (1931): 797–798; Clyde L. Randall, "Responsibility for Excellence," *Transactions of the American Association of Obstetrics and Gynecology* 75 (1964): 5–14.

152. "Minutes of Meeting of Council on Medical Education and Hospitals," *J. A. M. A.* 103 (1934): 48.

153. For a very detailed account of the convoluted negotiations that produced the collaboration in 1934, see Stevens, *American Medicine,* chapters 8, 10, and 11.

154. Brown, *Physicians and Medical Care,* p. 83; Stevens, *American Medicine,* p. 156; Mahlon Ashford, ed., *Trends in Medical Education* (New York: Commonwealth Fund, 1949), p. 174. The figures cited in these sources have been checked against the lists of residencies in approved hospitals published in the *Journal of the American Medical Association.*

155. George Rosen, *The Specialization of Medicine* (New York: Froben, 1944), pp. 78–79.

156. A. C. Christie, *Economic Problems of Medicine* (New York: Macmillan, 1935), p. 50.

157. Walter Bierring, "Consistency vs. Chaos in Medical Education and Licensure," *J. A. M. A.* 106 (1936): 1097–1099.

158. Cited in John S. Apperson, "Report on Advances in the Practice of Medicine," *Transactions of the Medical Society of Virginia*, 1886, p. 91.

159. Falk, Rorem, and Ring, *Costs of Medical Care*, pp. 205–207.

160. "Physicians' Incomes," *Medical Economics* 17 (September 1940): 38, 40 (chart 3A); "How Much the Doctor Collects," *Medical Economics* 18 (December 1940): 52.

161. Linsly R. Williams, "Present Status of the Practice of Medicine," *N. Y. State J. M.* 18 (1928): 1029.

162. Thorstein Veblen, *The Theory of the Leisure Class* (1899; reprint ed., New York: Modern Library, 1931), p. 115.

163. Williams, "Practice of Medicine," p. 1028. The "fifty-seven varieties" was an ironic reference to a contemporary advertising slogan of the H. J. Heinz Company about its pickles and other products.

164. Pusey, "Medical Education and Medical Service," p. 281.

165. *Ibid.*, p. 437.

166. Aura E. Sevringhaus, "Distribution of Graduates of Medical Schools in the United States and Canada according to Specialties, 1900–1964," *Journal of Medical Education* 40 (1965): 721–736.

167. Ashford, *Trends in Medical Education*, pp. 174, 279–282.

168. Apperson, "Advances in the Practice of Medicine," pp. 89–90.

169. "Proceedings of the San Francisco Session," *J. A. M. A.* 80 (1923): 1929; Pusey, "Medical Education and Medical Service," pp. 281, 437.

170. Allen Peebles, *A Survey of the Medical Facilities of Shelby County, Indiana: 1929* (Washington, D.C.: Committee on the Costs of Medical Care, No. 6, 1930), pp. 26 ff.

171. Pusey, "Medical Education and Medical Service," p. 439.

172. Williams, "Practice of Medicine," p. 1027.

173. Frank Billings, "The Resourceful General Practitioner of Modern Medicine," *J. A. M. A.* 80 (1923): 524.

174. Falk, Rorem, and Ring, *Costs of Medical Care*, p. 213.

175. Gerald Green, *The Last Angry Man* (New York: Charles Scribner's Sons, 1956), pp. 320–321.

176. Charles L. Dana, "The Doctor's Future," *New York Medical Journal* 97 (1913): 3.

177. *Ibid.*

178. George Rosen, "The First Neighborhood Health Center Movement—Its Rise and Fall," *Am. J. Pub. Health* 61 (1971): 1620–1637.

179. Frank Billings, "The Future of Private Medical Practice," *J. A. M. A.* 76 (1921): 353.

180. Winslow, *Life of Hermann M. Biggs*, pp. 345–355; Milton Terris, "Hermann Biggs' Contribution to the Modern Concept of Health Centers," *Bull. Hist. Med.* 20 (1946): 387–412; B. R. Richards, "What New York State Has Done in Health Centers," *Am. J. Pub. Health* 11 (1921): 214–216.

181. Billings, "Future of Private Medical Practice," p. 354.

182. Dana, "The Doctor's Future," p. 2 (italics added).

183. Nathan Van Etten, "Abuses of the Medical Charity and Free Service of Physicians," in *The Medical Profession and the Public* (Philadelphia: American Academy of Political and Social Science, 1934), p. 20.

184. Quoted in Edgar Sydenstricker, "Medical Practice and Public Needs," in *The Medical Profession and the Public,* p. 21.

185. Bonner, *Kansas Doctor,* p. 226.

186. J. L. Pomeroy, "County Health Administration in Los Angeles," *Am. J. Pub. Health* 11 (1921): 796–800; J. L. Pomeroy, "Health Center Development in Los Angeles County," *J. A. M. A.* 93 (1929): 1546–1550; D. L. W. Worcester, "Public Health and Private Doctors," *Survey Graphic* 23 (1934): 149–155, 202–204.

187. Samuel G. Busey, *Personal Reminiscences and Recollections of Forty-Six Years Membership in the Medical Society of the District of Columbia, and Residence in this City . . .* (Washington, D.C.: The Author, 1895).

188. *Trans. A. M. A.* 20 (1869): 41. It should be noted that on large plantations in the South it was customary to hire a practitioner on an annual basis to provide medical care for the family and the slaves. After the Civil War, physicians engaged in contract practices with benevolent societies and similar groups. See Duffy, *History of Medicine in Louisiana,* 2: 100–101, 339–402; and Joseph I. Waring, *A History of Medicine in South Carolina 1825–1900* (Columbia: South Carolina Medical Association, 1967), pp. 118, 329.

189. Henry Harris, *California's Medical Story* (San Francisco: J. W. Stacey, 1932), p. 263.

190. Cited in Albert T. Lytle, "Contract Medical Practice. An Economic Study," *N.Y. State J. M.* 15 (1915): 106.

191. S. S. Goldwater, "Dispensaries: A Growing Factor in Curative and Preventive Medicine," *Boston Medical and Surgical Journal* 172 (1915): 614.

192. George S. Mathews, "Contract Practice in Rhode Island," *Bull. Am. Acad. Med.* 10 (1909): 599–606. In addition to Mathews, the panelists included A. L. Benedict (Buffalo), Charles S. Sheldon (Madison, Wis.), H. T. Partree (Eatontown, N. J.), A. Ravagli (Cincinnati), and Woods Hutchinson (New York), all physicians, and John C. McManemin, past president of a lodge, the Fraternal Order of Eagles. The entire proceedings of the symposium appeared in the *Bulletin,* with discussion of the papers, on pp. 580–640.

193. Mathews referred specifically to the towns of Pawtucket, Central Falls, Woonsocket, and Lonsdale and to the working-class sections of Providence (Olneyville, Mount Pleasant, North End, and South Providence).

194. Mathews, "Contract Practice in Rhode Island," p. 601.

195. Irving Howe, *World of Our Fathers* (New York: Harcourt Brace Jovanovich, 1976), pp. 184–190.

196. Morris Joseph Clurman, "The Lodge Practice Evil of the Lower East Side," *Medical Record* 78 (1910): 717.

197. Howe, *World of Our Fathers,* p. 188.

198. *Ibid.*

199. Goldwater, "Dispensaries," p. 614.

200. J. L. White, "Hygiene and Doctors' Fees," *Transactions, Illinois State Medical Society,* 1889, pp. 382–393; Thomas N. Bonner, *Medicine in Chicago, 1850–1950: A Chapter in the Social and Scientific Development of a City* (Madison, Wis.: American History Research Center, 1957), p. 217.

201. Bonner, *Medicine in Chicago,* p. 217.

202. *Ibid.,* p. 218.

203. Robert A. Allen, "State Insurance Against Sickness," *J. A. M. A.* 63 (1914), 185–87. See also George Rosen, "Contract or Lodge Practice and Its Influence on Medical Attitudes to Heal the Insurance," *Am. J. Pub. Health* 67 (1977): 374–78.

204. Goldwater, "Dispensaries," p. 616.

205. Welch, "Medicine in Providence," p. 200.

206. Charles McIntire, "From the Field," *Bulletin Am. Acad. Med.* 10 (1909): 654.

207. Lytle, "Contract Medical Practice," p. 106.

208. *Ibid.*, p. 108.

209. For the early history of the health insurance movement, see Pierce Williams and Isabel Chamberlain, *The Purchase of Medical Care Through Fixed Periodic Payment* (New York: National Bureau of Economic Research, 1932), pp. 34–57; *A. M. A., Bulletin* 11 (1916): 250–354, 355–378, and 12 (1916–1917): 79–94, 315–418; Burrows, *A. M. A.*, pp. 134–148.

210. Williams and Chamberlain, *Purchase of Medical Care*, pp. 40–44.

211. The figures in different sources vary somewhat but do not contradict the basic point. See, e.g., Williams and Chamberlain, *Purchase of Medical Care*, p. 46; Elizabeth W. Wilson, *Compulsory Health Insurance* (New York: National Industrial Conference Board, Studies in Individual and Collective Security, No. 3, 1947), pp. 2, 15; John A. Lapp, "The Findings of Official Health Insurance Commissions," *American Labor Legislation Review* (March 1920): 27.

212. Bonner, *Kansas Doctor*, p. 224.

213. James P. Warbasse, "The Socialization of Medicine," *J. A. M. A.* 64 (1914): 266.

214. I. M. Rubinow, "Social Insurance and the Medical Profession," *J. A. M. A.* 64 (1915): 381–386.

215. "Report of Committee on Social Insurance," *A. M. A., Bulletin* 11 (1916): 250. Favill died of pneumonia on February 20, 1916; the House of Delegates of the A. M. A. convened in June 1916, at which time the committee presented its report. His death was a serious loss to the committee, since he had been active in organizations dealing with public health problems. Favill was influential in the National Association for the Study and Prevention of Tuberculosis, as well as in the National Committee for Mental Hygiene, of which he became the first president in 1909. See John Favill, ed., *Henry Baird Favill A. B., M. D., LL. D. 1860–1916. A Memorial Volume. Life, Tributes, Writings, Compiled by His Son* (Chicago: privately printed, 1917).

216. "Invalidity, Old Age and Unemployment Insurance, and a General Summary of Social Insurance," *A. M. A., Bulletin* 12 (1917): 418.

217. *Ibid.*, pp. 416–417.

218. *The Future of the Medical Practitioner, An Address Delivered before the Cleveland Academy of Medicine, January 16, 1920*, p. 1. See also David Reisman, *The Doctor of the New School, Address Delivered before the Ohio Society of Philadelphia, May 4th, 1923* (Philadelphia, 1923), pp. 14–15.

219. Bonner, *Medicine in Chicago*, p. 220.

220. *J. A. M. A.* 74 (1920): 1319. See also Odin W. Anderson, "Compulsory Medical Care Insurance, 1910–1950," *Annals of the American Academy of Political and Social Science* 273 (1951): 108.

221. Billings, "Future of Private Medical Practice," pp. 349–354; "Conference of Constituent State Medical Associations," *J. A. M. A.* 77 (1921): 1742–1744.

222. A. M. A., House of Delegates, *Proceedings, 73rd Annual Session*, May 1923, p. 44.

223. H. H. Shoulders, "An Insurance Plan of Benefits vs. a Hospital Plan of Benefits for Ex-Soldiers of the United States," *Bulletin, A. M. A.* 26 (1931): 146–147; Burrow, *A. M. A.*, pp. 158–160.

224. David R. Hyde et al., "The American Medical Association: Power,

Purpose and Politics in Organized Medicine," *Yale Law Journal* 63 (1954): 952–953.

225. Arthur M. Schlesinger, Jr., *The Crisis of the Old Order, 1919–1939* (Boston: Houghton Mifflin, 1957), pp. 11–45.

BIBLIOGRAPHICAL NOTE

Charles E. Rosenberg

Until comparatively recent years, the great majority of publications in the history of medicine concerned the development of ideas and techniques. Biographies of men and institutions were also produced in abundance, but the social and economic context of the protagonists (whether human or organizational) was generally ignored. Great doctors, hospitals, and medical schools tended to exist in a world motivated by abstract benevolence and the desire for intellectual achievement.

Since at least the 1920s, however, some medical men and medical historians have been writing a history informed by an awareness of the profession's social and economic context. The term "business" had, after all, been synonymous with "practice" throughout the nineteenth century. (For an attempt to reconstruct this thread of interest in medicine's social history, see Susan Reverby and David Rosner, "Beyond the 'Great Doctors'," in David Rosner and Susan Reverby, eds., *Health Care in America: Essays in Social History* [Philadelphia: Temple University Press, 1979].) Nineteenth- and twentieth-century

cut here

medical thought also found room for a continuing tradition of interest in "social medicine," a body of concepts and studies which emphasized the connections between particular incidences of disease and death and particular social and economic conditions and which saw the melioration of these debilitating conditions as a fundamental responsibility of the medical profession. (For a useful introduction to this literature, see George Rosen, "What is Social Medicine?," *Bull. Hist. Med.*, 21 [1947], 674–733. This and a number of related essays by Rosen are conveniently reprinted in his *From Medical Police to Social Medicine: Essays on the History of Health Care* [New York: Science History Publications, 1974].)

From the 1930s on, such scholars as Bernhard Stern, Henry Sigerist, George Rosen, and Erwin Ackerknecht addressed themselves to the social aspects of past medicine. But even among these contextually oriented historians, interest in the ordinary physician's *practice*, especially its economic dimensions, remained episodic. The development of public health concepts, for example, loomed much larger in this body of work. Here again, George Rosen was a pioneer, publishing in the 1940s two brief but influential monographs relating to the economics of medicine: *The Specialization of Medicine with Particular Reference to Ophthalmology* (New York: Froben, 1944), and *Fees and Fee Bills: Some Economic Aspects of Medical Practice in Nineteenth Century America* (Baltimore: Johns Hopkins University Press, 1946).

It was not until the late 1960s and the 1970s that this emphasis began to change. Economic issues in medicine, previously the concern of a small policy-oriented minority among historians, became a more general subject of debate. The reasons seem obvious enough. One was the anti-authoritarianism of the Vietnam and post–civil rights years. In the discipline of history generally, it was a period during which social history and especially the experience of ordinary men and women assumed an unaccustomed prominence; it thus became increasingly difficult to regard medicine as an essentially or exclusively intellectual pursuit. Concern with medicine's economic aspects was by no means limited to radicals, however. America's bitter and prolonged debate over modes of payment for increasingly costly medical care motivated an enormous diversity of scholarly investigation and polemic. It is obvious that George Rosen's evaluation of the structure of American medical practice will fit naturally into this well defined controversy.

Thus, within the past generation efforts have been undertaken to

inject a vigorous economic element into our understanding of the historical evolution of American medicine. Gerald E. Markowitz and David Rosner, for example, in a frequently cited article ("Doctors in Crisis: A Study of the Use of Medical Education Reform to Establish Modern Professional Elitism in Medicine," *American Quarterly* 25 [1973]: 83–107, saw the Flexner reforms as a way of solving the crisis of "oversupply," the goal being to limit access to the medical market to a smaller number of elite practitioners. E. Richard Brown's *Rockefeller Medicine Men: Medicine and Capitalism in America* (Berkeley and Los Angeles: University of California Press, 1979) provided a consistent but even sharper version, emphasizing the class interest latent in the overtly disinterested good works of the Rockefeller Foundation. The social aspects of medicine became not necessary and inevitable consequences of the profession's ability to mobilize an effective body of knowledge, but rather attempts to grasp social and economic authority through the promise of that knowledge. The role of the American Medical Association has also been much controverted; see, for example, Elton Rayack, *Professional Power and American Medicine: The Economics of the American Medical Association* (Cleveland: World, 1967); James G. Burrow, *AMA: Voice of American Medicine* (Baltimore: Johns Hopkins University Press, 1963); James G. Burrow, *Organized Medicine in the Progressive Era: The Move toward Monopoly* (Baltimore: Johns Hopkins University Press, 1977).

Much of this critical evaluation of the medical profession has been connected with a more general interest in understanding the role of the professions in modern society and their claims to social authority. Not surprisingly, sociologists have been leaders in shaping this ostensibly normative but necessarily political literature. For influential works in this area, see Eliot Freidson, *Profession of Medicine* (New York: Dodd, Mead, 1970); Magali Sarfatti Larson, *The Rise of Professionalism: A Sociological Analysis* (Berkeley and Los Angeles: University of California Press, 1977); Jeffrey L. Berlant, *Profession and Monopoly: A Study of Medicine in the United States and Great Britain* (Berkeley and Los Angeles: University of California Press, 1975); and Paul Starr, *The Social Transformation of American Medicine* (New York: Basic Books, 1982). Though these syntheses disagree in many respects, all unite in bringing medicine into a world of mundane social and economic relationships. In this connection, see also William Rothstein, *American Physicians in the Nineteenth Century: From Sects to Science* (Baltimore: Johns Hopkins University Press, 1972); Rothstein is partic-

ularly impressed, however, by the effects of "valid" therapeutic knowledge in reshaping the profession's social place. Another important overview of American medicine, emphasizing the twentieth century and the role of specialism, has been provided by Rosemary Stevens, *American Medicine and the Public Interest* (New Haven: Yale University Press, 1971). Stevens' study and the more recent analysis by Starr constitute together the most comprehensive and accessible overview of the public policy history of American medicine in the twentieth century.

The great bulk of the ever increasing material relating to the social, economic, and institutional context of American medicine has appeared in the form of articles and monographs. For a convenient overview of this literature, see Ronald L. Numbers, "The History of American Medicine: A Field in Ferment," *Reviews in American History* 10 (1982): 245–263. Numbers' footnotes provide a useful guide to the controversy occasioned by much of this new and often antagonistic scholarship. The growing interest in the social and economic history of American medicine has also resulted in the publication of a number of collections and symposia. Judith W. Leavitt and Ronald L. Numbers, eds., *Sickness and Health in America: Readings in the History of Medicine and Public Health* (Madison: University of Wisconsin Press, 1977) contains a particularly useful critical bibliography. Morris Vogel and Charles E. Rosenberg, eds., *The Therapeutic Revolution: Essays in the Social History of American Medicine* (Philadelphia: University of Pennsylvania Press, 1979), and Rosner and Reverby, *Health Care,* also provide useful leads to recent work in the field.

In recent years, there has been a good deal of historical attention paid to the longstanding debate over national health insurance and third-party payment. For more detailed discussion of the health insurance debate, see Daniel S. Hirshfield, *The Lost Reform: The Campaign for Compulsory Health Insurance in the United States from 1932 to 1943* (Cambridge: Harvard University Press, 1970); Ronald L. Numbers, *Almost Persuaded: American Physicians and Compulsory Health Insurance, 1912–1920* (Baltimore: Johns Hopkins University Press, 1978); Sylvia A. Law, *Blue Cross: What Went Wrong?,* 2nd ed. (New Haven: Yale University, 1976); and Ronald L. Numbers, ed., *Compulsory Health Insurance: The Continuing American Debate* (Westport, Conn.: Greenwood, 1982). Valuable background materials

are also to be found in the studies by Stevens, Starr, and Burrow cited above.

Another area in which there has been a burgeoning interest in recent years is the social and economic history of hospitals. Useful studies of local areas include Morris Vogel, *The Invention of the Modern Hospital: Boston 1870–1930* (Chicago and London: University of Chicago Press, 1980); David Rosner, *A Once Charitable Enterprise: Hospitals and Health Care in Brooklyn and New York, 1885–1915* (Cambridge: At the University Press, 1982); Leonard K. Eaton, *New England Hospitals, 1790–1833* (Ann Arbor: University of Michigan Press, 1957); and Charles E. Rosenberg, "From Almshouse to Hospital: The Shaping of Philadelphia General Hospital," *Health and Society* 60 (1982): 108–154. Rosner is particularly forceful in arguing for the economic impact of the Panic of 1893 and the consequent desire by hospitals and municipalities to rationalize costs. Vogel is particularly concerned to underline the social and political contrasts between Massachusetts General and Boston City hospitals and the shifting social functions of the hospital. In relation to the hospital, see also Harry Dowling, *City Hospitals: The Undercare of the Underprivileged* (Cambridge: Harvard University Press, 1982); Rosemary Stevens," 'A Poor Sort of Memory': Voluntary Hospitals and Government before the Great Depression," *Health and Society* 60 (1982): 551–584; Charles E. Rosenberg, "And Heal the Sick: The Hospital and Patient in 19th Century America," *J. Social Hist.*, 10 (1977): 428–447; and Charles E. Rosenberg, "Inward Vision and Outward Glance: The Shaping of the American Hospital, 1880–1914," *Bull. Hist. Med.* 53 (1979): 346–391.

Related to the interest in the social and economic history of hospitals has been a renewed interest in the history of nursing, a history now written from an increasingly political (and often feminist) perspective. For recent studies which provide an introduction to this literature, see Celia Davies, ed., *Rewriting Nursing History* (London: Croom Helm, 1980); Barbara Melosh, *"The Physician's Hand": Work, Culture and Conflict in American Nursing* (Philadelphia: Temple University Press, 1982); and Ellen Condliffe Lagemann, ed., *Nursing History: New Perspectives and Possibilities* (New York: Teachers College Press, 1983). Brian Abel-Smith, *A History of the Nursing Profession* (London: Heinemann, 1960) and Brian Abel-Smith, *The Hospitals in England and Wales, 1800–1948: A Study of Social*

Administration in England and Wales (Cambridge: Harvard University Press, 1964) provide valuable one-volume syntheses which have, thus far, no parallels for American events.

Another question relating particularly to the hospital and the economics of health care is the rise of an increasingly complex and capital-intensive technical armamentarium. For a convenient survey of events in this area, see Stanley Reiser, *Medicine and the Reign of Technology* (Cambridge: At the University Press, 1978). This field is still a much neglected one, however. A problem related both to the growth of public medicine and to the hospital is that of outpatient care. In this regard, see George Rosen, "The First Neighborhood Health Center Movement: Its Rise and Fall," *Am. J. Pub. Health* 61 (1971): 1620–1637, and Charles E. Rosenberg, "Social Class and Medical Care in Nineteenth-Century America," *J. Hist. Med.* 29 (1974): 32–54. The history of public medicine is a complex subject that, as Rosen emphasizes, in turn has a number of relationships with medical practice and the economic interests of physicians. See, in this connection, Barbara Rosenkrantz, *Public Health and the State: Changing Views in Massachusetts, 1842–1936* (Cambridge: Harvard University Press, 1972); Barbara Rosenkrantz, "Cart before Horse: Theory, Practice, and Professional Image in American Public Health," *J. Hist. Med.* 29 (1974): 55–73; and George Rosen, *Preventive Medicine in the United States 1900–1975: Trends and Interpretations* (New York: Science History Publications, 1975).

Much of the competitive pressure felt by regular physicians has come from so-called "sectarian" physicians—botanic and homeopathic physicians in the nineteenth century and osteopathic and chiropractic practitioners in the twentieth. These "deviant" theories of practice deserve more historical attention, but some work has already appeared. See, in this connection, Joseph Kett, *The Formation of the American Medical Profession: The Role of Institutions, 1780–1860* (New Haven: Yale University Press, 1968); Martin Kaufman, *Homeopathy in America: The Rise and Fall of a Medical Heresy* (Baltimore: Johns Hopkins University Press, 1971); Harris L. Coulter, *Divided Legacy: The Conflict between Homeopathy and the American Medical Association: Science and Ethics in American Medicine, 1800–1914* (1973; reprint ed., Richmond, Cal.: North Atlantic Books, 1982); and Norman Gevitz, *The D.O.s: Osteopathic Medicine in America* (Baltimore: Johns Hopkins University Press, 1982).

INDEX

JACKSON LIBRARY - LANDER UNIV.
3 6289 00211672 8

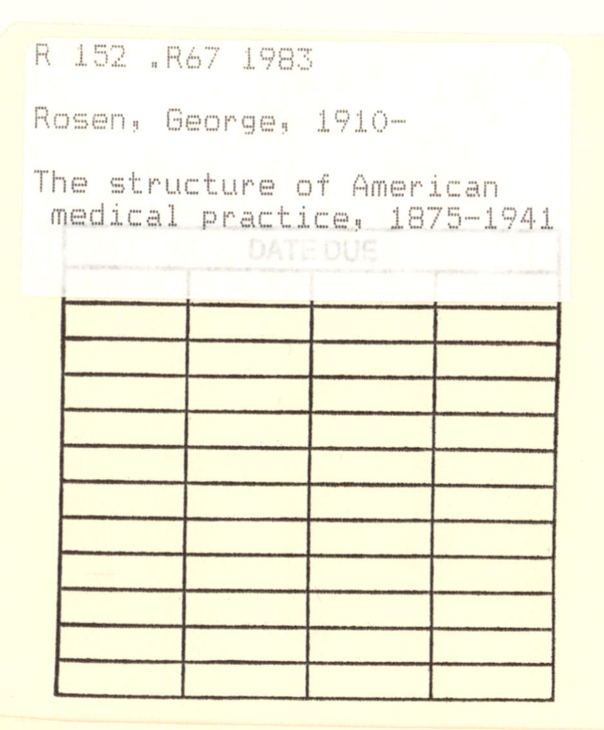
R 152 .R67 1983
Rosen, George, 1910-
The structure of American medical practice, 1875-1941
DATE DUE